Intrusion Detection and Prevention for Mobile Ecosystems

CRC Series in Security, Privacy and Trust

SERIES EDITORS

Jianying Zhou

Institute for Infocomm Research, Singapore

jyzhou@i2r.a-star.edu.sg

Pierangela Samarati

Università degli Studi di Milano, Italy

pierangela.samarati@unimi.it

AIMS AND SCOPE

This book series presents the advancements in research and technology development in the area of security, privacy, and trust in a systematic and comprehensive manner. The series will provide a reference for defining, reasoning and addressing the security and privacy risks and vulnerabilities in all the IT systems and applications, it will mainly include (but not limited to) aspects below:

- Applied Cryptography, Key Management/Recovery, Data and Application Security and Privacy;
- Biometrics, Authentication, Authorization, and Identity Management;
- Cloud Security, Distributed Systems Security, Smart Grid Security, CPS and IoT Security;
- Data Security, Web Security, Network Security, Mobile and Wireless Security;
- Privacy Enhancing Technology, Privacy and Anonymity, Trusted and Trustworthy Computing;
- Risk Evaluation and Security Certification, Critical Infrastructure Protection;
- Security Protocols and Intelligence, Intrusion Detection and Prevention;
- Multimedia Security, Software Security, System Security, Trust Model and Management;
- Security, Privacy, and Trust in Cloud Environments, Mobile Systems, Social Networks, Peer-to-Peer Systems, Pervasive/Ubiquitous Computing, Data Outsourcing, and Crowdsourcing, etc.

PUBLISHED TITLES

Intrusion Detection and Prevention for Mobile Ecosystems

Edited by
**Georgios Kambourakis, Asaf Shabtai,
Constantinos Kolias, and Dimitrios Damopoulos**

CRC Press
Taylor & Francis Group
Boca Raton London New York

CRC Press is an imprint of the
Taylor & Francis Group, an **informa** business

CRC Press
Taylor & Francis Group
6000 Broken Sound Parkway NW, Suite 300
Boca Raton, FL 33487-2742

First issued in paperback 2020

© 2018 by Taylor & Francis Group, LLC
CRC Press is an imprint of Taylor & Francis Group, an Informa business

No claim to original U.S. Government works

ISBN-13: 978-0-367-57306-5 (pbk)
ISBN-13: 978-1-138-03357-3 (hbk)

Visit the Taylor & Francis Web site at
http://www.taylorandfrancis.com

and the CRC Press Web site at
http://www.crcpress.com

Contents

SECTION III MOBILE NETWORK SECURITY AND INTRUSION DETECTION

SECTION IV INTRUSION DETECTION IN DYNAMIC AND SELF-ORGANIZING NETWORKS

Contents

SECTION I MOBILE PLATFORMS SECURITY, PRIVACY, AND INTRUSION DETECTION

SECTION II MALWARE DETECTION IN MOBILE PLATFORMS

v

SECTION III MOBILE NETWORK SECURITY AND INTRUSION DETECTION

SECTION IV INTRUSION DETECTION IN DYNAMIC AND SELF-ORGANIZING NETWORKS

Editors

Georgios Kambourakis, PhD, earned his diploma in applied informatics from the Athens University of Economics and Business and PhD in information and communication systems engineering from the Department of Information and Communications Systems Engineering of the University of the Aegean. He also earned Master of Education (EdM) from the Hellenic Open University. Currently, Georgios is an associate professor at the Department of Information and Communication Systems Engineering, University of the Aegean, Greece, and the director of Info-Sec-Lab. His research interests are in the fields of mobile and wireless networks security and privacy, VoIP security, public key infrastructure, IoT security, DNS security, and security education and he has more than 120 refereed publications in the above areas. In the fall semester of 2017, he conducted research on IoT Security and Privacy in George Mason University, Fairfax County, Virginia. He has guest edited special issues of several journals, including *ACM/Springer Mobile Networks and Applications, Computer Standards & Interfaces, IEEE Computer Magazine, Information Sciences, and Computer Communications*. He has been involved in several national and EU funded R&D projects in the areas of Information and Communication Systems Security. He is a reviewer for a plethora of IEEE and other international journals and has served as a technical program committee member for more than 220 international conferences in security and networking. More info at: www.icsd.aegean.gr/gkamb.

Asaf Shabtai, PhD, is a senior lecturer (assistant professor) in the Department of Software and Information Systems Engineering at Ben-Gurion University (BGU) of the Negev. Asaf is also a senior researcher at the Telekom Innovation Laboratories at BGU. Asaf is an expert in information systems security and has led several large-scale projects and researches in this field. His main areas of interests are security of computer networks and smart mobile devices, malware detection, fraud detection and credit risk prediction, data leakage detection, analysis of encrypted traffic, security awareness, IoT security, development of security mechanisms for avionic systems, as well as the application of machine learning in the cyber security domain. Asaf has published over 70 refereed papers in leading journals and conferences. In addition, he has coauthored a book on information leakage detection and prevention. Asaf earned a BSc in mathematics and computer science (1998); BSc in information systems engineering (1998); MSc in information systems engineering (2003); and PhD in information systems engineering (2011), all from Ben-Gurion University.

Constantinos Kolias, PhD, is a research assistant professor at the George Mason University, Virginia. He earned his diploma in computer science from the Technological Educational Institute of Athens, Greece. He also earned a MSc and PhD in information and communication system security, both from the Department of Information and Communication Systems Engineering, University of the Aegean, Greece. He is the creator of the AWID dataset, the first dataset benchmark for intrusion detection in wireless networks. Also, he is the lead engineer in the creation of the first Internet-of-Things laboratory at the National Institute of Standards and Technology (NIST). His primary research interests lie in the field of security for Internet of Things, next-generation wireless communication protocols, intrusion detection for wireless networks, and mobile surveillance.

Dimitrios Damopoulos, PhD, earned his BSc in industrial informatics from the Technological Educational Institute of Kavala, Greece. He also earned MSc in information & communication systems security and PhD in information and communication systems engineering, both from the Department of Information and Communication Systems Engineering, University of the Aegean, Greece. Currently, he is with Stevens Institute of Technology as an assistant professor. His research interests focus on smartphone security, mobile device intrusion detection and prevention systems, mobile malware, as well as mobile applications and services.

Contributors

Razan Abdulhammed
Department of Computer Science and
  Engineering
University of Bridgeport
Bridgeport, Connecticut

Abdullah J. Alzahrani
College of Computer Science and
  Engineering (CCSE)
University of Hail
Hail, Saudi Arabia

Ioannis Askoxylakis
Institute of Computer Science
Foundation for Research and
  Technology—Hellas
Heraklion, Crete, Greece

Konstantia Barmpatsalou
CISUC/DEI
University of Coimbra
Coimbra, Portugal

Thierry Benoist
European Commission
Joint Research Centre
Ispra, Italy

Laurent Beslay
European Commission
Joint Research Centre
Ispra, Italy

Gerardo Canfora
Department of Engineering
University of Sannio
Benevento, Italy

Ana R. Cavalli
Telecom SudParis and Montimage
France

Andrea Ciardulli
European Commission
Joint Research Centre
Ispra, Italy

Tiago Cruz
CISUC/DEI
University of Coimbra
Coimbra, Portugal

Eleni Darra
Department of Digital Systems
University of Piraeus
Piraeus, Greece

Khaled Elleithy
Department of Computer Science and
  Engineering
University of Bridgeport
Bridgeport, Connecticut

Yuval Elovici
Department of Software and
  Information Systems Engineering
Telekom Innovation Laboratories at
  Ben-Gurion University
Beer-Sheva, Israel

William Enck
Department of Computer Science
North Carolina State University
Raleigh, North Carolina

Miad Faezipour
Department of Computer Science and
    Engineering
University of Bridgeport
Bridgeport, Connecticut

Konstantinos Fysarakis
Institute of Computer Science
Foundation for Research and
    Technology—Hellas
Heraklion, Crete, Greece

Dimitris Geneiatakis
Cyber and Digital Citizens' Security
    Unit
European Commission
Joint Research Centre
Ispra, Italy

Luigi Gentile
Koine srl
Benevento, Italy

Ali A. Ghorbani
Canadian Institute for Cybersecurity
University of New Brunswick
Fredericton, New Brunswick, Canada

Mordechai Guri
Department of Software and
    Information Systems Engineering
Cyber-Security Research Center
Ben-Gurion University of the Negev
Beer-Sheva, Israel

Vinh Hoa La
Telecom SudParis and Montimage
France

Sotiris Ioannidis
Institute of Computer Science
Foundation for Research and
    Technology—Hellas
Heraklion, Crete, Greece

Roger Piqueras Jover
Bloomberg LP
New York, NY

Vasilios Katos
Department of Computing
Bournemouth University
Poole, Dorset, United Kingdom

Sokratis K. Katsikas
Department of Digital Systems
University of Piraeus
Piraeus, Greece

and

Center for Cyber and Information
    Security
Norwegian University of Science and
    Technology
Norway

Ioannis Kounelis
Cyber and Digital Citizens' Security
    Unit
European Commission
Joint Research Centre
Ispra, Italy

Lam-For Kwok
Department of Computer Science
City University of Hong Kong
Hong Kong

Apostolos Malatras
European Commission
Joint Research Centre
Ispra, Italy

Louis Marinos
European Union Agency for Network
    and Information Security (ENISA)
Heraklion, Crete, Greece

Fabio Martinelli
Istituto di Informatica e Telematica
Consiglio Nazionale delle Ricerche
Pisa, Italy

Ivan Martinovic
Department of Computer Science
University of Oxford
Oxford, United Kingdom

Charalabos Medentzidis
Electrical and Computer Engineering
    Department
Aristotle University of Thessaloniki
Thessaloniki, Greece

Weizhi Meng
Department of Applied Mathematics
    and Computer Science
Technical University of Denmark
Copenhagen, Denmark

Francesco Mercaldo
Institute for Informatics and
    Telematics
National Research Council of Italy
    (CNR)
Pisa, Italy

Edmundo Monteiro
CISUC/DEI
University of Coimbra
Coimbra, Portugal

Adwait Nadkarni
Department of Computer Science
North Carolina State University
Raleigh, North Carolina

Igor Nai Fovino
Cyber and Digital Citizens' Security
    Unit
European Commission
Joint Research Centre
Ispra, Italy

Yuri Poliak
Department of Software and
    Information Systems Engineering
Cyber-Security Research Center
Ben-Gurion University of the Negev
Beer-Sheva, Israel

Ignacio Sanchez
European Commission
Joint Research Centre
Ispra, Italy

Andrea Saracino
Istituto di Informatica e Telematica
Consiglio Nazionale delle Ricerche
Pisa, Italy

Asaf Shabtai
Department of Software and
    Information Systems Engineering
Ben-Gurion University of the Negev
Beer-Sheva, Israel

Bracha Shapira
Department of Software and
    Information Systems Engineering
Telekom Innovation Laboratories at
    Ben-Gurion University
Beer-Sheva, Israel

Nava Sherman
Department of Software and
    Information Systems Engineering
Ben-Gurion University of the Negev
Beer-Sheva, Israel

Paulo Simoes
CISUC/DEI
University of Coimbra
Coimbra, Portugal

Yannis Soupionis
European Commission
Joint Research Centre
Ispra, Italy

Gary Steri
Cyber and Digital Citizens' Security
    Unit
European Commission
Joint Research Centre (JRC)
Ispra, Italy

Vincent F. Taylor
Department of Computer Science
University of Oxford
Oxford, United Kingdom

Vasant Tendulkar
Department of Computer Science
North Carolina State University
Raleigh, North Carolina

Akash Verma
Department of Computer Science
North Carolina State University
Raleigh, North Carolina

Corrado Aaron Visaggio
Department of Engineering
University of Sannio
Benevento, Italy

Jianying Zhou
Information Systems Technology and
    Design
Singapore University of Technology
    and Design
Singapore

Introduction

During the past few years, mobile devices such as smartphones and tablets have penetrated the market at a very high rate. From an end-user perspective, the unprecedented advantages these devices offer revolve not only around their high mobility, but also extend to their ease of use and the plethora of their applications. As the permeation of mobile devices increases, the development of mobile apps follows this frantic rate, by being built in great numbers on a daily basis. On the downside, this mushrooming of mobile networks and portable devices has attracted the interest of several kinds of aggressors who possess a plethora of invasion techniques in their artillery. Such ill-motivated entities systematically aim to steal or manipulate users' or network data, and even disrupt the operations provided to legitimate users.

Their goal is assisted by the fact that while a continuously increasing number of users has embraced mobile platforms and associated services, most of them are not security-savvy and usually follow naive privacy preservation practices on their routine interaction with their devices. Until now, a great mass of research work and practical experiences have alerted the community about the nature and severity of these threats that equally affect end-users, providers, and even organizations.

One can identify several more reasons behind the proliferation of malware and the spanning of novel invasion tactics in mobile ecosystems. First, mobile devices are used extensively for sensitive tasks, including bank transactions and e-payments, private interaction such as engagement in social media applications, or even mission critical processes in healthcare. Second, smart, ultraportable, and wearable devices such as smartwatches and smartglasses are highly personal, and thus can be correlated with a single user; they embed several sensors and functionalities capable of collecting many details about the context of users, while they are constantly connected to the Internet. Third, numerous researches and case studies have shown that despite the ongoing progress, native security mechanisms of modern mobile operating systems or platforms can be outflanked. Even worse, most of the applied wireless communication technologies are eventually proven to be prone to numerous attacks. Admittedly, under this mindset, the attack surface for evildoers grows, further augmenting the expansion of volume and sophistication of malware apps.

To cope with this situation, defenders need to deploy smarter and more advanced security measures along with legacy ones.

The book at hand comprises a number of state-of-the-art contributions from both scientists and practitioners working in intrusion detection and prevention for mobile networks, services, and devices. It aspires to provide a relevant reference for students, researchers, engineers, and professionals working in this particular area or those interested in grasping its diverse facets and exploring the latest advances in intrusion detection in mobile ecosystems. More specifically, the book consists of 16 contributions classified into 4 pivotal sections:

- *Mobile platforms security, privacy, and intrusion detection*: Introducing the topic of mobile platforms security, privacy, and intrusion detection, and offering related research efforts on attacking smartphone security and privacy, a way to create reliable smartphone end-user apps in an ad hoc manner, a privacy risk assessment for Android apps, and an inference system for mobile forensics.
- *Malware detection in mobile platforms*: Investigating advanced techniques for malware and rootkit detection in the Android platform, and exploring the different kinds of intrusive apps and data leakage due to malware in the same platform.
- *Mobile network security and intrusion detection*: Experimentally exploring mobile botnets, demonstrating ways for attacking LTE by applying low-cost software radio, and an intrusion detection framework based on SMS.
- *Intrusion detection in dynamic and self-organizing networks*: Focusing on intrusion techniques in self-organizing networks, wireless sensor networks, 6LoWPAN-based wireless sensor networks, and co-operative intelligent transportation systems.

MOBILE PLATFORMS SECURITY, PRIVACY, AND INTRUSION DETECTION

Chapter 1

A Review of Intrusion Detection and Prevention on Mobile Devices: The Last Decade

Weizhi Meng, Jianying Zhou, and Lam-For Kwok

Contents

1.1 Introduction

Nowadays, mobile devices such as various phones are developing at a rapid pace and have become common in our daily lives. The worldwide smartphone market grew 0.7% year over year in 2016, with 344.7 million shipments, where Android dominated the market with an 87.6% share, according to data from the International Data Corporation (IDC) [1]. As a result, more users start utilizing mobile devices as a frequent storage medium for sensitive information (e.g., passwords, credit card numbers, private photos) [2], as well as use them for security-sensitive tasks due to their fast and convenient data connection [3]. Owing to this, smartphones have become an attractive target for hackers and malware writers [4,5].

As we know, mobile devices are easily lost, and the stored personal and sensitive information in those lost devices might be exploited for malicious use [6]. Therefore, it is unsurprising that designing secure solutions for mobile devices remains a topic of current interest and relevance. In addition to user authentication schemes [7], intrusion detection and prevention systems (IDSs/IPSs) are the most commonly used technology to protect the mobile environment.

Generally, based on specific detection approaches, intrusion detection systems (IDSs) can be generally classified into two types: *signature-based IDS* and *anomaly-based IDS*. For the former [8,9], it mainly detects a potential attack by examining packets and comparing them to known signatures. A signature (also called *rule*) is a kind of description for a known attack, which is usually generated based on expert knowledge. For the latter [10,11], it identifies an anomaly by comparing current events with preestablished normal profile. A normal profile often represents a normal behavior or network events. An alarm will be generated if any anomaly is detected. In addition, according to deployment, IDSs can be categorized into host-based systems and network-based systems.

As compared to IDSs, intrusion prevention systems (IPSs) are able to react and stop current adversary actions. Most IPSs can offer multiple prevention capabilities to adapt to various needs. IPSs usually allow security officers to choose the prevention capability configuration for each type of alert, such as enabling or disabling prevention, as well as specifying which type of prevention capability should be used. To control and reduce false actions, IPSs may have a learning or simulation mode that suppresses all prevention actions and instead indicates when a prevention action would have been performed. This allows security officers to monitor and fine-tune the configuration of the prevention capabilities before enabling prevention actions [8].

Motivation and focus. As mobile devices are often short of power and storage, traditional intrusion and prevention techniques are hard to deploy directly. However, in the last decade, with the rapid development of mobile devices, there is an increasing need to apply IDSs/IPSs to protect these devices. It is very critical to establish an appropriate defense mechanism on such resource-limited devices. Motivated by this, in this chapter, we aim to present a review, introducing recent advancement within the last decade regarding the development of mobile intrusion detection and prevention efforts in the literature, and providing insights about current issues, challenges, and future directions in this area. Our contributions of this chapter can be summarized as below:

■ First, we introduce the background of IDSs/IPSs in more detail and then investigate the development of IDS/IPS on mobile devices within the last decade, by examining notable work in the literature.
■ Then, we identify the issues and challenges of designing such defense mechanisms on mobile devices, describe several potential solutions, and analyze the future directions in this field.

The remaining parts of this chapter are organized as follows. Section 1.2 introduces the background of IDSs and IPSs. Section 1.3 surveys the recent work regarding IDSs/IPSs on mobile devices within the last decade. Section 1.4 identifies issues and challenges of designing an appropriate defense on mobile devices. Section 1.5 describes several potential solutions and points out future directions in this area, and Section 1.6 concludes this chapter.

1.2 Background

In this section, we introduce the background of IDSs and IPSs, respectively. The generic workflow of IDSs/IPSs is depicted in Figure 1.1.

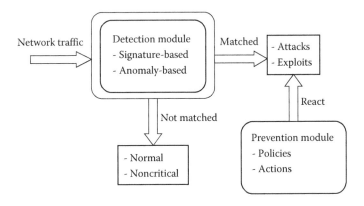

Figure 1.1   Generic workflow of IDSs and IPSs.

1.2.1 Intrusion Detection

Intrusion detection is the process of monitoring the events occurring in a local system or network and analyzing them for any sign of violations. An IDS is a software that automates the intrusion detection process. An IDS can typically provide several functions [8]:

- *Monitoring and recording information.* Targeted information can be usually recorded either locally or distributed according to concrete settings. For example, IDSs can store the information in an enterprise management server.
- *Notifying security officers.* This notification, also known as an alert, can be sent to security officers if any malicious event is identified. This alert has many forms such as emails, pages, and messages.
- *Generating reports.* Such reports often summarize the monitored events and provide details on particular interested events. Security officers can also track the historical status based on the reports.

Based on the deployment, there are two major types of IDSs: host-based IDS (HIDS) and network-based IDS (NIDS). An HIDS mainly monitors the events that occur in a local computer system, and then reports its findings. On the other hand, an NIDS aims to monitor current network traffic and detect network attacks through analyzing incoming network packets.

Moreover, according to the detection methods, an IDS can be primarily labeled as a signature-based, anomaly-based IDS. Figure 1.1 shows the generic workflow of intrusion detection.

- A signature is a pattern that corresponds to a known attack or exploit; thus, signature-based detection is the process of comparing signatures against observed events to identify possible incidents, for example, a telnet attempt with a username of "root," which is a violation of an organization's security policy [8].
- Anomaly-based detection is the process of comparing normal profiles against observed events to identify significant deviations. The profiles are developed by monitoring the characteristics of typical activity over a period of time. The system can use statistical methods to compare the characteristics of current activity to thresholds related to the profile.

Discussion. Signature-based detection is the simplest detection method, since it only compares the current unit of activity, such as a packet or a log entry, to a list of signatures using string comparison operations. Therefore, it is very effective at detecting known threats but largely ineffective at detecting previously unknown threats. By contrast, anomaly-based detection methods is good at detecting previously unknown threats, but may produce many false alarms. For example, if a particular maintenance activity that performs large file transfers occurs only once a month, it might not be observed during the training period; when the maintenance occurs, it is likely to be considered a significant deviation from the profile and trigger an alert [8].

In addition to the above two detection methods, another detection is called stateful protocol analysis, which is the process of comparing predetermined profiles to identify deviations for each protocol status. It relies on vendor-developed universal profiles that specify how particular protocols should and should not be used. In the current literature, hybrid IDSs are developing at a rapid pace.

1.2.2 Intrusion Prevention

An IPS is a software that has all the capabilities of an IDS and can also attempt to stop possible incidents. As compared with an IDS, IPS technologies can respond to a detected threat by attempting to prevent it from succeeding. They can provide the following functions:

- *Stopping the attack.* IPSs can react to existing attacks, such as terminating the network connection or user session that is being used for the attack, and blocking access to the target from the offending user account, IP addresses, etc.
- *Changing the security environment.* An IPS can change the configuration of other security controls to disable an attack (i.e., reconfiguration of a device), or even launch patches to be applied to a host if the host has detected vulnerabilities.
- *Changing the attack's content.* Some IPSs can remove or replace malicious portions of an attack/program to make it benign. For instance, an IPS can remove an infected file attachment from an email and then permit the cleaned email to reach its recipient.

Discussion. As depicted in Figure 1.1, IPSs rely on the detection of potential threats. Thus, intrusion prevention will perform behind IDSs. A typical IPS usually contains the basic functions of an IDS, or specified as IDPS (intrusion detection and prevention system). It is worth noting that IDSs/IPSs can be classified differently according to specific targets and focuses. Table 1.1 describes several classification categories accordingly.

1.3 Review on IDSs/IPSs on Mobile Devices

As the mobile environment is different from computers, appropriate security systems should consider both detection accuracy and resource consumption. In this section, our focus is the application of IDSs/IPSs on mobile devices.

Table 1.1 Classification Categories According to Different Focuses

	Description
By Detection Topology	
Network-based methods	IDS/IPS is mainly deployed in a network.
Host-based methods	IDS/IPS is mainly deployed in a local system.
Hybrid methods	IDS/IPS searches anomalies in both network and local systems.
By Source	
Application level	IDS/IPS examines events from application level.
Kernel level	IDS/IPS examines events from kernel level.
Hardware level	IDS/IPS examines events from hardware level.
Hybrid source	IDS/IPS provides an overall protection throughout various levels.
By Detection Approach	
Anomaly-based methods	IDS/IPS identifies anomalies from normal profiles.
Signature-based methods	IDS/IPS identifies threats through signature matching.
Behavior-based methods	IDS/IPS identifies anomalies based on predefined valid behavior.
Hybrid methods	IDS/IPS combines the above detection approaches.
By Focus	
Malicious apps	IDS/IPS focuses on the detection of malware.
Information leakage system	IDS/IPS focuses on the detection of information leakage.
Hybrid	IDS/IPS considers all current threats.

1.3.1 From 2004 to 2006

Jacoby and Davis [12] designed a warning system via a host-based intrusion detection, which could alert security administrators to protect smaller mobile devices. They operated through the implementation of battery-based intrusion detection (called *B-bid*) on mobile devices by correlating attacks with their impact on device power consumption using a rules-based host intrusion detection engine (HIDE). *B-bid* mainly measured the energy use over time to decide if an attack is present or not. As a result, probabilistic bounds for energy and time can have confidence intervals to measure abnormal behavior of power dissipation. In 2005, Nash et al. [13] also focused on battery exhaustion attacks, and designed an IDS, which takes into account the performance, energy, and memory constraints of mobile computing devices. Their system uses several parameters, such as CPU load and disk accesses, to estimate the power consumption using a linear regression model.

For general solutions, Jansen et al. [14] focused on mobile devices such as PDAs and proposed a framework for incorporating core security mechanisms in a unified manner, through adding improved user authentication, content encryption, organizational policy controls, virus protection, firewall and intrusion detection filtering, and virtual private network communication. This is a conceptual study; thus, it needs more tests in reality. Kannadiga et al. [15] discussed the characteristics of intrusion detection for pervasive computing environments (e.g., mobile, embedded, handheld devices) and described a mobile agent-based IDS to be deployed in a pervasive computing environment.

Miettinen et al. [16] advocated that mobile systems should be equipped with proper second-line defense mechanisms that can be used to detect and analyze security incidents. They proposed a framework for intrusion detection functionalities in mobile devices, which combines both host-based and network-based data collection. The data collection module is responsible for collecting and receiving the host-based monitoring data. Then, the data collection module forwards the monitoring data to the IDS module, which is responsible for the actual intrusion detection process. Then, Buennemeyer et al. [17] proposed a Battery-Sensing Intrusion Protection System (called *B-SIPS*) for mobile devices, which could alert on power changes detected on small wireless devices.

1.3.2 From 2007 to 2010

From 2007, more research has been done in protecting mobile devices against threats. Mazhelis and Puuronen [18] focused on personal mobile devices and designed a conceptual framework for mobile-user substitution detection. The framework was mainly based on the observation that user behavior and environment could reflect the user's personality in a recognizable way. More specifically, the framework could be decomposed into a descriptive and a prescriptive part. The former is concerned with the description of an object system (i.e., the user, his/her personality, behavior, and environment). The latter considers technical components (e.g., databases, knowledge bases, and processing units) that are needed to implement the UIV system based on the above object system. The proposed approach aimed at verifying the user's identity and detecting user substitution, which can be used in intrusion detection and fraud detection.

Venkataram et al. [19] proposed a generic model called Activity Event-Symptoms (AES) model for detecting fraud and attacks during the payment process in the mobile e-commerce environment. The model was designed to identify the symptoms and intrusions by observing various events/transactions that occur during mobile commerce activity. Niu et al. [20] studied the relationship between power consumption and parameters of system state on mobile computing devices by using genetic algorithm and artificial neural network. They found that an IDS may produce many false alarms, if it only relies on the CPU load to detect and identify a battery exhaust attack. It is suggested that an IDS should at least take hard-disk read and write operations and network transmission into account.

Later, Chung et al. [21] described an approach of Just-on-Time Data Leakage Protection Technique (JTLP), which aims to protect the user's private data of mobile devices from malicious activity and leakage. In a mobile device, all users' access can be monitored by JTLP monitor process (JTLPM) by using the access control method. When an unauthorized user or malicious code attempts to transmit some host of the outer network, JTLPM monitors all outbound suspicious packets to the external host. If a packet contains user's important data, this packet should be thrown out of the mobile device by the JTLPM. Brown and Ryan [22] argued that the current security model for a third-party application running on a mobile device requires its user to trust that application's vendor, but they cannot guarantee complete protection against the threats. Thus, they introduced a security architecture that prevents a third-party application deviating from its intended behavior, defending devices against previously unseen malware.

For anomaly detection, Schmidt et al. [23] demonstrated how to monitor a smartphone running Symbian OS to extract features for anomaly detection. They found that most of the top 10 applications preferred by mobile phone users could affect the monitored features in different ways. However, their features needed to be sent to a remote server, due to capability and hardware limitations.

Started from 2009, with the popularity of smartphones, most research has moved to malware detection. Shabtai et al. [24] evaluated a knowledge-based approach for detecting instances of known classes of mobile devices malware based on their temporal behavior. The framework relied on lightweight agent to continuously monitor time-stamped security data within the mobile device and to process the data using a light version of the Knowledge-Based Temporal Abstraction (KBTA) methodology. Then, Liu et al. [25] adopted power consumption analysis and proposed *VirusMeter*, a general malware detection method, to detect anomalous behaviors on mobile devices. The rationale underlying VirusMeter is the fact that mobile devices are usually battery powered and any malicious activity would inevitably consume some battery power. By monitoring power consumption on a mobile device, *VirusMeter* could catch misbehaviors that lead to abnormal power consumption.

Later, Xie et al. [26] proposed a behavior-based malware detection system named *pBMDS*, which adopts a probabilistic approach through correlating user inputs with system calls to detect anomalous activities in cellphones. In particular, *pBMDS* observes unique behaviors of the mobile phone applications and the operating users on input and output constrained devices, and leverages a hidden Markov model (HMM) to learn application and user behaviors from two major aspects: process state transitions and user operational patterns. Based on these, *pBDMS* identifies behavioral differences between malware and human users. Shabtai et al. [27] then proposed a behavioral-based detection framework for Android mobile devices. The proposed framework adopted an HIDS that could continuously monitor various features and events obtained from the mobile device, and apply machine learning methods to classify the collected data as normal (benign) or abnormal (malicious).

1.3.3 From 2011 to 2013

Smartphone vendors will ship more than 450 million devices in 2011. At the same period, mobile malware was propagated in a quick manner; therefore, many IDSs/IPSs were built aiming to perform the detection of malware. Chaugule et al. [28] found that mobile malware always attempts to access sensitive system services on the mobile phone in an unobtrusive and stealthy fashion. For example, the malware may send messages automatically or stealthily interface with the audio peripherals on the device without the user's awareness and authorization. They then presented *SBIDF*, a Specification Based Intrusion Detection Framework, which utilizes the keypad or touchscreen interrupts to differentiate between malware and human activity. In the system, they utilized an application-independent specification, written in Temporal Logic of Causal Knowledge (TLCK), to describe the normal behavior pattern, and enforced this specification to all third-party applications on the mobile phone during runtime by monitoring the intercomponent communication pattern among critical components.

Then, Burguera et al. [29] focused on detecting malware in the Android platform. They put the detector to embed into a framework for collecting traces from an unlimited number of real users based on crowdsourcing. The framework could analyze the data collected in the central server using two types of data sets: those from artificial malware created for test purposes and those from real malware found in the wild. They considered that monitoring system calls should be one of the most accurate techniques to determine the behavior of Android applications, since they provide detailed low-level information. Bukac et al. [30] then summarized the development of HIDS before 2012 regarding the detection of intrusions from a host network traffic analysis, process behavior monitoring, and file integrity checking. Damopoulos et al. [31] focused on anomaly detection on mobile devices and conducted an evaluation among different classifiers (i.e., Bayesian networks, radial basis function, K-nearest neighbors, and random Forest) in terms of telephone calls, SMS, and web browsing history.

La Polla et al. [32] later presented a survey on the security of mobile devices, and pointed out that mobile services had significantly increased due to the different form of connectivity provided by mobile devices, such as GSM, GPRS, Bluetooth, and Wi-Fi. They particularly described the state of the art on threats, vulnerabilities, and security solutions over the period from 2004 to 2011, by focusing on high-level attacks, like those to user applications. Roshandel et al. [33] argued that there is a fundamental difference in the attitude of a typical user when it comes to using their mobile device as compared to their personal computers. In addition, there is little by the way of security and privacy protection available on these mobile computing platforms. As a result, they developed a Layered Intrusion Detection and Remediation framework (LIDAR), which could automatically detect, analyze, protect, and remediate security threats on Android devices. More specifically, they developed several algorithms that may help detect abnormal behavior in the operation of Android smartphone and tablets that could potentially detect the presence of malware. Li and Clark [34] discussed that mobile devices were being rapidly integrated into enterprises, government agencies, and even the military, as these devices may hold valuable and sensitive content. They then provided an overview of the current threats based on data collected from observing the interaction of 75 million users with the Internet.

Curti et al. [35] investigated the correlations between the energy consumption of Android devices and some threats such as battery-drain attacks. They then described a model for the energy consumption of single hardware components of a mobile device during normal usage and under attack. Their model can be implemented in a kernel module and used to build up an energetic signature of both legal and malicious behaviors of Wi-Fi hardware component in different Android devices. The proposed Energy-Aware Intrusion Detection solutions were able to reliably detect attacks on mobile devices based on energetic footprints. To reach this, they adopted a step-wise approach: (1) they developed some built-in solutions allowing to measure and analyze energy consumption directly on each device, neglecting the usage of external hardware; (2) they performed measurements on some devices and benign/malicious applications and resulted in a database of consumption patterns for both benign smartphone activities and known attacks.

1.3.4 From 2014 to 2016

From 2014, most research studies aimed at developing a security mechanism to protect mobile devices against malware, while less work investigated how to build an IDS/IPS. Several surveys about the malware protection can be referred to in References 7 and 36.

For building IDSs/IPSs, Yazji et al. [37] focused on the problem of efficient intrusion detection for mobile devices via correlating the user's location and time data. They developed two statistical profiling approaches for modeling the normal spatiotemporal behavior of the users: one is based on an empirical cumulative probability measure and the other is based on the Markov properties of trajectories. An anomaly could be detected when the probability of a particular evolution (e.g., location, time) matching the normal behavior of a given user becomes lower than a certain threshold. Papamartzivanos et al. [38] identified that modern app markets had been flooded with applications that not only threaten the security of the OS, but also in their majority, trample on user's privacy through the exposure of sensitive information. They discussed and developed a cloud-based crowdsourcing mechanism that could detect and alert for changes in the app's behavior.

Later, Sun et al. [39] developed various host-based intrusion prevention systems (HIPS) on Android devices, in order to protect smartphones and prevent privacy leakage. In particular, they analyzed the implementations, strengths, and weaknesses of three popular HIPS architectures, and demonstrated a severe loophole and weakness of an existing popular HIPS product in which hackers can readily exploit. Based on this, they designed a more secure and extensible HIPS platform called *Patronus*. *Patronus* can dynamically detect existing malware based on runtime information, without the need to modify the hardware. Damopoulos et al. [40] then investigated two issues: the first one was how to define an architecture, which could be used for implementing and deploying a system in a dual-mode (host/cloud) manner and irrespectively of the underlying platform, and the second one was how to evaluate such a system. Their approach allows users to argue in favor of a hybrid host/cloud IDS arrangement and to provide quantitative evaluation facts on if and in which cases machine learning-driven detection is affordable when executed on devices such as smartphones. Damopoulos et al. [41] later described a tool that was able to dynamically analyze any iOS software in terms of method invocation (i.e., which API methods the application invokes and under what order), and produce exploitable results that can be used to manually or automatically trace software behavior to decide if it contains malicious code.

1.3.5 Discussion

Mobile devices have already become a part of people's lives. Once a mobile device is compromised, a wide range of threats may occur. For example, attackers might sell the uncovered personal data; they might leverage stored credentials to gain access to a device; or they might use the device as a gateway into enterprise data and resources by leveraging a trust relationship between the device and the IT infrastructure [34]. Even worse, the device could be put into a botnet or used to send unauthorized premium-rate SMS messages. Thus, how to manage these risks at scale and the problem is becoming more complex.

Intrusion detection and prevention techniques are one of the promising solutions to secure mobile devices. However, as compared to a desktop computer, mobile devices are often short of power and resources. Traditional IDS/IPS tools may not be applicable on mobile platforms. Therefore, there is a need for developing advanced and energy-ware intrusion detection and prevention solutions.

In addition to local mobile device protection, existing mobile networks usually consist of many mobile devices (e.g., medical smartphone network [42]), so it is also an important topic to secure the mobile environment, such as adding monitors [43], verifying locations [44], and performing deep packet inspection [45].

To compare the performance between IDSs/IPSs, three major factors should be considered: detection accuracy, time consumption, and CPU usage.

- *Detection accuracy.* This factor is extremely important for an IDS/IPS, where an ideal detection system should provide high accuracy and low false rates.
- *Time consumption.* Intrusion detection is a time-sensitive task, where a quick identification can reduce the damage (e.g., financial loss, data leakage). Generally, an ideal detection system should be able to identify threats in a faster manner.
- *CPU usage.* As mobile devices have limited resources, CPU usage becomes a critical factor to determine whether an IDS/IPS is feasible in real-world applications. An ideal detection system is expected to consume less CPU and ensure the availability of devices.

1.4 Issues and Challenges

As mentioned, mobile devices face the same (or a higher) level of malicious attacks that have plagued the desktop computing environments [34]. However, typical mobile devices are different from common computers:

- *Mobility.* Mobile devices are much more flexible due to their size and weight so that users can bring their devices everywhere. In comparison, a typical desktop computer is often deployed in a particular site. The mobility requires to design more dynamic security mechanisms on mobile devices.
- *Limited resources.* A typical mobile device (e.g., phones, iPads) has only limited power and computation capabilities, making it not powerful enough to implement traditional intrusion detection and prevention techniques. When designing a mobile security mechanism, there is a balance that should be considered between performance and energy.

Owing to these features, mobile IDS/IPS may suffer from many issues and challenges, which are the same as in traditional computing environments. The major issues and challenges are summarized in Figure 1.2.

- *Event overload.* Mobile devices can generate a massive amount of local and network events, with the rapid development of system computation. Such events may exceed the capability of a mobile IDS/IPS. For instance, a signature-based NIDS may drop lots of network packets if the incoming packets exceed their maximum processing capability.

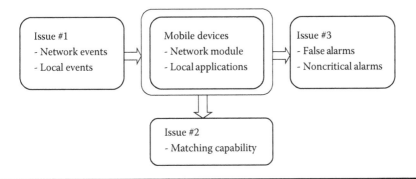

Figure 1.2   Issues and challenges of mobile IDSs/IPSs.

■ *Expensive signature matching.* For signature-based IDSs/IPSs, the signature matching module is often too expensive for resources that the computing burden is at least linear to the size of an incoming string [46]. Subsequently, the performance of IDS/IPS may be greatly degraded due to the heavy operational burden.

■ *Massive false alarms.* Both signature- and anomaly-based IDSs/IPSs may generate a large number of false alarms, which can significantly increase the difficulty in analyzing alarms and adversely affect the analysis results.

These issues and challenges can greatly degrade the performance of a mobile IDS/IPS, such as missing network and local events and dispersing analysis directions. In addition, these issues can increase the workload and burden of deploying IDSs/IPSs, causing mobile devices to be out of availability in a quick manner. Therefore, it is a critical topic for developing an appropriate security mechanism on the mobile environment.

1.5 Solutions and Future Trend

In this section, we describe several potential solutions for the above issues and challenges, and point out the future trend in this field.

1.5.1 Potential Solutions

In order to design a proper security mechanism on mobile devices, it is necessary to consider the above issues and make particular improvement with particular goals. According to the issues in Figure 1.2, it is promising to implement various modules or additional mechanisms (e.g., packet filter, alarm filter) to strengthen the IDS/IPS performance.

1.5.1.1 Packet Filter Development

The reduction of packet filter is a straightforward solution to improve the performance of intrusion detection and prevention on mobile devices. The idea of the packet filter is to reduce the number of target packets through filtering out certain packets early based on their IP confidence. To build an appropriate packet filtration mechanism, several factors should be considered:

- The filter should have a minimum impact on system and network performance.
- The filter should be efficient and provide a good filtration rate.
- The filter should not degrade the security level of IDSs/IPSs.

Based on these factors, there are several kinds of event/packet filters in the literature, such as blacklist-based, list-based, and trust-based filter.

- *Blacklist-based event/packet filter.* Blacklist is a common technique that is used in filtering events and packets. This type of filter can alleviate the burden of either a signature- or anomaly-based IDS/IPS in processing a massive number of target events. This filter can realize a weighted ratio-based method (statistic-based method) in the monitor engine to calculate the IP confidences and to generate the blacklist [47,48].
- *List-based event/packet filter.* In addition to blacklist, the whitelist can also be useful in real applications. Thus, it is a solution to combine the whitelist and the blacklist techniques in constructing a list-based event/packet filter [49].
- *Trust-based event/packet filter.* To leverage the blacklist generation, trust computation can be applied to such filters. For example, Bayesian inference model [50] can be used to enhance the computation of IP confidence and further improve the performance of filters in a large-scale network environment. One basic assumption is that all events/packets are independent of each other.

1.5.1.2 Alarm Filter Development

The large number of false alarms can greatly reduce the efficiency of an IDS and significantly increase the burden of analyzing these alarms. For example, thousands of alarms may be generated in one day, which are a big burden for a security officer. Even worse, false alarms may have a negative impact on the analysis of IDS outputs. Hence, false alarm reduction is an important issue for IDSs/IPSs. There are many techniques that can be considered:

- *Adaptive false alarm filter.* In real scenarios, machine learning is often applied to false alarm reduction. However, the filtration accuracy of an algorithm may be fluctuant. As a result, an adaptive false alarm filter is promising to select the best algorithm from a pool of algorithms [51]. Such filter enables the algorithm selection to be performed in an adaptive way with the purpose of maintaining a high and stable filtration accuracy.

■ *Knowledge-based alarm filter.* Expert knowledge is very crucial in deciding whether an alarm is critical or not. Therefore, knowledge-based alert verification can be combined to construct an alarm filter [52], that is, employing a rating mechanism to classify incoming alarms.

■ *Contextual alarm filter.* Many false alarms are produced, since the IDSs/IPSs are not aware of the contextual information of their deployed environment. Hence, considering contextual information is a promising method to improve the quality of output alarms [53].

1.5.1.3 Matching Capability Improvement

The expensive process of signature matching is a key limiting factor for deploying IDSs/IPSs on mobile devices. Therefore, there is a need for improving the matching capability. Besides traditional string matching algorithms, exclusive signature matching scheme is a promising solution.

The major difference between regular signature matching and exclusive signature matching is that the latter aims to identify a mismatch rather than to confirm an accurate match during the signature matching [46]. This scheme can be adaptive in selecting the most appropriate single character for exclusive signature matching in terms of different network environments. In particular, our scheme respectively calculates the character frequency of both stored NIDS signatures and matched signatures with the purpose of adaptively and sequentially determining the most appropriate character in the comparison with packet payload [54].

1.5.1.4 Overall Improvement

Moreover, the above solutions can be integrated into one comprehensive mechanism [55]. Taking EFM [56] as an example, this mechanism is composed of three major components: a *context-aware blacklist-based packet filter*, an *exclusive signature matching component*, and a *KNN-based false alarm filter*. In particular, the *context-aware blacklist-based packet filter* is responsible for reducing the workload of IDSs by filtering out network packets by means of IP reputation. The *exclusive signature matching component* is implemented in the context-aware blacklist-based packet filter aiming to speed up the process of signature matching. The *KNN-based false alarm filter* is responsible for filtering out false alarms (positives) that are produced by the packet filter and the IDS.

1.5.2 Future Trend

Intrusion detection and prevention is a basic solution to protect mobile devices against malicious use. With the development of mobile platforms, malware and theft use are the major threats. Additionally, several vulnerabilities in the operation system may threaten the phone security [57]. As a result, IDSs/IPS should be further enhanced through combining new features.

- *Behavioral-based detection.* The current smartphones often feature a touch-screen as the input method. As compared with the traditional button-based input, touchscreen enables more actions such as multitouch and touch movement. For instance, multitouch is a new feature, where users can touch the screen with multiple fingers at the same time [58,59]. The new feature may result in novel threats such as smudge attacks [60], but also enable behavioral-based detection (e.g., multitouch-included authentication [61–67]). With more biometrics implemented on mobile devices, biometric authentication should be given more attention in the future.
- *Graphical passwords.* There is an increasing number of applications installed on mobile devices such as graphical passwords; thus, it is necessary to apply IDS/IPS techniques to those passwords. Such combination can provide more comprehensive protection to the mobile environment [4,68]. Several research studies on graphical passwords can be referred to in References 69–79.
- *Cloud-based mechanism.* Computation resources are often a key limiting factor for deploying complex IDS/IPS techniques on mobile devices. With the advent of cloud, it is promising to offload expensive operations to the cloud side (e.g., offloading the signature matching process [80]).

1.6 Conclusions

Research in mobile device and smartphone security has been conducted for several years. Security solutions for mobile devices and smartphones must defend against viruses, malware, botnets, and attacks through the deployment of a wide spectrum of mobile applications. Intrusion detection and prevention techniques are one basic solution to protect mobile devices and users' privacy. However, it is not an easy task for building an appropriate defense mechanism on resource-limited mobile devices.

In this chapter, we present a review, introducing recent advancement within the last decade regarding the development of mobile intrusion detection and prevention efforts in the literature. Then, we give insights about current issues and challenges for mobile IDSs/IPSs such as overhead event/packets, massive false alarms, and matching bottleneck. By focusing on these issues, we introduce potential solutions in constructing event/packet alarm filter and improving signature matching. At last, we point out that future mobile IDSs/IPSs may cooperate with more applications such as behavioral-based detection and graphical passwords and utilize the resources from new environments such as cloud.

References

1. Smartphone Vendor Market Share, Q2. http://www.idc.com/prodserv/smartphone-market-share.jsp, 2016.

2. A. K. Karlson, A. B. Brush, and S. Schechter, Can I borrow your phone?: Understanding concerns when sharing mobile phones, in *Proceedings of the 27th International Conference on Human Factors in Computing Systems (CHI)*, ACM, New York, pp. 1647–1650, 2009.

3. P. Dunphy, A. P. Heiner, and N. Asokan, A closer look at recognition-based graphical passwords on mobile devices, in *Proceedings of the Sixth Symposium on Usable Privacy and Security (SOUPS)*, ACM, New York, pp. 1–12, 2010.

4. Y. Meng, W. Li, and L.-F. Kwok, Enhancing click-draw based graphical passwords using multi-touch on mobile phones, in *Proceedings of the 28th IFIP TC 11 International Information Security and Privacy Conference (IFIP SEC)*, Springer, Berlin, Heidelberg, Auckland, New Zealand, pp. 55–68, 2013.

5. A. Shabtai, Y. Fledel, U. Kanonov, Y. Elovici, S. Dolev, and C. Glezer, Google Android: A comprehensive security assessment, *IEEE Security Privacy*, **8**(2): 35–44, 2010.

6. Mobile and NCSA. Report on Consumer Behaviors and Perceptions of Mobile Security, January 2012. Available at: http://docs.nq.com/NQ_Mobile_Security_Survey_Jan2012.pdf.

7. W. Meng, D. S. Wong, S. Furnell, and J. Zhou, Surveying the development of biometric user authentication on mobile phones, *IEEE Communications Surveys Tutorials*, **17**: 1268–1293, 2015.

8. K. Scarfone and P. Mell, Guide to Intrusion Detection and Prevention Systems (IDPS), NIST Special Publication. http://csrc.nist.gov/publications/nistpubs/800-94/SP800-94.pdf, pp. 800–894, 2007.

9. G. Vigna and R. A. Kemmerer, NetSTAT: A network-based intrusion detection approach, in *Proceedings of the 1998 Annual Computer Security Applications Conference (ACSAC)*, IEEE Press, New York, pp. 25–34, 1998.

10. A. K. Ghosh, J. Wanken, and F. Charron, Detecting anomalous and unknown intrusions against programs, in *Proceedings of the 1998 Annual Computer Security Applications Conference (ACSAC)*, IEEE Computer Society, Scottsdale, AZ, USA, pp. 259–267, 1998.

11. A. Valdes and D. Anderson, Statistical methods for computer usage anomaly detection using NIDES, Technical Report, SRI International, January 1995.

12. G. A. Jacoby and N. J. Davis IV, Battery-based intrusion detection, in *Proceedings of IEEE Global Telecommunications Conference (GLOBECOM)*, IEEE, December Dallas, Texas, USA, pp. 2250–2255, 2004.

13. D. C. Nash, T. L. Martin, D. S. Ha, and M. S. Hsiao, Towards an intrusion detection system for battery exhaustion attacks on mobile computing devices, in *Proceedings of the 3rd IEEE International Conference on Pervasive Computing and Communications Workshops*, IEEE, Kauai Island, HI, USA, pp. 141–145, 2005.

14. W. Jansen, V. Korolev, S. Gavrila, T. Heute, and C. Séveillac, A unified framework for mobile device security, in *Proceedings of the International Conference on Security and Management (SAM)*, CSREA Press, Las Vegas, Nevada, USA, pp. 9–14, 2004.

15. P. Kannadiga, M. Zulkernine, and S. I. Ahamed, Towards an intrusion detection system for pervasive computing environments, in *Proceedings of the International Conference on Information Technology: Coding and Computing (ITCC)*, IEEE, Bangalore, India, pp. 277–282, 2005.

16. M. Miettinen, P. Halonen, and K. Hatonen, Host-based intrusion detection for advanced mobile devices, in *Proceedings of the International Conference on Advanced Information Networking and Applications (AINA)*, IEEE Computer Society, Vienna, Austria, pp. 72–76, 2006.

17. T. K. Buennemeyer, G. A. Jacoby, W. G. Chiang, R. C. Marchany, and J. G. Tront, Battery-sensing intrusion protection system, in *Proceedings of the 2006 IEEE Workshop on Information Assurance*, IEEE Computer Society, Egham, Surrey, UK, pp. 176–183, 2006.

18. O. Mazhelis and S. Puuronen, A framework for behavior-based detection of user substitution in a mobile context, *Computers & Security*, **26**(2): 154–176, 2007.

19. P. Venkataram, B. Sathish Babu, M. K. Naveen, and G. H. Samyama Gungal, A method of fraud & intrusion detection for e-payment systems in mobile e-commerce, in *Proceedings of the IEEE International Performance, Computing, and Communications Conference*, IEEE Computer Society, New Orleans, Louisiana, USA, pp. 395–401, 2007.

20. L. Niu, X. Tan, and B. Yin, Estimation of system power consumption on mobile computing devices, in *Proceedings of the International Conference on Computational Intelligence and Security (CIS)*, IEEE, Harbin, Heilongjiang, China, pp. 1058–1061, 2007.

21. B.-H. Chung, Y.-H. Kim, and K.-Y. Kim, Just-on-time data leakage protection for mobile devices, in *Proceedings of the International Conference on Advanced Communication Technology (ICACT)*, IEEE, Pyeongchang, Korea, pp. 1914–1915, 2008.

22. A. Brown and M. Ryan, Monitoring the execution of third-party software on mobile devices, in *Proceedings of the 11th International Symposium on Recent Advances in Intrusion Detection (RAID)*, Springer, Cambridge, MA, USA, pp. 410–411, 2008.

23. A.-D. Schmidt, F. Peters, F. Lamour, S. A. Camtepe, and S. Albayrak, Monitoring smartphones for anomaly detection, in *Proceedings of the 1st International Conference on MOBILe Wireless MiddleWARE, Operating Systems, and Applications*, ICST, Innsbruck, Austria, pp. 1–7, 2008.

24. A. Shabtai, U. Kanonov, and Y. Elovici, Detection, alert and response to malicious behavior in mobile devices: Knowledge-based approach, in *Proceedings of RAID 2009*, LNCS 5758, Springer, Saint-Malo, Brittany, France, pp. 357–358, 2009.

25. L. Liu, G. Yan, X. Zhang, and S. Chen, VirusMeter: Preventing your cellphone from spies, in *Proceedings of RAID 2009*, Springer, Saint-Malo, Brittany, France, pp. 244–264, 2009.

26. L. Xie, X. Zhang, J.-P. Seifert, and S. Zhu, PBMDS: A behavior-based malware detection system for cellphone devices, in *Proceedings of the 3rd ACM Conference on Wireless Network Security (WiSec)*, ACM, Hoboken, NJ, USA, pp. 37–48, 2010.

27. A. Shabtai and Y. Elovici, Applying behavioral detection on Android-based devices, in *Proceedings of Mobilware 2010*, Springer, Chicago, IL, USA, pp. 235–249, 2010.

28. A. Chaugule, Z. Xu, and S. Zhu, A specification based intrusion detection framework for mobile phones, in *Proceedings of ACNS*, Springer, Nerja, Spain, pp. 19–37, 2011.

29. I. Burguera, U. Zurutuza, and S. Nadjm-Tehrani, Crowdroid: Behavior-based malware detection system for Android, in *Proceedings of ACM Conference on Computer and Communications Security*, ACM, Chicago, IL, USA, pp. 15–25, 2011.

30. V. Bukac, P. Tucek, and M. Deutsch, Advances and challenges in standalone host-based intrusion detection systems, in *Proceedings of TrustBus 2012*, LNCS 7449, Springer, Vienna, Austria, pp. 105–117, 2012.

31. D. Damopoulos, S. A. Menesidou, G. Kambourakis, M. Papadaki, N. Clarke, and S. Gritzalis, Evaluation of anomaly-based IDS for mobile devices using machine learning classifiers, *Security and Communication Network*, **5**(1): 314, 2012.

32. M. La Polla, F. Martinelli, and D. Sgandurra, A survey on security for mobile devices, *IEEE Communications Surveys and Tutorials*, **15**(1): 446–471, 2013.
33. R. Roshandel, P. Arabshahi, and R. Poovendran, LIDAR: A layered intrusion detection and remediation framework for smartphones, in *Proceedings of the 4th ACM Sigsoft International Symposium on Architecting Critical Systems*, ACM Press, Vancouver, British Columbia, Canada, pp. 27–32, 2013.
34. Q. Li and G. Clark, Mobile security: A look ahead, *IEEE Security & Privacy*, **11**: 78–81, 2013.
35. M. Curti, A. Merlo, M. Migliardi, and S. Schiappacasse, Towards energy-aware intrusion detection systems on mobile devices, in *Proceedings of the 2013 International Conference on High Performance Computing and Simulation*, IEEE, Helsinki, Finland, pp. 289–296, 2013.
36. P. Faruki, A. Bharmal, V. Laxmi, G. Vijay, S. G. Manoj, M. Conti, and M. Rajarajan, Android security: A survey of issues, malware penetration, and defenses, *IEEE Communications Surveys and Tutorials*, **17**(2): 998–1022, 2015.
37. S. Yazji, P. Scheuermann, R. P. Dick, G. Trajcevski, and R. Jin, Efficient location aware intrusion detection to protect mobile devices, *Personal and Ubiquitous Computing*, **18**(1): 143–162, 2014.
38. D. Papamartzivanos, D. Damopoulos, and G. Kambourakis, A cloud-based architecture to crowdsource mobile app privacy leaks, in *Proceedings of the 18th Panhellenic Conference on Informatics (PCI)*, ACM, Athens, Greece, pp. 1–6, 2014.
39. M. Sun, M. Zheng, J. C. S Lui, and X. Jiang, Design and implementation of an Android host-based intrusion prevention system, in *Proceedings of the 30th Annual Computer Security Applications Conference (ACSAC)*, ACM Press, New Orleans, LA, USA, pp. 226–235, 2014.
40. D. Damopoulos, G. Kambourakis, and G. Portokalidis, The best of both worlds: A framework for the synergistic operation of host and cloud anomaly-based IDS for smartphones, in *Proceedings of the Seventh European Workshop on System Security (EuroSec)*, ACM, New York, Article 6, 2014.
41. D. Damopoulos, G. Kambourakis, S. Gritzalis, and S. O. Park, Exposing mobile malware from the inside (or what is your mobile app really doing?), *Peer-to-Peer Networking and Applications*, **7**(4): 687–697, 2014.
42. W. Meng, W. Li, X., Yang, and K. K. R Choo, A Bayesian inference-based detection mechanism to defend medical smartphone networks against insider attacks, *Journal of Network and Computer Applications*, **78**: 162–169, 2017.
43. J. S. Ransbottom and G. A. Jacoby, Monitoring mobile device vitals for Effective Reporting (ER), in *Proceedings of the IEEE Conference on Military Communications*, IEEE, Washington, DC, USA, pp. 329–335, 2006.
44. R. A. Malaney, Wireless intrusion detection using tracking verification, in *Proceedings of the IEEE International Conference on Communications*, IEEE, Glasgow, Scotland, pp. 1558–1563, 2007.
45. G. A. Jacoby and S. Mosley, Mobile security using separated deep packet inspection, in *Proceedings of the 5th IEEE Consumer Communications and Networking Conference (CCNC)*, IEEE, Las Vegas, NV, USA, pp. 482–487, 2008.
46. Y. Meng, W. Li, and L.-F. Kwok, Single character frequency-based exclusive signature matching scheme, in *The 11th IEEE/ACIS International Conference on Computer and Information Science (ICIS 2012)*, Studies in Computational Intelligence, Springer, Shanghai, China, pp. 67–80, 2012.

47. Y. Meng and L.-F. Kwok, Adaptive context-aware packet filter scheme using statistic-based blacklist generation in network intrusion detection, in *Proceedings of the 7th International Conference on Information Assurance and Security (IAS)*, IEEE, Melacca, Malaysia, pp. 74–79, 2011.

48. Y. Meng and L.-F. Kwok, Adaptive blacklist-based packet filter with a statistic-based approach in network intrusion detection, *Journal of Network and Computer Applications*, **39**: 83–92, 2014.

49. Y. Meng and L.-F. Kwok, Enhancing list-based packet filter using ip verification mechanism against ip spoofing attack in network intrusion detection, in *Proceedings of the 6th International Conference on Network and System Security (NSS)*, Lecture Notes in Computer Science 7645, Springer, Wuyishan, Fujian, China, pp. 1–14, 2012.

50. Y. Meng, L.-F. Kwok, and W. Li, Towards designing packet filter with a trust-based approach using Bayesian inference in network intrusion detection, in *Proceedings of the 8th International Conference on Security and Privacy in Communication Networks (SECURECOMM)*, Lecture Notes in ICST 106, Springer, Padua, Italy, pp. 203–221, 2012.

51. Y. Meng and L.-F. Kwok, Adaptive false alarm filter using machine learning in intrusion detection, in *Proceedings of the 6th International Conference on Intelligent Systems and Knowledge Engineering (ISKE)*, Advances in Intelligent and Soft Computing, Springer, Shanghai, China, pp. 573–584, 2011.

52. Y. Meng, W. Li, and L.-F. Kwok, Intelligent alarm filter using knowledge-based alert verification in network intrusion detection, in *The 20th International Symposium on Methodologies for Intelligent Systems (ISMIS)*, Lecture Notes in Artificial Intelligence 7661, Springer, Macau, China, pp. 115–124, 2012.

53. Y. Meng and L.-F. Kwok, Adaptive non-critical alarm reduction using hash-based contextual signatures in intrusion detection, *Computer Communications*, **38**: 50–59, Elsevier, 2014.

54. Y. Meng, W. Li, and L.-F. Kwok, Towards adaptive character frequency-based exclusive signature matching scheme and its applications in distributed intrusion detection, *Computer Networks*, **57**(17): 3630–3640, Elsevier, 2013.

55. W. Meng and L.-F. Kwok, Enhancing the performance of signature-based network intrusion detection systems: An engineering approach, *HKIE Transactions*, **21**(4): 209–222, Taylor & Francis, 2014.

56. W. Meng, W. Li, and L.-F. Kwok, EFM: Enhancing the performance of signature-based network intrusion detection systems using enhanced filter mechanism, *Computers & Security*, **43**: 189–204, Elsevier, 2014.

57. W. Meng, W. H. Lee, S. R. Murali, and S. P. T Krishnan, JuiceCaster: Towards automatic juice filming attacks on smartphones, *Journal of Network and Computer Applications*, **68**: 201–212, 2016.

58. D. Fiorella, A. Sanna, and F. Lamberti, Multi-touch user interface evaluation for 3D object manipulation on mobile devices, *Journal on Multimodal User Interfaces*, **4**(1): 3–10, 2010.

59. M. Frank, R. Biedert, E. Ma, I. Martinovic, and D. Song, Touchalytics: On the applicability of touchscreen input as a behavioral biometric for continuous authentication, *IEEE Transactions on Information Forensics and Security*, **8**(1): 136–148, 2013.

60. T. Kwon and S. Na, TinyLock: Affordable defense against smudge attacks on smartphone pattern lock systems, *Computers & Security*, **42**: 137–150, 2014.

61. D. Damopoulos, G. Kambourakis, and S. Gritzalis, From keyloggers to touchloggers: Take the rough with the smooth, *Computers & Security*, **32**: 102–114, 2013.

62. A. De Luca, A. Hang, F. Brudy, C. Lindner, and H. Hussmann, Touch Me Once and I Know It's You!: Implicit authentication based on touch screen patterns, in *Proceedings of the 2012 ACM Annual Conference on Human Factors in Computing Systems (CHI)*, ACM, New York, pp. 987–996, 2012.

63. Y. Meng, D. S. Wong, and L.-F. Kwok, Design of touch dynamics based user authentication with an adaptive mechanism on mobile phones, in *Proceedings of the 29th Annual ACM Symposium on Applied Computing (SAC)*, ACM Press, Gyeongju, Korea, pp. 1680–1687, 2014.

64. Y. Meng, D. S. Wong, R. Schlegel, and L.-F. Kwok, Touch gestures based biometric authentication scheme for touchscreen mobile phones, in *Proceedings of the 8th China International Conference on Information Security and Cryptology (INSCRYPT)*, Springer, Beijing, China, pp. 331–350, 2012.

65. I. Oakley and A. Bianchi, Multi-touch passwords for mobile device access, in *Proceedings of the 2012 ACM Conference on Ubiquitous Computing (UbiComp)*, ACM Press, Pittsburgh, PA, USA, pp. 611–612, 2012.

66. N. Sae-Bae, K. Ahmed, K. Isbister, and N. Memon, Biometric-rich gestures: A novel approach to authentication on multi-touch devices, in *Proceedings of the SIGCHI Conference on Human Factors in Computing Systems (CHI)*, ACM Press, Austin, Texas, USA, pp. 977–986, 2012.

67. N. Sae-Bae, N. Memon, K. Isbister, and K. Ahmed, Multitouch gesture-based authentication, *IEEE Transactions on Information Forensics and Security*, **9**(4): 568–582, 2014.

68. J. Thorpe and P. C. Van Oorschot, Human-seeded attacks and exploiting hotspots in graphical passwords, in *Proceedings of 16th USENIX Security Symposium on USENIX Security Symposium*, USENIX Association, Boston, MA, USA, pp. 1–16, 2007.

69. W. Meng, W. Li, L. F. Kwok, and K. K. R Choo, Towards enhancing click-draw based graphical passwords using multi-touch behaviours on smartphones, *Computers & Security*, **65**: 213–229, 2017.

70. T. Takada and Y. Kokubun, MTAPIN: Multi-touch key input enhances security of PIN authentication while keeping usability, *International Journal of Pervasive Computing and Communications*, **10**(3): 276–290, 2014.

71. F. Tari, A. A. Ozok, and S. H. Holden, A comparison of perceived and real shoulder-surfing risks between alphanumeric and graphical passwords, in *Proceedings of the 2nd Symposium on Usable Privacy and Security (SOUPS)*, ACM, New York, NY, USA, pp. 56–66, 2006.

72. D. Van Thanh, Security issues in mobile eCommerce, in *Proceedings of the 11th International Workshop on Database and Expert Systems Applications (DEXA)*, IEEE, USA, pp. 412–425, 2000.

73. P. C. Van Oorschot, A. Salehi-Abari, and J. Thorpe, Purely automated attacks on PassPoints-style graphical passwords, *IEEE Transactions on Information Forensics and Security*, **5**(3): 393–405, 2010.

74. P. C. Van Oorschot and J. Thorpe, Exploiting predictability in click-based graphical passwords, *Journal of Computer Security*, **19**(4): 699–702, 2011.

75. S. Wiedenbeck, J. Waters, J.-C. Birget, A. Brodskiy, and N. Memon, Passpoints: Design and longitudinal evaluation of a graphical password system, *International Journal of Human-Computer Studies*, **63**(1–2): 102–127, 2005.

76. S. Wiedenbeck, J. Waters, J.-C. Birget, A. Brodskiy, and N. Memon, Authentication using graphical passwords: Effects of tolerance and image choice, in *Proceedings*

of the 2005 Symposium on Usable Privacy and Security (SOUPS), ACM, Pittsburgh, Pennsylvania, USA, pp. 1–12, 2005.

77. T.-S. Wu, M.-L. Lee, H.-Y. Lin, and C.-Y. Wang, Shoulder-surfing-proof graphical password authentication scheme, *International Journal of Information Security*, **13**(3): 245–254, 2014.

78. B. B. Zhu, J. Yan, G. Bao, M. Yang, and N. Xu, Captcha as graphical passwords—A new security primitive based on hard AI problems, *IEEE Transactions on Information Forensics and Security*, **9**(6): 891–904, 2014.

79. Y. Meng, Designing click-draw based graphical password scheme for better authentication, in *Proceedings of IEEE International Conference on Networking, Architecture, and Storage (NAS)*, IEEE Press, Xiamen, China, pp. 39–48, 2012.

80. Y. Meng, W. Li, and L.-F. Kwok, Design of cloud-based parallel exclusive signature matching model in intrusion detection, in *The 15th IEEE International Conference on High Performance Computing and Communications (HPCC)*, IEEE, pp. 2013.

Chapter 2

Attacking Smartphone Security and Privacy

Vincent F. Taylor and Ivan Martinovic

Contents

2.1 Introduction

The smartphone landscape continues to grow at an explosive pace as devices become more powerful, feature-rich, and more affordable to the average consumer. Gartner reports smartphone sales as exceeding 1.4 billion units in 2015, up 14.4% over the previous year [1]. Smartphones offer greatly increased functionality over traditional feature phones due to the availability of full-blown operating systems providing advanced APIs to third-party app developers. Smartphones are predominantly powered by Android or iOS, with Android maintaining a commanding lead of the market with 84.7% market share as of 2015 Q3 with iOS in a far second at 13.1% [2]. Other operating systems represented in the landscape include Windows Phone and BlackBerry among others. On top of the operating system, smartphones offer a variety of network interfaces for connectivity, multitasking facilities, and open application programming interfaces (APIs) for supporting third-party app development. Android and iOS have rich app marketplaces, each offering access to approximately 1.5 million apps [3,4] that add additional functionality to the smartphone. The always-connected, extremely extensible nature of smartphones exposes a large footprint on the device where weaknesses in the underlying hardware or software may be exploited by an attacker.

    The smartphone landscape is very large, and has a number of layers of software, protocols, and services that work together to deliver an experience to the consumer. The interaction between consumer, apps, smartphone, service provider, and the wider Internet is supported by various wireless protocols that provide connectivity. Thus, a smartphone may be vulnerable to attacks coming from installed apps, wireless interfaces, running services, and the underlying configuration of the device. We are motivated to systematize this knowledge of attacks and attack vectors, as this will provide a compendium to security researchers intending to develop intrusion detection and prevention systems* (IDS) for the smartphone ecosystem. We do this by comprehensively enumerating the ways in which the security and privacy of a smartphone can be attacked. By understanding the ways in which smartphones can be attacked, we obtain a mechanism to compare them to traditional workstations, giving useful insight into the additional or varied risks that need to be addressed when building technology to secure smartphones.

---

* For brevity, we refer to intrusion detection and prevention systems as simply IDS for the remainder of this chapter.

2.2 Background

To understand the ways smartphones are attacked, we first need to understand the operating systems that run on these devices and the security models and features they employ. There are four major smartphone operating systems: Android, iOS, Windows Phone, and BlackBerry [5]. In addition to typical low-level tasks such as memory management and process scheduling, smartphone operating systems provide features critical in today's smartphone landscape, such as allowing access to a touchscreen, camera, Bluetooth, Wi-Fi, NFC, GPS, microphone, and other such hardware. Aside from providing access to the typical smartphone hardware, the operating system also mediates access to the underlying cellular radio, enabling communication with a mobile network carrier.

2.2.1 iOS

iOS is a mobile operating system developed by Apple Inc. It has a healthy app ecosystem that surrounds it with over 1.4 million iOS applications available for download. The operating system itself is proprietary, closed source, and written in C, C++, Objective-C, and Swift. It is a Unix-like operating system and features a hybrid kernel that runs on 64/32-bit ARM processors. Before iOS apps are made available to the public in the Apple App Store, they must undergo a thorough vetting process by Apple. Apps must pass reliability testing and other analysis to ensure that they are not malicious or otherwise unsavory. Apple's vetting process includes manual testing and static analysis to determine whether an app tries to perform actions outside of what it claims to do [6]. This vetting process is not always perfect and indeed security researchers have uncovered ways of circumventing the protections put in place by Apple [7]. In the case of Jekyll [8], the malicious app passed the vetting process by rearranging its code to add new, malicious functionality, after passing the approval process. The iOS kernel uses code signing to ensure that all apps running on a device come from an approved source and have not been tampered with [9]. Additionally, all third-party apps are sandboxed by iOS to prevent them from accessing data stored by other apps and modifying the system. However, Han et al. described how to "break out" of the iOS sandbox by leveraging dynamically loaded, private APIs in malicious apps [10]. Finally, iOS enforces a secure boot chain and file encryption using a per-file key.

2.2.2 Android

Android is a mobile operating system developed by Google and the Open Handset Alliance. Android devices are powered by a healthy app ecosystem providing access to over 1.6 million apps. The core of the operating system is written in C, with additional components written in C++, and the user interface portions written in Java. Like iOS, it is also a Unix-like operating system; however, it features a monolithic kernel, designed to run on a number of processor platforms such as ARM, x86, and MIPS. In stark contrast to the Apple App Store, the Google Play Store does not require an exhaustive app vetting process before an app is admitted to the store. In general, apps are dynamically tested with a Google security service known as Bouncer [12]. Google automatically scans apps using dynamic analysis and combines the results of this analysis with signals from its reputation engine after it has analyzed the account of the app developer themselves. Security researchers John Oberheide and Charlie Miller demonstrated techniques that could be used to fingerprint Bouncer [13]. They identified unique characteristics of the Bouncer emulation framework such as the hostname, phone number, and Android ID. By checking for these fingerprints, malware can pretend to be benign when being tested by Bouncer and then become malicious when installed on victim devices. In an early study [14], Enck et al. analyzed the source code of 1100 Android apps and found no evidence of malware or exploitable vulnerabilities. Unfortunately, the landscape has deteriorated since then. Indeed, Zhou and Jiang [15] provide a characterization of the evolution of Android malware. On the Android platform, every app runs in its own sandbox by default. As a result of this, each app is isolated from other apps and the system itself, except by using well-defined APIs and system services such as interprocess communication (IPC). However, researchers have found ways for apps to break out of their sandbox and read arbitrary files using symbolic links [16]. Android uses a Linux-like user approach, where each app is executed as a different user and thus inherits the security provided by the operating system in protecting its resources and files. In addition to sandboxing and permissions, Android is also designed to prevent platform modification by malware and also has the capability of remotely removing malware from a device if required [17].

Comparison of Smartphone Operating Systems: Table 2.1 shows a comparison of the similarities and differences between the four most popular smartphone operating systems, and summarizes our effort in distilling this information from the literature [18–23]. For brevity, we do not compare an exhaustive set of features for these operating systems. Instead, we target the main characteristics of the operating systems that contribute the most to vulnerabilities, and thus are most interesting to IDS developers. We look at the OS family, CPU architectures supported, source code model, programming languages used, and reverse engineering tools that are available. Android is based on the Linux family of operating system, while iOS' Darwin and BlackBerry's QNX are Unix-like operating systems. The outlier here is the Windows Phone operating system, which is built around the Windows family of operating systems. All four of these smartphone operating systems are built to run on ARM processors, with Android offering the capability to run on x86 and MIPS processors as well. Android dominates the market, being delivered on 82.8% of smartphones, iOS on 13.9%, and Windows Phone and BlackBerry trailing distantly with 2.6% and 0.3% deployment, respectively, as of 2015 Q3 [2]. Android is the only open-source operating system on the list and all are written in C/C++

Table 2.1 Summary of the Main Characteristics of Android, iOS, Windows Phone, and BlackBerry That Contribute to Their Attack Surface

Operating System	Android	iOS	Windows Phone	BlackBerry
OS family	Linux	Darwin	Windows CE-7, Windows NT-8	QNX
Vendor	Google Inc., Open Handset Alliance (and OEMs)	Apple Inc.	Microsoft (and OEMs)	BlackBerry Ltd.
CPU architecture	ARM, ARM64, x86, MIPS	ARM, ARM64	ARM (ARM64 upcoming [11])	ARM
Market share (2015 Q3) [2]	84.7%	13.1%	1.7%	0.3%
Source code	Open	Closed	Closed	Closed
Programming language	C, C++, Java	C, C++, Objective-C, Swift	C, C++	C++
Application store	Google Play Store	App Store	Windows Phone Store	BlackBerry World
Reverse engineering tools	apktool, dex2jar, JD-Compiler, XDA auto tool	iRET toolkit, Windows Explorer, oTool, iExplorer, Class-dump-z	Decompresser, Visual Studio, .NET Decompiler	JD-GUI, VSMTool, COD extractor

or other variants of C. Each of the operating systems is supported by a single official application marketplace, which provides third-party apps to users. Importantly, Android and Windows Phones have OEMs that (sometimes) modify the standard operating systems and unwittingly introduce vulnerabilities [24]. Finally, all four operating systems have a suite of reverse engineering tools available to assist with vulnerability analysis.

Definitions: We now define key terms that are used throughout the chapter.

- *Attack vector*: The means by which an attack is carried out against a system.
- *Exploit*: The method used to take advantage of a vulnerability.
- *Vulnerability*: Any weakness in a system that exposes it to risk.

Threat model: In evaluating the landscape as it concerns intrusion detection and prevention, we systematize adversaries based on their capabilities, goals, and relationship to the smartphone under attack:

- *Local adversary (active/passive)*: This attacker is present on or controls the local network. A local attacker may also be logically adjacent to the device (e.g., spoofing a cell tower) or have close physical access to a device (e.g., in close physical contact with the victim).
- *Remote adversary (active/passive)*: This attacker is present outside of the local network and may control segments of the network between the victim and the destination of their traffic.

Passive adversaries may eavesdrop on and observe traffic from the communication channels in the network. They may also observe data from *side channels*, such as device sensors [25], power consumption [26,27], and wireless transmissions [28,29]. Conversely, active adversaries are able to read, modify, or inject data into a communication channel. Note that malicious app developers (or adversaries who modify/repackage apps) fall into the category of active remote adversary. Our types of adversary are not necessarily mutually exclusive. Indeed, adversaries may change position in the network and more than one adversaries may collude to achieve a more complex objective. The specific target of the adversary may be one or more of

- *The victim themselves*: The adversary is intending to cause harm to the victim and does this by attacking their smartphone to cause loss of data or perform denial-of-service (DoS).
- *The device itself*: The adversary may be intent on exfiltrating personal data from the device such as contacts, credit card information, social security numbers, pictures, or videos. In the case of corporate espionage, the adversary may be targeting the employee of a company to obtain intellectual property, unpublished reports, or other sensitive business data.
- *Device resources*: Data on a device may be immaterial to an adversary who is targeting smartphones to exploit their resources such as storage, processing power, and bandwidth. This is especially common for adversaries interested in "recruiting" devices for a botnet.

For the remainder of this chapter, we frame the attacks and attack vectors in relation to the position and intent of the adversary. This is summarized by the flowchart of the decision-making process of an attacker shown in Figure 2.1. In general, an adversary has one of three objectives when attacking a smartphone:

■ *Perform DoS*: The adversary is concerned with preventing the device from performing its prescribed functionality. This attack is fairly noticeable, since the user will perceive degradation in performance or a missing device (in the case of theft).

■ *Utilize device resources*: The adversary is concerned with leveraging the resources (CPU, memory, network access) of a device to further their own goals, for example, recruiting devices for a botnet or as a proxy for launching further attacks.

■ *Steal data*: The adversary is concerned with obtaining sensitive data from a device such as user account information, credit cards, multimedia, and sensor data. Note that the sensitive data the adversary is interested in may not yet exist, so the adversary may plant a backdoor, for example, when spying on a spouse.

2.3 Smartphone Attack Vectors

Developing IDSs for smartphones is complicated by the fact that smartphones are devices that communicate over a variety of wireless interfaces/networks and provide a highly customizable and extensible platform. Thus, smartphones will necessarily have a number of areas that must be exposed in order for them to provide their stated functionality. Moreover, what really distinguishes smartphones from other computing platforms is the multitude of sensors they contain and their ultra-high mobility, which makes them susceptible to loss/theft/physical access. The following list distils [30–32] 13 vulnerable areas (or "weak points)" on typical smartphones that will continue to be targets for delivering exploits:

■ *Browser*: May contain vulnerabilities in parsing web pages, processing Javascript, or providing WebView functionality to apps.

■ *Baseband processor*: Smartphones can be tricked into connecting to rogue base stations, which can then attack the mobile radio interface.

■ *Messaging services*: Short message service (SMS)/multimedia messaging service (MMS) messages may be used to deliver malicious payloads.

■ *Wireless interfaces*: Attackers can attempt to attack a smartphone from any one of the myriad of (noncellular) wireless interfaces.

■ *SIM card*: Attackers may be able to manipulate SIM cards to attack a device or steal data.

■ *Memory card*: Many smartphones provide slots for external memory cards. These are frequently unencrypted and data can be retrieved if the smartphone or memory card itself is misplaced.

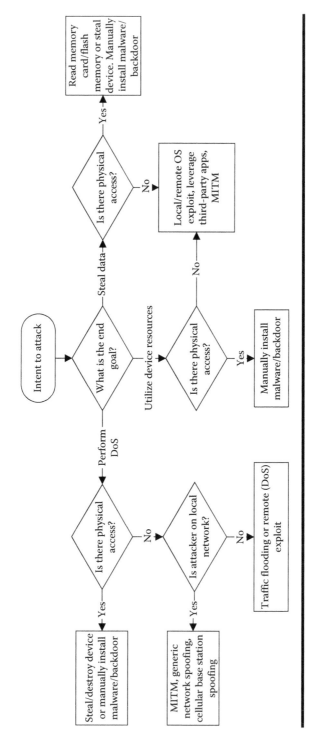

Figure 2.1 Flowchart of a typical smartphone attack from the attacker's point of view.

- *Hardware interfaces/ports*: Smartphones may be vulnerable to attacks coming over their exposed ports, such as USB ports. By opening a device, an attacker can also try to pilfer data through low-level circuitry such as JTAG ports.
- *Operating system*: Adversaries can attack typical weaknesses found in operating systems. In some cases, smartphone operating systems may not be as mature (or robust) as desktop operating systems.
- *Third-party apps*: Third-party apps can access any resource that they have been given permission to access. Additionally, apps can attempt to break out of the OS-provided sandbox. Attackers can also use vulnerable third-party apps as a proxy for conducting further attacks on a smartphone through privilege escalation.
- *Users*: Users can advance attacks if they make bad device configuration choices or are victims of social engineering.
- *Memory*: Physical memory on the device can be modified to remove protective mechanisms from the system.
- *Firmware*: An attacker may target submodule (such as Wi-Fi interface cards) firmware to obtain long-term elevated privileges on the device.
- *Device itself*: An attacker may target any one of the side channels coming from the device itself, for the purposes of device fingerprinting or recovering sensitive data such as encryption keys or screen-lock codes.

We systematize attack vectors as belonging to either of four categories: drive-by attacks, app ecosystems, physical attacks, and social engineering [31,33,34]. In detailing attack vectors, we use *italicized text* to denote the vulnerable areas affected. Also, we only briefly identify some attacks to put the attack vectors into perspective. We leave a detailed treatment of all the major attacks against smartphones for Section 2.4.

2.3.1 Drive-by Attacks

Vulnerable areas affected: Browser, baseband processor, messaging services, wireless interfaces, operating system, third-party apps, users.

In the case of drive-by attacks (or watering hole attacks), an attacker attempts to exploit existing bugs in the software running on the smartphone that processes external data. A common method of delivering drive-by attacks involves exploiting vulnerabilities in the *browser* on the smartphone to make it execute a malicious payload. These attacks can be carried out *en masse* since popular web pages can be compromised and laced with malicious payloads. Alternately, links to malicious pages can be sent to *users* through traditional channels such as email, *messaging services*, and social media. An attacker can also manipulate unencrypted HTTP pages to insert malicious payloads or take advantage of improperly handled SSL/TLS (in *third-party apps*) to perform a man-in-the-middle (MITM) attack [35] and insert the malicious payload that way. Note, however, that attacks targeting smartphone browsers have more limited potential than desktop browsers due to application isolation via sandboxing.

Drive-by attacks can also leverage any one of a device's *wireless interfaces*. Various spoofing attacks can be performed against Wi-Fi and Bluetooth, and indeed, base station spoofing can used against the *baseband processor*. Other targets of drive-by attacks may include WebViews (*operating system*), a user interface component in smartphone development frameworks that allows apps to easily render web pages from within the app. A WebView can be used to provide an interface to a Javascript component and thus external Javascript can be executed on a device. Additionally, WebViews can also allow a website to access data stored on devices. If an attacker is able to intercept or modify the content of a URL that is loaded by a WebView (using an MITM attack or cross-site scripting), they can use functions from within the WebView framework to access data from the device. Worryingly, privilege escalation exploits have been published that allow arbitrary code execution on vulnerable devices through WebViews [36]. Another class of attacks, known as component hijacking attacks [37], leverage the drive-by attack principle to access private data and spoof intents.

Drive-by attacks may exploit network services, pieces of software running on a device that open ports to listen for incoming connections. Traditionally, network services only run on devices acting as servers, such as web (HTTP) servers listening on port 80. In the smartphone landscape, however, to satisfy the great need for interconnectivity, mobile devices can be found running network services such as Android Debug Bridge (ADB) [38], Virtual Private Network (VPN), Virtual Network Computing (VNC), Remote Desktop (RDP), and Secure Shell (SSH) services. In the case where smartphones are configured to share their Internet connection through a mobile hotspot, they can be expected to also run Dynamic Host Configuration Protocol (DHCP) services and act as a default gateway. These additional services all increase the number of avenues for exploit. Network services offer an attractive interface for attackers to attempt to exploit, since they provide a (usually) always open entrance that is accessible via the network. Exploiting network services is also particularly attractive to an adversary since no user intervention is typically required to allow the exploit to take place and after successful exploitation there may be no immediate indication to the user that an attack has indeed happened. By default, smartphones may not have any network services installed, but there are a wide variety of third-party apps that users install, which offer additional functionality that requires the use of network services. Indeed, Nielson reports that the average user uses 26.8 different apps per month [39], and any of these could potentially leverage network services.

Drive-by attacks, while successful on traditional workstations, may have more limited impact when translated to the smartphone arena. On Android, most software is implemented in Java and executed by the Dalvik Virtual Machine. This mitigates some of the typical attack strategies (such as buffer overflows) since low-level data structures are protected by boundary checks. However, many Android apps also leverage libraries implemented in native code; thus, some parts of many apps continue to be susceptible to traditional attacks against memory corruption bugs. These attacks are quite dangerous since they can lead to code execution on the device [40], with the user not necessarily knowing that they have been compromised.

2.3.2 App Ecosystems

Vulnerable areas affected: Third-party apps.

By and large, user installation of grayware/malware is limited due to the use of "trusted" software repositories such as the official app stores. App stores and smartphone operating systems utilize strong technical mechanisms to ensure a restriction on the *third-party apps* that can be installed on a device. Attacks coming from app ecosystems leverage the fact that if grayware/malware can be placed in a marketplace, it can quite quickly be available for infecting the entire ecosystem. Additionally, grayware/malware authors are incentivized by the fact that users are typically more trusting of apps if they find them in the official app marketplaces.

As mentioned in Section 2.2, app stores employ various degrees of vetting before allowing an app to become available for the general public to download. Additionally, smartphone operating systems restrict, by default, the "sideloading" of apps, that is, installing apps to a device through unofficial channels. For this reason, most grayware/malware affecting smartphones are delivered as Trojan-horse apps via an app store. Thus, malicious authors must develop apps with some functionality, but containing malicious payloads hidden from app store vetting using timebombs, dynamic code loading, reflection, code obfuscation, and/or IP address checking (to determine whether the app is being run through an app store's vetting engine). Recently, the BrainTest trojan [41] utilized all the aforementioned strategies to evade app store detection.

One strategy used by grayware/malware authors, and most commonly observed in third-party app marketplaces, involves the repackaging of legitimate apps to inject malware [15], which can then attempt to exploit the *operating system/firmware* or steal data from the *memory card*. Fraudsters have also been known to modify the advertising portions of legitimate apps to insert their own code. This allows them to fraudulently obtain revenue from a legitimate app [42]. Other less malicious apps (and their included libraries) have been known to leverage additional and unnecessary dangerous permissions, ostensibly to have greater access to sensitive data and resources, which can then be used for profiling a user [43–45] for reasons such as better advertisement targeting, or more maliciously, selling user data directly to other third parties.

Smartphone worms are much more limited than Trojan-horse apps but may begin to see wider adoption with the availability of operating system exploits propagated by modern smartphone connectivity features such as portable hotspots/NFC and even older channels such as Bluetooth/SMS/MMS/WAP. Operating system protection mechanisms, such as SELinux, offer mitigation for system exploits by enhancing the boundaries of app sandboxes [46] and thus worms may have more limited success.

2.3.3 Physical Attacks

Vulnerable areas affected: SIM card, memory card, hardware interfaces/ports, memory, firmware, device itself.

One class of physical attacks come about from dismantling the *device itself* and/or being able to connect to and interface directly with the *hardware interfaces/ports* on the device; we call these *physical (tampering)* attacks. Physical attacks may also make use of side channels that enable the inference of private data located on a device; we call these *physical (general)* attacks. We expand on specific physical attacks in Section 2.4.1, but right now we enumerate general physical attack approaches:

1. Accessing a device that does not use a screen-lock and transferring the data from the device using copy/paste/attach features within the operating system.
2. Accessing *memory cards* within the device itself and removing them to obtain data that was stored on the device.
3. Inferring PIN/screen-lock codes from smudges on a smartphone touch-screen [47].
4. Leveraging ports, such as USB ports, on the device to perform further attacks [48–50].
5. Modifying physical *memory* chips on the circuit board to introduce new software and/or affect the *firmware* of low-level hardware (such as Wi-Fi adaptors).
6. The *SIM card(s)* in a device can be removed to retrieve sensitive data such as messages and phone numbers. Malicious payloads can also be written to a SIM card.

An attacker can leverage their access to the *hardware interfaces/ports* of a device to place malware or other data on the device or execute commands. Lau et al. demonstrated how it was possible to install arbitrary apps on an iOS device through the USB port [49]. The ADB can also be used to launch attacks. The ADB is a command line tool that can be used to connect to and run commands on Android devices using a desktop.

Another class of physical attacks comes from leveraging the physical state of a device or physical access to the device to attack it. Leveraging the physical sensors on a smartphone is an example of utilizing the physical state of a device to enable attacks. The literature exemplifies using the accelerometer/gyroscope [25,51], and light sensor [52] to steal device credentials/passwords.

Attackers can also leverage physical access to a device to attempt to pass the "lock screen," provided that screen-locking is enabled in the first place [53]. A locked device is usually guarded by PINs, patterns, and passwords, and more recently, using biometrics such as fingerprints. An adversary being able to success-fully unlock a device depends on the complexity of the credential used to lock the device. Approaches as rudimentary as looking at screen smudges have demonstrated potential in assisting attackers to bypass locked screens [47]. Biometric approaches, which show much promise, have been shown to be dangerous if implemented incor-rectly [54], with the end result being a potential compromise of a user's biometric data such as a fingerprint. Needless to say, the compromise of a user's biometric is a serious problem, as by nature it cannot be replaced.

2.3.4 Social Engineering

Vulnerable areas affected: Browser, operating system, users.

With social engineering, the user of a smartphone is tricked into revealing cre-dentials or performing actions that assist the attacker in furthering their attack. These attacks are dangerous in that they employ nontechnical strategies to elicit private information from users and, as such, generic IDS solutions to address social engineering are not available. The problem of social engineering is exacerbated by the fact that users may not know that they have been successfully attacked until long after the fact, if at all. Three common social engineering attacks specific to smartphones are

1. *Making malicious apps look like legitimate apps*: Malware/grayware authors typ-ically build clones of popular applications to trick a user into installing their version because it has a name and description very similar to the app they actually want [42,55].
2. *Enticing users using device-specific details*: Smartphone users may be tricked by ads and web pages that give them advice specific to their device make and model. Attackers commonly use the User-Agent sent by a browser/app to identify the device before sending customized messages to the user about faults with their specific device such as poor battery or malware infections. The users are then led to download malware, which supposedly solves their "problems" [56].
3. *Malware pretending to be a second factor of authentication*: Desktops infected by the Zeus malware may instruct users to download an authentication compo-nent to their smartphone as a second factor of authentication when the user attempts to log in to their online bank [57]. The malware then captures a user's bank login credentials.

Aside from social engineering that leads to malware installation, other typical social engineering attacks that result in the user giving away their credentials are just as detrimental as on traditional desktops. Especially considering that a user may be logged into several services from their smartphone at the same time, social engineering presents a high-reward attack vector to adversaries.

Table 2.2   Attack Vectors and What Vulnerable Area on the Smartphone They Target

Attack Vector	Vulnerable Area Affected
Drive-by attacks	Browser, messaging services, wireless interfaces, SIM card, memory card, operating system, third-party apps, users, memory, firmware
App ecosystems	Third-party apps
Physical attacks	Baseband processor, SIM card, memory card, hardware interfaces, USB, memory, firmware
Social engineering	Browser, operating system, users

Table 2.2 summarizes the relationship between the attack vectors and the vulnerable areas that they target. From the table, it can be seen that drive-by attacks have the potential to affect the most areas on a smartphone. This is perhaps unsurprising, as drive-by attacks are made possible by bugs/vulnerabilities in the software on the smartphone itself, and thus there is a rich attack surface that can be targeted. Physical attacks have the second largest number of vulnerable areas and target weaknesses in the physical hardware/characteristics of the smartphone. App ecosystems and social engineering target fewer vulnerable areas directly, but can be used as a proxy for delivering more dangerous drive-by exploits if users are tricked into installing apps or performing particular actions on their device.

Table 2.3 shows common attack vectors, the level of sophistication required to achieve success, and the potential effect of device compromise. The level of sophistication refers to the technical expertise required from the attacker and ranges from low (minimal technical ability required), medium (moderate technical ability required, with published exploits easily available/adaptable), to high (advanced technical ability required, usually requiring the development of zero-day exploits or advanced reverse-engineering skills). The effect of compromise ranges from low (information disclosure or minor annoyance), medium (low + greater annoyance

Table 2.3   Attack Vectors and Their Main Characteristics

Attack Vector	Level of Sophistication	Effect of Compromise
Drive-by attacks	Medium/high	Low/medium/high
App ecosystems	Low	Low/medium
Physical (tampering)	High	High
Physical (general)	Low/medium	Low/medium
Social engineering	Low	Low/medium

and potentially costing the user money, e.g., premium rate SMS/calls), to high (full compromise of the device with unfettered access by the adversary). The main insight from Table 2.3 is that social engineering requires minimal skill and has the potential to affect many victims, but the effect of the attack is typically low. Drive-by attacks, on the other hand, can prove to be very effective to attackers since published exploits are available (especially for older devices that are still widely used [58]) and can yield good returns in terms of the effect of compromise while targeting a moderate number of victims. Physical tampering of devices requires high sophistication by adversaries but may yield significant rewards and are usually employed at the nation-state/law-enforcement level. Worryingly, unsophisticated attackers can combine social engineering with published drive-by exploits to obtain a significant return on investment, especially if attacks target users with older devices.

2.4 Smartphone Attack Hierarchy

We classify smartphone attacks based on the position of the attacker in the "space" relative to the smartphone under attack as follows:

- *Physical versus nonphysical*: As shown in Figure 2.2, the first level of differentiation is whether the attack is performed by physically accessing the device. This is a logical separation of attacks as it broadly divides attacks into those that require tangible access to a device as opposed to those that access the device in an intangible way. IDS developers will typically focus on nonphysical attacks. Nonphysical (or intangible) attacks can be further separated into two categories: local and remote.
- *Local versus Remote*: Local and remote refer to the logical proximity of the attacker to the victim device in terms of location on the network. Broadly speaking, local attacks are carried out by attackers that are on the current local

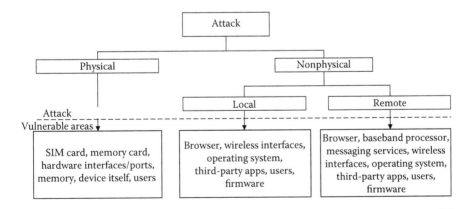

Figure 2.2   Taxonomy of smartphone attacks.

area network (LAN) segment (or otherwise logically adjacent to the victim) and includes well-known attacks such as Address Resolution Protocol (ARP) spoofing, MITM, and traffic analysis attacks. Remote attacks are carried out by adversaries who are able to launch attacks from beyond the local network segment, that is, more than one network hop away.

■ *Interactive versus noninteractive*: Attacks can also be categorized into whether they are interactive or noninteractive. By interactive and noninteractive, we mean whether the smartphone user is required to perform a particular action on their smartphone for the attack to be successful. In general, noninteractive attacks are more attractive (and more difficult to exploit) since no user intervention is required and, as a result, may be more stealthy.

In Sections 2.4.1 and 2.4.2, we compare typical attacks that fall into the categories of physical and nonphysical, to assess their characteristics relative to each other and gain an understanding of the motivations of attackers for using one type of attack over another. Table 2.4 provides a compendium of examples of attacks for all the exploits mentioned in this section.

2.4.1 Physical Attacks

Physical attacks are carried out by attackers that target the hardware of the device itself. In other words, these attacks require attackers to physically touch the device in order to carry out their malfeasance. The main classes of physical attacks are: hardware tampering, attacking the device over its built-in ports, and leveraging physical sensors on the device to garner data.

2.4.1.1 Hardware Tampering

Hardware tampering attacks are directed at the physical circuitry of the device itself. Security of hardware is often an afterthought since many manufacturers consider the device hardware secure through obscurity. Tampering with hardware requires esoteric knowledge and a particular skillset, but common low-level interfaces and circuitry on most electronic devices allow attacks to be performed against a wide range of devices. For example, many electronic devices, when disassembled to the circuit board level, will have exposed serial and JTAG ports. These ports can be used to intercept debug messages, send commands, or flash the firmware of the device. Serial and JTAG interfaces are widely used for communication between submodules in embedded systems and an attacker with reasonable skill and patience can usually find ways of accessing these buses. By being in physical possession of or in close proximity to a device, an attacker may also leverage data gleaned from any of the physical side channels on the device such as power consumption or electromagnetic emanations. By leveraging side-channel information, attackers can cheaply [59] determine secret keys from a device's embedded circuitry [27].

Table 2.4   Examples of Common Attacks against Smartphones and Their Characteristics

Name	Vector	Vulnerable Area	Local/Remote	Interactivity	Examples
Man-in-the-middle attacks	Drive-by	Browser, third-party apps, baseband processor, wireless interfaces	Both	Noninteractive	SSL MITM on apps [35,66]
Network spoofing	Drive-by	Wireless interfaces, browser, operating system	Local	Noninteractive	DNS spoofing [67], DHCP spoofing [68], ARP spoofing [69,70], Wi-Fi attacks [67,71]
Base station spoofing	Drive-by	Baseband processor, messaging services	Local	Noninteractive	Cell tower spoofing [72]
Network service attacks	Drive-by	Operating system, network services	Both	Noninteractive	ADB exploits [38], SSH exploits [73]
SMS/MMS/WAP attacks	Drive-by	Messaging services	Remote	Both	Stagefright [74–76], iOS SMS bugs [77], sending USSD codes via WAP [78]

(Continued)

Table 2.4 (*Continued*)   Examples of Common Attacks against Smartphones and Their Characteristics

Name	Vector	Vulnerable Area	Local/Remote	Interactivity	Examples
Software attacks	Drive-by, app ecosystem	Browser, operating system, third-party apps	Remote	Interactive	iOS browser weaknesses [79,80], WebView attacks [81], Jekyll App Store vetting evasion [8], integer overflow attack [40], bypassing sandbox [16], malicious iOS apps [10], malicious Android apps [41]
NFC attacks	Physical	Wireless interfaces	Local	Both	Android NFC exploit [82,83]
OS attacks	Drive-by	Operating system, third-party apps	Remote	Noninteractive	Generic iOS vulnerabilities [84], generic Android vulnerabilities [85], generic Windows Mobile vulnerabilities [86], generic BlackBerry vulnerabilities [87], GTalkService exploits [88,89], LibTIFF buffer overflow [90]
Bluetooth attacks	Drive-by	Wireless interfaces	Local	Both	Bluejacking, blusnarfing, bluebugging [91–93]
Physical attacks	Physical	Hardware interfaces/ports, memory, memory card, firmware, SIM card, device itself	Local	Noninteractive	Android forensics and scraping [60–62], theft, cloning SIM cards [26], deriving secret keys [27]

(Continued)

Table 2.4 (*Continued*) Examples of Common Attacks against Smartphones and Their Characteristics

Name	Vector	Vulnerable Area	Local/Remote	Interactivity	Examples
USB attacks	Physical, drive-by	Hardware interfaces/ports	Local	Noninteractive	Juice Jacking [48], Juice Filming [50], Mactans: installing apps silently [49]
Side-channel attacks	Physical, drive-by	Device itself, wireless interfaces, SIM card, users	Local	Noninteractive	Inferring app usage [29], device fingerprinting [28,94,95], deriving secret keys [27], cloning SIM cards [26], detecting installed apps [96], inferring app usage patterns [29], inferring user mood [97]
Social engineering attacks	Social engineering	Users, third-party apps	Both	Interactive	App cloning [42,55], arousing curiosity [56], Trojan-horse attack [57]
Sensor attacks	Physical	Device itself	Both	Noninteractive	Leveraging accelerometer/gyroscope [25], using light sensor [52], using microphone [64]
Authentication attacks	Physical	Device itself, users	Local	Noninteractive	Smudge attack [47], attacks against biometric authentication [54]

Even if the attacker cannot leverage the aforementioned interfaces in a reasonable way, they can attack other components of the device such as the flash memory. The attacker can potentially de-solder flash memory from the circuit board, and use other tools and hardware to read and modify the filesystem, bootloader, or other sensitive configuration or software [60,61]. Munro [62] showed how to leverage hardware tampering to read data from a device, crack PIN codes, boot a smartphone from an operating system installed on a memory card, read from and write to a UART, access flash memory, and wipe a device.

2.4.1.2 Ports

Smartphones can also be attacked using their built-in ports. Of all the physical ports on a device, the standard USB port is most commonly used. The attack to be launched at a device over USB depends on the USB mode that the device is in. Common USB modes include mass storage, media device, tethering, fastboot, and ADB. Devices running iOS were shown to be vulnerable to arbitrary app installation over USB [49]. Android devices that have USB debugging enabled have the ADB daemon running on the device. The ADB daemon allows the running of commands with special system privileges. Using ADB, an attacker can bypass some of the security features of the operating system. It is important to note, however, that the majority of Android devices will not have ADB enabled by default (since it is a feature mostly used by developers) and later ($\geq 4.2.2$) Android devices have key-based authentication of desktops. However, vulnerabilities found in Android 4.2.2–4.4.2 allowed attackers to bypass ADB authentication [63]. An interesting avenue for attacking a locked ADB-enabled device is to obtain access to a desktop that is already permitted to connect to the device, and using ADB to execute commands on the device via a terminal on that desktop. It is expected that only few devices will be vulnerable to attacks using ADB; however, the attractiveness of this approach is increased when considering the power gained from attacking a device that is ADB-accessible.

In a popular attack called "juice-jacking," attackers can steal a victim's data or install software on their device via the USB port [48]. USB ports on a smartphone are not the only ports that are vulnerable. Indeed, other physical ports that are ripe for attack are the SIM card ports, SD card ports, HDMI ports, and docking connectors, to name a few.

2.4.1.3 Physical Sensors

Smartphones have a plethora of built-in sensors that enable rich interactions between users and apps. By leveraging sensors such as the accelerometer/gyroscope, researchers showed how it was possible to infer keystrokes on a smartphone based on how the device moved as credentials were being typed in References 25 and 51. Other researchers showed that minor movements when entering credentials are sufficient to change the ambient light hitting the light sensor in a way that can be exploited to reduce the search space required for guessing a PIN [52]. The microphone on a smartphone has also been used to extract high-value audio data from smartphones [64]. Using power side channels, researchers were able to clone SIM cards [26] and retrieve cryptographic keys [27]. Other researchers used imperfections in smartphone sensors for device fingerprinting and tracking [65]. One of the big differentiators between smartphones and traditional computers is the myriad of built-in sensors that they contain. Thus sensor side channels contribute a novel attack surface to smartphones and is an open area for further IDS research.

2.4.2 Nonphysical Attacks

Nonphysical attacks on a smartphone do not require physical manipulation of the device. Thus these attacks will predominantly come from across a communication channel, be it Wi-Fi, 34/4G, Bluetooth, etc. As mentioned earlier, local attacks come from the local network segment from physically (or logically) adjacent devices. These attacks are predominantly launched by attackers on the same Wi-Fi network as the victim. Remote attacks come from attackers further removed than the local network segment, that is, more than one hop away. Remote attacks can come from hosts on the Internet and may be delivered by malicious web pages, email, or by exploiting apps installed on a device. Potential attack vectors also include malicious SMS/MMS messages carefully crafted in a way that exploits a vulnerability on the system [74]. Some attacks can also be categorized as being "local or remote" in cases where they can be launched from either location.

2.4.2.1 Local

On the local network segment, smartphones are susceptible to the typical spoofing attacks that traditional workstations are also susceptible to. With ARP spoofing [69,70], the attacker convinces the victim device that it is the gateway. The same principle applies for DNS [67] and DHCP spoofing [68] attacks. In DNS spoofing, the attacker responds to the victim's DNS queries. Usually, the responses point a domain name to an IP address under the control of the attacker. By doing this, the attacker receives traffic from the victim that was intended for a different recipient. Similarly, with DHCP spoofing, the attacker responds with false DHCP responses, usually telling the victim to use the attacker's IP address as a default gateway. Once again, this tricks the victim into sending their traffic to the attacker. Once the attacker receives the traffic, they are free to inspect, modify, and/or forward the packets to some destination.

Alternately, the attacker can simply drop the packets and cause a DoS attack. It is worth noting that the attacker need not be present on the local area network as a client to perform local attacks. The attacker can use additional hardware such as femtocells or wireless access points to trick the victim into connecting to cellular [98] and Wi-Fi networks directly under their control [72]. If successful, the attacker has full domain over all traffic from the victim and is free to manipulate it as they see fit. The level of effort and sophistication required to carry out attacks over cellular networks is currently high but falling rapidly due to the availability of picocells/femtocells and open-source cellular infrastructure software [98]. Picocells can typically supply coverage within a range of 200 m and can thus target thousands of potential victims if placed in a crowded area. By spoofing the cellular network, attackers essentially become a man-in-the-middle and can attack devices either by sending low-level commands, or perform more traditional attacks against Internet connectivity like a typical man-in-the-middle. Once the MITM attack is underway, the attacker can potentially insert malicious payloads into the content being received by users using their choice of drive-by attack.

2.4.2.2 Local or Remote

Certain attacks can be carried out whether the attacker is on the local network segment or somewhere on the path to the destination. Two notable examples of attacks in this category are traffic analysis attacks and MITM attacks.

- *Traffic analysis*: By doing traffic analysis, an attacker is able to glean additional information from network traffic. Traffic analysis is usually done on encrypted traffic since the payload cannot be examined directly. By leveraging pattern matching or statistical analysis on captured traffic, the attacker can determine things such as the apps that a user has installed on their device [95,96], specific actions that the user is doing within these apps [29,94], or identify devices on the network based on their unique traffic fingerprint [28].
- *MITM*: If an attacker successfully launches an MITM attack against a victim, they have the power to intercept, analyze, modify, and forward the traffic as they see fit. If transport layer encryption, such as TLS, is used, an attacker may be limited in what they can do with the traffic. Assuming that the victim properly validates and handles TLS certificates, the attacker would no longer be able to read or modify (without detection) the victim's traffic. However, the attacker would still be able to drop, delay, or route the traffic as they see fit. All of this assumes that the client handles the certificates properly. Indeed, it has been shown that on smartphones, many apps do not handle certificates properly and, as a result, a number of apps are still vulnerable to MITM attacks even when the connection is "secured" with TLS [35,99].

2.4.2.3 Remote

Remote attacks are aimed at the victim device from over the network, with the attacker usually being many hops away. Attackers can target network services running on a smartphone, SMS/MMS handlers in the operating system, client-side web browsers, email client applications, and other apps running on the smartphone. Each of these pieces of software can be exploited in different ways.

■ *Network services*: By default, many smartphones do not have network services installed, but various third-party apps open ports and actively listen for incoming connections. A simple feature in Android, called the ADB, can allow access to the device via TCP [30]. Other apps that listen for connections over TCP include RDP, SSH, VNC, and other similar protocols. Each of these technologies exposes a potential area for exploit as they offer a potential means of entry into the smartphone. Open ports can be enumerated by using port-scanning tools such as Nmap to probe the device. Once the list of listening services is obtained, the attacker can then proceed to fingerprint the actual software that is running behind the port before attempting to exploit it. Network services have not received great attention in the literature, and thus network service attacks on smartphones continues to be an open area of research.

■ *Messaging*: In addition to the traditional weak points that smartphones have as a result of them being Internet-enabled, there are other weak points that come about from their connectivity to the cellular network. Three notable examples of such weak points are the SMS [77] system, the MMS [74–76] system, and the wireless application protocol (WAP) [78] system. The WAP Push message feature implements a Service Loading (SL) request. This request may cause a smartphone to request a URL. By carefully crafting a web page, Ravishankar Borgaonkar [78] demonstrated how a smartphone can be directed, via Unstructured Supplementary Service Data (USSD) codes, to do things such as lock the SIM card or perform a factory restore. More recently, the attacks leveraging vulnerabilities in MMS handling came to prominence with the announcement of the "Stagefright" exploit [74]. This exploit leverages the fact that some devices automatically process a video received by MMS so that it is ready for viewing when the user opens the message. By sending a carefully crafted MMS message, an attacker could potentially perform arbitrary actions on a victim's device through remote code execution, without any interaction required from the victim.

■ *Browser*: The web browsers installed on smartphones present a rich and dynamic area of potential vulnerability since, in addition to executing their own application logic, they support a number of technologies and protocols as well, such as Javascript. Javascript itself is an entire scripting language and contributes to the complexity of the code that underlies the web browser. Indeed, an entire spectrum of Javascript attacks exist for exploiting browser weaknesses. The most common way to exploit a browser is the "drive-by" or "watering hole" attack whereby an attacker causes the user to visit a malicious URL. This is often done through social engineering. The URL is usually laced with malicious payloads that attack vulnerabilities in the browser. Since a browser is a necessarily complex piece of software handling many technologies, it is a very susceptible piece of the stack that has been exploited several times in the literature [79–81]. Worryingly, smartphone browsers may even be at more risk since they may not be a mature as their desktop counterparts. However, owing to the strict nature of app sandboxing, smartphone browser exploitation may have limited return on investment after successful exploitation.

■ *Email clients*: Email clients are another potential vulnerable area on smartphones. Attacks exploiting vulnerabilities in email clients can be delivered en masse, for example, by using unsolicited email as a vector for delivery. Since many email clients typically render email based on the HTML tags and rich multimedia that they contain, attacks on email clients are possible by exploiting the improper handling of malformed messages. Additionally, email clients usually pass on downloaded attachments to other software for handling, which make the email client itself a potential vector for malicious email attachments to be passed on to other software running on the system. Attacks on smartphone email clients are not common and may be an interesting area for future research.

■ *Third-party apps*: Third-party apps contribute in several ways to allowing attackers to breach security or invade privacy on smartphones. While the apps available in the official app stores are vetted to varying extents, malicious apps sometimes slip past these protection mechanisms [8]. Moreover, users sometimes opt to install apps from third-party marketplaces, which are known to contain greater amounts of grayware/malware [42]. But malware is not the only issue; legitimate third-party applications can be exploited by an adversary and caused to perform malicious actions. For example, by performing an MITM attack on the connection between an app and its server [35], an adversary can potentially instruct the app to perform tasks that are outside of its design and leak sensitive data from a smartphone. Mitigating attacks that leverage third-party apps may be hard given that many of these third-party apps have a legitimate reason to access the sensitive APIs and data that they do on a smartphone. By exploiting weaknesses in these apps, adversaries can obtain access to all the data that the app has permission to access. Alternately, adversaries can also make Trojan-horse apps that actually perform a legitimate task but also pilfer data behind the scenes. These types of apps are not easily identified since they may have legitimate reasons to access the APIs that they do and there is no straightforward way to differentiate them from nonmalicious apps. This is an open research problem.

Less-privileged smartphone apps can leverage confused deputy attacks to obtain sensitive data and send it from a device. In a confused deputy attack [37], a rogue app without a particular permission tricks one or more apps with the required permissions into carrying out the task that it is not permitted to do [100]. For example, an Android app that did not have permission to connect to the Internet* could use an "Intent" to make the browser app

---

* This example is an oversimplification. We note that in iOS/Android 6, permission to use the Internet is granted to apps by default.

make the connection for it. Colluding apps can also combine each of their sets of granted permissions to achieve a greater overall goal [101]. For example, one app with access to the address book may collude with another app with access to the Internet, to achieve the overall goal of surreptitiously sending a user's address book over the Internet. Colluding apps can communicate directly or by using covert channels, further adding to the complexity of detecting them. Identifying and taming one or more colluding apps is an interesting area of research for intrusion detection and prevention.

2.5 Smartphone App Marketplaces and Malware

Adversaries do not necessarily need to perform elaborate exploits if their only aim is to pilfer sensitive data from a device. This can be done in a straightforward way with the "consent" of the user, if the adversary creates a Trojan-horse application and entices the user to install it. Many apps have a legitimate reason to access data and sensitive APIs on a device to provide their functionality. For example, a navigation app has a reason to retrieve a user's geographical location and also requires access to the Internet to load maps. If this navigation app had a secondary purpose of tracking users, it would be very difficult, if at all possible, to identify this sort of malfeasance. To this end, malware authors are developing and publishing grayware apps with dubious behavior to further their malevolent goals.

This worrying trend has not gone unnoticed and indeed the literature [44,102–104] is replete with examples of malicious apps making it into both the Google Play Store and Apple App Store. Much harder to detect and remove are grayware apps such as Trojan-horse apps that are not outrightly malicious but instead mask their malfeasance under the guise of providing some reasonable functionality [43,45]. Malicious (or Trojan-horse apps) are able to perform a variety of actions on a smartphone and are only limited by sandboxing and the permissions that they are allowed to perform. However, some malicious apps also contain operating system exploits, which they can deploy to break out of their sandbox and/or perform privilege escalation [105]. Many authors have proposed solutions to identify malware in app marketplaces. For example, Chakradeo et al. [106] used statistical methods to measure correlations in app characteristics, which reduced the time taken to scan a marketplace.

Repackaged apps are variants of legitimate apps that are reverse-engineered and modified to add or change app behavior. A common tactic is to edit variables in advertisement code or programmatically click ads to route advertising revenue fraudulently to the attacker [42,107]. Additionally, adversaries sometimes rebrand apps (by modifying the app name and icon) and pass them off as their own. This worrying trend underscores the fact that app marketplaces themselves are a large attack vector that are difficult to police. Moreover, users tend to misplace trust in apps coming from app marketplaces [108,109], both official and unofficial, and unwittingly expose themselves to greater risk than on their desktop computers. This problem is compounded by the fact that anti-malware solutions are less widely deployed on smartphones [109]. Thus, adversaries have significant motivation to deliver attacks via app marketplaces.

2.6 Attack Vector Mitigation Using IDS/IPS

We now summarize the mitigation strategies that are used to counter attacks against smartphones. This analysis is summarized in Table 2.5. As discussed in Section 2.3, drive-by (or watering hole) attacks are a popular attack vector given that these attacks can (usually) be deployed remotely and target a moderate number of users without much additional effort required. Drive-by attacks typically target vulnerabilities in the operating system or other software running on smartphones. Unsurprisingly then, most drive-by attacks can be thwarted by simply keeping the smartphone operating system and other third-party software up-to-date. However, this is more easily said than done since many smartphones, especially older Android devices, are still widely used, but no longer receive updates from their vendor. Any useful IDS will need to be able to protect older unpatched devices. We elaborate on the phenomena of unpatched devices in Section 2.8.

As discussed in Section 2.5, third-party apps offer a low-investment avenue for attackers to get their code running on users' devices. The potentially exploitable userbase can be quite significant considering that publishers have turned to buying app reviews [110], to make their app seem more legitimate in app marketplaces. Attackers can opt to use the permissions their apps have been granted to pilfer as much data from the device as possible, or, with some more investment, can embed exploits into their app to elevate the app's privileges while using techniques [12] to ensure that their app is not removed from the app marketplace. Successfully deploying such an app can lead the attacker to obtaining anything from sensitive data on the device to total compromise of the device. The level of effort required varies, depending on whether the attacker just wants to get some personal information or fully compromise devices. These attacks are hard to target specific users, though

Table 2.5 Common Attack Vectors and Their Mitigation Strategies

Attack Vector	Mitigation Strategy
Drive-by attack	Software updates to patch existing vulnerabilities
App ecosystems	Whitelist of apps that can be installed or more intensive app vetting procedures
Physical attacks	Encrypt all data on a smartphone and use tamper-proof hardware
Social engineering	User education and software updates (to limit privilege escalation in case of successful attack)

most times, attackers do not have specific targets in mind when leveraging third-party apps to distribute their malfeasance. The danger from third-party apps can be mitigated, in the first instance, by more intensive app vetting in official market-places. IDS employed by app stores need to be able to detect dynamic code loading, logic bombs, and the like, and they need to be able to do it at scale while not being detectable by the malware itself.

An adversary having physical access through a victim misplacing their device or having it stolen opens up significant avenues of attack for that device. The low investment required to access misplaced devices makes theft attractive to unskilled adversaries. However, skilled and resourceful adversaries may also target specific devices, such as company-issued devices, in an effort to obtain valuable intellectual property or inside information. If a device is lost/stolen, an adversary has immediate access to all unencrypted data on that device and may access it by directly connecting to memory cards. In the case where the device was not sufficiently secured using screen-lock codes, the adversary would have access to all the other features of the device and could impersonate the owner of the device. Researchers found, in one study, that 29% of participants failed to use a screen-lock on their devices and, in general, underestimate how much personal information is stored on their device [53]. Many physical attacks can be easily mitigated by user training, use of full device encryption, and use of screen-lock functions. However, other attacks that target the underlying circuitry to circumvent protection mechanisms must be remedied using alternative strategies such as tamper-proof hardware.

Finally, social engineering can be used to gain entry into a smartphone. Since modern smartphone operating systems have several robust security features, attackers have turned to manipulating the user of a device to further their goals. By tricking a user into installing third-party apps or leading them to spoofed web pages, an attacker can obtain access to sensitive device APIs (if a user is tricked into granting an app sensitive permissions) or other privileged information. Social engineering requires varying effort depending on the nature of the attack and how elaborate it is, with the potential gain often being proportional to the social engineering effort required. These nontechnical attacks are mitigated by user education (about social engineering and how to remain safe) and software updates to mitigate privilege escalation if the attacker successfully gained an entry point into the smartphone. Intrusion detection and prevention tackling social engineering attacks on smartphones is an open research area.

2.7 Inherited Weak Points and Countermeasures

A critical part of any analysis of smartphone attacks and attack vectors has to do with understanding the similarities and differences in attacking smartphones as opposed to traditional desktops/workstations. In some ways, attacks are of a similar form and are merely adapted to work on smartphones. In other ways, idiosyncrasies of smartphones make attacks on them easier to happen and harder to defend against. For example, in contrast to the typical desktop, the ability to monitor and disrupt attacks on smartphones is reduced since they have multiple points for traffic ingress and egress. Additionally, high mobility (and consequent ease of misplacement) and the plethora of sensors that make smartphones unique also contribute to the rich attack surface they contain.

On a workstation, it usually suffices to install security software such as a firewall and anti-malware solution. Smartphones have many more ways of connecting to the outside world using various wireless technologies and ports and thus a typical firewall approach is no longer sufficient. Trivially, smartphones are more easily lost or misplaced and this offers unique attack vectors to an adversary who stole or otherwise happened upon a device. Full device encryption and screen-locks reduce the potential profit to an adversary that steals or otherwise obtains physical access to a device.

2.7.1 Built-in Mitigation Strategies

Modern smartphone platforms often contain advanced features that mitigate the likelihood of a successful compromise by an adversary. Indeed, smartphones utilize features such as sandboxing, Data Execution Prevention (DEP), Address Space Layout Randomization (ASLR), and verified boot that add to the complexity of the task of an adversary intent on exploiting a smartphone. Table 2.6 summarizes the most common attack mitigation technologies used on smartphones as well as the attack surface that they defend. These technologies are all inherited from modern desktop/server operating systems.

Sandboxing is a well-known security mechanism that is also used on smartphone operating systems for separating running programs. Apps running in a sandbox may only access a tightly controlled set of resources as arbitrated by the operating system. Any additional resources required are accessed through well-defined

Table 2.6 Mitigation Features Used to Increase the Level of Investment Required by an Attacker to Successfully Exploit a Smartphone

Mitigation Feature	Weak Point Defended
Sandboxing	Browser, operating system, third-party apps
DEP	Browser, operating system, third-party apps
ASLR	Browser, operating system, third-party apps
Verified boot	Operating system, firmware
Cryptography	SIM card, memory card, firmware, operating system

APIs and, in many cases, apps need to have their intentions to access "third-party" resources declared, *a priori*, to the operating system. DEP is another feature borrowed from modern computer operating systems for use on smartphones. DEP demarcates areas of memory as containing data that is executable or nonexecutable. This protects against malicious exploits such as buffer overflow attacks that store executable instructions in a data area of memory. ASLR is typically combined with DEP for even greater security. ASLR randomizes the addresses for key memory areas such as the base of the executable file as well as the stack, heap, and relevant libraries. This makes it very difficult for an attacker to correctly jump to an exploited function in memory and protects against buffer overflow attacks. Verified boot is a hardware and/or software technique concerned with restricting the software that can run on the device during boot up. Verified boot typically only allows software cryptographically signed by the manufacturer to run on the device. This provides an additional layer of security since it detects and prevents potentially compromised software from running on critical parts of the system. In iOS, a secure boot chain ensures that low-level software has not been tampered with and that the iOS will only run on validated Apple devices [9].

2.7.2 Attack Vectors and Attack Surfaces on Workstations

A breadth of knowledge already exists with regard to securing the attack surfaces of desktops/workstations. By understanding the similarities and differences with smartphones and workstations, we have an effective foundation from which to understand how best to engineer approaches to secure the attack surfaces on smartphones. Smartphones, in general, contain all the attack surfaces that typical desktops and workstations contain. The smartphone also contains additional attack surfaces, which come from the fact that it is also a mobile phone that connects to a cellular network. As a result, smartphones have additional attack surfaces coming from their messaging capability (SMS/MMS that are delivered over the cellular network), their cellular interface (the physical hardware and firmware that is responsible for providing connectivity to cellular networks), and other artifacts of mobile connectivity such as various ad hoc communication technologies such as NFC, Bluetooth, and the like. These additional attack surfaces, originating from the idiosyncrasies of smartphone technology, are interesting areas of further research since they do not exist on desktops/workstations and, as such, their protective technologies may not be as mature as those of other attack mitigation systems.

2.8 Related Work and Open Research Problems

By design, smartphones are portable, high connectivity devices with a variety of sensors, technologies, ports, and interfaces. On desktop computers, a firewall can effectively mitigate many attacks since most nonphysical attacks come from services exposed to the network. Conversely, on a smartphone, simply installing a firewall can reduce attacks coming from over the network, but those attacks only target a fraction of the vulnerable points present. There are still many non-Internet technologies that run on smartphones, such as NFC, Bluetooth, SMS/MMS, and these continue to provide vulnerable points that cannot be readily "firewalled." Engineering IDS solutions for protecting these weak points is an open research challenge.

As we saw in Table 2.4, drive-by attacks are a popular vector of delivering a malicious payload to a smartphone. From Table 2.5, we also saw that the most common mitigation strategy for drive-by attacks is software updates. This is not surprising because most drive-by attacks aim to exploit some software vulnerability on the smartphone. However, getting software patches to end-users is easier said than done. As Thomas et al. [111] discovered, some vulnerabilities will not have been deployed to 95% of vulnerable devices until more than 5 years after the release of the fix. This leaves a very large window of opportunity for attackers to continue to exploit unpatched devices and this is often no fault of the end-user. Indeed, the Android landscape is highly fragmented [112], meaning that there is no unified way of pushing fixes to Android smartphones automatically. On the other hand, vulnerabilities on Apple and BlackBerry devices can be patched more easily since these vendors have greater control over the operating system and update channels. Thus, one of Android's greatest advantages, its open-source nature, is also one of its biggest disadvantages from a security standpoint, in that most Android devices cannot get a security update as soon as it is available. This problem is also made worse on both Android/Windows Phones because of OEMs that modify the operating systems to add their own features, hampering the upgrade process, and sometimes introducing new vulnerabilities themselves [24]. Thus, any IDS should be able to work even on devices with outdated operating systems.

2.8.1 Related Work

Shabtai et al. [113] propose a methodology for evaluating the effectiveness of security solutions on Android. The authors propose evaluation criteria such as visibility, security solution administration, inherent cost, security level, and other miscellaneous artifacts. Given the many security solutions that have been proposed, this work provides an important mechanism for comparing the utility of one solution to another. Along similar lines, Louk et al. [114] argue for the use of monitoring, detecting, tracking, and notification (MDTN) as a means of securing against intrusions in smartphone environments. The authors demonstrate the utility of this approach in identifying malware and show that it outperforms some existing approaches.

Shabtai et al. [115] describe a method for intrusion detection on mobile devices using the knowledge-based temporal abstraction (KBTA) methodology. This approach is suitable for identifying previously unknown malware on mobile devices. Their system works by polling the device for particular metrics such as number of SMSs sent, along with critical system events. This data is combined with a knowledge-base that allows an abstraction into high-level patterns. The authors deploy their proposed system as a host-based intrusion detection system (HIDS) and show that it yields a performance rate above 94%. The authors further argue for the utility of their system on battery-constrained devices such as smartphones, by showing an average CPU consumption of only 3%.

Houmansadr et al. [116] was one of the first to leverage the cloud for intrusion detection. Zounouz et al. [117] later followed a similar approach and proposed Secloud, a cloud-based security solution for smartphones. Secloud works by emulating a smartphone using cloud resources and sending device input from the device to be protected to this virtual device. In this way, Secloud can perform resource-intensive security analysis without burdening the actual physical device. Secloud's emulator performs virus-scanning, file-integrity checking, system-call monitoring and intrusion detection and response. The authors validate the utility of Secloud by showing that it accurately detects intrusions while consuming negligible resources.

Along similar lines, Shabtai et al. [118] present Andromaly, a malware detection framework for Android. Andromaly is host-based and feeds continuously collected features and events from the target device to anomaly detectors. The anomaly detectors are built around machine learning classifiers. The authors evaluate several classification algorithms and feature selection methods. They show that malware can be detected in a way that is both lightweight and accurate. Shabtai et al. [119], in a related work, propose a behavior-based anomaly detection system that identifies anomalies based on traffic patterns.

More recently, Ariyapala et al. [120] combine host and network metrics to build an intrusion detection system for smartphones. The authors capture metrics such as CPU utilization, energy consumption, running processes, user activity and network traffic. Damopoulos et al. [121] propose a framework that unifies host- and cloud-based intrusion detection systems. The authors validate their system by showing that it can deliver quick and accurate results using computations that are affordable on an iPhone. Papamartzivanos et al. [122] crowdsource information on privacy leaks from smartphones using a cloud-based architecture. Finally, Damopoulos et al. [123] propose one of the first anomaly-based intrusion detection systems for mobile devices and use iPhone user data to show that their system can detect intrusions with up to a 99.8% true-positive rate.

2.8.2 *Open Research Problems*

Smartphones are devices designed with seamless connectivity in mind. Thus, these devices are shipped with a wide variety of wireless interfaces powered by various technologies. Since smartphones are utilized by users of varying technical skills, manufacturers may be tempted to sacrifice security to appease the consumer. Although there is the "walled-garden" approach being taken with the app repositories, apps of dubious intent, but seemingly legitimate, are an increasing problem. Indeed, third-party apps that access sensitive data on a device are free to package it and send it over the Internet. The destination or reason for doing this may not be immediately clear after static or dynamic analysis so there is still a risk of data theft by third-party apps that were entrusted to carry out a specific task. This problem is further complicated by confused deputy attacks and apps that collude to avoid detection. IDS solutions that identify and mitigate covert channels for app collusion remains a challenge.

From our survey of the literature, we uncovered that email client apps and network service apps (those that open ports) have not received wide attention from adversaries or the research community. These categories of apps may potentially be more vulnerable because of their distinctive characteristics. For email apps, the rending of HTML email or automatic downloading of attachments can provide unique access to the system if malicious emails/attachments are not handled properly. Network service apps that open ports on a smartphone may also introduce vulnerabilities if they are not designed properly. Even very mature network services on desktops/workstations/servers contain vulnerabilities, so it would be no surprise if the less mature smartphone versions of these services also contain vulnerabilities. This fact becomes worrying when one considers that many apps are developed by small teams or individuals with potentially little knowledge or concern for security. Submodule firmware (such as Wi-Fi) has also received little attention from attackers. This may be because it requires esoteric knowledge to actually interface with hardware. If successfully exploited, submodule firmware can provide long-term, almost undetectable, elevated privileges on a smartphone. IDS approaches to protecting hardware are a welcome area for future research.

More recent additions to smartphones such as NFC communication and the use of biometrics for authentication introduce new areas of potential vulnerability to be exploited. Given that these features are increasingly being used for making purchases using a smartphone, it seems natural that the attention of adversaries would shift in this direction. Indeed, the literature shows researchers who were capable of intercepting data from contactless payment cards [124]. Biometric authentication on mobile devices is also an area worth exploring since keeping credentials safe while allowing secure authentication on devices that are easily lost is an ongoing challenge.

Rogue cellular base stations are also a challenge due to the availability and rapidly falling prices of cellular infrastructure such as picocells. Combined with the prevalence of unpatched devices with gaping security holes, rogue base stations have the potential to quickly and silently compromise many devices at once. Out-of-date devices also increase the likelihood of smartphone worms that propagate by exploiting wireless interfaces for transmission. As mentioned earlier in this chapter, IDS systems will need to continue to offer protection even on devices that are no longer supported by their manufacturer.

2.9 Conclusion

Smartphones are poised to take over from desktops and workstations as the device of choice for communication, shopping, banking, and web browsing. As these devices become pervasive, the interest of adversaries naturally turns in that direction, as the adversaries look at ways of exploiting users and stealing data to make a profit. In this chapter, we enumerated the various assets on a smartphone, as well as the attack surfaces that are present on these devices that can be used by an adversary to gain entry into the system. Until now, it was not well understood by a nonexpert how the various components of a smartphone collectively contributed to its overall attack surface. We showed that smartphones have all the attack surfaces that desktops do, with additional attack surfaces coming from the fact that they contain additional hardware for mobility, sensing, and ad hoc connectivity. We discuss the various attack mitigation features in use on smartphones, some directly borrowed from desktops/workstations, and highlight the various weak points that are defended using these technologies. We also analyzed, in general, how vulnerabilities on smartphones can be reduced, but also demonstrate how some operating systems, like Android, will naturally have more challenges in getting software updates to users. As smartphones become a more tightly knit part of the average person's life, it is important to have a solid grasp of the ways that these devices are vulnerable, so that we can continue to develop ways of keeping end-users safe, now and into the foreseeable future.

References

1. Gartner. Gartner Says Worldwide Smartphone Sales Grew 9.7 Percent in Fourth Quarter of 2015, http://www.gartner.com/newsroom/id/3215217, February 2016.
2. Gartner. Gartner Says Emerging Markets Drove Worldwide Smartphone Sales to 15.5 Percent Growth in Third Quarter of 2015, http://www.gartner.com/newsroom/id/3169417, November 2015.
3. Statista. Number of Available Applications in the Google Play Store from December 2009 to July 2015, http://www.statista.com/statistics/266210/number-of-available-applications-in-the-google-play-store/, 2015.
4. Statista. Number of Available Apps in the Apple App Store from July 2008 to June 2015, http://www.statista.com/statistics/263795/number-of-available-apps-in-the-apple-app-store/, 2015.
5. S. J. Vaughan-Nichols, Smartphone Operating Systems: The Rise of Android, the Fall of Windows, http://www.zdnet.com/article/smartphone-operating-systems-the-rise-of-android-the-fall-of-windows/, February 2013.
6. Apple Inc. App Review, https://developer.apple.com/app-store/review/.
7. Georgia Institute of Technology. Georgia Tech Uncovers iOS Security Weaknesses, http://www.news.gatech.edu/2013/07/31/georgia-tech-uncovers-ios-security-weaknesses, July 2013.
8. T. Wang, K. Lu, L. Lu, S. Chung, and W. Lee, Jekyll on iOS: When benign apps become evil, in *Proceedings of the 22Nd USENIX Conference on Security*, SEC'13, USENIX Association, Berkeley, CA, USA, pp. 559–572, 2013.

9. Apple Inc. iOS Security—White Paper, https://www.apple.com/business/docs/iOS_Security_Guide.pdf, September 2015.
10. J. Han, S. Kywe, Q. Yan, F. Bao, R. Deng, D. Gao, Y. Li, and J. Zhou, Launching generic attacks on iOS with approved third-party applications, in M. Jacobson, M. Locasto, P. Mohassel, and R. Safavi-Naini, editors, *Applied Cryptography and Network Security, vol. 7954 of Lecture Notes in Computer Science*, Springer, Berlin, Heidelberg, pp. 272–289, 2013.
11. Windows 10 "Redstone": 64-Bit ARM Support Could Deliver Fastest Windows Phones Ever, http://www.ibtimes.com/windows-10-redstone-64-bit-arm-support-could-deliver-fastest-windows-phones-ever-2267877, January 2016.
12. O. Hou, A Look at Google Bouncer, http://blog.trendmicro.com/trendlabs-security-intelligence/a-look-at-google-bouncer/, 2012.
13. J. Oberheide, Dissecting the Android Bouncer, https://jon.oberheide.org/blog/2012/06/21/dissecting-the-android-bouncer/, July 2012.
14. W. Enck, D. Octeau, P. McDaniel, and S. Chaudhuri, A study of Android application security, in *Proceedings of the 20th USENIX Conference on Security*, SEC'11, USENIX Association, Berkeley, CA, USA, pp. 21–21, 2011.
15. Y. Zhou and X. Jiang, Dissecting Android malware: Characterization and evolution, in *2012 IEEE Symposium on Security and Privacy (S&P)*, San Francisco, CA, USA, pp. 95–109, May 2012.
16. MWR InfoSecurity. Sandbox Bypass through Google Admin WebView, August 13, 2015. https://labs.mwrinfosecurity.com/advisories/sandbox-bypass-through-google-webview/, accessed June 5, 2017.
17. H. Lockheimer, Android and Security, http://googlemobile.blogspot.co.uk/2012/02/android-and-security.html, February 2012.
18. K. Bala, S. Sharma, and G. Kaur, A study on smartphone based operating system, *International Journal of Computer Applications*, 121(1): 17–22, 2015.
19. R. M Dabhi and S. K. V. Nakum, A paper on latest and upcoming smartphone OS, *International Journal of Advanced Research in Computer Science and Software Engineering*, 4(4): 219–222, 2014.
20. P. Kaur and S. Sharma, Google Android. A mobile platform: A review, *Recent Advances in Engineering and Computational Sciences (RAECS) 2014*, 1–5, March 2014.
21. M. Nosrati, R. Karimi, and H. A. Hasanvand, Mobile computing: Principles, devices and operating systems, *World Applied Programming*, 2(7): 399–408, 2012.
22. O. O. Okediran, O. T. Arulogun, R. A. Ganiyu, and C. A. Oyeleye, Mobile operating systems and application development platforms: A survey, *International Journal of Advanced Networking and Applications*, 6(1): 2195–2201, 2014.
23. T. N. Sharma, M. K. Beniwal, and A. Sharma, Comparative study of different mobile operating systems, *International Journal of Advancements in Research & Technology*, 2: 1–2, 2013.
24. L. Wu, M. Grace, Y. Zhou, C. Wu, and X. Jiang, The impact of vendor customizations on Android security, in *Proceedings of the 2013 ACM SIGSAC Conference on Computer & Communications Security (CCS '13)*, CCS '13, ACM, New York, NY, USA, pp. 623–634, 2013.
25. L. Cai and H. Chen, TouchLogger: Inferring keystrokes on touch screen from smartphone motion, in *Proceedings of the 6th USENIX Conference on Hot Topics in Security*, HotSec'11, USENIX Association, Berkeley, CA, USA, pp. 9–9, 2011.

26. Y. Yu, J. Liu, F-X Standaert, Z. Guo, D. Gu, S. Wei, Y. Ge, and X. Xie, Cloning 3G/4G SIM cards with a PC and an oscilloscope: Lessons learned in physical security, in *Black Hat USA*, Las Vegas, NV, USA, 2015.

27. P. Kocher, J. Jaffe, and B. Jun, Differential power analysis, in M. Wiener, editor, *Advances in Cryptology— CRYPTO' 99, vol. 1666 of Lecture Notes in Computer Science*, Springer, Berlin, Heidelberg, pp. 388–397, 1999.

28. T. Stöber, M. Frank, J. Schmitt, and I. Martinovic, Who do you sync you are?: Smartphone fingerprinting via application behaviour, in *Proceedings of the Sixth ACM Conference on Security and Privacy in Wireless and Mobile Networks*, WiSec '13, ACM, New York, NY, USA, pp. 7–12, 2013.

29. M. Conti, L. V. Mancini, R. Spolaor, and N. V. Verde, Can't you hear me knocking: Identification of user actions on Android apps via traffic analysis, in *Proceedings of the 5th ACM Conference on Data and Application Security and Privacy*, CODASPY '15, ACM, New York, NY, USA, pp. 297–304, 2015.

30. J. J. Drake, Z. Lanier, C. Mulliner, P. Oliva Fora, S. A. Ridley, and G. Wicherski, *Android Hacker's Handbook*, John Wiley & Sons, Indianapolis, IN, USA, 2014.

31. Fraunhofer Institute for Secure Information Technology. Smartphone Attack Vectors, https://www.sit.fraunhofer.de/fileadmin/dokumente/poster/Smartphone-Attack-Vectors_Poster_Fraunhofer-SIT.pdf

32. C. Miller, D. Blazakis, D. DaiZovi, S. Esser, V. Iozzo, and R.-P. Weinmann, *iOS Hacker's Handbook*, John Wiley & Sons, 2012.

33. National Security Agency. New Smartphones and the Risk Picture, https://www.nsa.gov/ia/_files/factsheets/mobilerisks.pdf, April 2012.

34. P. Schulz and D. Plohmann, Rheinische Friedrich-Wilhelms-Universitat Bönn, Germany, Technical Report, 2013.

35. S. Fahl, M. Harbach, T. Muders, L. Baumgärtner, B. Freisleben, and M. Smith, Why Eve and Mallory love Android: An analysis of Android ssl (in)security, in *Proceedings of the 2012 ACM Conference on Computer and Communications Security*, CCS '12, ACM, New York, NY, USA, pp. 50–61, 2012.

36. Rapid7. Android Browser and WebView addJavascriptInterface Code Execution, https://www.rapid7.com/db/modules/exploit/android/browser/webview_addjavascriptinterface

37. L. Lu, Z. Li, Z. Wu, W. Lee, and G. Jiang, CHEX: Statically vetting Android apps for component hijacking vulnerabilities, in *Proceedings of the 2012 ACM Conference on Computer and Communications Security*, CCS '12, ACM, New York, NY, USA, pp. 229–240, 2012.

38. T. Vidas, D. Votipka, and N. Christin, All your Droid are belong to us: A survey of current Android attacks, in *WOOT*, pp. 81–90, 2011.

39. Nielson. Smartphones: So Many Apps, So Much Time, http://www.nielsen.com/us/en/insights/news/2014/smartphones-so-many-apps–so-much-time.html, July 2014.

40. G. Gong, Exploiting Heap corruption due to Integer Overflow in Android libcutils, in *Black Hat USA*, Las Vegas, NV, USA, 2015.

41. A. Polkovnichenko, BrainTest—A New Level of Sophistication in Mobile Malware, http://blog.checkpoint.com/2015/09/21/braintest-a-new-level-of-sophistication-in-mobile-malware/, September 2015.

42. W. Zhou, Y. Zhou, X. Jiang, and P. Ning, Detecting repackaged smartphone applications in third-party Android marketplaces, in *Proceedings of the Second ACM Conference*

on Data and Application Security and Privacy, CODASPY '12, ACM, New York, NY, USA, pp. 317–326, 2012.

43. P. H. Chia, Y. Yamamoto, and N. Asokan, Is this app safe?: A large scale study on application permissions and risk signals, in *Proceedings of the 21st International Conference on World Wide Web*, WWW '12, ACM, New York, NY, USA, pp. 311–320, 2012.

44. A. Porter Felt, M. Finifter, E. Chin, S. Hanna, and D. Wagner, A survey of mobile malware in the wild, in *Proceedings of the 1st ACM Workshop on Security and Privacy in Smartphones and Mobile Devices*, ACM, Chicago, IL, USA, pp. 3–14, 2011.

45. Y. Zhou, Z. Wang, W. Zhou, and X. Jiang, Hey, you, get off of my market: Detecting malicious apps in official and alternative Android markets, in *NDSS*, San Diego, CA, USA, 2012.

46. Android. Security-Enhanced Linux in Android, https://source.android.com/security/selinux/index.html

47. A. J. Aviv, K. Gibson, E. Mossop, M. Blaze, and J. M. Smith, Smudge attacks on smartphone touch screens, *WOOT*, 10: 1–7, 2010.

48. Krebson Security. Beware of Juice-Jacking, http://krebsonsecurity.com/2011/08/beware-of-juice-jacking/, August 2011.

49. B. Lau, Y. Jang, C. Song, T. Wang, and P. H. Chung, Mactans: Injecting malware into iOS devices via malicious chargers, in *Black Hat USA*, Las Vegas, NV, USA, 2013.

50. W. Meng, W. Hao Lee, S. R. Murali, and S. P. T. Krishnan, Charging me and I know your secrets!: Towards juice filming attacks on smartphones, in *Proceedings of the 1st ACM Workshop on Cyber-Physical System Security*, CPSS '15, ACM, New York, NY, USA, pp. 89–98, 2015.

51. Z. Xu, K. Bai, and S. Zhu, TapLogger: Inferring user inputs on smartphone touchscreens using on-board motion sensors, in *Proceedings of the Fifth ACM Conference on Security and Privacy in Wireless and Mobile Networks*, WISEC '12, ACM, New York, NY, USA, pp. 113–124, 2012.

52. R. Spreitzer, Pin skimming: Exploiting the ambient-light sensor in mobile devices, in *Proceedings of the 4th ACM Workshop on Security and Privacy in Smartphones & Mobile Devices*, ACM, Scottsdale, AZ, USA, pp. 51–62, 2014.

53. S. Egelman, S. Jain, R. S. Portnoff, K. Liao, S. Consolvo, and D. Wagner, Are you ready to lock? in *Proceedings of the 2014 ACM SIGSAC Conference on Computer and Communications Security*, CCS '14, ACM, New York, NY, USA, pp. 750–761, 2014.

54. H. Xue, T. Wei, Y. Zhang, and Z. Chen, Fingerprints on mobile devices: Abusing and leaking, https://www.blackhat.com/docs/us-15/materials/us-15-Zhang-Fingerprints-On-Mobile-Devices-Abusing-And-Leaking-wp.pdf

55. C. Gibler, R. Stevens, J. Crussell, H. Chen, H. Zang, and H. Choi, AdRob: Examining the landscape and impact of Android application plagiarism, in *Proceeding of the 11th Annual International Conference on Mobile Systems, Applications, and Services*, MobiSys '13, ACM, New York, NY, USA, pp. 431–444, 2013.

56. J. Goodchild, Social Engineering: 3 Mobile Malware Techniques, http://www.csoonline. com/article/2129129/social-engineering/social-engineering–3-mobile-malware-techniques.html, July 2011.

57. P. Collinson, Don't Bank on Your Phone, It Could Be Hacked by Zeus "Trojan Horse", http://www.theguardian.com/money/2011/jul/22/smartphones-hacked-zeus-malware, July 2011.

58. D. R. Thomas, A. R. Beresford, T. Coudray, T. Sutcliffe, and A. Taylor, The lifetime of Android API vulnerabilities: Case study on the Javascript-to-Java interface, in *Revised*

Selected Papers of the 23rd International Workshop on Security Protocols XXIII, Vol. 9379, Springer-Verlag New York, Inc., New York, NY, USA, pp. 126–138, 2015.

59. G. Goller and G. Sigl, Side channel attacks on smartphones and embedded devices using standard radio equipment, in S. Mangard and A. Y. Poschmann, editors, *Constructive Side-Channel Analysis and Secure Design, vol. 9064 of Lecture Notes in Computer Science*, Springer International Publishing, New York, USA, pp. 255–270, 2015.

60. A. Hoog. *Android Forensics: Investigation, Analysis and Mobile Security for Google Android*, Elsevier, Amsterdam, The Netherlands, 2011.

61. N. Son, Y. Lee, D. Kim, J. I. James, S. Lee, and K. Lee, A study of user data integrity during acquisition of Android devices, *Digital Investigation*, 10: S3–S11, 2013.

62. K. Munro, Android scraping: Accessing personal data on mobile devices, *Network Security*, 2014(11): 5–9, 2014.

63. MWR Labs. Android 4.4.2 Secure USB Debugging Bypass, https://labs.mwrinfosecurity.com/advisories/android-4-4-2-secure-usb-debugging-bypass/

64. R. Schlegel, K. Zhang, X.-y. Zhou, M. Intwala, A. Kapadia, and X.F. Wang, Soundcomber: A stealthy and context-aware sound trojan for smartphones, in *NDSS*, vol. 11, San Diego, CA, USA, pp. 17–33, 2011.

65. S. Dey, N. Roy, W. Xu, R. Roy Choudhury, and S. Nelakuditi, AccelPrint: Imperfections of accelerometers make smartphones trackable, in *NDSS*, San Diego, CA, USA, 2014.

66. J. Hubbard, K. Weimer, and Y. Chen, A study of SSL proxy attacks on Android and iOS mobile applications, in *11th IEEE Consumer Communications and Networking Conference (CCNC) 2014*, IEEE, Las Vegas, NV, USA, pp. 86–91, 2014.

67. M.-W. Park, Y.-H. Choi, J.-H. Eom, and T.-M. Chung, Dangerous Wi-Fi access point: Attacks to benign smartphone applications, *Personal Ubiquitous Computing*, 18(6): 1373–1386, 2014.

68. G. Edgecombe, Detection of SSL-related security vulnerabilities in Android applications, https://grahamedgecombe.com/talks/android-ssl.pdf, 2014.

69. J. G. Beekman and C. Thompson, Man-in-the-middle attack on T-Mobile Wi-Fi calling, Technical Report, University of California at Berkeley, 2013.

70. R. Siles, Real world ARP spoofing, Technical Report, SANS Institute, 2003.

71. E. Dondyk, L. Rivera, and C. C. Zou, Wi–Fi access denial of service attack to smartphones, *International Journal of Security and Networks*, 8(3): 117–129, 2013.

72. C. Rose, Ubiquitous smartphones, zero privacy, *Review of Business Information Systems (RBIS)*, 16(4): 187–192, 2012.

73. N. Seriot, iPhone privacy, in *Black Hat DC*, Arlington, VA, USA, p. 30, 2010.

74. P. Nickinson, The "Stagefright" Exploit: What You Need to Know, http://www.androidcentral.com/stagefright, Aug 2015.

75. Z. Team, Experts Found a Unicorn in the Heart of Android, https://blog.zimperium.com/experts-found-a-unicorn-in-the-heart-of-android/, July 2015.

76. zLabs. Zimperium zLabs Is Raising the Volume: New Vulnerability Processing MP3/MP4 Media, https://blog.zimperium.com/zimperium-zlabs-is-raising-the-volume-new-vulnerability-processing-mp3mp4-media/, October 2015.

77. N. Golde and C. Mulliner, Countering SMS attacks: Filter recommendations, Technical Report, Technische Universität Berlin, 2011.

78. R. Lardner, Ravi Borgaonkar Says Android Phones Vulnerable to Wipeout Attack, http://www.huffingtonpost.com/2012/09/30/ravi-borgaonkar-says-andr_n_1923867.html, Sep 2012.

79. T. Dullien, Ralf-Philipp Weinmann & Vincenzo Iozzo Own the iPhone at PWN2OWN, http://blog.zynamics.com/2010/03/24/ralf-philipp-weinmann-vincenzo-iozzo-own-the-iphone-at-pwn2own/, March 2010.

80. C. Miller and V. Iozzo, Fun and games with Mac OS X and iPhone payloads, in *Black Hat Europe*, Amsterdam, The Netherlands, 2009.

81. T. Luo, H. Hao, W. Du, Y. Wang, and H. Yin, Attacks on WebView in the Android system, in *Proceedings of the 27th Annual Computer Security Applications Conference*, ACM, Orlando, FL, USA, pp. 343–352, 2011.

82. MWR InfoSecurity. MWR Labs expose vulnerabilities in Samsung Galaxy S5 and Amazon Fire Phone, https://www.mwrinfosecurity.com/media/press-releases/mwr-labs-expose-vulnerabilities-in-samsung-galaxy-s5-and-amazon-fire-phone/, 2014.

83. C. Miller, Exploring the NFC attack surface, in *Black Hat USA 2012*, Las Vegas, NV, USA, 2012.

84. CVE Details. Apple Iphone OS: List of Security Vulnerabilities, https://www.cvedetails.com/vulnerability-list/vendor_id-49/product_id-15556/Apple-Iphone-Os.html

85. CVE Details. Google Android: List of Security Vulnerabilities, https://www.cvedetails.com/vulnerability-list/vendor_id-1224/product_id-19997/Google-Android.html

86. CVE Details. Microsoft Windows Mobile: List of Security Vulnerabilities, https://www.cvedetails.com/vulnerability-list/vendor_id-26/product_id-9709/Microsoft-Windows-Mobile.html

87. CVE Details. Blackberry OS: List of Security Vulnerabilities, https://www.cvedetails.com/vulnerability-list/vendor_id-8356/product_id-25529/Blackberry-Blackberry-Os.html

88. J. Oberheide, Android hax, in *Proceedings of SummerCon 2010*, New York, USA, June 2010.

89. J. Oberheide, When Angry Birds Attack: Android Edition, https://www.duosecurity.com/blog/when-angry-birds-attack-android-edition, May 2011.

90. Rapid7. Apple iOS MobileSafari LibTIFF Buffer Overflow, https://www.rapid7.com/db/modules/exploit/apple_ios/browser/safari_libtiff

91. J. Thom-Santelli, A. Ainslie, and G. Gay, Location, location, location: A study of bluejacking practices, in *CHI '07 Extended Abstracts on Human Factors in Computing Systems*, CHI EA '07, ACM, New York, NY, USA, pp. 2693–2698, 2007.

92. G. Legg, The Bluejacking, Bluesnarfing, Bluebugging Blues: Bluetooth Faces Perception of Vulnerability, http://www.eetimes.com/document.asp?doc_id=1275730, 2005.

93. A. J. Solon, M. J. Callaghan, J. Harkin, and T. M. McGinnity, Case study on the Bluetooth vulnerabilities in mobile devices, *IJCSNS International Journal of Computer Science and Network Security*, 6(4): 125–129, 2006.

94. M. Conti, L. V. Mancini, R. Spolaor, and N. V. Verde, Analyzing Android encrypted network traffic to identify user actions, *IEEE Transactions on Information Forensics and Security*, 11(1): 114–125, 2016.

95. V. F. Taylor, R. Spolaor, M. Conti, and I. Martinovic, AppScanner: Automatic fingerprinting of smartphone apps from encrypted network traffic, in *1st IEEE European Symposium on Security and Privacy (Euro S&P)*, Saarbrucken, Germany, 2016.

96. S. Dai, A. Tongaonkar, X. Wang, A. Nucci, and D. Song, NetworkProfiler: Towards automatic fingerprinting of Android apps, in *Proceedings of the 32nd IEEE International*

Conference on Computer Communications (INFOCOM), IEEE, Turin, Italy, pp. 809–817, April 2013.

97. R. LiKamWa, Y. Liu, N. D. Lane, and L. Zhong, MoodScope: Building a mood sensor from smartphone usage patterns, in *Proceeding of the 11th Annual International Conference on Mobile Systems, Applications, and Services*, MobiSys '13, ACM, New York, NY, USA, pp. 389–402, 2013.

98. OpenBTS. OpenBTS—Open Source Cellular Infrastructure, http://openbts.org/

99. M. Georgiev, S. Iyengar, S. Jana, R. Anubhai, D. Boneh, and V. Shmatikov, The most dangerous code in the world: Validating SSL certificates in non-browser software, in *Proceedings of the 2012 ACM Conference on Computer and Communications Security*, CCS '12, ACM, New York, NY, USA, pp. 38–49, 2012.

100. A. Lineberry, D. Luke Richardson, and T. Wyatt, These aren't the permissions you're looking for, *Def Con 18*, Las Vegas, NV, USA, 2010.

101. S. Bugiel, L. Davi, A. Dmitrienko, T. Fischer, A.R. Sadeghi, and B. Shastry, Towards taming privilege-escalation attacks on Android, in *NDSS*, San Diego, CA, USA, 2012.

102. K. Bell, Everything You Need to Know about App Store Malware, http://mashable.com/2015/09/21/ios-app-store-malware/, September 2015.

103. G. Cluley, Malware Hits the Google Play Android App Store Again (and Again), https://grahamcluley.com/2015/09/video-malware-hits-google-play-android-app-store/, September 2015.

104. A. Mylonas, S. Dritsas, B. Tsoumas, and D. Gritzalis, Smartphone security evaluation. The malware attack case, in *Proceedings of the International Conference on Security and Cryptography (SECRYPT) 2011*, IEEE, Seville, Spain, pp. 25–36, 2011.

105. X. Jiang, GingerMaster: First Android Malware Utilizing a Root Exploit on Android 2.3 (Gingerbread), http://www.csc.ncsu.edu/faculty/jiang/GingerMaster/, 2011.

106. S. Chakradeo, B. Reaves, P. Traynor, and W. Enck, MAST: Triage for market-scale mobile malware analysis, in *Proceedings of the Sixth ACM Conference on Security and Privacy in Wireless and Mobile Networks*, WiSec '13, ACM, New York, NY, USA, pp. 13–24, 2013.

107. J. Crussell, R. Stevens, and H. Chen, MAdFraud: Investigating ad fraud in Android applications, in *Proceedings of the 12th Annual International Conference on Mobile Systems, Applications, and Services*, MobiSys '14, ACM, New York, NY, USA, pp. 123–134, 2014.

108. D. Pramod and R. Raman, A study on the user perception and awareness of smartphone security, *International Journal of Applied Engineering Research*, 9(23): 19133–19144, 2014.

109. Z. Benenson and L. Reinfelder, Should the users be informed? On differences in risk perception between Android and iPhone users, in *Symposium on Usable Privacy and Security (SOUPS)*, Newcastle, UK, 2013.

110. Z. Xie and S. Zhu, Appwatcher: Unveiling the underground market of trading mobile app reviews, in *Proceedings of the 8th ACM Conference on Security & Privacy in Wireless and Mobile Networks*, WiSec '15, ACM, New York, NY, USA, pp. 10:1–10:11, 2015.

111. D. R. Thomas, A. R. Beresford, T. Coudray, T. Sutcliffe, and A. Taylor, The lifetime of Android API vulnerabilities: Case study on the JavaScript-to-Java interface.

112. S. Kovach, Android Is Facing a Security Crisis, http://uk.businessinsider.com/stagefright-vulnerability-is-bad-news-for-android-2015-8, August 2015.

113. Y. Elovici, A. Shabtai, and D. Mimran, Evaluation of Security Solutions for Android Systems, https://arxiv.org/abs/1502.04870v1

114. M. Louk,H. Lim, and H. Lee, An analysis of security system for intrusion in smartphone environment, *The Scientific World Journal*, 2014: 2014.
115. A. Shabtai, U. Kanonov, and Y. Elovici, Intrusion detection for mobile devices using the knowledge-based, temporal abstraction method, *Journal of Systems and Software*, 83(8): 1524–1537, 2010.
116. A. Houmansadr, S. A. Zonouz, and R. Berthier, A cloud-based intrusion detection and response system for mobile phones, in *Proceedings of the 2011 IEEE/IFIP 41st International Conference on Dependable Systems and Networks Workshops*, DSNW '11, IEEE Computer Society, Washington, DC, USA, pp. 31–32, 2011.
117. S. Zonouz, A. Houmansadr, R. Berthier, N. Borisov, and W. Sanders, Secloud: A cloud-based comprehensive and lightweight security solution for smartphones, *Computers & Security*, 37: 215–227, 2013.
118. A. Shabtai, U. Kanonov, Y. Elovici, C. Glezer, and Y. Weiss, "Andromaly": A behavioral malware detection framework for Android devices, *Journal of Intelligent Information Systems*, 38(1): 161–190, 2012.
119. A. Shabtai, L. Tenenboim-Chekina, D. Mimran, L. Rokach, B. Shapira, and Y. Elovici, Mobile malware detection through analysis of deviations in application network behavior, *Computers & Security*, 43: 1–18, 2014.
120. K. Ariyapala, H. G. Do, H. N. Anh, W. K. Ng, and M. Conti, A host and network based intrusion detection for android smartphones, in *30th International Conference on Advanced Information Networking and Applications Workshops (AINA)*, Crans-Montana, Switzerland, pp. 849–854, March 2016.
121. D. Damopoulos, G. Kambourakis, and G. Portokalidis, The best of both worlds: A framework for the synergistic operation of host and cloud anomaly-based IDS for smartphones, in *Proceedings of the 7th European Workshop on System Security*, EuroSec '14, ACM, New York, NY, USA, pp. 6:1–6:6, 2014.
122. D. Papamartzivanos, D. Damopoulos, and G. Kambourakis, A cloud-based architecture to crowdsource mobile app privacy leaks, in *Proceedings of the 18th Panhellenic Conference on Informatics*, PCI '14, ACM, New York, NY, USA, pp. 59:1–59:6, 2014.
123. D. Damopoulos, S. A. Menesidou, G. Kambourakis, M. Papadaki, N. Clarke, and S. Gritzalis, Evaluation of anomaly-based IDS for mobile devices using machine learning classifiers, *Security and Communication Networks*, 5(1): 3–14, 2012.
124. T. P. Diakos,J. A. Briffa, T. W. C. Brown, and S. Wesemeyer, Eavesdropping near-field contactless payments: A quantitative analysis, *The Journal of Engineering*, January 2013.

Chapter 3

Reliable Ad Hoc Smartphone Application Creation for End Users

Adwait Nadkarni, Akash Verma, Vasant Tendulkar, and William Enck

Contents

3.1 Introduction

Smartphones are now commonplace in much of the developed world, and their popularity continues to rise. A key feature of smartphones is the wide variety of available third-party applications, commonly known as "apps." Users can find apps to enhance nearly any daily activity and provide entertainment during idle periods. Indeed, the official application markets for Android and iOS both contain over 700,000 applications [1,2].

Privacy is a significant problem for smartphone consumers. In the past several years, a number of research groups have identified widespread privacy concerns with smartphone apps in both Android [3–8] and iOS [9,10]. Popular media investigators such as the *Wall Street Journal* have made similar independent findings [11]. Smartphone apps leak a range of privacy-sensitive information, from seemingly innocent phone identifiers to geographic location to entire address books [3,5,9]. Researchers often speculate that such data are collected and sold to data brokers that perform analytics for selling advertisements. Regardless of the actual use, it is clear that privacy-sensitive data are being leaked by smartphone apps, often without user consent or information.

The current state of the smartphone application ecosystem leaves privacy-conscious consumers with a dilemma: either use the app while being aware of the privacy risks, or do not install the app. Many privacy-conscious consumers (including the authors) occasionally decide that an application's benefit outweighs its privacy risks. While recent research has proposed fine-grained privacy controls, none are likely to go mainstream. Solutions that modify the OS to allow finer-grained permission control [12–14], return fake values [7,14,15], or limit network connections with sensitive values [3,7] require significant technical expertise to build and install the custom OS for a specific device. Furthermore, these research prototypes have not undergone rigorous testing, nor are they frequently updated to new OS versions that contain new features and security patches. More recently, an array of solutions have proposed adding inline reference monitors to applications [16–19] rather than modifying the OS. Unfortunately, statically modifying an application package either results in a painful install process for the user, or requires an online trusted third-party to host modified apps (which to date does not exist). Finally, all of these solutions risk breaking applications in unknown ways, as developers frequently assume permissions are granted if the app is installed.

Privacy-conscious consumers sometimes have a third choice: use a mobile website in the phone's web browser. Many applications are simply a convenient way to access a popular website from a mobile device. Increasingly, website owners are developing and maintaining mobile versions of their websites, which can be recognized by the *m.* or *mobile.* domain prefix. Frequently, the mobile website functions very similar to the mobile app. However, there are security and privacy drawbacks to accessing the app through its corresponding mobile website. First, authentication tokens are stored in the web browser's cookie store, which has a larger attack surface than if they are stored in an app's private data storage. That is, mobile web browsers (e.g., Opera, Dolphin) can be considered to be operating systems in themselves, as they have to serve a diverse array of web applications, and hence must implement a large number of APIs to maintain general compatibility with web applications. Thus, the attack surface of a mobile web browser, if measured by the codebase size and number of app-facing APIs, is certainly larger than an ordinary native application that is more purpose-specific. Second, the shared cookie store allows advertisers and social networking sites to track users [20].

In this chapter, we propose *NativeWrap* as a new alternative model for privacy-conscious consumers to use web-based applications on smartphones. NativeWrap balances the security and privacy risks of using the smartphone application and the phone's web browser. When a user is visiting a website in the phone's browser that he/she would like to run as a native app, he/she "shares" the URL with NativeWrap. NativeWrap then "wraps" the URL into a native platform app while configuring best-practice security options. In effect, NativeWrap removes the third-party developer from the platform code, placing the user in control.

Specifically, NativeWrap provides the following properties:

■ *Isolated cookie store*: Web browsers have one cookie store and mediate access based on the same origin policy (SOP). Unfortunately, SOP is insufficient to prevent privacy loss when the same advertisement firm (e.g., DoubleClick) is used on many websites. SOP also does not prevent large social networking sites (e.g., Facebook) from identifying user browser habits by simply encouraging website owners to include social networking integration [20]. NativeWrap prevents such privacy loss by ensuring a separate cookie store for each wrapped website. It also prevents a compromised browser from leaking authentication cookies for multiple websites.

■ *Phishing prevention*: Phishing attacks are successful when the user clicks on a link and is fooled into entering sensitive information into a fake website. On smartphones, phishing attacks are aided by web browsers that remove the address bar to maximize the viewing area [21]. By using a native platform app, the user can be trained to always use the phone's application launcher to access security-sensitive services (e.g., banking). NativeWrap provides the native platform app experience to any website. It also pins the wrapped website to a specific domain to ensure embedded elements (e.g., ads) do not redirect the user to a malicious site.

■ *Correct SSL configuration*: Recent research has identified widespread misconfiguration of SSL in smartphone apps [22,23]. NativeWrap not only ensures proper SSL verification, but it also can pin the website to a certificate authority to remove dependence on a large root CA list. Furthermore, NativeWrap uses the HTTPS Everywhere [24] approach to provide the option of forcing SSL within the wrapped website [25].

■ *Limited, user-controlled permissions*: Developers of native mobile applications frequently include extra functionality that impinges on user privacy. NativeWrap defaults to Internet-only permission, with the ability for the user to add several common functional permissions when wrapping the website. Note that permissions are customized while wrapping the website into the wrapped app, and cannot be changed after the wrapped app has been created.

Our contribution: The primary contribution of this chapter is a new approach for privacy-concerned consumers to access web content from smartphones and mobile devices. To demonstrate the approach, we describe the prototype implementation of NativeWrap for Android, and show its compatibility with the top 500 websites from Alexa.com. NativeWrap has been deployed on the Google Play Store and installed more than 10,000 times since August 2014, with generally positive user reviews.[*]

3.2 Motivation

Before describing NativeWrap, we must first understand how and why many applications are developed. We begin with a short history of mobile application development while defining several key terms used throughout the chapter. We then provide a survey of mobile apps from the Google Play Store to better characterize the significance of the problem.

3.2.1 Background

The first feature-enhanced mobile phones provided an Internet connection and a web browser. Early users visited the same websites as provided for personal

---

[*] 4.3 star user rating (out of 5) as seen on August 2016.

computers; however, it quickly became clear that mobile versions of these websites were required to cater to the small display sizes on mobile phones. These websites, commonly known as *mobile WebApps* (or simply *WebApps*), are front ends developed specifically to suit the display and user interface (UI) aesthetics of mobile phones, and can be accessed by nearly any smartphone with a web browser.

As mobile phone platforms with native application environments emerged, developers began porting WebApp functionality to the popular platforms. These native applications (or *native apps* for short) are platform-specific, and are hosted on application markets such as the Google Play Store or the Apple App Store, from which users discover, download, and install them to their devices.

Native apps possess the ability to closely interact with the user and use the phone's hardware features such as accelerometers and GPS receivers to provide a rich user experience. As the usefulness of native apps grew, so did their popularity, ultimately leading users to frequently choose a native app over visiting the corresponding WebApp in the phone's web browser. In turn, more and more companies and organizations felt compelled to provide native app versions of their websites to stay up-to-date and maintain company image.

Developing and maintaining native apps requires significant resources. First, the application must be developed for each popular platform. Android and iOS use vastly different programming languages and design abstractions. Second, native app updates must occur via the platform's application market, which can include timely review processes (e.g., iOS) or at minimum user annoyance when apps are updated frequently. As a consequence, hybrid applications began to emerge. These hybrid applications are essentially WebApps "wrapped" in a "WebView" class within a native app. Both Android and iOS provide WebView primitives; therefore, only a very small amount of code needs to be written for each platform, and updates only need to occur at the web server. Toolkits such as PhoneGap simplify this process even further by providing a common template. To simplify discussion, this chapter terms these hybrid applications as *WebView apps*.

There are both security and privacy benefits and drawbacks to WebView apps versus using WebApps in the web browser. On the positive side, WebView apps are treated as security principals within their native platforms. This separation provides extra protection of user credentials and other sensitive data. WebView apps can also deter phishing. Once a user downloads a native app (e.g., a banking app), he/she becomes implicitly trained to access the service through the phone's launcher, and potentially less likely to be fooled by a link in an email. Note that since the WebView app simply displays the bank's website, making the user follow a bad link through the WebView app should not be possible unless the bank's website is compromised. Finally, WebView apps have separate cookie storage, which limits cross-site privacy concerns. For example, if a user is logged into Facebook in the web browser, whenever the users visits a website with a Facebook "like" button, Facebook is notified. In contrast, if the user accesses Facebook via a native or WebView app, the user's authenticated Facebook cookies are not present in the web browser. Similar privacy concerns with website advertisements are also mitigated.

WebView apps also have security and privacy drawbacks. WebView apps are generally relatively simple and their core functionality requires little more than permission to access the Internet. However, WebView apps often contain extra permissions. Many recent studies [3,4,7,9] have identified privacy leaks of geographic location and phone identifiers, often by advertisement libraries [6]. Finally, WebView apps with extra permissions can potentially do more harm if exploited [26].

3.2.2 Application Survey

NativeWrap is an alternative to any mobile website or native app that has a mobile website. However, our primary target is to replace WebView applications, as they are little more than a WebView widget rendering the mobile version of a website. To estimate a lower bound on the need for NativeWrap, we performed a survey of popular Android applications. Specifically, we sought to better understand (1) *what percentage of apps are WebView apps?* and (2) *what is the permission request profile of WebView apps?*

Our survey includes the top 500 free applications from each of the 25 application categories on the Google Play Store, as of January 2013. We excluded game and widget categories, as they are rarely full-screen WebView apps. We disassembled the applications using baksmali [27] and extracted the `AndroidManifest.xml` file for each app using AXMLPrinter2 [28]. We then used lightweight static code analysis heuristics to classify the apps (described below). Our survey results are summarized in Table 3.1.

Counting WebView apps: We identified WebView apps in two steps. First, we used grep on the dissembled code to identify all applications that create or initialize WebView objects with URLs. We found that roughly 81% of applications used WebViews. However, upon closer inspection of randomly chosen applications, we found many apps use WebViews for extra functionality such as displaying company policies or advertisements. To estimate the lower bound of WebView apps, we identified the applications that use WebViews within the file

Table 3.1   Application Survey Results

Characteristic	Number of Apps	Percentage (%)
Total apps	12,500	100.00
Apps that use WebViews	10,165	81.32
WebView apps	1066	8.52
Potentially overprivileged WebView apps	999	7.99
Apps potentially requiring location	630	5.04

that contains its main activity class. The main activity is specified in the application's `AndroidManifest.xml` and defines the first activity component started when the application is launched. If an app uses a WebView in its main activity class, it is highly likely that WebViews are core to the app's functionality. However, we stress that this is a lower bound, because developers may place WebView initializers in other classes called by the main activity class. This second search strategy identified 1066 apps, or 8.52% of our sample set, which is a significant-enough percentage of applications to be concerned about.

Permission use by WebView apps: Having identified a lower bound on the percentage of WebView apps, we turned to their security and privacy implications. Ideally, a WebView app should only require the Internet permission. However, we found that nearly all of the identified WebView apps ($\approx 93\%$) required more permissions. These applications are called "potentially overprivileged WebView apps" in Table 3.1. Users installing these WebView apps have no way to deny specific undesired permissions.

The WebView apps requested a total of 436 unique extra permissions, of which 333 were custom permissions declared by the applications themselves. Figure 3.1 further breaks down the frequency of popularly requested permissions. The figure

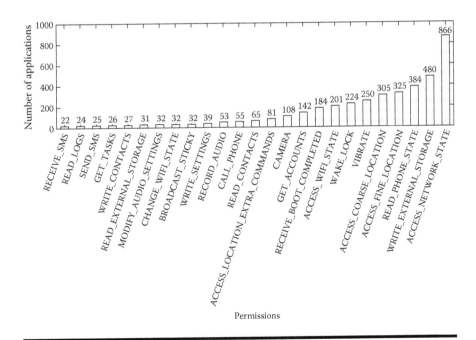

Figure 3.1 **Permissions frequently requested by WebView apps, in the ascending order of the number of applications that request them. Permissions requested by <20 applications are omitted.**

shows that most WebView apps request ACCESS_NETWORK_STATE, which is used to determine if the phone has a data connection, and can be used to differentiate cellular and WiFi connections. Additionally, WRITE_EXTERNAL_STORAGE is reasonable for WebView apps storing caches on the SD card. However, Figure 3.1 shows a wide variety of privacy- and security-relevant permissions. We note that the phone state and location permissions are the next highest requested permissions. These results clearly indicate WebView apps present privacy concerns.

Stowaway [26]: To characterize how many requested permissions are actually used by WebView app code, we analyzed 50 randomly selected applications with Stowaway [26]. We only analyzed 50 applications because Stowaway is not a stand-alone application and required us to manually upload the applications to a website. The Stowaway results are useful as they help describe the potential for increased damage if the WebView app is compromised (e.g., due to a vulnerability in WebKit). We found that half of the 50 apps requested permissions that are never used. This result indicates that NativeWrap can also help increase application security.

3.2.3 Threat Model

A fundamental premise behind our work is that both apps and mobile websites have advantages and disadvantages with respect to security and privacy. Our NativeWrap solution is designed to leverage the advantages of each while removing the disadvantages.

Mobile applications are written by potentially untrusted third-party developers. Recent studies have clearly demonstrated that many legitimate (i.e., nonmalware) apps leak privacy-sensitive values such as phone identifiers, location, and address books [3,7]. Often, these privacy leaks are a result of advertising and other nonrequired functionality. We seek to eliminate privacy loss due to nonrequired functionality.

Accessing mobile websites through the device's web browser also has security and privacy threats. We summarize these threats as follows.

Cross-site attacks: WebApps contain web elements from different origins. These elements can store cookies within the web browser's cookie store, and are frequently aware of the WebApp they are embedded within. By storing and retrieving cookies, the owners of these elements can track user's browsing habits. For example, consider a user logged into Facebook. Whenever the user visits a website that embeds a Facebook "like" button, Facebook is notified that the user visited the page, even if the user does not click the button [20]. Further investigations found that logging out of Facebook is not enough [29,30]. To regain privacy, the user must clear the cookie store. Similar privacy concerns arise with web advertisements that store cookies, that is, a privacy concern DoubleClick is infamously known for. Browser state, including a range of browser cache methods, can be used to track the user [31]. By having per-WebApp cookie stores and state, NativeWrap significantly mitigates, if not removes, such privacy threats.

Phishing: Phishing attacks commonly trick users into clicking on URLs that direct them to a website pretending to be the original (e.g., a bank website). Web browsers on smartphones often make this easier, because the browser hides the address bar to maximize the page viewing area [21]. An example of such an attack is "Tabnabbing" [32], wherein the attacker loads a fake page resembling some recently used website's login page into a browser tab that has been open, but inactive for a while. If the user is convinced the page is authentic, he/she may enter her credentials. NativeWrap seeks to mitigate such attacks by always clearly displaying the WebView app's name. NativeWrap further pins the WebView app to a domain to ensure phishing does not inadvertently originate from the domain, for example, via advertisements that hijack the screen [33].

Browser compromise: Upon compromising the web browser, an attacker potentially gains access to all of the user's cookies, including those that are used for authentication. The compromise could also result in a man-in-the-browser attack [34], wherein the compromised browser logs all user activity and input. NativeWrap mitigates these threats by treating each WebApp as a different security principal in the host operating system. This includes separate cookie stores and separate runtime principals for each WebApp. We note that newer web browser architectures such as Chrome for Android also provide defenses against such attacks. A more detailed comparison is provided in Section 3.7.

As described previously, adversaries and misbehaving corporations generally gain access to personal user data and cookies through cross-site attacks. The Facebook "like" button described previously is a great example. That said, prior work has built models to prevent leakage of private data through cross-site attacks. For example, Bauer et al. [35] provide an information flow control model that tracks the flow of sensitive information in the Chromium web browser and prevents leakage of sensitive information. Yet, such systems are unavailable to users unless ported into the web browser (e.g., Chrome for mobile).

As NativeWrap is deployed as an application, the OS and all its services are a part of NativeWrap's trusted computing base (TCB). Therefore, NativeWrap cannot provide security guarantees against an attacker with *root* privileges. For instance, the root user can entirely replace any installed legitimate app (including NativeWrap and its WebView apps) with an identical trojan. If NativeWrap is included with the OS distribution, as we describe in Section 3.6.3, an SEAndroid policy may provide some protection.

3.3 NativeWrap Design

NativeWrap provides an alternate model for accessing web-based content by providing a balance between installing a third-party application and using the phone's web

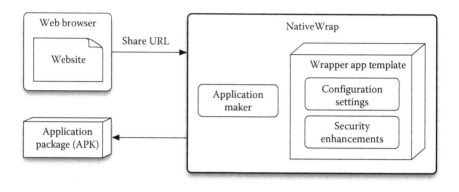

Figure 3.2   NativeWrap architecture.

browser. NativeWrap seamlessly allows end users to create safe and privacy-friendly applications for any website. To do this, the user must first visit the desired website in the phone's web browser. Once loaded, the user selects the "share" action that is often used to share a URL with messaging and social networking applications. When the user shares the URL, NativeWrap is available as a share target. Once NativeWrap receives a URL, it presents a configuration screen to the user. NativeWrap uses the URL to specify best practices defaults (e.g., forcing SSL, CA pinning). Once the configuration is confirmed, NativeWrap parameterizes a premade WebView wrapper template and installs the newly created application package. This architecture is shown in Figure 3.2.

The remainder of this section describes the objectives and design of NativeWrap. We note that while many parts of the discussion are Android-specific, NativeWrap is more general. We use Android where necessary to provide simplified and concrete discussion. Android also allows us to build and distribute a working NativeWrap prototype. We did not consider the other smartphone platforms for the prototype, because they cannot install applications without distributing them through the official application market. However, this need not necessarily be a limitation of NativeWrap. Other smartphone platforms (e.g., iOS) could easily include NativeWrap as part of the OS and provide it the ability to install the created applications.

3.3.1 Design Objectives

The primary objective of NativeWrap is to provide the user with a secure alternative to using WebView apps provided by third parties or accessing a WebApp via the browser. As such, NativeWrap seeks to achieve the following design objectives:

1. *Regulated permission set*: The WebView app should operate with the bare minimum privileges, that is, network access. If additional privilege is required (e.g., to access external storage to upload photographs), the user should be provided the option to grant it at the time of creating the WebView app. We rely on the user's discretion in determining what extra permissions to grant. However, only network access should be enabled by default, and the WebView app should operate correctly with only network access (with the exception of the function requiring more privilege).

   Again, more explanation is required regarding the mechanism for dynamically granting permissions during runtime. Also, how it is possible to know what kind of permissions a web app requires to run in its full potential and how this knowledge is transferred to NativeWrap?

2. *Separate WebApp-specific resources*: In the browser, WebApps share a cookie store, bookmarks, and history. If the browser is compromised, the authentication cookies of all WebApps may be compromised. Furthermore, the SOP is insufficient to prevent privacy loss when a cookie provider is included as a page element on many websites. Therefore, NativeWrap seeks to ensure separation of these resources. The resources should be specific to the WebApp; other WebApps should not be loaded into the original WebApp's container.

3. *Application-specific SSL configuration*: Web browsers must support the SSL needs of all websites. In contrast, a NativeWrap app needs only to support the SSL needs of one website. This feature must be leveraged to ensure the best possible SSL configuration for the app, including pinning the app to a CA certificate and forcing SSL if possible.

4. *Execution of trustworthy code*: The created WebView app should be free from known vulnerabilities and execute in a predictable manner. It should also prevent malicious arbitrary code from executing, and should be resistant to confused deputy attacks.

5. *Protection against unauthorized updates*: An attacker with physical possession of a locked device should not be able to update the WebView app (e.g., through the Android debug bridge [adb]).

3.3.2 Design Elements

We fulfill these design objectives on Android in four parts: a secure configurable wrapper, domain pinning, SSL pinning, and forcing HTTPS where possible.

3.3.2.1 Secure Configurable Wrapper

In order to keep the resources of WebApps isolated, we wrap WebApps into native Android applications. Each Android application has a unique Linux UID, making it a unique security principal. Therefore, native Android apps cannot access the private storages of other apps. By using this separation, we ensure protection for resources such as the cookie stores, saved passwords, etc.

Our native application template is actually an Android application built using a WebView as its primary layout view. The WebView is configured to display the WebApp associated with the URL supplied by the user. An alternate approach would have been modifying the default Android Open Source Project (AOSP) browser to support a single WebApp. After briefly considering this option, we determined that refactoring the browser app is generally a complex and error-prone process that may leave unknown vulnerabilities. Therefore, we opted for a clean design.

We configure our wrapper template to request only the Internet permission. While studying WebApps, we recognized that some websites allow users to upload files (e.g., photographs). WebViews can be programmed to relay file upload events to the Android OS. This feature will require the READ_EXTERNAL_STORAGE permission in future Android releases. Therefore, NativeWrap offers the user the option to add this permission while configuring the wrapper. Furthermore, the wrapper template is configured to only upload a file via the Android OS. Hence, the resulting app cannot directly access the external storage without the user's knowledge.

From our experience of deploying NativeWrap on the Google Play Store (described in Section 3.4), we found that NativeWrap could be too restrictive for some applications that genuinely require certain permissions (e.g., location) to execute their major functionality. We validated this hypothesis through user reviews and a study of popular WebApps. As our goal behind NativeWrap is to put the user in control, we added the location permission as a configuration option to NativeWrap, in a manner similar to the external storage permission. Other optional permissions can be added in the future if necessary.

3.3.2.2 Domain Pinning

The wrapper template is a native Android app that ensures that other native applications do not have access to the private resources of the WebApp wrapped in the template. To describe domain pinning, we call this wrapped WebApp the "primary WebApp" and the corresponding URL the "primary URL." Domain pinning only affects the primary URL and not resources referenced by that page. That is, domain pinning only allows the primary WebApp to load in the WebView, although inline content can be fetched from different sources. To simply state, only the primary WebApp and its subdomains may load in the main body of the WebView app.

If the user navigates outside the primary WebApp, he/she may be exposed to phishing or cross-site attacks. These attacks often rely on the browser's ability to load multiple WebApps, which then share the same resources such as cookie stores, history, and bookmarks. To prevent these attacks, we make the wrapper WebApp-specific by configuring the WebView to only work with the primary domain.

Requests to load or redirect to a WebApp outside this domain are forwarded to the phone's default web browser. To ensure the user is aware of this transition, we always display the name of the WebView app at the top of the screen. We also display a nonintrusive toast message when transitioning to the web browser. For example, in our experience with a native-wrapped Facebook WebView app, Facebook can display its content as well as ads, but once the user clicks on an advertisement, the request to load the new page is forwarded to the phone's web browser.

NativeWrap identifies the domain for the primary WebApp from the URL specified by the user. During our experimentation with initial versions of NativeWrap, we found that the full domain is not always appropriate. For example, www.bestbuy.com redirects to www-ssl.bestbuy.com for user login. Therefore, pinning the WebApp to www.bestbuy.com will not allow the user to log in, because the authentication cookies will be stored in the phone's browser. In this case, it is better to pin the WebView to bestbuy.com and allow all subdomains.

Pinning the WebApp to the second-level domain (e.g., bestbuy.com) is not always appropriate. For example, if the user is wrapping foo.blogspot.com, blogspot.com is too broad. However, we anecdotally observed that pinning the third-level domain is required significantly less frequently than the second-level domain. Therefore, we use the second-level domain as the default configuration, but also display the third-level domain as a clear option. We believe the cases when the third-level domain is needed will be obvious to most users.

Our experimentation with NativeWrap also uncovered redirection to other second-level domains. For example, blogspot.com redirects to accounts.google.com for authentication. Many websites use third parties such as Google and Facebook to authenticate. To address third-party authentication services, we suggest a whitelist solution. There are a relatively small number of authentication providers, which can be easily enumerated within the template. Furthermore, these domains generally are not the source of phishing attacks. Our current implementation only includes accounts.google.com and facebook.com, but additional entries can be easily added.

We note that including Facebook as trusted domain does not introduce privacy concerns unless the user actually logs into the WebApp via Facebook. In this case, Facebook may be notified of page visits within the primary WebApp if those pages contain Facebook "like" buttons.

3.3.2.3 SSL Pinning

Recent CA compromises have confirmed worst fears about the flaws of the CA model. An attack on Comodo in March 2011 resulted in it issuing nine fake certificates for websites, including Google, Microsoft, and Skype [36]. DigiNotar was compromised several months later [37], with the attacker(s) being able to issue over 500 fraudulent certificates, including a wildcard certificate for Google.

Fake SSL certificates are not limited to adversarial CA compromises. Nation states and other governing bodies can also force CAs to issue fake certificates. According to the Electronic Frontier Foundation's (EFF) SSL Observatory, there are about 650 odd organizations that function as CAs [38]. An Android version ships hundreds of such trusted CA certificates in its KeyStore, 140 for Android 4.2 [39]. If any one of these CAs is compromised, a fake SSL certificate for any website can be created, allowing the holder of the fake certificate to perform DNS redirection or MITM attacks.

In the wake of the CA compromises and growing cyber-political tension, researchers have given increased attention to the CA model. Convergence [40] is a promising solution resulting from this discourse. Convergence is based on the idea of "trust agility," where the user chooses a set of notaries to validate certificates, and multiple notaries can be added or removed as needed. Notaries situated in different geographic areas can further reduce the possibility of an attacker fooling all notaries. One option is to integrate functionality to use Convergence into NativeWrap, which would allow wrapped WebView applications to rely on notaries using the Convergence infrastructure. Note that this would need to be coupled with defining an initial set of notaries, as well as allowing the user to configure the notary template used for all newly created applications. However, we currently use a simpler, and perhaps more appropriate mechanism: SSL CA pinning.

Creating WebApp-specific native applications makes NativeWrap suitable for using SSL CA pinning. Individual WebApps commonly only use one CA; therefore, it becomes possible to pin a root CA certificate to a particular wrapper application. SSL CA pinning significantly reduces the attack surface for many WebApps. For example, since Google uses Equifax as a CA, a compromise of Comodo would not affect the created WebView app. In fact, many third-party developers have begun using SSL pinning for their native apps. Unfortunately, doing so has proved to be error prone [22].

NativeWrap uses a first-use approach to acquire the CA certificate for the WebApp loaded in the native wrapper, that is, we extract the CA certificate associated with the URL passed to NativeWrap. We then configure a TrustManager for the WebView class that only allows that root CA for SSL verification. We note that this approach is less flexible than Convergence, as the WebApp may wish to change its CA, which would require the WebView app to be recreated. This is not a problem for WebView apps created by third parties, as they could simply distribute an updated version in the application market. The first-use approach is also subject to compromise during acquisition of the CA certificate used for the pinning. Finally, WebApps that use multiple CAs may nondeterministically fail. However, we did not experience any such problems during our compatibility study described in Section 3.6.1.

3.3.2.4 Force HTTPS

Many websites provide both HTTP and HTTPS versions of their content. Unfortunately, URL references in content do not always use the HTTPS version of a URL when the user is visiting the HTTPS version of the site. ForceHTTPS [25] is a solution that allows website owners to configure the site to inform the browser that HTTPS should be used for all connections. However, to take advantage of ForceHTTPS, the user must be aware that an HTTPS version of the site is available. For example, Google Search provides both HTTP and HTTPS versions, and until only recently, the user would need to type "`https://`" to visit the HTTPS version. To take advantage of the optional HTTPS versions of websites, the EFF created the HTTPS Everywhere project [24]. This project provides an extension for Firefox and Chrome that consults a regular expression-based rule set identifying websites that have an HTTPS version. Users using the extension can ensure that they visit the HTTPS version of a website whenever possible, without the need to type "`https://`."

We have incorporated the HTTPS Everywhere concept into NativeWrap. When the user shares a URL with NativeWrap, NativeWrap consults the HTTPS Everywhere rule set to determine if an HTTPS version of the website is available. If so, the NativeWrap configuration template includes a "ForceHTTPS" checkbox, with the value selected by default.

If the user creates the app with the ForceHTTPS option enabled, the matched rule is included in the created WebView app. When the user uses the app, the rule is matched against every visited URL, substituting the HTTPS version whenever possible. Packaging a single rule works, since the wrapper is pinned to a single domain. This also works if the user selects the option to pin the wrapper to pin to the domain of the origin (e.g., `*.google.com` instead of `images.google.com`). In this case, the rule for `*.google.com` is applied, covering all its sub-domains.

We know that there are multiple ways to maintain the HTTPS Everywhere rule set. One option is to hard-code the rule set into the NativeWrap app, and update it by distributing a new version through the application market. However, this method is slow and potentially annoying for users. Therefore, NativeWrap currently retrieves the rule set by making a secure connection to our remote server, where the rules are stored and regularly updated as soon as the EFF git repository is updated.

3.3.2.5 Update Protection

Android requires applications to be signed with the developer's certificate for installation. The signature controls access to the app's resources, as well as the app itself. For instance, an update to the app can only be applied if the update and the app are signed with the same certificate [41].

While possession of the certificate is necessary, it is not sufficient to be able to update the app. The update package must also have the *same Android package name* as the installed application, for the attempt to succeed. Disassembling the WebView app created by NativeWrap running on an attacker's device is sufficient to extract the package name, or a static prefix if one is used.

Thus, an attacker can successfully install a malicious update via adb, *without root privileges*, on a *locked device* if the following conditions are satisfied: (1) the package name is constant, or sufficiently predictable and (2) a known certificate (e.g., the debug certificate available in the Android SDK) is used to sign the app. NativeWrap addresses both of these conditions to prevent malicious updates, as follows:

Random package name generation: For every new WebView app being created, NativeWrap generates a random 64-character package name suffix. For an attacker using adb, the package name will not give any information about the website wrapped in the WebView app. A random package name also makes it very difficult to launch a large-scale attack, as the attack would have to be targeted to a specific app on a specific device. Using random package names makes the attack difficult, but a targeted attack by a determined adversary is still possible. NativeWrap eliminates this possibility through dynamic certificate generation.

Dynamic certificate generation: If the debug keys from the SDK are used, or a keystore is hard-coded in the NativeWrap apk, an attacker would be easily able to get the signing certificate by disassembling the apk, which can in turn be easily obtained from the device. Thus, NativeWrap adopts an approach of generating keys on the device itself. When the user is about to create her first WebApp, NativeWrap initializes a key store, creates a signing certificate, and stores it inside its private, internal storage. This certificate is used to sign all the WebApps created on that device, and is unique to that particular installation of NativeWrap. Further, it cannot be accessed from NativeWrap's internal storage without root access. Thus, an adversary has no way of getting the certificate used to sign the WebView apps on a device. NativeWrap does not generate a new certificate for each WebView app created, as that adds no further protection, and instead adds a significant overhead to the WebView app generation process.

Finally, we note that an adversary with root privileges may not need to update an app at all, but may uninstall it entirely and replace it with an identical trojan. The damage that an adversary with root can cause extends beyond just NativeWrap, to all apps and services on the device. Thus, as described in the threat model in Section 3.2, NativeWrap does not defend against an adversary with root privileges.

3.4 NativeWrap Deployment

We deployed NativeWrap on the Google Play Store in August 2014, from where it was downloaded over a 1000 times in the first week. NativeWrap also received news coverage [42–44] that attracted privacy-conscious users to install and try it. As of August 2016, the application has been downloaded more than 10,000 times, and has a 4.3 star user rating, as shown in Figure 3.3.

Figure 3.3 NativeWrap's listing on the Google Play Store.

We now describe the enhancements and changes made to NativeWrap based on user reviews. Most requests were straightforward to incorporate, except for one: the addition of the location access permission to the wrapped application. In this section, we also describe the study we performed to understand the requirement of location by popular WebApps, the results of which justify the inclusion of a configuration option for location.

3.4.1 Overview of User Requests and Improvements

Based on user requests and reviews, we made the following improvements to NativeWrap:

1. *User-friendly setup page*: The WebView app setup page now uses terms understandable to nontechnical users, as shown in Figure 3.5a. A tutorial video on wrapping a simple WebApp has also been provided with the application on the Google Play Store.

2. *Favicon support*: NativeWrap can now extract the *favicon* from the URL of the web application to be wrapped, and use it as the icon of the native WebView app. This feature was incorporated on user demand, and is being enhanced with every update.

3. *Optional full-screen support*: NativeWrap intentionally displays the WebView app's name in the title bar, as an additional defense against phishing attacks. Based on user requests, we made the title bar optional by providing an additional "full-screen" mode of application creation.

4. *Location support*: Finally, NativeWrap users demanded an option to allow applications to access the user location. As allowing location access goes against NativeWrap's primary goal of preserving user privacy, this request was not immediately incorporated. Instead, we performed a study to determine if popular web applications truly require location, as we describe in Section 3.4.2. Based on the results of the study, we added an option to enable the location permission for the WebView application, as shown in Figure 3.5a.

3.4.2 WebApp Location Requirement Study

Based on user reviews, our hypothesis behind this study is that a significant number of popular WebApps require location information for their main or necessary add-on functions. We picked the top 250 unique English-language websites from Alexa.com as our dataset. We now describe the methodology and our findings.

Methodology: We performed the experiment on a Nexus 4 device running Android version 5.1.1. We manually navigated each website for a duration of 5 minutes, following the links that were most likely to prompt users for permission to access location (e.g., "Store Locator" in *lowes.com*). Note that the WebApp has to request access from the user (via the mobile web browser) before it can access location data. If a permission prompt was encountered, we noted the main purpose of the WebApp as well as the function performed by the WebApp at the point of location access. At the end of our study, we classified applications that requested for location into two categories, namely, (1) major functions, that is, location requested for the primary function of the application (e.g., Google Maps) or user-requested add-ons (e.g., the *Lowes.com Store Locator* function), or (2) minor functions, that is, location requested for nonessential (or unspecified) tertiary functions (e.g., baidu.com, which asks for location at the root page itself).

Findings: We found that 51 out of the 250 (i.e., 20.4%) of the WebApps prompted for location access. Additionally, all of the prompts were reached before we navigated to a link depth of 3, which suggests that location access was caused by a link that was a part of the root page or one of the pages directly reachable from the root page. Based on our classification criteria, we discovered that 46 out of those 51 apps, that is, 92% of the apps that used location or 18.4% of the entire set, used location for their main function or a user-requested add-on. This finding justifies the inclusion of location access as an option for the user on NativeWrap's configuration screen, similar to the way external storage access is already included.

3.5 Implementation

In this section, we describe the implementation of NativeWrap for the Android OS. We describe the basic flow of events that takes place when a URL is native-wrapped. The core NativeWrap logic is implemented as an Android application that can be installed on any Android phone. The application includes a wrapper template that is in and of itself an Android APK package. An example execution using Facebook is shown in Figure 3.4. The source code for NativeWrap can be found at https://wspr.csc.ncsu.edu/nativewrap/.

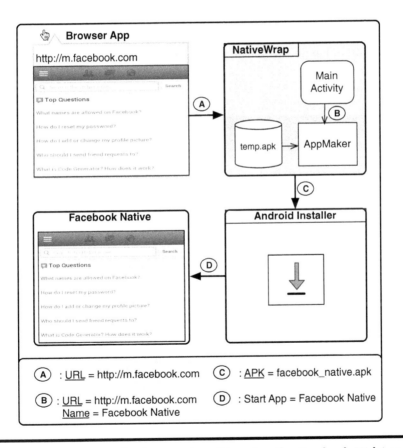

Figure 3.4   NativeWrap implementation: Wrapping the Facebook WebApp to create the *Facebook Native* application.

1. *Sharing the URL*: The process begins with the user visiting the target URL in the phone's web browser. Web browsers commonly have a "share" function that calls `startActivity` with an intent addressed to the ACTION_SEND action string and a data field containing the URL string of the current page. When Android resolves ACTION_SEND, multiple targets are available; therefore, it opens a chooser dialog that allows the user to choose the target. NativeWrap defines an intent filter for ACTION_SEND on its main activity. As such, NativeWrap is started automatically by Android, and there is no need for a persistent service.

2. *Customizing the wrapper*: Once NativeWrap receives the intent, it extracts the URL and populates the configuration template with defaults, as described in Section 3.3. At this point, the user can modify the URL, the pinned domain, specify an application name, enable additional permissions, etc., as shown in Figure 3.5a. The user then chooses to "Make the APK," which sends the customized parameters to *AppMaker*, which is a private activity component.

3. *AppMaker*: When AppMaker receives the customized parameters, it copies the default wrapper APK file to `temp.apk`. This APK is already configured to support SSL pinning, domain pinning, and some usability features to support

Figure 3.5 Configuration and installation of a *Facebook Native* app. (a) NativeWrap configuration screen. (b) Android installer.

a maximum number of web applications. It is also designed to retrieve the URL from an XML file in the /`assets` directory within the APK.

AppMaker first extracts the `AndroidManifest.xml` from `temp.apk`. We parse and modify the manifest file using AXML [45], as it is in a binary XML format. We change the package to one with a prefix name to "`edu.ncsu.nativewrap`," and a random and unique 64-character suffix, as described in Section 3.3.2.5. AppMaker changes the package name only in the manifest file. It does not rebuild the application. To ensure that the application executes correctly, we use the full classname of activity components specified in the manifest. Using the default relative class names attempts to call a nonexistent class, since the package name in the manifest no longer matches the prefix on the Java classes.

Next, AppMaker modifies the `label` attribute of the main activity. This is the activity started by the phone's application launcher, and changing its `label` to the application name specified by user ensures the user can easily find the WebView app in the list of icons and in the settings menus. Additionally, if the user chooses external storage read or write access, or location access, AppMaker adds a < `uses-permission` > specification for READ_EXTERNAL_STORAGE, WRITE_EXTERNAL_STORAGE, and the ACCESS_FINE_LOCATION permissions, respectively.

If the configuration indicates that the user wants to include the favicon as the icon of the WebView app, NativeWrap downloads the `favicon.ico` file from the root URL, if possible. The launcher icon in the WebView app's package is then replaced with the acquired favicon.

Further, NativeWrap also provides the optional functionality of creating full-screen WebView apps. The user makes the choice on the configuration screen, and the main activity of the WebView app is modified to make the app full screen. For user security, we ensure that when the user clicks a text input within the WebView app, the title of the app is shown as long as the text input is in focus. This prevents phishing attacks, and keeps the user informed of the identity of the WebView app. The user can simply flick the title in the upward direction to make it disappear again.

Finally, AppMaker creates a new XML file for the website URL, adds that file and the modified manifest file to `temp.apk`. The resulting package is signed with a prespecified key and renamed to < `application-name` > `.apk`. In order to install the `.apk`, the installer must be able to read the file. The most obvious place to store the `.apk` is the SD card, which is effectively readable by all applications. However, the SD card is also effectively writable by all applications. If the `.apk` is writable, a malicious application may exploit a race condition by modifying the file before it is installed. To avoid this race condition, we place the `.apk` in the root of NativeWrap's /`data` directory and make the file world-readable. Passing the full file path to the installer allows the package to be installed.

4. *Installing the APK*: Once the APK is created, AppMaker sends an intent message to the system with the full path to the APK to initiate its installation. As shown in Figure 3.5b, this intent invokes the Android's installer, which presents the user with a screen to install the application. Once the user approves the permission list, the WebView app is available in the phone's application launcher.

3.6 Evaluation

We begin the evaluation by comparing the HTML5 compatibility of NativeWrap with Google Chrome for Android, and studying how NativeWrap affects the compatibility of popular WebApps. Then, we describe two case studies to demonstrate the functionality and security benefits of NativeWrap.

3.6.1 Compatibility

We test NativeWrap's compatibility in two ways. First, we test the raw HTML5 compatibility using a standard benchmark. We then manually evaluate the top 500 Alexa websites.

3.6.1.1 HTML5 Compatibility Test

We performed a compatibility test for HTML5 support using `html5test.com`, on a Nexus 4 running Android 5.1.1. This test evaluates a web browser on how well it supports the HTML5 standard and new features, and generates a cumulative score chart for each aspect examined. Table 3.2 gives a comparison of the performance of Chrome for Android and NativeWrap's wrapper in the HTML5 compatibility test. NativeWrap performs almost as well as Chrome, only marginally lagging behind in a few features.

Our wrapper, and in turn the Android WebKit, supports almost all the features equally well to what Chrome browser does on the same platform. Although it does not fully or partially support features like Microdata and Streams, the Chrome browser lacks support for these features as well. However, we do support most other aspects of the standard, including form elements, essential parsing rules, audio, and video. NativeWrap's wrapper generally scores similar to Google Chrome for almost all of the features. Chrome scores better only in the audio (speech recognition) and output (web notifications) categories and marginally in web applications category.

Further, NativeWrap performed exactly as well as the stock Android 5.1 browser, and much better than the reported values for the stock Android 4.4 browser (428 points as per `html5test.com`[*]) and far better than that for stock Android 4.0

---

[*] Results accessed May 14, 2014.

Table 3.2   HTML5 Compatibility Score Comparison

Feature (Max Points)	Google Chrome	NativeWrap
Parsing rules (5)	5	5
Elements (30)	26	26
Forms (75)	73	73
Microdata (5)	0	0
Location and orientation (20)	20	20
Output (10)	10	5
Input (25)	25	25
User interaction (20)	18	18
Performance (25)	21	21
Security (40)	35	35
History and navigation (10)	10	10
Communication (35)	35	35
Video (35)	31	31
Audio (30)	30	25
Peer to peer (20)	15	15
2D graphics (25)	23	23
3D graphics (25)	20	20
Animation (5)	5	5
Responsive images (15)	15	15
Web applications (25)	21	20
Storage (35)	35	35
Files (15)	15	15
Streams (5)	3	3
Web components (10)	10	10
Other (20)	17	17
Total (555)	518	507

browser (272 points as per `html5test.com`[*]), which confirms our choice to build the wrapper from scratch rather than refactoring the AOSP browser.

3.6.1.2 Alexa Top 500 Study

To further verify our results on NativeWrap's compatibility, we tested NativeWrap with the top 500 websites in the United States from Alexa.com. We used a

---

[*] Results accessed May 14, 2014.

Google Nexus 4 device running Android version 5.1.1 for this experiment. It is worth mentioning that as of November 2015, 60.4% of the Quantcast [46] top 10,000 and 46% of top 100,000 web applications used HTML5 Doc-Type [47]. Even by a conservative estimate, the number is likely to have gone higher since.

We made a WebView app using NativeWrap for each website, and simultaneously tested the WebView app and the website in Chrome for Android. We tested the hypertext content as well as interactive multimedia content such as HTML5 audio and video tags, and also the intrawebsite navigation. We also tested the extra features such as location access and full-screen applications. None of the websites crashed or exhibited broken functionality during our tests. We infer the following from our results:

1. *NativeWrap supports most HTML5 features* commonly used by WebApps. This inference is supported by NativeWrap's HTML5 test compatibility scores, and is significant in the face of rising HTML5 use by WebApps.
2. *Websites detect browser compatibility* and present only compatible features. Websites could also redirect the user to an HTML4 version, though we did not observe any redirection on our native wrappers, possibly because it is compatible with most required HTML5 features that most websites currently use.
3. *Websites handle errors* and exceptions silently and transparently from the user, especially when they are related to HTML5, which is still not supported completely by most browsers.
4. *NativeWrap supports HTML4 content* well, and is completely compatible with websites that still work on HTML4.

3.6.2 Case Studies

3.6.2.1 Slick Deals

The Slick Deals WebApp keeps the user updated with the latest information on deals and offers on various products and services. The Android app for Slick Deals is a WebView application, and does not use the native Android UI to a great extent. It is a fairly popular application installed in around 100,000–500,000 devices, with a four star ranking on the Google Play Store. The app loads a WebView with the web address of the mobile WebApp, that is, `http://m.slickdeals.net`.

Slick Deals was one of the overprivileged applications obtained from our application survey described in Section 3.2. An analysis with Stowaway detected that the app requests the Android location permissions (both coarse and fine locations), but does not use any API that require these permissions. Even if it did call API that requested location, its purpose of displaying online deals would not justify the need for location information.

We created a new Slick Deals app using NativeWrap for this case study. The Slick Deals mobile website worked just as well on the new app as it did in the browser. At the same time, the original Slick Deals app did not offer any more functionality than the native-wrapped app, apart from a different font and color combination, but was in fact vulnerable to activity hijacking attacks when scanned with ComDroid [48].

3.6.2.2 Facebook for Android

Facebook tops the Alexa rankings as the most visited website worldwide as of April 2013. The Facebook app is also the most popular free Android app based on the number of installs from the Google Play Store, somewhere between 100 and 500 million as of April 2013. Based on the sheer number of users whose privacy depends on Facebook, it is an ideal candidate for a case study.

We compared three methods of accessing Facebook from an Android device: (1) the Facebook WebApp accessed via the phone's web browser shown in Figure 3.6a, (2) the Facebook for Android native app (version 3.1) shown in Figure 3.6b, and (3) the native-wrapped version of the Facebook app shown in Figure 3.6c. We evaluate each approach on two main factors: *usability*, which measures the convenience and features offered to the user, and *security*, which is based on the vulnerabilities in the approach, possible attack surfaces, and potential privacy violations.

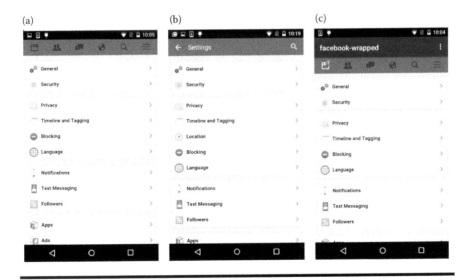

Figure 3.6 Facebook privacy settings page. (a) Facebook WebApp in Chrome browser. (b) Facebook for Android app. (c) Facebook-wrapped.

Accessing Facebook via the browser: As described in Section 3.2.3, using the Facebook app in the web browser exposes the user to various privacy and security problems, for example, the Facebook "like" button privacy issue or phishing attacks like the "tabnabbing." The other two approaches do not face such problems as they are directly installed as independent native applications on the smartphone, and have their own separate resources.

The browser-based approach also lacks the convenience of using a native app, as the user has to go through an additional step, that is, the web browser. The other two approaches provide dedicated apps for Facebook, and the native Facebook for Android app also utilizes some of the smartphone's resources and UI elements to provide a more immersive experience. Therefore, the web browser-based approach clearly does not measure up to other two approaches, in terms of both usability and security. Hence, we now only focus on the remaining two approaches.

Facebook for Android versus Facebook-wrapped: For this evaluation, we created a native-wrapped Facebook application with the URL m.facebook.com. We call it "Facebook-wrapped." We compare both the approaches on the basis of usability and security.

Facebook-wrapped and the Facebook for Android app are identical in terms of performing all of the core Facebook functionality, such as browsing pages and profiles, liking and sharing objects, uploading pictures, managing the user's account and privacy settings, etc. Facebook-wrapped lacks three primary features that Facebook for Android provides: (1) Android notifications, (2) contacts integration, and (3) geo-location check-in. However, users willing to sacrifice these features can benefit from privacy advantages.

Both the Facebook for Android and the Facebook-wrapped app are installed as native applications, and hence are not affected by the threats faced by the web browser-based approach. The Facebook-wrapped app can perform all of the core Facebook functionality. Therefore, ideally, Facebook for Android should also not require more than the Internet permission. This is not the case because Facebook for Android has many value-add features such as taking pictures and geo-location check-in. However, Facebook for Android also requests a number of nonobvious permissions. For example, it can access call logs, contacts, and recently added a permission, allowing it to track what applications the user is currently running [49]. While there are likely reasonable justifications for all of Facebook for Android's permission requests based on various integration features, the functionality is not required by all users.

Facebook-wrapped on the other hand does not require any special privilege other than network access and the permission to read external storage (API 17 onwards, optional). The primary observed drawback was the inability to use geo-location check-in. However, we view Facebook-wrapped as a privacy-friendly alternative to Facebook for Android. Users interested in these privacy benefits are less likely to use the location feature.

3.6.3 Deployment Trade-Offs

When designing NativeWrap, we debated between bundling it with a custom Android and creating a stand-alone third-party application that can be downloaded from the Google Play Store. Clearly, a stand-alone third-party application is more desirable and will reach a wider audience. Unfortunately, this deployment approach requires the user to modify the "unknown sources" application side-loading security setting. That is, the user has to choose to allow apps from unknown sources to install on the phone. Considering that most users are not security experts, allowing side-loading of apps from unknown sources may make the user vulnerable to attacks by malicious applications. Expert users can reduce their vulnerability time frame by checking the option immediately before using NativeWrap, and unchecking it immediately afterward. Testing showed that "unknown sources" was the only Android security option that needed to be disabled. NativeWrap was successfully tested with the "Verify Apps" feature activated.

The "unknown sources" limitation can be eliminated by making NativeWrap part of the Android OS. For example, NativeWrap could be deployed as a preinstalled system application and configured with the `ApplicationInfo.FLAG_PRIVILEGED` set in the package manager service. Doing so would inform the system package installer that NativeWrap install requests are not from an unknown source.

3.7 Related Work

Web browser hardening: Web browsers are the central aspect of our Internet use. Anupam et al. analyzed JavaScript- and VBScript-based attacks on the web application data in 1998 [50], and their work was one of the first to note how operating systems security primitives (e.g., "ACL" [51], "capabilities" [52–54]) apply to the multiapplication environment in the browser. Since then, many approaches based on standard OS primitives have been proposed for enhancing the browser's security.

Tahoma [55] treats web applications as first class objects, and uses virtual machines (VMs) to isolate web applications from each other and the browser from the underlying operating system. Each web application instance starts in a new VM and has its own virtual disk space, screen, input devices, etc. A key difference with respect to NativeWrap is that Tahoma allows the web application to specify the domains that will run in its VM instance in a manifest file. Delegating the browser configuration (domains to pin, security enhancements, etc.) to the web application exposes the user to cross-site attacks and to some extent phishing attacks described in Section 3.2.3. App Isolation [56] similarly allows web developers to configure domain pinning, and to optionally select isolated storage.

The OP Browser [57] splits the browser design into distinct function-specific components (e.g., web page, storage, UI) and makes the communication between these subsystems explicit, trusting the underlying operating system and the Java Virtual Machine (JVM) to maintain isolation between components. Such a model makes browser compromise difficult to achieve through exploits in individual subsystems, and provides strong isolation guarantees. Although OP Browser starts web applications in new instances (processes), it still has a common cookie store for all web application instances in the *storage* component. Although the reference monitor will follow the SOP, the common cookie store will lead to privacy issues such as the Facebook "like" button problem. Instead of simply starting a new process, NativeWrap leverages the UID-based separation provided by the underlying Android OS and ensures complete isolation between wrappers.

Google Chrome for Android also leverages the UID-based sandboxing provided by the Android OS. Every new browser "tab" is started in a new principal instance, that is, a process, and every such process has a different UID. This allows Chrome to regulate permissions allocated to each such principal, and provides isolation with respect to resources and data for each principal. A major limitation of the Chrome for Android browser is that it puts content from various origins in the same tab, that is, in the same principal instance, meaning that the privileges allocated to a tab may still be accessible to the content from a different origin than the main content of the tab, leading to cross-site attacks.

The Gazelle web browser [58] recognizes the need for isolating web application principals into separate instances. Content from different domains, even if accessed in the same tab or embedded in the same web page, is put in separate principal instances. Therefore, Gazelle prevents embedded content of one principal executing code in another principal's context. In spite of such protections, principals in Chrome as well as Gazelle share common resources like cookie stores, which can result in privacy problems, some of which are described in this chapter. The fundamental reason behind this difference is that NativeWrap's wrapper provides a single web application environment, while Chrome for Android, Gazelle, OP browser, and other similar approaches [56,59–61] attempt to achieve complete app-specific isolation in a multiapp environment.

Privacy violations by native apps: Most web browsers available today are vulnerable to many of the attacks described in our threat model (Section 3.2.3). Google Chrome for Android is relatively resistant to browser compromise due to its UID-based sandboxing, but still vulnerable to phishing and cross-site attacks. Native WebView apps defined in Section 3.2.1 by default do not share browser state and cookie stores, and hence are not vulnerable to cross-site or browser phishing attacks. Nevertheless, native WebView apps that are overprivileged cause privacy concerns [3–8].

There are different strategies for preventing privacy violations by such applications. Aurasium [19] repackages Android apps to make them policy compliant and to prevent privilege escalation attacks. A similar approach is taken by Dr. Android [18] and RetroSkeleton [16]. TISSA [14] allows the user to manage the private information granted to the app both during and after installation. It also has a provision to supply applications fake information. Apex [13] retrofits the Android package installer to install an application with custom policies. TaintDroid [3] uses taint tracking to alert the user when an application tries to export private data off the device. AppFence [7] and MockDroid [15] give the user a choice to provide fake information to apps that demand private data. In case the user needs to divulge information, AppFence prohibits the receiving app from exporting the data off the device.

Modifying an application package or its functionality may cause an application to break. Therefore, NativeWrap instead takes the control out of the hands of the developer, and packages a reliable template according to the security settings configured by the user.

Other WebApp wrappers: PhoneGap [62] allows developers to create native wrappers for HTML5 WebApps, and also provides JavaScript API to access the phone's resources. Thus, PhoneGap-based applications can potentially be just as privacy invasive as other native applications. PhoneGap is also only used by developers to wrap their HTML5 apps in native wrappers, and cannot be used by the user without the source code for the HTML5 app.

Finally, close in implementation, but drastically different in motivation, is the Fluid app [63]. Fluid is designed to create a native version of any website for Mac OS X for user convenience. NativeWrap is designed specifically to address the security and privacy needs of smartphone users and is proposed as an alternate model for accessing web content on smartphones. As such, Fluid does not provide the best practices security configuration provided by NativeWrap, nor does it provide the basic facility, that is, a separate cookie store per wrapped WebApp in its free version.

3.8 Summary

Third-party native applications have become the *de facto* way for users to access web content on smartphones. In this chapter, we argued that native applications offer many security and privacy benefits over accessing the web content using the phone's web browser. Unfortunately, many of the native applications provided by third parties hold privacy concerns in and of themselves. To resolve this tension, we proposed NativeWrap as an alternative approach for smartphone users to access web content. NativeWrap "wraps" a given URL into a native application and applies security best practices configuration. In doing so, NativeWrap removes third-party developers from platform code and places users in control of privacy-sensitive operation.

Acknowledgments

This work was funded in part by the National Security Agency, and NSF grants CNS-1222680 and CNS-1253346. Any opinions, findings, and conclusions or recommendations expressed in this chapter are those of the authors and do not necessarily reflect the views of the funding agencies.

References

1. Apple, Apple Updates iOS to 6.1, March 2013, http://www.apple.com/pr/library/2013/01/28Apple-Updates-iOS-to-6-1.html
2. B. Womack, Google says 700000 applications available for Android. *Bloomberg Businessweek*, October 2012, http://www.businessweek.com/news/2012-10-29/google-says-700-000-applications-available-for-android-devices
3. W. Enck, P. Gilbert, B.-G. Chun, L. P. Cox, J. Jung, P. McDaniel, and A. N. Sheth, TaintDroid: An information-flow tracking system for realtime privacy monitoring on smartphones, in *Proceedings of the 9th USENIX Symposium on Operating Systems Design and Implementation (OSDI)*, USENIX Association, Vancouver, Canada, pp. 393–407, October 2010.
4. W. Enck, D. Octeau, P. McDaniel, and S. Chaudhuri, A study of android application security, in *Proceedings of the 20th USENIX Security Symposium*, USENIX Association, San Francisco, CA, pp. 315–331, August 2011.
5. C. Gibler, J. Crussell, J. Erickson, and H. Chen, AndroidLeaks: Automatically detecting potential privacy leaks in android applications on a large scale, in *Trust and Trustworthy Computing, Lecture Notes in Computer Science* vol. 7344, 2012.
6. M. Grace, W. Zhou, X. Jiang, and A.-R. Sadeghi, Unsafe exposure analysis of mobile in-app advertisements, in *Proceedings of the ACM Conference on Security and Privacy in Wireless and Mobile Networks (WiSec)*, ACM, Tuscon, AZ, pp. 101–112, 2012.
7. P. Hornyack, S. Han, J. Jung, S. Schechter, and D. Wetherall, These aren't the Droids you're looking for: Retrofitting android to protect data from imperious applications, in *Proceedings of the ACM Conference on Computer and Communications Security (CCS)*, ACM, Chicago, IL, pp. 639–652, 2011.
8. R. Stevens, C. Gibler, J. Crussell, J. Erickson, and H. Chen, Investigating user privacy in android ad libraries, in *IEEE Mobile Security Technologies (MoST)*, 2012.
9. M. Egele, C. Kruegel, E. Kirda, and G. Vigna, PiOS: Detecting privacy leaks in iOS applications, in *Proceedings of the ISOC Network and Distributed System Security Symposium (NDSS)*, The Internet Society, San Diego, CA, pp. 177–183, February 2011.
10. J. Han, Q. Yan, D. Gao, J. Zhou, and R. Deng, Comparing mobile privacy protection through cross-platform applications, in *Proceedings of the Annual Network and Distributed System Security Symposium (NDSS)*, The Internet Society, San Diego, CA, 2013.
11. S. Thurm and Y. I. Kane, Your apps are watching you, http://online.wsj.com/article/SB10001424052748704694004576020083703574602.html
12. M. Conti, V. T. N. Nguyen, and B. Crispo, CRePE: Context-related policy enforcement for android, in *Proceedings of the 13th Information Security Conference (ISC)*, Springer, Boca Raton, FL, pp. 331–345, October 2010.
13. M. Nauman, S. Khan, and X. Zhang, Apex: Extending android permission model and enforcement with user-defined runtime constraints, in *Proceedings of ASIACCS*, ACM, Beijing, China, pp. 328–332, 2010.

14. Y. Zhou, X. Zhang, X. Jiang, and V. W. Freeh, Taming information-stealing smart-phone applications (on android), in *Proceedings of the International Conference on Trust and Trustworthy Computing (TRUST)*, Springer, Pittsburgh, PA, pp. 93–107, June 2011.

15. A. R. Beresford, A. Rice, N. Skehin, and R. Sohan, MockDroid: Trading privacy for application functionality on smartphones, in *Proceedings of the 12th Workshop on Mobile Computing Systems and Applications (HotMobile)*, ACM, Phoenix, AZ, pp. 49–54, 2011.

16. B. Davis and H. Chen, RetroSkeleton: Retrofitting android apps, in *Proceedings of the International Conference on Mobile Systems, Applications, and Services (MobiSys)*, ACM, Taipei, Taiwan, pp. 181–192, 2013.

17. H. Hao, V. Singh, and W. Du, On the effectiveness of API-level access control using bytecode rewriting in android, in *Proceedings of the ACM SIGSAC Symposium on Information Computer and Communications Security (ASIACCS)*, ACM, Hangzhou, China, pp. 25–36, 2013.

18. J. Jeon, K. K. Micinski, J. A. Vaughan, A. Fogel, N. Reddy, J. S. Foster, and T. Millstein, Dr. Android and Mr. Hide: Fine-grained permissions in android applications, in *Proceedings of the ACM Workshop on Security and Privacy in Smartphones and Mobile Devices (SPSM)*, ACM, Raleigh, NC, pp. 3–14, 2012.

19. R. Xu, H. Saidi, and R. Anderson, Aurasium: Practical policy enforcement for android applications, in *Proceedings of the USENIX Security Symposium*, USENIX Association, Bellevue, WA, pp. 539–552, 2012.

20. A. Efrati, "Like" button follows web users, http://online.wsj.com/article/SB1000 142405274870428150457632944143299 5616.html?mod = WSJ˙Tech˙LEADTop, 2011.

21. A. P. Felt and D. Wagner, Phishing on mobile devices, in *Proceedings of the Workshop on Web 2.0 Security and Privacy (W2SP)*, IEEE, Oakland, CA, 2011.

22. S. Fahl, M. Harbach, T. Muders, L. Baumgartner, B. Freisleben, and M. Smith, Why Eve and Mallory love Android: An analysis of Android SSL (in)security, in *Proceedings of the 2012 ACM Conference on Computer and Communications Security (CCS)*, ACM, Raleigh, NC, pp. 50–61, 2012.

23. M. Georgiev, S. Iyengar, S. Jana, R. Anubhai, D. Boneh, and V. Shmatikov, The most dangerous code in the world: Validating SSL certificates in non-browser software, in *Proceedings of the ACM Conference on Computer and Communications Security (CCS)*, ACM, Raleigh, NC, pp. 38–49, 2012.

24. Electronic Frontier Foundation. HTTPS Everywhere, https://www.eff.org/https-everywhere. Accessed April 2013.

25. C. Jackson and A. Barth, ForceHTTPS: Protecting high-security web sites from network attacks, in *Proceedings of the 17th International ACM Conference on World Wide Web*, ACM, Beijing, China, pp. 525–534, 2008.

26. A. P. Felt, E. Chin, S. Hanna, D. Song, and D. Wagner, Android permissions demystified, in *Proceedings of the ACM Conference on Computer and Communications Security (CCS)*, ACM, Chicago, IL, pp. 627–638, 2011.

27. smali—An assembler/disassembler for Android's dex format, https://code.google.com/p/smali/. Accessed April 2013.

28. android4me—J2ME port of Google's Android, https://code.google.com/p/android4me/. Accessed August 2012.

29. N. Cubrilovic, Logging out of Facebook is not enough, http://www.nikcub.com/posts/logging-out-of-facebook-is-not-enough, 2011.

30. B. Slawski, Facebook patent application describes receiving data from logged-out users to target ads, http://www.seobythesea.com/2011/09/facebook-patent-application-target-ads/, 2011.
31. C. Jackson, A. Bortz, D. Boneh, and J. C. Mitchell, Protecting browser state from web privacy attacks, in *Proceedings of the 15th International Conference on World Wide Web*, ACM, Edinburgh, Scotland, pp. 733–744, 2006.
32. A. Raskin, Tabnabbing: A new type of phishing attack, http://www.azarask.in/blog/post/a-new-type-of-phishing-attack/, 2010.
33. C. Amrutkar, K. Singh, A. Verma, and P. Traynor, VulnerableMe: Measuring systemic weaknesses in mobile browser security, in *Proceedings of the International Conference on Information Systems Security (ICISS)*, Springer, Guwahati, India, pp. 16–34, 2012.
34. P. Guhring, Concepts against man-in-the-browser attacks, http://www.cacert.at/svn/sourcerer/CAcert/SecureClient.pdf. Accessed December 2012.
35. L. Bauer, S. Cai, L. Jia, T. Passaro, M. Stroucken, and Y. Tian, Run-time monitoring and formal analysis of information flows in chromium, in *Proceedings of the ISOC Network and Distributed Systems Security Symposium (NDSS)*, The Internet Society, San Diego, CA, Feb 2015.
36. W. Leonhard, Weaknesses in SSL certification exposed by Comodo security breach, https://www.infoworld.com/t/authentication/weaknesses-in-ssl-certification-exposed-comodo-security-breach-593, 2011.
37. D. Kaplan, DigiNotar breach fallout widens as more details emerge, http://www.scmagazine.com/diginotar-breach-fallout-widens-as-more-details-emerge/article/211349/, 2011.
38. The Electronic Frontier Foundation. EFF SSL Observatory, https://www.eff.org/observatory. Accessed October 2012.
39. N. Elenkov, Certificate pinning in Android 4.2, http://nelenkov.blogspot.com/2012/12/certificate-pinning-in-android-42.html, 2012.
40. M. Marlinspike, Convergence, http://convergence.io/. Accessed March 2013.
41. Android Developers, Signing considerations, https://developer.android.com/studio/publish/app-signing.html#considerations, August 2016.
42. Help Net Security, Control Android app permissions with NativeWrap, http://www.net-security.org/secworld.php?id = 17283, August 2014.
43. R. Lemos, Android web apps get extra security with privacy wrapper, http://www.wired.co.uk/news/archive/2014-08/22/privacy-wrapper-android, August 2014.
44. R. Lemos, Researchers create privacy wrapper for Android web apps, http://arstechnica.com/security/2014/08/researchers-create-privacy-wrapper-for-android-web-apps/, August 2014.
45. axml—Read write Android binary XML files, https://code.google.com/p/axml/. Accessed January 2013.
46. quantcast. Top sites, https://www.quantcast.com/top-sites. Accessed November 2015.
47. builtWith. HTML5 DocType usage statistics, https://trends.builtwith.com/docinfo/HTML5-DocType. Accessed November 2015.
48. E. Chin, A. Porter Felt, K. Greenwood, and D. Wagner, Analyzing inter-application communication in android, in *Proceedings of the 9th Annual International Conference on Mobile Systems, Applications, and Services (MobiSys)*, ACM, Washington, DC, USA, pp. 239–252, 2011.
49. E. Protalinski, Facebook's Android app can now retrieve data about what apps you use, http://thenextweb.com/facebook/2013/04/13/facebooks-android-app-can-now-retrieve-data-about-what-apps-you-use/, 2013.

50. V. Anupam and A. Mayer, Security of web browser scripting languages: Vulnerabilities, attacks, and remedies, in *Proceedings of the 7th USENIX Security Symposium*, USENIX Association, San Antonio, TX, pp. 187–200, 1998.
51. G. Fernandez and L. Allen, Extending the Unix protection model with access control lists, in *Proceedings of the USENIX Summer Symposium*, USENIX Association, Berkeley, CA, pp. 119–132, 1988.
52. P. A. Karger and A. J. Herbert, An augmented capability architecture to support lattice security and traceability of access, in *Proceedings of the IEEE Symposium on Security and Privacy*, IEEE, Oakland, CA, pp. 2–12, May 1984.
53. J. S. Shapiro, *EROS: A Capability System*, PhD thesis, University of Pennsylvania, 1999.
54. W. Wulf, E. Cohen, W. Corwin, A. Jones, R. Levin, C. Pierson, and F. Pollack, HYDRA: The kernel of a multiprocessor operating systems, *Communications of the ACM*, 17(6): 1974.
55. R. S Cox, J. G. Hanson, S. D. Gribble, and H. M. Levy, A safety-oriented platform for web applications, in *2006 IEEE Symposium on Security and Privacy*, IEEE, Oakland, CA, pp. 350–364, 2006.
56. E. Y. Chen, J. Bau, C. Reis, A. Barth, and C. Jackson, App Isolation: Get the security of multiple browsers with just one, in *Proceedings of the 18th ACM Conference on Computer and Communications Security*, ACM, Chicago, IL, pp. 227–238, 2011.
57. C. Grier, S. Tang, and S. T. King, Secure web browsing with the OP web browser, in *Proceedings of the 2008 IEEE Symposium on Security and Privacy*, IEEE, Oakland, CA, pp. 402–416, 2008.
58. H. J. Wang, C. Grier, A. Moshchuk, S. T. King, P. Choudhury, and H. Venter, The multi-principle OS construction of the gazelle web browser, in *Proceedings of the USENIX Security Symposium*, USENIX Association, Montreal, Canada, pp. 417–432, 2009.
59. L.-S. Huang, Z. Weinberg, C. Evans, and C. Jackson, Protecting browsers from cross-origin CSS attacks, in *Proceedings of the 17th ACM Conference on Computer and Communications Security*, ACM, Chicago, IL, pp. 619–629, 2010.
60. K. Jayaraman, W. Du, B. Rajagopalan, and S. J. Chapin, ESCUDO: A Fine-grained protection model for web browsers, in *Proceedings of the 2010 IEEE 30th International Conference on Distributed Computing Systems (ICDCS)*, IEEE Computer Society, Genova, Italy, pp. 231–240, 2010.
61. S. Tang, H. Mai, and S. T. King, Trust and protection in the illinois browser operating system, in *Proceedings of the 9th USENIX Conference on Operating Systems Design and Implementation*, USENIX Association, Vancouver, BC, Canada, pp. 17–32, 2010.
62. PhoneGap, http://phonegap.com/about/, 2012. Accessed May 5, 2013.
63. T. Ditchendorf, Turn your favorite web apps into real Mac apps. http://fluidapp.com/about, 2012. Accessed May 5, 2013.

Chapter 4

Android Applications Privacy Risk Assessment

Dimitris Geneiatakis, Charalabos Medentzidis,
Ioannis Kounelis, Gary Steri, and Igor Nai Fovino

Contents

4.1 Introduction

Forecasts for mobile Internet penetration show that its end-users' basis will grow at a pace of 25% in the coming years, while currently mobile users are approximated to two billion as reported in [1]. This means that in the next few years, mobile end-users' basis will be greater than landline Internet end-users. Further, the evolution of mobile software and the underlying mobile infrastructures, for example, the augmentation in mobile Internet speeds and the one-stop shop model in which the various official or third-party app stores are based on can validate this trend.

In this direction, device manufacturers and software houses produce smartphones capable of processing information similar to personal computers (PCs). To do so, these devices incorporate a high-level operating system (OS) easily managed by end-users, while they provide access to different built-in sensors offering opportunities for new advanced services. This way, these OSs provide a fine-grained access to personal data, for example, GPS, Camera, and Contacts.

On the one side, these advances are on the benefit of the end-users, while on the other side, they increase even more the attack surface against them. This means that an adversary, in the era of mobile world toward a unified communication model, is presented with much more opportunities and capacities to gain access to sensitive user or network data. Even worse, having in mind that smartphones are nothing less than a mobile personal inventory, they essentially become a valuable target for adversaries.

So, while the majority of end-users may have developed a certain degree of trust to centralized software stores, for example, Google Play Store and Apple App Store, it is rather improbable for one to be totally sure about the quality of any given application from a security and privacy point of view. For example, spying applications can collect end-users' geographical position or steal personal information for their coders to sell them, for instance, to marketing companies [2]. Not only different "families" of malware [3], but even popular and well-established applications, named as goodware, especially on software stores, take advantage of their access to sensitive resources for stealthily manipulating and/or stealing personal information as demonstrated in various research works so far [4–6].

Primarily, as mentioned previously, this is because mobile OSs have a more granular approach for providing access to personal data through third-party software, which is naturally not the case for PC OSs. For instance, third-party Android OSs software have direct access to end-user's private storage, and consequently adversaries can easily retrieve personal information without the end-user's consent. Examples of mobile software invasive behavior are, for instance, games that request access to unique identifiers or user location that are not needed by the app to function. Ultimately, it is up to each mobile device end-user to judge if software behavior is invasive according to his/her personal perception.

So, a major open question is whether end-users are in a position to become informed and cope with the various forms of software intrusiveness threatening their private sphere as a consequence of software usage, especially when using their mobile devices as they provide a fine-grained access to private data sources. Toward this direction, Theoharidou et al. [7] and Mylonas et al. [8] introduce a risk management approach in order to assess end-users' impact of a privacy breach, while in [9] automatically assesses the review-to-behavior fidelity of applications in terms of security and privacy. Although these works might be beneficial, they are based on end-user-centric reviews, without taking into consideration software innate properties.

In this chapter, we introduce a risk management approach to assess mobile application's intrusiveness to end-users' private sphere. In this work, we focus on Android applications, aka apps, as Android is the most employed OS in mobile devices. However, a similar approach can also be used for other existing platforms. Our approach relies on app's static features, for example, permissions and running services, to compute its intrusiveness using an entropy-based heuristic metric. To the best of our knowledge, this is the first work that quantifies app's intrusiveness based on its features themselves. In this way, end-users will be able to understand how intrusive an app is. We evaluated our methodology using the top hundred apps of Google Play Store and well-known malware.

The rest of this chapter is structured as follows. In Section 4.2, we report on our motivation and we introduce background information with reference to this work in Section 4.3. In Section 4.4, we present our methodology for grading Android's mobile application risk, while in Section 4.5, we evaluate it using different mobile applications. In Section 4.6, we overview and discuss the related work. Finally, in Section 4.7, we conclude this chapter and we give some ideas for future work.

4.2 Motivation

Software security, especially at the operating system (OS) level, has been greatly enhanced during the last few years. So far, various protection solutions at the OS level have been proposed in the literature, including Address Space Layout Randomization (ASLR) [10,11], Control-Flow Integrity (CFI) [12], and canaries [12]. These solutions are mostly considered to work jointly toward reducing the OS attack surface. However, adversaries do not solely focus on OS-level vulnerabilities, but also target known or unknown application flaws to somehow gain access to otherwise private user space.

Hence, to enhance security at the application level, several additional countermeasures and alternative approaches have been proposed as well. Kc et al. [13] introduce an Instruction Set Randomization (ISR) approach to protect applications against the different types of injection attacks. Moreover, the feasibility of the employment of ISR on commodity systems, for example, x86 is demonstrated in Reference 14, while ASLR for mobile OS has been studied in Reference 10. In addition, Provos et al. [15] spawn a new process for each user connected to an OpenSSH service running without administrator privileges. This happens in a memory area separated from the one of the main process, in order to eliminate the chances for a malicious entity to acquire unauthorized root access to the system.

Such an approach for instance is incorporated by Android OS in which end-users' applications are executed on separated processes.

Despite the advances on securing software running on commodity hardware as well as proposals targeting on enhancing end-users' security and privacy, one can safely argue that they are still quite far from being complete if we consider that data leakages and data manipulation occur in real services. At the same time, very little attention is paid to software intrusiveness when it comes to preserving end-users' security and privacy, especially in the mobile world. For instance, we believe that end-users' attack surface could be minimized if they install less intrusive software. To do so, there is a need of employing the appropriate technique of informing end-users about software intrusiveness in a quantitative way before software installation takes place.

4.3 Android Architecture Overview

Android is a multilayer architecture OS. Its basis relies on a Linux kernel and supports a custom virtual machine suitable for mobile resource constraint devices. Android apps are executed in an isolated environment in order to be protected from other services flaws. In such an environment, communication among apps and OS services is restricted, and could be accomplished only through Interprocess Communication (IPCs). In Android, apps code, configuration data, and other required resources for app's execution are contained in a single file, named as Android application package (APK).

In this architecture, the manifest holds essential information for app's configuration. In a nutshell, the manifest describes from an implementation point of view all the components required for its normal execution. For instance, it contains (a) the permissions required by the app during execution for getting access to the corresponding "sensitive resources," (b) message receivers that provide to the given app the ability to read messages broadcasted either by other apps or by the OS, and (c) the definition of app's services running in the background for the app's needs. However, these are only a few of the manifest's elements. A detailed analysis for the app's manifest is beyond the scope of this chapter, and can be found in the Android documentation.*

4.4 Application Risk Assessment

In this section, we introduce our approach for assessing app's intrusiveness based on entropy theory and app's innate features. This is because, risk is associated with the

---

* https://developer.android.com/guide/topics/manifest/manifest-intro.html

lack of knowledge about a future event, while entropy is a measure for uncertainty. Further, we believe that this approach should be automated, so that end-users can use it as is, without requiring any specific knowledge. To do this, we build a custom-made architecture for automatically extracting mobile app's features and assess its risk in terms of intrusiveness. Briefly, we rely on app's permissions and its APIs as its innate features, since these data complement each other. Note that the permission provides very high-level information about applications behavior but it is not useful enough alone. However, in mobile apps, sensitive APIs are connected with permissions to grant access during execution.

In the following subsections, we report on background information with regard to entropy information, and to our approach for computing the app's intrusiveness.

4.4.1 Entropy

Information theory entropy, introduced by Shannon [16], models a system's uncertainty of symbols with regard to the expected value. That is, the predictability of the symbol assuming that the corresponding probabilities are known. In other words, reduced uncertainty corresponds to a lower entropy and vice versa. Consequently, symbol repetition can contribute to the identification of hidden redundancy in the information handled by the corresponding system.

Briefly, considering that a symbol S in a set L has probability $P_S(i)$, the entropy of the set L is calculated based on the formula (4.1).

$$H(S) = - \sum_{i=1}^{n} P(i) * \log_b p(i) \tag{4.1}$$

It should be noted that the entropy of a set L maximizes when all instances (i.e., symbols) that consist the set are equiprobable. In that case, the uncertainty of the outcome maximizes as the repetition of symbols in the set L is "minimized." That is, the higher the symbol repetition in a specific set, the lesser the entropy. In case in which two sets are independent, the entropy of both sets can be computed using the formula (4.2).

$$H(A, B) = H(A) + H(B) \tag{4.2}$$

4.4.2 Android Software Intrusiveness

We believe that to assess any software's intrusiveness and understand better possible consequences on end-users' private sphere, among others, its innate properties should be taken into account. So, in this work, we rely on app's permissions and sensitive APIs that are contained in the APK. More specifically, "uses-permissions" tag defines in a coarse-grained approach the sensitive resources that the app might need to access during its execution, while sensitive APIs are highly connected with access to sensitive resources, for example, network and camera. Even though additional

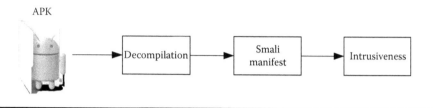

Figure 4.1 **High-level architecture for assessing app's intrusiveness.**

components can be used toward assessing app's intrusiveness in this work, we study only two main sensitive resources such as permissions and APIs; we are planning to extend our approach by including additional features in a future work.

To extract such information, that is, uses-permission and code APIs, we relied on apktool.[*] Briefly, we reverse-engineered the app (APK) to gain access to its manifest data and smali source code. Then, using custom-made tools, we extracted the permissions that are defined in the manifest, and the app's smali source code from which we rebuilt the method's original signatures. This information is used to compute a given app's intrusiveness. Our approach's high-level architecture is depicted in Figure 4.1.

In our approach, we assume that there is a set P of n permissions in mobile OS as well as a set M of k APIs. An app $A(i)$ requests a subset of P permissions and M APIs to perform its activity. We use two binary variables $X(i,j)$ and $Y(i,m)$ to represent the status of each permission $P(j)$ and of each method $M(m)$ (API), whether it exists or not, correspondingly. In this way, we build two one-dimensional vectors that are used to compute $A(i)$ app's intrusiveness according to formula (4.2). Note that in our case, the vectors consist of 170 elements for modeling app's permissions based on Google's documentation[†] for API level 21, and 1310 elements for "sensitive" methods as determined by Felt et al. [17].

So, consider for instance an app $A(k)$ that after the extraction of permissions and sensitive methods from the APK leads to the identification of 5 permissions and 10 sensitive APIs. Then we generate the vectors $P(k)$ and $M(k)$ that correspond to the existence or not of the specific feature in the app and then we compute the overall app's entropy. So, in this specific case, permissions and methods entropy are 0.0656 and 0.0237 and consequently its final value is 0.893.

Overall, Figure 4.2 reports on the risk trend of a given app considering different numbers of permissions and APIs incorporated in it according to formula (4.2). This risk trend corresponds to the entropy values of permissions and APIs, respectively, which in both cases have linear function form. In this theoretical analysis, we assume that there are no feature repetitions among the available features; however, this is not

---

[*] https://ibotpeaches.github.io/Apktool/
[†] http://developer.android.com/reference/android/Manifest.permission.html

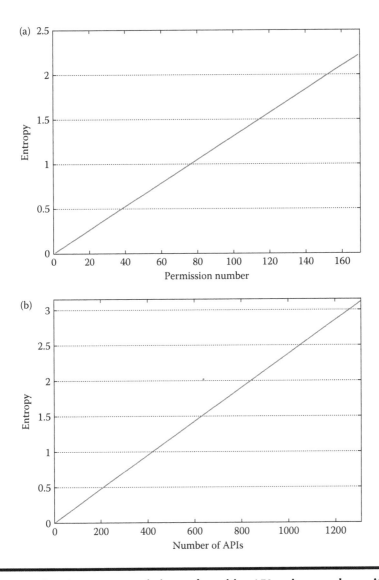

Figure 4.2 Apps' entropy permission and sensitive APIs privacy and security risk assessment. (a) Entropy risk trend for different number of permissions. (b) Entropy risk trend for different number of sensitive APIs.

always the case in the real world, especially for APIs that an app might use. Without loss of generality, the less sensitive features an app might rely on the less intrusive behavior has, and vice versa. Obviously, an app that has no sensitive features and neither a permission nor an API has no entropy and consequently no risk according to our model. On the other hand, if an app has, for example, 170 permissions and 1310 APIs, its risk reaches the maximum value of intrusiveness, that is, 5.31.

4.5 Evaluation

We evaluate our approach using two different sets of mobile apps: (a) Google Play Store's 100 top apps and (b) 100 malware apps randomly selected. All the apps were collected during August 2016. We should mention that our approach's aim is not to classify apps into different categories, that is, goodware and malware, but to assess their intrusiveness in terms of sensitive requested resources. This way, end-users can be informed in a quantitative approach about an app's intrusiveness without the need to install and use it beforehand, and we can compare different apps under their intrusiveness perspective.

4.5.1 Goodware

Figure 4.3a reports on the intrusiveness score of the top 100 apps of Google Play Store. According to our approach, apps' intrusiveness values range between 0 and 0.76 (for additional stats, refer to Table 4.1). As mentioned earlier, the less intrusive the score of a given app, the less offensive the behavior of the app. Recall that an app's intrusiveness is directly related with its incorporated functionality. So, if an app declares that it accomplishes a task, let us say X, while it hides other additional tasks, we can identify it; we do not identify the task itself, but the app's high entropy could be used to deduce this fact.

Based on our outcomes, 20% of the top 100 apps score is less than 0.1, while 40% of them are between 0.1 and 0.2. Only 9% of the apps reach values greater than 0.6; however, 15% of the examined apps have values between 0.3 and 0.5. Figure 4.4 illustrates the probability density function of goodware intrusiveness. At this point, we should mention that apps with high score do not mean that they act maliciously, and vice versa, but it indicates the app's high intrusiveness on end-users' private sphere. Interestingly, in our analysis, the most intrusive app is a security-related app, while communication and social media apps also reach very high. We are planning to study different categories of intrusiveness in a future work.

4.5.2 Malware

We also report on malware intrusiveness in Figure 4.3b in order to represent the whole picture of software intrusiveness. This is because unfortunately malware is part of the software that end-users might use (usually without noticing). So, it is interesting to study this kind of software, over our approach, as well.

Briefly, malware app's intrusiveness values range between 0.15 and 0.45 (for additional stats, refer to Table 4.1). According to our analysis, 20% of the malware apps have intrusiveness values less than 0.1. 40% of them have a score between 0.1 and 0.2, while 23% have values between 0.2 and 0.3. Surprisingly, only 1% have intrusiveness score greater than 0.5 (Figure 4.5).

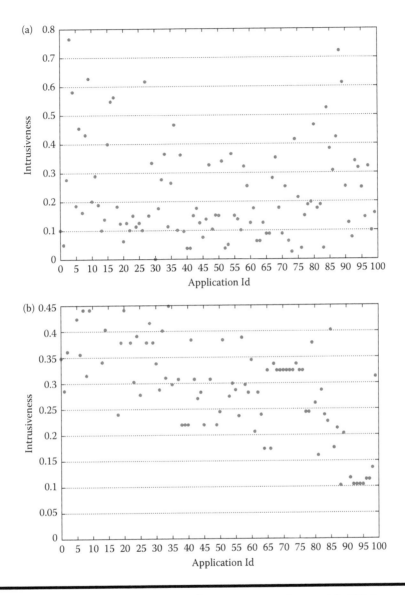

Figure 4.3   Apps' intrusiveness: Goodware versus malware. (a) 100 top apps' intrusiveness. (b) Malware intrusiveness.

4.5.3 *Discussion*

The empirical analysis demonstrates that our approach provides promising results with regard to software intrusiveness of Android apps to end-users' private sphere. Results show that malware does not have a higher intrusiveness score compared to goodware. From one side, this fact might be considered expected as malware relies

Table 4.1   Goodware and Malware Intrusiveness Score Statistics Summary

Category	Average	Maximum	Minimum	Standard Deviation
Goodware	0.22	0.765	0	0.168
Malware	0.15	0.42	0.02	0.08

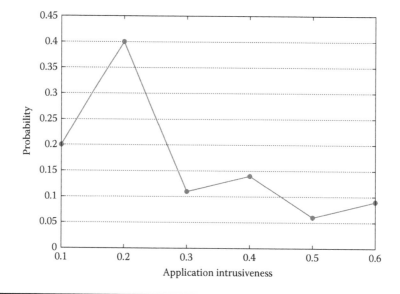

Figure 4.4   Probability density function for goodware intrusiveness.

on sophisticated approaches to hide its intrusiveness. On the other side, this could also be a limitation of our approach as we rely on sensitive APIs as determined by Felt et al. [17], which currently might be outdated. Moreover, in the current version of our approach, we assume that all permissions and APIs have the same significance on the intrusiveness score. Thus, we are planning for our future extensions to also use other research works identifying these sensitive APIs, like the research work presented in [18], and weighted the different features in order to achieve higher accuracy on app's risk estimation. In any case, as mentioned earlier, our goal is to provide a framework that enable end-users to easily assess app's intrusiveness instead of classifying whether an app is a malware or not.

4.6 Related Work

In this section, we overview related works that deal with end-users' privacy. A detailed security and privacy analysis of these works is beyond the scope of this

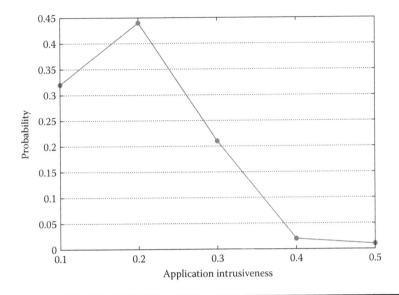

Figure 4.5   Probability density function for malware intrusiveness.

chapter, and can be found in Reference 19. In brief, we distinguish related works in two main categories based on the main characteristics of the solutions that are used to assess end-users' privacy:

1. *Static and dynamic monitoring*: Monitoring data flows and access to other sensitive resources, either statically or during application execution, to detect possible privacy leakages.
2. *Privacy risk evaluation*: Assessing either the risk of using an application or the impact that data privacy leakages can have to end-users, without the need to execute an app.

An overview of these solutions can be found in Table 4.2. The next two subsections analyze the above two categories more in detail.

4.6.1 Static and Dynamic Monitoring

TaindDroid [20] is among the very first works demonstrating privacy issues of third-party apps in Android through dynamic execution analysis on a virtualized environment. To do this, the authors extended Android OS to make it able to track sensitive data flows between sources and sinks. Using this solution, end-users could identify possible data leakages of the apps they are willing to use before installing them in the phone. In this direction, PasDrdoid [21] builds on the advantages of TaintDroid, and enables end-users to determine on the fly whether a transaction

Table 4.2 Overview of the Related Work

Solution	Data Tracking	Privacy Risk Evaluation	Characteristics
[20]	Yes	No	Dynamic—Functions at OS level
[21]	Yes	No	Dynamic—Functions at OS level
[22]	Yes	No	Dynamic—Controls access at data sources
[23]	Yes	No	Static—Data flow correlation
[24]	Yes	No	Dynamic—Cloud-based solution
[25]	No	Yes	Static—Permission-based analysis
[26]	No	Yes	Dynamic—Privacy behavior analysis
[27]	No	Yes	Static—Machine learning-based solution
[28]	No	Yes	Dynamic—Develops privacy profiles
[29]	No	Yes	Introduces risk indicators
[30]	No	Yes	Dynamic—Monitor app's behavior
[31]	No	Yes	Dynamic—A cooperative cloud-based app's privacy evaluation approach
[9]	No	Yes	Assess app's reviews for assessing its risk
[7,8]	No	Yes	Assess the impact of app's security breach on end-users
[32]	No	Yes	Measuring users' perception with regard to security and privacy
[33]	No	Yes	Probabilistic generative models for risk scoring
[34]	No	Yes	Risk assessment based on permissions
[35]	No	Yes	Symbolic execution—Identifies unintended data transactions

that includes sensitive data should be allowed or blocked. Orthogonally, Aspect-Droid [36] focuses on monitoring sensitive data on a modified instrumented version of the original app, without the need to modify the underlying infrastructure, that is, OS.

DroidJust [23] is an automated static solution that justifies whether an access to sensitive data is accomplished for the purposes of the app. This is achieved through a correlation between incoming and outgoing data flows based on static analysis. In this way, end-users can assess how offensive a given app is based on the DroidJust justification for the access to sensitive data performed by the app. However, privacy leakages might be connected with end-users' intention to accomplish a specific operation, so the analysis of sensitive data transmission in Android for privacy leakage detection presented in AppIntent [35] is an important complementary function.

In an alternative approach, Beresford et al. [22] in their solution enable end-users to control access to sensitive data by granting access either to mock or to actual (sensitive) data. However, this solution requires the use of a customized version of the Android OS.

AntMon [24] introduces a service, which is completely transparent to end-users, and records apps' mobile traffic to a cloud service for analysis in order to deduce whether a privacy leakage exists or not.

4.6.2 *Privacy Risk Evaluation*

Theoharidou et al. [7] and Mylonas et al. [8] introduce a risk management approach in order to assess end-users' impact of a security breach in their mobile device. In these works, the authors take into consideration end-users' expectations with regard to personal and sensitive data disclosure and assess possible impacts on a qualitative scale from zero to four. In this direction, Jorgensen et al. [29] develop risk indicators to study mobile app's risk as perceived by end-users.

Complementary, Kong et al. [9] automatically assess the review-to-behavior fidelity of apps in terms of security and privacy. Although these works might be beneficial, they are based on end-user-centric reviews and their subjective opinion about privacy leakages, without taking into consideration software innate properties.

Close to our approach, Agrawal et al. [25] focus on apps' intrusiveness. Their framework analyzes permissions and classifies them into different categories assigning a specific score to each of them, that is, from 1 to 5. So, the permissions of a given app are exported in order to calculate app's intrusiveness using a maximum function of permissions' score.

AppProfiler [26] develops a knowledge base of mapping between APIs and fine-grained privacy behaviors, which is used to analyze an app in terms of privacy behavior and privacy report easily understandable by end-users. In the direction of identifying privacy leakages on apps for kids, Liu et al. [27] introduce a machine learning approach using app's static features, that is, category to classify whether an app violates kids' privacy.

In a postactive approach Roshandel et al. [28] collect information during apps' execution in order to develop their privacy profile and inform end-users about apps' privacy behavior. This approach helps users to better calibrate their own preference, and determine if they would like to continue using the app. Similarly, RiskMon [30], having in mind also end-users' diverse perception for privacy risk, introduces a solution for a continuous monitoring of app's behavior. This solution consists of building a specific level of privacy of a set of apps that end-users trust in order to develop the expected behavior and compare it with the behavior of other untrusted apps. This is done during app's execution, and it identifies the ones diverging from the expected behavior. In the same direction, Papamartzivanos et al. [31] introduce a cloud-based architecture that enables end-users to share privacy-related information for a specific app, which is logged locally. This way, authors build a cooperative privacy awareness system for end-users that have installed and use the same application.

Using probabilistic generative models for risk scoring, Hao Peng et al. [33] introduce the notion of risk scoring and risk ranking for Android applications in order to improve risk communication. The results, based on real dataset extracted from Google Play Store, show that the Naive Bayes with Informative Priors (PNB) has performance equivalent or better than other more sophisticated approaches, and that it gives an accurate indication about why an application has a high risk and how the latter can be reduced.

Sarma et al. [34] propose an Android apps' risk assessment that takes into account also the benefits associated with a permission and a comparison with the permissions requested by similar apps. They conduct an analysis on a large dataset of applications and malware, categorizing the permissions according to their level of criticality and obtaining a signaling mechanism triggered when the risk is above a defined threshold.

A complementary and different approach is the work of Chin et al. [32], a study aimed at measuring the perception users have about smartphone's privacy and security in order to guide the design of possible solutions. The results highlight users' concerns on running sensitive tasks on smartphones, mainly for the fear of data loss and to a low level of security they associate to wireless channels. The conclusion drawn up by the authors is to embed, directly on the devices, data backup and wipe functionalities as well as the adoption of security indicators in the applications store.

4.7 Conclusions and Future Work

Familiarizing end-users with mobile app's privacy and security risk is a challenging task. This is mainly due to the fact that end-users have different technical backgrounds, perspectives, and notions about the terms of privacy and security. Consequently, any approach aimed at identifying and disseminating the risks of an app is always insufficient for some end-users. As a consequence, those approaches should always be orthogonal to each other in order to complement their limitations.

In this chapter, we introduced a practical methodology for assessing mobile app's intrusiveness, with regard to end users' privacy, taking into consideration apps' innate characteristics. Our approach can be considered as a standalone solution that is able to quantify any app's intrusiveness. Outcomes show that app's intrusiveness relies on the functionality incorporated in its code. Currently, we use only permissions and APIs as features for calculating app's intrusiveness. Thus, we are planning to extend our methodology to include other useful features existing in the app, such as interprocess communications activities, data sources, etc., and weighting the different features appropriately in order to develop a more fine-grained intrusiveness assessment methodology.

References

1. The International Data Corporation. Mobile Internet users to top 2 billion worldwide in 2016, according to IDC, December 2015.
2. FBI warns Loozfon, Finfisher mobile malware hitting Android phones, http://www.networkworld.com/article/2223327/security/fbi-warns-loozfon-finfisher-mobile-mal ware-hitting-android-phones.html, October 2012.
3. Mobile malware evolution 2015, https://securelist.com/analysis/kaspersky-security-bulletin/73839/mobile-malware -evolution-2015/, February 2016.
4. J. Chen, H. Chen, E. Bauman, Z. Lin, B. Zang, and H. Guan, You shouldn't collect my secrets: Thwarting sensitive keystroke leakage in mobile IME apps, in *Proceedings of the 24th USENIX Conference on Security Symposium*, SEC '15, USENIX Association, Berkeley, CA, USA, pp. 675–690, 2015.
5. Z. Deng, B. Saltaformaggio, X. Zhang, and D. Xu, Iris: Vetting private API abuse in iOS applications, in *Proceedings of the 22nd ACM SIGSAC Conference on Computer and Communications Security*, CCS '15, ACM, New York, NY, USA, pp. 44–56, 2015.
6. W. Enck, P. Gilbert, S. Han, V. Tendulkar, B.-G. Chun, L. P. Cox, J. Jung, P. McDaniel, and A. N. Sheth, Taintdroid: An information-flow tracking system for realtime privacy monitoring on smartphones, *ACM Transactions on Computer Systems*, **32**(2): 5:1–5:29, 2014.
7. M. Theoharidou, A. Mylonas, and D. Gritzalis, *A Risk Assessment Method for Smartphones*, Springer, Berlin, Heidelberg, pp. 443–456, 2012.
8. A. Mylonas, M. Theoharidou, and D. Gritzalis, *Assessing Privacy Risks in Android: A User-Centric Approach*, Springer International Publishing, Cham, pp. 21–37, 2014.
9. D. Kong, L. Cen, and H. Jin, Autoreb: Automatically understanding the review-to-behavior fidelity in Android applications, in *Proceedings of the 22nd ACM SIGSAC Conference on Computer and Communications Security*, CCS '15, ACM, New York, NY, USA, pp. 530–541, 2015.
10. H. Bojinov, D. Boneh, R. Cannings, and I. Malchev, Address space randomization for mobile devices, in *Proceedings of the Fourth ACM Conference on Wireless Network Security*, WiSec '11, ACM, New York, NY, USA, pp. 127–138, 2011.

11. H. Shacham, M. Page, B. Pfaff, E.-J. Goh, N. Modadugu, and D. Boneh, On the effectiveness of address-space randomization, in *Proceedings of the 11th ACM Conference on Computer and Communications Security*, CCS '04, ACM, New York, NY, USA, pp. 298–307, 2004.

12. M. Abadi, M. Budiu, Ú. Erlingsson, and J. Ligatti, Control-flow integrity, in *Proceedings of the 12th ACM Conference on Computer and Communications Security*, CCS '05, ACM, New York, NY, USA, pp. 340–353, 2005.

13. G. S. Kc, A. D. Keromytis, and V. Prevelakis, Countering code-injection attacks with instruction-set randomization, in *Proceedings of the 10th ACM Conference on Computer and Communications Security*, CCS '03, ACM, New York, NY, USA, pp. 272–280, 2003.

14. G. Portokalidis and A. D. Keromytis, Fast and practical instruction-set randomization for commodity systems, in *Proceedings of the 26th Annual Computer Security Applications Conference*, ACSAC '10, ACM, New York, NY, USA, pp. 41–48, 2010.

15. N. Provos, M. Friedl, and P. Honeyman. 2003. Preventing privilege escalation. In *Proceedings of the 12th conference on USENIX Security Symposium—Volume 12 (SSYM'03)*, Vol. 12. USENIX Association, Berkeley, CA, USA, pp. 16–16.

16. E. S. Claude, A mathematical theory of communication, *The Bell System Technical Journal*, **27**(3): 379–423, 1948.

17. A. P. Felt, E. Chin, S. Hanna, D. Song, and D. Wagner, Android permissions demystified, in *Proceedings of the 18th ACM Conference on Computer and Communications Security*, CCS '11, ACM, New York, NY, USA, pp. 627–638, 2011.

18. K. W. Y. Au, Y. F. Zhou, Z. Huang, and D. Lie, Pscout: Analyzing the Android permission specification, in *Proceedings of the 2012 ACM Conference on Computer and Communications Security*, CCS '12, ACM, New York, NY, USA, pp. 217–228, 2012.

19. Sufatrio, D. J. J. Tan, T.-W. Chua, and V. L. L. Thing, Securing Android: A survey, taxonomy, and challenges, *ACM Computing Surveys*, **47**(4): 58:1–58:45, 2015.

20. W. Enck, P. Gilbert, B.-G. Chun, L. P. Cox, J. Jung, P. McDaniel, and A. N. Sheth, Taintdroid: An information-flow tracking system for realtime privacy monitoring on smartphones, in *Proceedings of the 9th USENIX Conference on Operating Systems Design and Implementation*, OSDI '10, USENIX Association, Berkeley, CA, USA, pp. 393–407, 2010.

21. S.-H. Hung, S.-W. Hsiao, Y.-C. Teng, and R. Chien, Real-time and intelligent private data protection for the Android platform, *Pervasive and Mobile Computing*, **24**: 231–242, 2015. Special Issue on Secure Ubiquitous Computing.

22. A. R. Beresford, A. Rice, N. Skehin, and R. Sohan, Mockdroid: Trading privacy for application functionality on smartphones, in *Proceedings of the 12th Workshop on Mobile Computing Systems and Applications*, HotMobile '11, ACM, New York, NY, USA, pp. 49–54, 2011.

23. X. Chen and S. Zhu, Droidjust: Automated functionality-aware privacy leakage analysis for Android applications, in *Proceedings of the 8th ACM Conference on Security & Privacy in Wireless and Mobile Networks*, WiSec '15, ACM, New York, NY, USA, pp. 5:1–5:12, 2015.

24. A. Shuba, A. Le, M. Gjoka, J. Varmarken, S. Langhoff, and A. Markopoulou, Antmonitor: Network traffic monitoring and real-time prevention of privacy leaks in mobile devices, in *Proceedings of the 2015 Workshop on Wireless of the Students, by the Students, for the Students*, S3 '15, ACM, New York, NY, USA, pp. 25–27, 2015.

25. A. Agrawal, B. Sodhi, and T. V. Prabhakar, A multi-dimensional measure for intrusion: The intrusiveness quality attribute, in *Proceedings of the 9th International ACM Sigsoft*

Conference on Quality of Software Architectures, QoSA '13, ACM, New York, NY, USA, pp. 63–68, 2013.

26. S. Rosen, Z. Qian, and Z. M. Mao, Appprofiler: A flexible method of exposing privacy-related behavior in Android applications to end users, in *Proceedings of the Third ACM Conference on Data and Application Security and Privacy*, CODASPY '13, ACM, New York, NY, USA, pp. 221–232, 2013.

27. M. Liu, H. Wang, Y. Guo, and J. Hong, Identifying and analyzing the privacy of apps for kids, in *Proceedings of the 17th International Workshop on Mobile Computing Systems and Applications*, HotMobile '16, ACM, New York, NY, USA, pp. 105–110, 2016.

28. R. Roshandel and R. Tyler, User-centric monitoring of sensitive information access in Android applications, in *Proceedings of the Second ACM International Conference on Mobile Software Engineering and Systems*, MOBILESoft '15, IEEE Press, Piscataway, NJ, USA, pp. 144–145, 2015.

29. Z. Jorgensen, J. Chen, C. S. Gates, N. Li, R. W. Proctor, and T. Yu, Dimensions of risk in mobile applications: A user study, in *Proceedings of the 5th ACM Conference on Data and Application Security and Privacy*, CODASPY '15, ACM, New York, NY, USA, pp. 49–60, 2015.

30. Y. Jing, G.-J. Ahn, Z. Zhao, and H. Hu, Riskmon: Continuous and automated risk assessment of mobile applications, in *Proceedings of the 4th ACM Conference on Data and Application Security and Privacy*, CODASPY '14, ACM, New York, NY, USA, pp. 99–110, 2014.

31. D. Papamartzivanos, D. Damopoulos, and G. Kambourakis, A cloud-based architecture to crowdsource mobile app privacy leaks, in *Proceedings of the 18th Panhellenic Conference on Informatics*, PCI '14, ACM, New York, NY, USA, pp. 59:1–59:6, 2014.

32. E. Chin, A. P. Felt, V. Sekar, and D. Wagner, Measuring user confidence in smartphone security and privacy, in *Proceedings of the Eighth Symposium on Usable Privacy and Security*, SOUPS '12, ACM, New York, NY, USA, pp. 1:1–1:16, 2012.

33. H. Peng, C. Gates, B. Sarma, N. Li, Y. Qi, R. Potharaju, C. Nita-Rotaru, and I. Molloy, Using probabilistic generative models for ranking risks of Android apps, in *Proceedings of the 2012 ACM Conference on Computer and Communications Security*, CCS '12, ACM, New York, NY, USA, pp. 241–252, 2012.

34. B. P. Sarma, N. Li, C. Gates, R. Potharaju, C. Nita-Rotaru, and I. Molloy, Android permissions: A perspective combining risks and benefits, in *Proceedings of the 17th ACM Symposium on Access Control Models and Technologies*, SACMAT '12, ACM, New York, NY, USA, pp. 13–22, 2012.

35. Z. Yang, M. Yang, Y. Zhang, G. Gu, P. Ning, and X. S. Wang, AppIntent: Analyzing sensitive data transmission in Android for privacy leakage detection, in *Proceedings of the 2013 ACM SIGSAC Conference on Computer and Communications Security*, CCS'13, ACM, New York, NY, USA, pp. 1043–1054, 2013.

36. A. Ali-Gombe, I. Ahmed, G. G. Richard III, and V. Roussev, Aspectdroid: Android app analysis system, in *Proceedings of the Sixth ACM Conference on Data and Application Security and Privacy*, CODASPY '16, ACM, New York, NY, USA, pp. 145–147, 2016.

Chapter 5

From Fuzziness to Criminal Investigation: An Inference System for Mobile Forensics

Konstantia Barmpatsalou, Tiago Cruz,
Edmundo Monteiro, and Paulo Simoes

Contents

5.1 Foreword

It is a generally accepted truth that as soon as the popularity of mobile devices started increasing, security incidents also followed the same path. More specifically, in 2015, an increase of 38% in overall "detected information security incidents"

[1] was observed. Cases concerning attacks on mobile devices showed a 12% increase in comparison to the equivalent findings of 2014. Meanwhile, 32% of the attacks on corporate assets involved computers and mobile devices [2]. The aforementioned numbers correspond solely to incidents that unfolded online. If one takes into account the traditional crime scene investigation cases, which, among others, involve owners of mobile devices and their usage, then the rates grow even more.

Investigators are not only responsible for managing an increased number of incidents but also for handling growing amounts of data, since contemporary mobile devices are used in data-intensive use cases. This combination of facts creates a new challenge in the field of MF in front of a new main challenge. How will the investigators' work be facilitated in terms of data off-loading and automatic or semi-automatic crime recognition?

Usually, in a field such as online criminology, where a piece of evidence cannot be directly associated to either being suspicious or not and uncertainty is a reality, hard computing methods, such as Naive Bayes, Support Vector Machine (SVM) or Random Forest Classification (RFC) perform in a less efficient way [3]. Their two-valued logic is rather restricting when the environment under research is multivariate. On the other hand, soft computing methods such as fuzzy logic, artificial neural networks (ANNs), genetic algorithms, and evolutionary computing show better results when managing data based on approximation and nonlinearity.

Proper evidence data handling can not only give a new perspective in the investigation process, but also complement highly correlated information security principles, such as intrusion detection. This chapter aims to validate the correlation between the forensic science in mobile devices and mobile intrusion detection systems, by simultaneously respecting the fundamental boundaries between them. It also discusses the potential of automated investigation processes and proposes an equivalent high-level schema.

5.2 Related Work

The current section discusses the advances in the aforementioned disciplines of MF and mobile IDSs and provides a background for the correlation validation that follows.

5.2.1 Mobile Forensics

After the "Big Bang" of the smartphone era (beginning of widespread commercial use), the mobile forensics (MF) universe also started expanding. The majority of the first research papers were exclusively dedicated to various acquisition types and techniques, interacting either with the hardware (physical acquisition) or the software (logical and pseudo-physical acquisition) parts of a target mobile device [4].

Various new theoretical and practical models were presented and the evolution of the discipline was rapid in the timespan of less than a decade.

However, new research concerns emerged. The field of acquisition methodology became more complex, since the device memory is no longer the unique part under investigation. Researchers are facing the challenges of acquisition in various environments, such as the cloud storage services, which are defined by different principles and pose new restrictions [5]. Additionally, a proper investigation process does not result in a flawless acquisition stage. Investigators need a detailed and structured event presentation, so as to achieve time-efficient correlation, especially when handling acquired data in bulk [6].

Data retrieved by forensic acquisition techniques in compromised devices or systems form a discipline examining the impact of threats and attacks on them and can also be used as a testing ground for identification, detection and prevention of malware, and other suspicious entities [7]. Parsing of retrieved data is also an emerging subdiscipline. Its span varies from the development of new scripting methods solely dedicated to evidence parsing in mobile technologies to the implementation of novel techniques, such as the post-cold-boot attack RAM parsing described by Hilgers et al. [8].

Data generated from mobile devices have a nonlinear relationship with the functionality these devices serve. As an immediate result, closing cases becomes a rather difficult task for the investigators. Acquired data classification is time-consuming, whereas evidence correlation to other cases is either impossible or very complicated. Even though evidence classification, correlation, and crime identification would be highly estimated research priorities, they are some of the least developed disciplines in the field of MF. Nonetheless, the particular fields have a very high potential and can lead to impressive results in digital investigation, especially when interoperating with other security mechanisms, such as intrusion detection and prevention.

5.2.2 Intrusion Detection

The increasing amount of malicious activity on smartphones is a proof that sophisticated mechanisms have to be implemented toward that direction. One of these mechanisms are mobile intrusion detection systems (IDSs), defined as security systems that monitor computer (and recently mobile) systems and network traffic and analyze the available data for security incidents such as external invasions or internal misuse [9,10]. Moreover, an IDS is designed in such a way that it can foresee and prevent the system from future attacks, by gathering and analyzing data from already performed ones. Lately, the use of IDSs has expanded to the mobile discipline, where their functionality is related also to detection of malicious activity and applications, as claimed by Shabtai et al. [11]. Currently, there are two major categories in which an IDS can be classified: host-based (HIDS) and network-based (NIDS), as well as the third, a hybrid version of the two previously mentioned types.

Even though the smartphones discipline is relatively new, research on mobile IDSs started way before their era. The very first attempts in the field date back to 1997, but this survey incorporates research papers published within the past 8 years, in order to focus on more up-to-date techniques.

A behavioral detection solution for various malware types (viruses, worms, and Trojans) was implemented by Bose et al. [12]. The main idea behind the detection mechanism was the creation of a database containing various behavioral patterns of devices becoming infected by malware. Afterward, SVM was used in order to classify actual patterns encountered. Battery-sensing intrusion protection system (B-SIPS), introduced by Buennemeyer et al. [13], carried out the task of notifying the users when changes related to resources drain occurred.

One of the relatively early efforts in the field of soft computing for IDSs was carried out by Fries [14]. The author used a combination of fuzzy systems and genetic algorithms, as a means of solving the lack of adaptability in detecting mutant threats by "providing near optimal solutions for NP-complete problems" [14]. Even though it is not a piece of research entirely dedicated to mobile devices, its approach is flexible enough to be used in a mobile environment.

"A power-aware, malware detection framework that monitored, detected and analyzed previously unknown energy-depletion threats" was presented by Kim et al. [15]. An HIDS operating at the application level and within the resources limitations of a mobile device was proposed by Schmidt et al. [16]. The presented framework performed analysis of static function calls. A remote anomaly detection system (RADS) was the functionality core in the paper by Schmidt et al. [17]. Features related to the state of a monitored device that would prove helpful for anomaly detection were acquired and forwarded to the RADS. Collected data then would be useful in order to distinguish the cases of normal and anomalous behavior.

The solution provided by Schmidt et al. [18] was an HIDS performing static analysis and using signature-based detection. Executables were parsed for specific system calls, which were then compared with those of malware executables. Sandboxing, the default feature in the Android OS when an application runs in an isolated environment, was used by Blasing et al. [19]. In their research, the authors performed static and dynamic analysis in a sandboxed environment. According to Damopoulos et al. [20], TaintDroid, the solution implemented by Enck et al. [21], was the first to run directly on the device.

However, it did not use pure anomaly detection but taint tracking, that is, tracing "the flow of sensitive data through third-party applications" [22]. Its functions are resumed in sensitive information labeling and alert creation when those data were used. One major drawback of the system was the initial consideration of all the applications as not trusted. Application filtering or exceptions would have increased the efficiency of the system.

Paranoid Android by Portokalidis et al. [23] is a hybrid IDS with cloud support, which exploited the advantages of remote analysis without the restrictions of a mobile environment. Knowledge-based temporal abstraction (KBTA) was used by Shabtai et al. [24], so as to recognize temporal behavioral patterns. Crowdroid, a hybrid IDS stated in [25] used crowdsourcing, so as to collect accurate information about users' devices behavioral profiles. Then it compared them with patterns of known and self-written malware.

In "Andromaly," Shabtai et al. [11] developed a live event monitoring IDS and used machine learning techniques in order to trace malware existence. Damopoulos et al. [20] implemented an IDS with complementary features of host and cloud-based detection so as to profit from both by maximizing performance. Papamartzivanos et al. [26] created a cloud based, crowdsourcing, behavioral detection mechanism, so as to identify data leak patterns from applications that could compromise the end users' privacy. Lastly, Ariyapala et al. [27] implemented a hybrid IDS in the form of a mobile application which collects host and network data and offloads the task of anomaly detection on the cloud.

5.3 Intrusion Detection and MF: Two Worlds Apart?

The correlation between IDSs and forensic evidence has been widely disputed throughout the literature, with many different opinions appearing. More specifically, Stephenson [28] exposes the two controversial opinions. On the one hand, usage of forensic evidence for intrusion detection support is considered unsuitable for the systems' characteristics. On the other hand, forensic data are the perfect solution for collection of live or postmortem evidence. Sommer [29] claims the inverse, that logs deriving from IDSs are actual forensic evidence. This assumption was also validated by Arasteh et al. [30], who designed a model based on it.

Judging by the conclusions drawn by the aforementioned research papers, the two disciplines' compatibility is rather controversial. However, the terms overlap and the two worlds are intersecting in many points. In this section, we aim to enumerate the differences, the bisection points, and clarify the role of each discipline. Firstly, an IDS examines incidents that occur mainly within a short time frame; it can be adjustable but needs to be short for system efficiency and for immediate decision on countermeasures. On the contrary, a forensic system has no time limitations concerning the time span of an incident. It tracks activity within a bigger, flexible, and adjustable time frame, which can be configured by the investigators according to their requirements.

In terms of operational timing, IDSs are used as preventive measures. Their scope is to protect a system while a malicious action is occurring, so they have to be previously loaded with knowledge (in the form of signatures or patterns). Forensic methods are used after a crime is conducted, in order to figure out which entity was responsible for it. However, this does not prohibit the forensic analysis results from being used as learning patterns in the future.

An IDS receives a respectful quantity of data when operating. As a result, and for the sake of efficiency, not every alert should be considered as a major system emergency in the need of countermeasures. If that case would occur, then the resulting systems would provide considerably higher FP rates. During a forensic investigation, the very same alert can be an emergency event and should be regarded as one, according to the case under investigation. The results and the interpretation of data vary according to the scenario under examination. The same data may have different interpretations and weight (importance) for different phenomena. Thus, a forensic environment is more context-sensitive than an IDS.

Lastly, an IDS has to be as autonomous as possible, without the need for human intervention. Besides, forensic systems will always be in need of human expertise. This, however, does not prevent forensic intelligence and automation from offloading many investigators' tasks, especially when they are handling big data.

Therefore, there are no bounds that would restrict the interoperability of systems of the two disciplines, at least in theory. They can coexist in the same system and data flow between them can be cyclic. For example, IDS logs can serve as subjects for forensic investigations and vice versa. However, this interaction needs to be properly contextualized and the roles of each discipline have to be defined a priori, so as to avoid conflicts related to their significance. Figure 5.1 describes the relationship between an IDS and a forensic mechanism coexisting in the same system. As a system operates, it is continuously protected by the IDS, which also performs logging and provides feedback to the forensic mechanism for evidence investigation. Simultaneously, the forensic mechanism investigates and audits the function of the system, based on past evidence and intrusion detection logs. The investigation results can feed future patterns to the IDS.

In the next section, we will analyze the role and operation of such a forensic mechanism within a protected mobile environment.

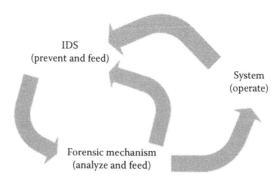

IDS
(prevent and feed)

System
(operate)

Forensic mechanism
(analyze and feed)

Figure 5.1   IDS and forensic mechanisms correlation in a system.

5.4 A Fuzzy Inference System for MF

A forensic mechanism integrated in a mobile environment is a characteristic that cannot be ignored when it comes to the aforementioned system's analysis and design, because its existence can improve the overall quality of service in terms of security, as well as actively contribute to intrusion prevention. It can provide a source of expert knowledge to the investigators as a complementary means of prediction and serve as a background for pattern creation to be used as feeding material for the IDS, especially with the appropriate data analysis and one-time configuration. For such a procedure to take place, there are a few key points that have to be clarified in order to avoid future contextualization conflicts and dead-ends.

- Even though mobile devices have capabilities which resemble the ones of the high-end desktop systems, especially in terms of software performance and functionality, their hardware (low-level) status, energy efficiency, and storage space capacity differ significantly and need an alternative investigation approach, which takes their particularities into account without causing a resource drain.
- Moreover, data types handled by mobile devices are more specific and limited, comparing to the vastness of computer-manipulated data. This fact facilitates a potential investigative procedure.
- Acquired data from mobile devices do not completely belong to the devices, since the latter interoperate with many different systems. One of the most common examples are cloud storage services. Owners of mobile devices usually store data in cloud repositories, other than using the limited handset storage means. There is also a variety of gadgets (smart watches, wristbands, and professional equipment, such as wearables with sensors, electronic stethoscopes, etc.) that synchronize with mobile devices and their data are also stored in the equivalent applications. However, when it comes to criminal investigation, every data source is valuable and needs to be taken into consideration.
- The systemic environment in which a device operates is also an important factor concerning the data interpretation. Forensic investigation is highly context-sensitive and different patterns may lead to different assumptions in one environment and have a completely opposite meaning in another. A detailed system description and scope approach must precede every investigation.

When all the aforementioned prerequisites are assured, one can proceed to the system design and the definition of its basic functionalities. To the best of our knowledge, this is the first research approach toward logically induced rule generation and a fuzzy system implementation for pure crime behavioral identification. Previous work in collaboration between the disciplines of fuzzy systems and MF is mainly related to pure quantitative metrics for network forensics purposes [31] or to semantics [32] and are elaborated in the following paragraphs.

A methodology capable of creating "incremental fuzzy decision trees based on network service type to reduce the human intervention and time-cost" was implemented by Liu and Feng [33]. In the same line of thought, [31] highlighted the network traffic—big data related upcoming issue—and presented the combination of fuzzy logic and expert systems in order to classify different attack types and subsequently filter the important ones. Rostamipour and Sadeghiyan [34] implemented an automated fuzzy expert system for network forensics, which was not in need of human involvement in order to create evidence from the information feed it had. The system was able to pinpoint toward attack patterns and entities.

An approach rather oriented toward the semantic universe, the paper by Stoffel et al. [35], suggested an "automatic procedure for expert-system-like rule generation" [35] by the use of fuzzy clustering in actual evidence datasets consisting of phrases or words used during crime reporting (such as trespassing, theft, murder), provided by law enforcement agencies. Other works in the field mainly involve plaintext document parsing for specific terms and clustering with appropriate algorithms such as "K-means, K-medoids, Single Link, Complete Link, and Average Link" [32] and using "self-knowledge algorithms" as a means of classification between suspicious terms in data retrieved from mobile devices [36].

A crime is not an action that can be characterized by strict norms and evidence cannot be strictly classified as suspicious or not. As a result, it cannot be represented by binary metrics, but by a rather complex set of rules, leading to values corresponding to probability levels. Moreover, forensic intelligence is not in such a developed level so as to claim that it can replace human expertise. Actually, its role is more suitable for complementary support by providing insights and guidance.

In a system consisting of multiple variables-factors with variant weights according to the case under examination, traditional hard computing approaches of crisp classification are not proven efficient. On the contrary, accepting that an action or a piece of evidence can have different degrees of belonging to a potential suspiciousness level output approaches the description of a more realistic phenomenon. Moreover, evidence as data can show a certain level of uncertainty, varying from moderate to high. This fact makes hard computing a less appropriate solution. Fuzzy systems as a subset of soft computing show high tolerance levels to imprecision, uncertainty, partial truth, nonlinearity, and approximation [37].

The proposed system aims to calculate the overall probability of suspiciousness level in specific, predefined criminal investigation scenarios, while receiving forensically acquired data as inputs. It consists of a hybrid mechanism working on three different levels; firstly, data are processed by a fuzzy system, which calculates the suspiciousness level per category according to logical IF–THEN rules constructed by previous expert knowledge in the area of information security [38]. An example of such a rule can be written in a simplified version as follows: "If a caller's appearance frequency is high and a call duration is long, then the suspiciousness level for

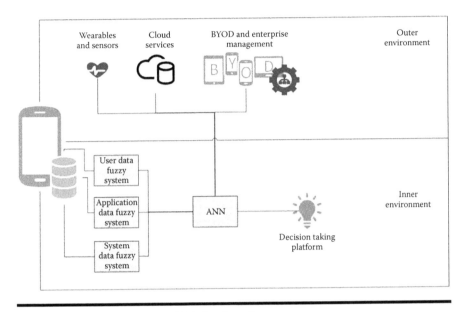

Figure 5.2   Proposed system high-level outline.

d given criminal activity is low." Secondly, the initial assumptions and the fuzzy system efficiency are evaluated by an ANN (cooperative neuro-fuzzy system), and lastly, the results are sent and depicted in an observer-friendly format in a properly formed decision-taking platform. The extended mechanism based on mobile forensic principles is depicted in Figure 5.2.

The schema consists of two major parts, which are distinguished by the fact of being or not being components of the mobile device under investigation. The first part, known as the "inner environment," concerns the data an individual can acquire from a mobile device by different acquisition methods (physical, logical, or pseudo-physical) [39], in a human interpretable format and preferably already structured. Data are then split into three different categories, grouped by their particular characteristics, as described by Barmpatsalou et al. [4].

As already mentioned at the beginning of this section, auditing of mobile devices is no longer limited to the handset environment, but expands to more sources of information [40]. External factors, such as data from wearable devices and sensors, cloud services, and enterprise mobility management with BYOD support, constitute the second part, named "Outer Environment." They also (when existing) play a role in the overall suspiciousness level calculation. Alongside with the fuzzy system derived data, they contribute to the learning process conducted by the ANN and the final outputs are projected on a decision-taking platform. The current paper focuses on examining the performance and the efficiency of the fuzzy systems.

The authors provided a forensic data grouping according to the degree of membership to the three following sets: user, application, and system data. More precisely, the user data group refers to information "imported and edited by users, such as text messages, contact lists, pictures and all sorts of customized application data," whereas the system data concern information manipulated mainly by mobile operating systems, such as connection handlers (GPS, WiFi), network usage statistics, operating systems "defaults and structural elements (IMEI, IMSI)." Finally, the third category comprises "data used by applications as background procedures and other similar entries handled by applications" [4]. Each data subcategory is further fragmented to its structural elements. For example, an SMS message is broken down to characteristics such as the length, the appearance frequency, the sender's device mobility state (whether it is a mobile device, a fixed line, or a service). An entry in the GPS-location dataset can consist of the respective latitude, longitude, and timestamp, whereas an entry corresponding to an installed application may contain elements such as the installation timestamp, the application name, the process ID (PID), etc. The authors used the "Cambridge Device Analyzer" [41] dataset, an actual log collection from mobile devices which contains various data types. Owing to space limitations, SMS logs are the only category under examination present in the paper.

The first step concerns the creation of a use case. According to the aforementioned expert knowledge source [38], the authors decided to use a protest that escalated from a peaceful event to rioting due to officers' infiltration to the protesters' side as an experimental background. The memory of the officers' devices is forensically acquired or logged so as to be used for suspicious pattern identification. An SMS logged entry consists of the following attributes: (appearance frequency of each sender's name or number, message length, and sender's country of origin).

Every subcategory characteristic is actually a linguistic variable, which can receive values in different scales (very low to very high, very short to very long, local, or foreign), according to the existing attribute type. After the variables' definition and scaling are complete, the rules can be generated. Expert knowledge is used in order to create a list of assumptions, which are used as guidelines for further rule generation.

- The officers are using dedicated devices only during their service.
- The protesters are more likely to use unknown sources (one-time payphones), street phones, or the Internet for communication), but not foreign numbers.
- A longer message signifies more time availability, so it is a less suspicious pattern.
- The lower the appearance frequency of a message is, the more suspicious the exchange becomes (it might be an encoded message, and the officer would not use the dedicated phone for infiltration purposes but only in the heat of the moment).

1. If (Length is very_short) and (ApFreq is very_low) and (CountrySource is abroad) then (Suspiciousness is medium) (1)
2. If (Length is very_short) and (ApFreq is very_low) and (CountrySource is less_local) then (Suspiciousness is high) (1)
3. If (Length is very_short) and (ApFreq is very_low) and (CountrySource is local) then (Suspiciousness is very_high) (1)
4. If (Length is very_short) and (ApFreq is very_low) and (CountrySource is abroad) then (Suspiciousness is medium) (1)
5. If (Length is short) and (ApFreq is very_low) and (CountrySource is less_local) then (Suspiciousness is high) (1)
6. If (Length is short) and (ApFreq is very_low) and (CountrySource is local) then (Suspiciousness is very_high) (1)
7. If (Length is medium) and (ApFreq is very_low) and (CountrySource is abroad) then (Suspiciousness is medium) (1)
8. If (Length is medium) and (ApFreq is very_low) and (CountrySource is less_local) then (Suspiciousness is high) (1)
9. If (Length is medium) and (ApFreq is very_low) and (CountrySource is local) then (Suspiciousness is high) (1)
10. If (Length is long) and (ApFreq is very_low) and (CountrySource is abroad) then (Suspiciousness is low) (1)
11. If (Length is long) and (ApFreq is very_low) and (CountrySource is less_local) then (Suspiciousness is low) (1)
12. If (Length is long) and (ApFreq is very_low) and (CountrySource is local) then (Suspiciousness is medium) (1)
13. If (Length is very_long) and (ApFreq is very_low) and (CountrySource is abroad) then (Suspiciousness is low) (1)
14. If (Length is very_long) and (ApFreq is very_low) and (CountrySource is less_local) then (Suspiciousness is low) (1)
15. If (Length is very_long) and (ApFreq is very_low) and (CountrySource is local) then (Suspiciousness is medium) (1)
16. If (Length is very_short) and (ApFreq is low) and (CountrySource is abroad) then (Suspiciousness is medium) (1)
17. If (Length is very_short) and (ApFreq is low) and (CountrySource is less_local) then (Suspiciousness is high) (1)
18. If (Length is very_short) and (ApFreq is low) and (CountrySource is local) then (Suspiciousness is very_high) (1)
19. If (Length is short) and (ApFreq is low) and (CountrySource is abroad) then (Suspiciousness is low) (1)
20. If (Length is short) and (ApFreq is low) and (CountrySource is less_local) then (Suspiciousness is medium) (1)
21. If (Length is short) and (ApFreq is low) and (CountrySource is local) then (Suspiciousness is high) (1)
22. If (Length is medium) and (ApFreq is low) and (CountrySource is abroad) then (Suspiciousness is very_low) (1)
23. If (Length is medium) and (ApFreq is low) and (CountrySource is less_local) then (Suspiciousness is low) (1)
24. If (Length is medium) and (ApFreq is low) and (CountrySource is local) then (Suspiciousness is medium) (1)

Figure 5.3 **"IF–THEN" rules definition.**

■ The less an exchange appears and the shorter a message is, the higher the probability of suspiciousness becomes.

Figure 5.3 contains a snapshot of the "IF–THEN" rules generated.

After defining the appropriate system rules, the system has to be parameterized. The limits of each input variable are defined and the membership function is chosen and adapted according to the needs of the system. Figure 5.4 represents the triangular and trapezoidal membership functions for the input variables of the SMS entries: appearance frequency, length, and country source.

The variables are then used as inputs in each fuzzy system and the final output is calculated as the total suspiciousness level of each data type combination. The suspiciousness level is higher when the output membership function value is higher than 0.9. The next step, known as the fuzzy system evaluation, involves the testing of the system with actual data. For the evaluation procedure, three datasets from different mobile devices were used. Since no suspicious patterns are known to exist in the datasets, the most appropriate membership function approach is the one which detects the less—but not equal to zero—potentially suspicious entries. In Table 5.1, we present the percentages of potentially suspicious patterns encountered in each device sample when triangular and trapezoidal membership functions were used.

It is noticeable that both of the membership functions are performing similarly. However, the triangular membership function leads to a smaller number of potentially suspicious patterns; thus, it is considered the best approach. Nevertheless, more membership function types have to be evaluated and compared, so

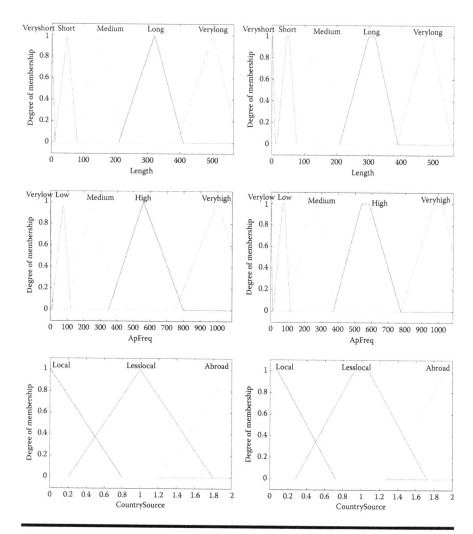

Figure 5.4   Triangular and trapezoidal membership functions.

Table 5.1   Percentages of Potentially Suspicious
SMS Patterns per Membership Function

Dataset/Function	Triangular		Trapezoidal	
Device 1	49/1311	3.7%	50/1311	3.8%
Device 2	77/1138	6.7%	87/1138	7.6%
Device 3	31/947	3.2%	40/947	4.2%

as to select the most appropriate one. Afterward, the fuzzy evaluation should be completed for other data types, such as calls, Wi-Fi data, geolocation coordinates, etc. Since fuzzy systems do not have learning capabilities, it is rather difficult to adapt them and their parameters for different datasets. This is where the cooperative nature of fuzzy systems and neural networks provides further assistance. The neural network adjusts the parameters according to the datasets and optimizes their values, with a future aim to identify not only the suspicious pattern, but also the crime that it belongs to.

5.5 Conclusion

Automation in mobile criminal investigation is a rather challenging task and this assumption escalates more since there is a respectable research gap in the particular area and existent works are relatively few. However, with the appropriate triggering provided by surveys and some very first implementations, the field has the potential to show substantial development. The current chapter worked toward this direction by proposing a self-evaluating and learning schema of criminal activity recognition. Its element of success depends highly on the detailed description of its structural parts, that is, the logical rules which characterize each action and data type involved. Its aims is to offload the tasks of investigators handling digital crime cases and also provide some useful feedback and logging as prior knowledge against security breached in systems or stand-alone mobile devices.

References

1. PwC, Turnaround and transformation in cybersecurity: Key findings from The Global State of Information Security* Survey 2016. Retrieved from http://www.pwc.com/gx/en/issues/cyber-security/information-security-survey/download.html, 2016.
2. PwC, Global Economic Crime Survey 2016. Retrieved from http://www.pwc.com/gx/en/economic-crime-survey/pdf/GlobalEconomicCrimeSurvey2016.pdf, 2016.
3. R. A. R. Ashfaq, X. Z. Wang, J. Z. Huang, H. Abbas, and Y. L. He, Fuzziness based semi-supervised learning approach for intrusion detection system, *Information Sciences*, 378: 484–497, 2017.
4. K. Barmpatsalou, D. Damopoulos, G. Kambourakis, and V. Katos, A critical review of 7 years of mobile device forensics, *Digital Investigation*, 10(4): 323–349, 2013.
5. S. Zawoad, A. K. Dutta, and R. Hasan, SecLaaS: Secure logging-as-a-service for cloud forensics, in *The Proceedings of the 8th ACM SIGSAC Symposium on Information, Computer and Communications Security*, ACM, Hangzhou, China, pp. 219–230, 2013.
6. D. Kasiaras, T. Zafeiropoulos, N. Clarke, and G. Kambourakis, Android forensics: Correlation analysis, in *The 2014 9th International Conference for Internet Technology and Secured Transactions (ICITST)*, IEEE, London, United Kingdom, pp. 57–162, 2014.

7. E. Casey, *Smartphone Forensics and Mobile Malware Analysis*. Retrieved from http://cas eite.drupalgardens.com/content/smartphone-forensics-and-mobile-malware-analysis, 2013.

8. C. Hilgers, H. Macht, T. Müller, and M. Spreitzenbarth, Post-mortem memory analysis of cold-booted android devices, in *Proceedings of the 2014 Eighth International Conference on IT Security Incident Management & IT Forensics (IMF '14)*, IEEE Computer Society, Washington, DC, pp. 62–75, 2014.

9. SANS Institute, *Intrusion Detection Systems: Definition, Need and Challenges*. Retrieved from: https://www.sans.org/reading-room/whitepapers/detection/intrusion-detection-systems-definition-challenges-343, 2001.

10. A. S. K. Panthan, *The State of the Art in Intrusion Prevention and Detection*, Auerbach Publications, New York, NY, 2014.

11. A. Shabtai, U. Kanonov, Y. Elovici, C. Glezer, and Y. Weiss, Andromaly: A behavioral malware detection framework for android devices, *Journal of Intelligent Information Systems*, 38(1): 161–190, 2011.

12. A. Bose, X. Hu, K. G. Shin, and T. Park, Behavioral detection of malware in mobile handset, in *Proceedings of the International Conference on Mobile Systems, Applications, and Services (MobiSys)*, New York, NY, pp. 225–238, 2008.

13. T. K. Buennemeyer, T. M. Nelson, L. M. Clagett, J. P. Dunning, R. C. Marchany, and J. G. Tront, Mobile device profiling and intrusion detection using smart batteries, in *HICSS '08: Proceedings of the 41st Annual Hawaii International Conference on System Sciences*, IEEE Computer Society, Washington, DC, p. 296, 2008.

14. T. P. Fries, A fuzzy-genetic approach to network intrusion detection, in *Proceedings of the 10th Annual Conference Companion on Genetic and Evolutionary Computation (GECCO '08)*, ACM, New York, NY, pp. 2141–2146, 2008.

15. H. Kim, J. Smith, and K. Shin, K, Detecting energy-greedy anomalies and mobile malware variants, in *Proceedings of the 6th International Conference on Mobile Systems, Applications, and Services (MobiSys '08)*, ACM, New York, NY, pp. 239–252, 2008.

16. A. D. Schmidt, H. G. Schmidt, J. Clausen, K. A. Yksel, O. Kiraz, A. Camtepe, and S. Albayrak, Enhancing security of linux-based android devices, in *Proceedings of 15th International Linux Kongress*, Lehmann, Hamberg, Germany, 2008.

17. A. D. Schmidt, F. Peters, F. Laamour, and S. Alabayrak, Monitoring smartphones for anomaly detection, *Mobile Networks and Applications (MONET)*, 14(1): 92–106, 2009.

18. A. D. Schmidt, A. Camptepe, and S. Albayrak, Static smartphone malware detection, in *5th Security Research Conference (Future Security 2010)*, Berlin, Germany, 2010.

19. T. Bläsing, L. Batyuk, A. D. Schmidt, S. A. Camtepe, and S. Albayrak, An android application sandbox system for suspicious software detection, in *5th International Conference on Malicious and Unwanted Software (MALWARE)*, Nancy, Lorraine, pp. 55–62, 2010.

20. D. Damopoulos, G. Kambourakis, and G. Portokalidis, The best of both worlds: A framework for the synergistic operation of host and cloud anomaly-based IDS for smartphones, in *Proceedings of the 7th European Workshop on System Security*, ACM, New York, NY, pp. 6:1–6:6, 2014.

21. W. Enck, P. Gilbert, B. G. Chun, L. P. Cox, J. Jung, P. McDaniel, and A. N. Sheth, Taintdroid: An information-flow tracking system for realtime privacy monitoring on smartphones, in *Proceedings of the 9th USENIX Conference on Operating Systems Design and Implementation, OSDI'10*, USENIX Association, Berkeley, CA, pp. 1–6, 2010.

22. M. La Polla, F. Martinelli, and D. Sgandurra, A Survey on security for mobile devices, *IEEE Communications Surveys and Tutorials*, 15(1): 446–471, 2013.

23. G. Portokalidis, P. Homburg, K. Anagnostakis, and H. Bos, Paranoid android: Versatile protection for smartphones, in *Proceedings of the 26th Annual Computer Security Applications Conference (ACSAC '10)*, ACM, New York, NY, pp. 347–356, 2010.

24. A. Shabtai, U. Kanonov, and Y. Elovici, Intrusion detection for mobile devices using the knowledge-based, temporal abstraction method, *Journal of System and Software*, 83: 1524–1537, 2010.

25. I. Burguera, U. Zurutuza, and S. Nadjm-Tehrani, Crowdroid: Behavior-based malware detection system for Android, in *Proceedings of the 1st ACM Workshop on Security and Privacy in Smartphones and Mobile Devices*, ACM, New York, NY, pp. 15–26, 2011.

26. D. Papamartzivanos, D. Damopoulos, and G. Kambourakis, A cloud-based architecture to crowdsource mobile app privacy leaks, in *Proceedings of the 18th Panhellenic Conference on Informatics (PCI '14)*, ACM, New York, NY, pp. 59:1–59:6, 2014.

27. K. Ariyapala, H. G. Do, H. N. Anh, W. K. Ng, and M. Conti, A host and network based intrusion detection for android smartphones, in *2016 30th International Conference on Advanced Information Networking and Applications Workshops (WAINA)*, Crans-Montana, Switzerland, pp. 849–854, 2016.

28. P. Stephenson, The application of intrusion detection systems in a forensic environment, in *Third International Workshop on the Recent Advances in Intrusion Detection (RAID 2000)*, Toulouse, France. Retrieved from http://www.raidsymposium.org/raid2000/program.html, 2000.

29. P. Sommer, Intrusion detection systems as evidence, *Computer Networks*, 31(23): 2477–2487, 1999.

30. A.R. Arasteh, M. Debbabi, A. Sakha, and M. Saleh, Analyzing multiple logs for forensic evidence, digital investigations, in *Proceedings of the 7th Annual Digital Forensic Research Workshop (DFRWS'07)*, Digital Investigation, 4(Supplement 1): S82–S91, 2007.

31. N. Liao, S. Tian, and T. Wang, Network forensics based on fuzzy logic and expert system, *Computer Communications*, 32(17): 1881–1892, 2009.

32. L. F. da Cruz Nassif and E. R. Hruschka, Document clustering for forensic computing: An approach for improving computer inspection, in *The 10th International Conference on Machine Learning and Applications and Workshops (ICMLA)*, Honolulu, HI, pp. 265–268, 2011.

33. Z. Liu and D. Feng, Incremental fuzzy decision tree-based network forensic system, *Lecture Notes in Computer Science*, 3802: 995–1002, 2005.

34. M. Rostamipour and B. Sadeghiyan, Network attack origin forensics with fuzzy logic, in *The 2015 5th International Conference on Computer and Knowledge Engineering (ICCKE)*, Mashhad, pp. 67–72, 2015.

35. K. Stoffel, P. Cotofrei, and D. Han, Fuzzy methods for forensic data analysis, in *SoCPaR*, pp. 23–28, 2010, http://ieeexplore.ieee.org/document/5685848/

36. F. Marturana, G. Me, R. Berte, and S. Tacconi, A quantitative approach to triaging in mobile forensics, in *2011 IEEE 10th International Conference on Trust, Security and Privacy in Computing and Communications*, Changsha, pp. 582–588, 2011.

37. A. Gegov, *Fuzzy Networks for Complex Systems*, Springer, Berlin, Germany, 2010.

38. SALUS, *Deliverable 2.1: SALUS PPDR Use Cases—Intermediate*. Retrieved from https://www.sec-salus.eu/wp-content/uploads/2014/05/SALUS_D2.1_v1.0.pdf, 2014.

39. K. Barbatsalou, E. Monteiro, and P. Simoes, Mobile forensics: Evidence collection and malicious activity identification in PPDR systems, in *Proceedings of the International Conference in Information Security and Digital Forensics*, SDIWC, Thessaloniki, Greece, pp. 42–48, 2014.
40. S. Siboni, A. Shabtai, N. O. Tippenhauer, J. Lee, and Y. Elovici, Advanced security testbed framework for wearable IoT devices, *ACM Transactions on Internet Technology*, 16(4), 2016.
41. D. Wagner, A. Rice, and A. Beresford. Device analyzer: Understanding smart-phone usage, in *The 10th International Conference on Mobile and Ubiquitous Systems: Computing*, Networking and Services, Tokyo, Japan, 2013.

MALWARE DETECTION IN MOBILE PLATFORMS

Chapter 6

Function-Based Malware Detection Technique for Android

Nava Sherman and Asaf Shabtai

Contents

6.1 Introduction

The popularity and adoption of smartphones has fostered the rapid proliferation of mobile malware, especially on Android. According to Trend Micro predictions, the number of high-risk malicious applications for Android will reach two million in 2016 [1]. The open-market nature of Android applications and the lack of awareness of smartphone users to potential risks [2] make the Android operating system (OS) susceptible to malware attacks over the Internet. A comprehensive survey on the current state of the threats and vulnerabilities of mobile devices was done in Reference 3.

Most current antivirus software uses signature-based detection. The use of signatures makes antivirus applications vulnerable to malware obfuscation methods that can be easily applied to Android applications. This is due to tools such as the "Proguard," which is used for obfuscating applications and is included in the Android SDK. As a case in point, Rastogi et al. [4] applied several simple transformations to malware applications that are detected by the most commonly used (Android) antivirus software. None of these transformations affected the behavior of the application; however, the changes they made dramatically decreased the detection rate of the malware by various antivirus software. A similar evaluation conducted by Fedler et al. [5] also showed that most antivirus products could be evaded by making minor alterations to malware. Their recommendation was to apply static and dynamic analysis methods for comprehensive and scalable detection of malware applications.

The prevailing analysis methods for mobile malware detection found in the literature use static analysis [6–8], dynamic analysis [9,10], and hybrid techniques [11,12].

An advantage of dynamic analysis techniques, in which information gathered by executing the application is the basis for malware detection, is that they can detect the download and execution of native code by a malware application or the use of encryption and obfuscation methods that can bypass static code analysis. The downside of dynamic analysis is its lack of code coverage and the fact that it is time and resource consuming.

Static analysis methods usually apply reverse engineering to the Android application code to extract the properties and functionalities of the applications.

This study proposes a static analysis approach for detecting malicious Android applications. The approach focuses on the detection of malicious code writers' attempts to avoid detection mechanisms (such as signature-based detectors). The approach is based on the fact that attackers use obfuscation techniques in an attempt to hide malicious code. As described in Reference 13, these malware obfuscation techniques include dead code insertion, register reassignment, subroutine reordering, instruction substitution, code transposition, code integration, and more. Wong et al. [14] suggested that these techniques cause alterations to the file structure, making it distinct from benign code structure, and they introduce a malware detection technique based on this observation. The suggested method profiles the applications by extracting features derived from the structural properties of the functions in the applications. The extracted features consist of three groups of features from three different relevant domains: (1) geological analysis extracting structural-based features derived from the coding time and the length of the functions (e.g., measured by the number of opcodes), (2) textual analysis computing the *tf-idf* measure (term frequency-inverse document frequency) [15] for prominent functions, and (3) graph theory analysis representing functions' concurrence as a graph and extracting structural features from the graph.

We combined the proposed features from different domains for the following two reasons. First, combining the proposed features may prevent malicious code writers from bypassing the detection. Since the detection is based on different and independent groups of features, it will be more complicated to develop malicious code capable of evading all types of features. Second, the learning algorithm can benefit from a larger set of features, especially from features that provide different views on the instances (i.e., Android applications), thus creating a more robust and accurate predictor. We conducted an experiment using 60,000 benign and 10,000 malicious Android applications. Using the proposed method, we were able to classify Android applications with an area under the ROC curve (AUC) of 0.97.

6.2 Related Work

Most of the research in the field of Android malware detection uses static analysis, dynamic analysis, and hybrid techniques to detect new malware.

Dynamic analysis is performed by monitoring the application at runtime, either on the endpoint or on an emulator/sandbox. An advantage of the dynamic detection method is that it can detect the download and execution of native code by the malware or the use of encryption, both of which can bypass static code analysis. In addition, dynamic analysis is immune to code obfuscation. Some previous works on dynamic analysis proposed analyzing the behavior of applications by monitoring network traffic. For example, Shabtai et al. [16] used a cross-feature analysis approach to detect self-updating malware applications, and Wei et al. [17] dynamically analyzed the network traffic of applications running in a sandbox in order to detect the DNS resolution behavior of Android malware. Others aimed at tracking known malware behaviors, including Bose et al. [9] who dynamically traced sequences of API function calls. These sequences were matched to sequences that represent known malware behaviors. Mohd et al. [18] dynamically traced the executed system calls (using an emulator) for detecting SMS exploitation. Shamili et al. [19] used an SVM classifier based on features collected from phone calls, SMS records, and data communication to detect specific malware types. Papamartzivanos et al. [20] designed a cloud-based architecture in order to detect privacy violation, create a collaborative infrastructure for exchanging information related to applications privacy exposure level, and supply a behavior-driven detection mechanism.

While the studies mentioned above are aimed at detecting specific and predefined malware behavior, the following studies examined the behavior of applications on several levels of the Android platform in order to detect malware without predefining the specific behavior of the application. Burguera et al. [21] monitored the Linux kernel system call usage as a basis for malware classification. Reina et al. [22] inspected not only the system calls but also the interactions between the Android components (binder) and the underlying Linux system to characterize low-level OS-specific and high-level Android-specific behaviors. Zhao et al. [23] analyzed the level above and based their classifier on the recorded system resources' access requests for each process. Zhang et al. [24] combined dynamic tracing of the permission requests for resources by applications, with tracking sensitive operations on the granted resources (using taint tracking). Rastogi et al. [4] presented the "AppsPlayground" framework that uses dynamic analysis on different Android platform levels: Linux system call tracing, API call tracing, and taint tracking. A thorough classification of the proposed solutions, according to the information they use (analyze) and the methods applied (e.g., signature vs. machine learning solutions), is presented in Reference 3.

The downside of the dynamic analysis techniques used in all of the above works is that it is complicated to simulate the exact trigger condition, which causes the malicious behavior to be executed. There are many types of trigger conditions; for example, the malware may be activated only at a certain time or when connected to a specific network. The trigger condition may also be a certain user or network input. Furthermore, malware code writers often use antiforensic techniques that can detect being monitored (running in a sandbox or in a debugger environment) and hide their activity. The need to pinpoint the exact conditions required to activate the malicious behavior makes the dynamic analysis techniques time and resource consuming. In order to address this problem, Rastogi et al. [4] used a special type of sandbox emulation with automatic exploration strategies in their "AppsPlayground" framework. Tam et al. [25] presented a different approach to this problem in the framework "CopperDroid." They implemented a technique that artificially simulates the analyzed malware with events, based on the malware's manifest file.

Complementary to the dynamic analysis are the static analysis methods, which usually use reverse engineering on an Android application to extract its properties in order to detect malware. One static analysis strategy is analyzing the way malware works and trying to detect these behaviors in an application's code. For example, Grace et al. [12] statically detected certain behaviors (e.g., encryption and dynamic code loading) as a basis for selecting suspicious applications for further analysis. Zhou et al. [26] hashed small sequences of the Dalvik byte code to create small "fingerprints" in order to detect repackaged applications.

Other approaches use information extracted from different levels of the Android platform to identify malicious applications. Yuksel et al. [27] classified the Android-based security solutions into four groups according to the information they analyze: information extracted from the OS level, permissions requested by the application, source code, and application/service behavior. An example an OS-based solution is provided by Alam et al. [6] who used sequences of control flow patterns (derived from the machine-level native code) for malware detection. Permission-based solutions use the manifest file, particularly its permission request, as information for static analysis. Permissions[*] are an Android platform security mechanism that allows or restricts application access to protected APIs and resources (see the *Android User Guide*). By default, Android applications have no permissions, and their access to protected APIs or resources on the device is prohibited. Permissions are requested by the application through the manifest file. During installation, the user is asked to grant the permissions, and the application will contain the permissions that the user gives. Since permissions serve as the gate used to access critical system resources, many methods use them to track malware.

Debelo et al. [28] classified Android applications using an Support Vector Machine (SVM) classifier, based on requests for high-risk permissions. In their framework "DroidMat," Wu et al. [29] expanded upon Debelo's work and extracted features from both the manifest file (permissions and intent) and API calls (extracted from the binary dex file). They then used a clustering algorithm to detect malware families. Yerima et al. [30] utilized parallel classifiers built on a common set of features extracted from API calls, permissions, and commands to detect malicious Android applications. Ali-Gombe et al. [8] combined opcode sequences found in sensitive functional modules with permission file information to detect malicious applications. Chuang et al. [31] used the APIs that are more often used in normal applications than in malicious applications as features for an SVM classifier, Similarly, Arp et al. [32] combined features extracted from the manifest file (e.g., requested hardware component) and from the Dalvik executable (e.g., restricted API calls) file to construct a joint vector space. This vector space was used not only for the detection process but also for analyzing the malware behavior.

One of the problems with using the permissions as a basis for malware detection is that the request of certain permissions in the application manifest file does not necessarily mean that it is actually used within the code. Thus, researchers have tried to reconstruct the application call and data flow in their static analysis. Elish et al. [33] developed an assurance score for applications by calculating the percentage of sensitive operations (data or system resource access) triggered by user activity for each application. Other methods statically track the combination of call and data flow to reconstruct the application's behavior [34].

Some works combine both dynamic and static analysis methods. For example, Luoxu Min et al. [35] proposed a system that dynamically simulates suspected

---

[*] http://developer.android.com/guide/topics/security/permissions.html

actions, which are chosen according to the result of static analysis of the application.

In this study, we attempt to explore a novel static analysis approach for detecting malicious Android applications. The proposed approach is based on the assumption that malware application writers employ evasion techniques to avoid detection. These techniques tend to change the code structure, and the new method aims at detecting these changes. Several works used structural changes to detect malware. These works used the function's coding time [36] or the function's length [37,38] on Windows executable files. Our research differs from the previous works in three ways. First, the detection method utilizes features that are generic and not platform-specific, and therefore it can be applied to any platform (e.g., Windows and iOS). Second, the proposed features are based on the structure and distribution of the code's functions. The advantage of using functions is that a function contains more semantic information than a sequence of bytes, and thus may be more effective (as shown in the work of Shabtai et al. [39]), the accuracy increases with features that contain more semantic information. Finally, we combine features from different domains, thus preventing malicious code writers from bypassing the detection.

6.3 Method

Static analysis of Android applications using machine learning techniques is applied in order to detect malicious applications. This analysis uses three types of features that are extracted from the functions that constitute the application's code:

- *Geological features*—Extracted based on the estimated coding time of functions and the functions' length distribution (measured in Dalvik opcodes)
- *Textual analysis features*—Extracted based on the calculation of the *tf-idf* value [15] of prominent functions
- *Graph-based features*—Structural features extracted from the graph that is generated from the functions' concurrence

For extracting the three groups of features, we utilize a database of functions that are extracted from a known set of benign applications. This database is termed the *reference database*. The *reference database* contains the functions that are identified and extracted from that set of benign files, as well as the estimated coding date of each function, structural properties of the function such as length (in opcode), and the function's frequency in the corpus and cooccurrence with other functions in files (i.e., functions that exist in the same file).

Therefore, the proposed method is composed of three main phases as presented in Figure 6.1: *reference function database creation* (indicated by the light gray arrows), *training phase* (indicated by the dark gray arrows), and *evaluation/detection phase* (indicated by the black arrows).

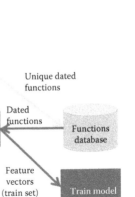

Figure 6.1 **The proposed method's three main phases: (1)** *reference function database creation phase*, **indicated by the light gray arrows (extracting functions from the reference database dataset and building a** *reference database* **that will be used in all of the next phases); (2)** *training phase* **indicated by the dark gray arrows (in this phase, features are extracted and used to build and train a classifier); and (3)** *evaluation/detection phase* **indicated by the black arrows (the classifier's performance in classifying Android applications is evaluated).**

6.3.1 Generating the Reference Database

The first phase (indicated in Figure 6.1 by the light gray arrows) is generating the *reference database*. The input to this phase is a set of known benign Android applications denoted by *R*. In the first step, the functions residing within an Android application are identified and extracted by the function extractor module. The functions are extracted from the .dex (Dalvik executable) file, which is a part of the Android application package (i.e., .apk file). The .dex file is compiled Java bytecode executed by the Android's virtual machine. To extract the functions, each .dex file of an Android application package (.apk file) is disassembled. The Dedexer application[*] is used for the disassemble process. We then identify and extract all functions

---

[*] http://dedexer.sourceforge.net/

in the .dex file. Each function is represented as a sequence of Dalvik opcodes, which constitute the function.

Then, the extracted functions are processed by the function analyzer module. The function analyzer first identifies and merges identical functions. Two functions are considered identical if they have the same length and share the exact Dalvik opcode sequence. Next, the creation date of each function is estimated by the function analyzer module. As each function may appear in more than one file, we set the creation date of a function as the creation date of the oldest file it appeared in. In order to determine the creation date of an .apk file, we could use the following heuristics: (1) creation date of the .dex file; (2) publishing date of the file in the Internet; or (3) creation date from the .apk signature. We opted to use the creation date from the .dex file, because it is assumed to be more accurate than the other two sources although it still can be manipulated by application developers (especially malicious application developers). To partially overcome the problem of forgery, we removed applications with implausible/anomalous file creation dates (such as 1.1.1970) from our datasets. Algorithm 6.1 summarizes the process of setting the creation date of functions. Finally, the function analyzer module stores the functions in the *reference database* along with the function's metadata: creation date, size (i.e., number of Dalvik opcodes), and the identifiers to the files in which the function appears.

Algorithm 6.1 Setting the creation date of functions

Input:	$R = \{a_1, \ldots, a_n\}$
	P – The platform of applications in R
Output:	DF_{APP} dated set of functions containing a list of functions and their estimated creation date $\langle f, date \rangle$

1. $DF_{APP} = \emptyset$
2. **For each** $a \in R$
3.   *tempFunc* $\leftarrow$ *extractFunctions*(a, P)
4.   *date* $\leftarrow$ *extractFileDate*(a, P)
5.   **For each** *fempFunc*
6.   *similar* = '*false*'
7.   **For each** $df \in DF_{APP}$
8.   *if is Similar*(f, df, P)*then*
9.   *similar* = '*true*'
10.   *updateFunctionDate*$(df, date)$
11.   *if similar* = = '*false*' *then*
13. $DF_{APP} \leftarrow DF_{APP} \cup \langle f, date \rangle$
14. return DF_{APP}

6.3.2 Training Phase

In the training phase (indicated in Figure 6.1 by the dark gray arrows), the *reference database* is used to extract features from a given labeled set of benign and malicious applications, that is, the training set. This set does not include the set of benign applications used for generating the *reference database*. The training set and extracted features are used for building a classification model.

First, the functions are extracted from the .apk files in the training set. This is done using the function extractor module described in Section 6.3.1. Then, for each file, three sets of features are extracted using the *reference database*. A description of the three classes of extracted features follows. These features are used to build and train a classifier.

6.3.2.1 Geological Features

This class of features focuses on the code's structure. It combines two feature groups: The first group of features is based on statistical information regarding the length of the functions (in opcodes). The assumption behind the choice of feature group is that in order to avoid detection, malware writers tend to apply obfuscation techniques such as using wrappers, function outlining, and adding null code, which eventually change the distribution of the functions' lengths. The second group of features is based on statistical information regarding the estimated age (i.e., coding time) of the functions contained in the application. Our choice of this group of features is based on the assumption that evasion techniques applied by malware writers result in a much larger amount of new code, compared to benign programs that tend to maximize code reuse.

To extract the first group of features, we sort the functions in the *reference database* according to their length. We then divide the functions into l equal-sized bins; the number of bins l is provided as a parameter. The smallest function and the largest function in each bin determine the bin's range. Then for each file in the training set we extract the following features:

- *Percentage of functions in lengthBin[i]*—For each *lengthBin[i]*, $i = 1 \ldots l$, we calculate the percentage of the application's functions whose length matched the range of the respective bin.
- *Longest function*—The length (in opcodes) of the longest function normalized by the average and standard deviation of the lengths of the functions in the *reference database*.
- *Shortest function*—The length of the shortest function normalized by the average and standard deviation of the lengths of the functions in the *reference database*.

■ *Length stdev*—The standard deviation of the length of the functions in the application file normalized by the average and standard deviation of the lengths of the functions in the *reference database*.

■ *Length average*—The average length of the functions in the application file normalized by the average and standard deviation of the lengths of the functions in the *reference database*.

To extract the second group of features, we sort the functions in the *reference database* according to their creation date. We then divide the functions into t equalsized bins; the number of bins t is provided as a parameter. The oldest function and the newest function of each bin determines the bin's coding time interval. Then, for each file in the training set, we determine the creation date of its functions by searching each of its functions in the *reference database*. If a match is found, the function is assigned with the date of the function in the *reference database*; or else, it is considered a function without a date. Finally, after dating all of the functions in the training set applications, we extract the following features:

■ *Percentage of functions in creationDateBin[i]*—For each *creationDateBin[i]*, $i = 1 \ldots t$, we calculate the percentage of the application's functions for which the creation time matched the range of the respective bin.

■ *Percentage no coding time*—The percentage of functions with no coding time.

■ *Oldest coding time*—The oldest function's coding time (in seconds from 1970) normalized by the average and standard deviation of the coding times of the functions in the *reference database*.

■ *Newest coding time*—The newest function's coding time (in seconds from 1970) normalized by the average and standard deviation of the coding time on the functions in the *reference database*.

■ *Average coding time*—The average coding time of functions normalized by the average and standard deviation of the coding times of the functions in the *reference database*.

■ *Median coding time*—The median of function's coding time normalized by the average and standard deviation of the coding times of the functions in the *reference database*.

■ *Standard deviation coding time*—The standard deviation of the functions' coding time normalized by the average and standard deviation of the coding times of the functions in the *reference database*.

6.3.2.2 Text Analysis Features

We apply this class of features from the text categorization domain to our functionbased malware detection task. For each function, we calculate its *tf-idf* measure, which is a well-known measure in the text categorization field, often used as a weighting factor in information retrieval and text mining [40]. The acronym *tf-idf* is short for term frequency-inverse document frequency. It is a numerical statistic intended to reflect how important a word (i.e., term) is to a document in a collection or corpus. The *tf-idf* value increases proportionally to the number of times a word appears in the document, but is offset by the frequency of the word in the corpus, which helps to adjust for the fact that some words appear more frequently in general. In addition to the many uses of this measure for search engines as a central tool in scoring and ranking documents, it is also used for classification and malware detection [39].

We refer to each function as a "term," and an application is treated as a "document." Each file in the *training set* is then represented using a set of "terms" (functions) denoted by M. This set of functions is selected from the *reference database*. We chose the $|M|$ functions with the highest *document frequency (df)* value (note that in our experiments, we chose the top 10,000 functions with the highest *df* value; thus in total we had 10,000 features). Then, for each function f in M and for each file in the training set, we compute its *tf-idf* value as follows:

1. Compute:

$$idf\left(f,r\right) = \frac{\log|M|}{1 + \left|app \in R : f \in app\right|}$$

where R is the set of benign applications used for generating the *reference dataset*, and $|app \in R: f \in app|$ is the number of documents (applications) in the corpus (R), where function f appears. We adjust the denominator by adding 1 in order to avoid division by zero.

2. Compute the term frequency (*tf*) value of a function in an application (*d*) as the frequency of function f in M normalized by the maximum frequency of a function in the application *d*.

3. Calculate the *tf-idf* measure by $tfidf\left(f,d,R\right) = tf\left(f,d\right) \times idf\left(f,R\right)$.

6.3.2.3 Graph-Based Features

This set of features is based on the assumption that the cooccurrence of functions differs between benign and malicious applications. For example, two functions may tend to appear together more in benign applications than in malicious applications. Based on this hypothesis, we extract a third set of graph-based features and use these features as a basis for classifying applications.

In order to extract the graph-based features, we first generate the functions' cooccurrence graph based on the functions stored in the *reference database*. This *reference graph* is a weighted graph denoted by $G = (V, E)$, where the vertices of the graph (V) are the functions in the *reference database*, and E is the set of edges connecting the functions. The weight of an edge $e \in E$ that connects two vertices (i.e., two functions) is set to be the number of applications in which both functions appear. In cases in which two functions do not appear together in any application file, there is no edge connecting the two related vertices. The weight is normalized by the sum of applications in which each function (individually) appears.

For each application *i* in the *training set*, we extract the graph-based features as follows: First, we identify the largest subset of vertices $V_i \subseteq V$ in the *reference graph* G such that each $v \in V_i$ represents a function common to the *reference database* and the application *i*. Next, we calculate the following set of graph vertex measures for each $v \in V_i$ on the *reference graph*:

- *Degree (or valency)* of a vertex in a graph is the number of edges, which are connected to the vertex.
- *Closeness centrality* of a vertex in a graph is the reciprocal of the sum of the shortest path distances from vertex *u* to all $n - 1$ other vertices. Since the sum of the distances depends on the number of vertices in the graph, closeness is normalized by the sum of the possible distances $(n - 1)$. It is defined as $C(u) = \frac{(n-1)}{\sum_{(v=1)}^{(n-1)} d(v,u)}$, where $d(v,u)$ is the shortest path distance between *v* and *u*, and *n* is the number of vertices in the graph.
- *PageRank* is a measure that ranks the vertices in the graph based on the structure of the incoming links. It was originally used to rank web pages.
- *Shortest path* computes the shortest path length from a vertex to all reachable vertices.
- *Eigenvector centrality* is a measure of the influence of a vertex in a network. It assigns relative scores to all vertices in the network based on the concept that connections to high-scoring vertices contribute more to the score of the vertex in question than equal connections to low-scoring vertices. For a given graph $G = (V,E)$ with $|V|$ number of vertices, let $A = (a_{v,t})$ be the adjacency matrix, that is, $a_{v,t} = 1$ if vertex *v* is linked to vertex *t*, and $a_{v,t} = 0$ in case it is not. The centrality score of vertex *v* can be defined as

$$x_v = \frac{1}{\lambda} \sum_{t \in M_v} x_t = \frac{1}{\lambda} \sum_{t \in G} A_{v,t} x_t$$

where M_v is the set of the neighbors of *v* and λ is a constant.

Finally, for each of the vertex measures above, we compute the following six statistics (computed from the values of the measure over all vertices in set V_i: minimum, maximum, average, median, kurtosis, skewness, and standard deviation.

The same set of features was also computed for the subgraph induced from the *reference graph* (G) by $V_i \subseteq V$. A vertex-induced subgraph (sometimes simply called an "induced subgraph" or "overlay graph)" is a subset of the vertices of graph G, along with all the edges in which both endpoints are in the subset. Figure 6.2 illustrates the subgraph induced from the complete graph k_{10} by the vertex subset (1, 2, 3, 5, 7, 10).

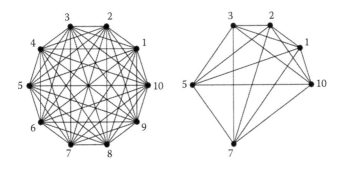

Figure 6.2 Induced graph.

6.3.3 Evaluation (Detection) Phase

In the third phase (indicated in Figure 6.1 by the black arrows), which comes after evaluating the classifier, we evaluate the classifier's performance using a test set (applying the classifier on new unlabeled application files). We extract the proposed features (in the same way as described in Section 6.3.2) for each file in the test set and apply the model that was generated during the training phase in order to classify the file as malicious or benign.

6.4 Evaluation

6.4.1 Dataset

For the experiments, we used a dataset that was composed of 60,000 benign Android applications and 10,000 malicious Android applications. The applications were supplied and labeled by a well-known antivirus company. A classification of the malware applications in the dataset is provided in Appendix 6A. The collection of benign files includes applications from the Google Play Store that were identified as the most popular applications among the antivirus company users. The labeling was verified by the platform "VirusTotal," which classifies the applications according to a number of the most popular antivirus engines. The dataset contained files from 2012 to 2013. The application's date was determined by the creation date of its .dex file.

6.4.2 Evaluation

In order to simulate a real-life environment where a model built from known (labeled) applications is used to classify new applications, we used a chronological evaluation to test and compare our methods. In the chronological evaluation, a

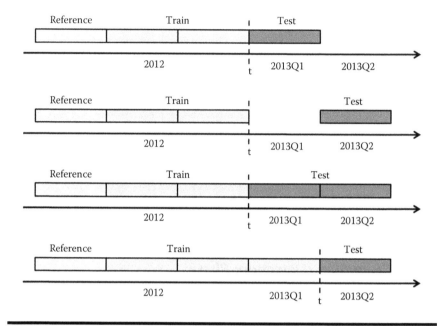

Figure 6.3   Chronological evaluation.

classifier is trained on files created before time *t* and tested on files created at a later time (see Figure 6.3).

As presented in Figure 6.3, we used different sets of applications for the training and *reference database* creation. This was done in order to avoid overfitting. It also gave us an opportunity to repeat the experiment several times and ensure the validity of our results. We repeated the following process three times. Each time, we divided the benign applications in the training set randomly into three folds. Then for each division, we had three iterations. As illustrated in Figure 6.3 in each iteration, a different fold was used for generating the *reference database*, and the other two folds were used for training a model. We sorted the applications according to their creation date into three groups:

1. Android applications created in 2012
2. Android applications created from January through April 2013 (first four months of 2013)
3. Android applications created from May through August 2013 (second four months 2013)

We then ran a series of four tests:

1. Two-thirds of the 2012 applications were used to train the classifier (the features were extracted using the *reference database* built from the remaining third

of the 2012 applications). The classifier was evaluated using a dataset built from the January through April 2013 applications (the first timeline in Figure 6.3 illustrates this test).

2. Two-thirds of the 2012 applications were used to train the classifier. The classifier was evaluated using May through August 2013 applications (the second timeline in Figure 6.3).

3. Two-thirds of the 2012 applications were used to train the classifier. The classifier was evaluated using January through August 2013 applications (the third timeline in Figure 6.3).

4. In order to simulate the updating of the classifier, in the fourth test, the classifier was trained on a dataset containing applications from 2012 and January through April 2013, and it was evaluated on applications from May through August 2013 (the fourth timeline in Figure 6.3).

We executed the four tests mentioned above on each combination of reference dataset, training dataset, and test set. In total we had 36 (3*3*4) tests for each set of features. We conducted all of the tests for each set of features and for a dataset that combined the features of all of the sets (combined feature group). The complete list of computed features is presented in Appendix 6B.

In machine learning applications, feature selection is often applied in order to reduce the number of extracted features, some of which may be redundant or irrelevant. This is done in order to prevent overfitting, enhance the model's generality, and reduce the model's complexity and processing runtime. Since two of our feature sets (the combined feature group and the text analysis feature group) included a large number of features, we applied feature selection to these sets. We compared the results of three feature selection methods: Info Gain (top 50 and top 100) and CFS.

6.5 Results

In the experiments, we attempted to answer the following research questions:

Research Question 1. Is it possible to detect unknown malicious applications on Android devices using the proposed feature groups?

In order to answer the first research question and identify the most effective feature set (geological analysis features, textual analysis features, graph theory analysis features) for detecting malware on Android, we compared the average ROC value and average TPR (for FPR = 5%) of the different feature sets. The results were averaged over all of the tests, classifiers, and feature selection methods. The results are presented in Table 6.1. From the results, we conclude that the best feature group was the combined feature group, which benefits from the three types of feature sets and improves the performance of the textual analysis feature set by 7.2%.

Research Question 2. Which classifier is the most accurate at detecting malware on Android devices?

Table 6.1 Feature Group Comparison

Feature Group	Average ROC	Average TPR (FPR = 5%)
Geological analysis	0.859	0.616
Textual analysis	0.861	0.628
Graph theory	0.785	0.440
Combined	0.923	0.765

Table 6.2 Classifier Comparison

Classifier	Average ROC	Average TPR (FPR = 5%)
Rotation Forest	0.909	0.726
Random Forest	0.868	0.671
Logistic Regression	0.862	0.556
AdaBoost	0.859	0.663

We evaluated the following learning algorithms with Weka's default parameter configuration: Logistic Regression, Rotation Forest (Random Forest with 10 trees as the base classifier), Random Forest (10 trees), and AdaBoost (Random Forest with 10 trees as the base classifier), in order to determine the classifier with the best performance. We compared the classifiers' average ROC values and average TPRs (for FPR = 5%). The results were averaged over all tests, feature groups, and feature selection methods. The results are presented in Table 6.2. The results indicate that Rotation Forest outperformed all of the others classification algorithms.

Research Question 3. For the textual analysis feature set and the combined feature set, which method of feature selection yields the most accurate detection results? Info Gain top 50, Info Gain top 100, or CFS? We compared the performance (average ROC and average TPR for FPR = 5%) of the feature selection methods (third research question). The results were averaged over all tests, both feature sets (the textual analysis feature group and the combined feature group), and each of the classifiers. The results are presented in Table 6.3.

The results show that CFS had the best performance. However, we can see that the performance of the feature selection method interacts with the used classifier. As shown in Table 6.4 (the best results are in bold), for our best classifier, Rotation Forest, the best feature selection method is Info Gain (top 100).

Research Question 4. With regard to geological features, which number of bins performs better, six or nine bins?

To answer this research question, we compared the average ROC and average TPR (for FPR = 5%) of the geological analysis features using six and nine bins (for

Table 6.3   Comparison of the Feature Selection Methods

Feature Selection	Average ROC	Average TPR (FPR = 5%)
CFS	0.897	0.738
Info Gain 100	0.891	0.680
Info Gain 50	0.887	0.672

Table 6.4   Comparison of the Feature Selection Methods by Classifier

Algorithm	CFS	Info Gain 100	Info Gain 50
AdaBoost (using Random Forest)	0.889	0.874	0.869
Logistic Regression	0.892	0.874	0.870
Random Forest	0.890	0.885	0.879
Rotation Forest	0.918	**0.934**	0.932

both size and coding time). The ROC and TPR were averaged over all of the tests and classifiers. Table 6.5 presents the results.

We can see that changing the number of bins affects the results and that nine bins perform better than six bins.

Research Question 5. What level of performance can we achieve when we select the best feature set (combined feature group), the best classifier (Rotation Forest), and the best feature selection method for the classifier (Info Gain top 100)? How does this performance change over time? Can updating the model help to deal with the concept drift?

To answer this research question, we measured the AUC and average TPR (for FPR = 5%) of the optimal model. The results are shown in Table 6.6. To test the performance change over time, we used the model trained on a dataset of 2012 applications and compared the results when tested on the applications from the first four months of 2013 (the first timeline in Figure 6.1) with the results when tested on the applications created in the second four months of 2013 (the second timeline in Figure 6.1). The results show that the performance deteriorate if we

Table 6.5   Comparison of the Number of Bins (Geological Features)

Feature Number of Bins	Average ROC	Average TPR (FPR = 5%)
6	0.847	0.730
9	0.859	0.750

Table 6.6   Optimized Results

Train/Test	Average ROC	Average TPR (FPR = 5%)
2012/First four months of 2013	0.970	0.866
2012/Second four months of 2013	0.952	0.799
2012 + First four months of 2013/Second four months of 2013	0.967	0.846

use the same model on newer files, which might be a sign of a concept drift. By observing the results of a *training set* composed of 2012 and applications from the first four months of 2013 and tested on a set built from applications created in the second four months of 2013 (the fourth timeline in Figure 6.1), we can see that this concept drift can be overcome if we update the training set with new applications. By doing so, we can achieve a result of AUC 0.97 for both tests.

Figure 6.4 depicts the ROC graph of the best results (for both tests: tests one and four in Figure 6.1). The graph shows that the performance of both tests are similar and therefore that updating the model overcomes the deterioration of performance over time.

Research Question 7. When using an old model to classify new applications, will updating the *reference database*, in addition to updating the model, contribute to the performance more than just updating the model?

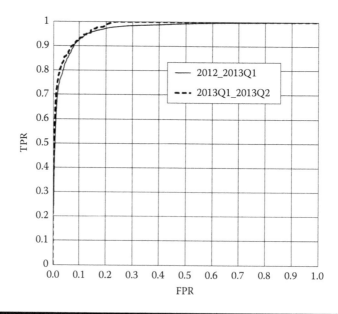

Figure 6.4   Optimized results of ROC graph.

Table 6.7   Results Obtained by Updating the Model Compared with Those Obtained by Updating the Model and the *Reference Database*

Reference Dataset Files	Average ROC	Average TPR (FPR = 5%)
2012	0.90	0.73
2012/First four months of 2013	0.93	0.77

In order to answer this research question, we divided the benign applications from both 2012 and the first four months of 2013, randomly to three folds. We repeated this process three times. For each division, we ran three iterations. In each iteration, a different fold was used for generating the *reference database* and the other two folds were used for training a model. The model was tested on a test set built of applications from the second four months of 2013. In total, we ran all of our tests nine times. We compared the performance (average ROC and average TPR for FPR = 5%) of these tests, to those of the fourth test in Figure 6.1 (testing a model trained on 2012 and first four months of 2013 applications on a test set of applications from of second four months of 2013). In these tests, the *reference database* was built only from 2012 files. The results were averaged over all iterations, classifiers, feature groups, and feature selection methods. The results are presented in Table 6.7.

From the results, we can see that updating the *reference database*, in addition to updating the model, may contribute to the accurate classification of new files.

6.6 Discussion and Future Work

In this work, we proposed a static analysis method that uses function-based features for malware detection. The features were taken from three different domains: geological analysis, textual analysis, and graph theory analysis. From the evaluation that we conducted, we can infer that these sets of features may be a good basis for classifying Android applications.

Our results show that by combining these groups of features and using the Rotation Forest classifier, we can achieve an AUC of 0.97. We also showed that in the case in which the model built is updated with new applications, these strong results are sustained over time.

Thus, we may conclude that combining different function-based features from different domains contributes to the creation of a robust and effective classifier.

We identified several limitations of the proposed method that should be addressed in future work. First, since the proposed method analyzes the Java code, it cannot detect malware written in a native language or encrypted viruses (similar to static analysis approaches that are also based on information extracted from the Java code). In addition, our geological features are based on the creation date of the applications .dex file. Malicious code writers may forge the creation time in order to bypass detection. However, we assume that since the assignment of the creation date is based on a large repository of (legitimate) files, it will not have any effect on the creation date of legitimate functions, and therefore the impact on detection accuracy will be limited. We also propose combining features from different domains to make it even harder to evade the proposed detection approach. Another possible way to try to evade the proposed method is to manipulate the malware code by adding benign functions. This way the malicious code writer can change the length and age statistics and the concurrence of the functions. These evasion methods should be investigated in future work in order to better assess the robustness of the proposed method and improve it (e.g., by using a reference database of malware functions).

In the future, we would like to further explore and enhance the set of features (e.g., the number of bins or more graph-based features).

As our features are universal in nature and are not platform-specific, a possible direction will be to explore these features as a basis for the classification of files from other platforms or malware. We may investigate this direction further and test the option to apply transfer learning techniques, which aim to transfer the knowledge achieved by previous tasks to a different target task by using data from different domains [41]. *Transfer learning* is mostly beneficial in situations where there are insufficient data in the target domain. In such cases, we can use the existing data from source domains to solve problems in the target domain. Using transfer learning techniques, we can attempt to train a model on one platform (e.g., Android) or malware family and try to use it for classification of files (after applying transfer leaning) on another platform (e.g., Windows) or other malware families.

Glossary

Android SDK: A software development kit that enables developers to create applications for the Android platform. The Android SDK includes sample projects with source code, development tools, an emulator, and required libraries to build Android applications.

API: Application program interface (API) is a set of routines, protocols, and tools for building software applications.

Dalvik Executable File: Dalvik is an open-source, register-based virtual machine (VM) that is part of the Android OS. The Dalvik VM executes files in the Dalvik Executable (.dex) format and relies on the Linux kernel for additional functionality like threading and low-level memory management.

DNS: Short for Domain Name System (or Service or Server), an Internet service that translates domain names into IP addresses.

Overfitting: A modeling error that occurs when a function is too closely fit to a limited set of data points. Overfitting the model generally takes the form of making an overly complex model to explain idiosyncrasies in the data under study. In reality, the data being studied often have some degree of error or random noise within it. Thus, attempting to make the model conform too closely to slightly inaccurate data can infect the model with substantial errors and reduce its predictive power.

Virtual Machine: Self-contained operating environment that behaves as if it is a separate computer. For example, Java applets run in a Java virtual machine (VM) that has no access to the host operating system.

Appendix 6A: Malware Dataset

Malware Family	*Number of Applications*
Backdoor	
Andup	36
Anserver	26
Basebrid	1210
Basebridge	39
Bgserv	10
Cawitt	2
Fakengry	57
Fjcon	42
Gingermaster	148
GinMaster	707
Glodream	68
Golddream	48
GoYear	35
Kmin	162
KungFu	680
Qdplugin	2
Xsider	19

Continued

Malware Family	Number of Applications
Yzhc	1
Others	149
Trojan	
BadNews	31
Boqx	129
DorDrae	19
Fakedoc	76
Fakeplayer	7
FakeRun	6
Faketoken	23
Fav	234
Gamex	46
Gappusin	60
Jsmshider	7
Ksapp	282
Meds	209
Mmarketpay	163
Mseg	123
MTK	364
Trojan Downloader	
Ddlight	14
Updtkill	7
UpdtKiller	52
Vdloader	5
Wroba	10
Trojan SMS	
Boxer	192

Continued

Malware Family	Number of Applications
DontLookBack	14
Droidap	10
Elpso	19
FakeLogo	71
Fakeinst	1643
FakePlayer	8
FakeStud	24
Hippo	18
Hispo	3
Ikangoo	104
Jifake	13
Koomer	6
Kyview	37
Opfake	1296
RuSMS	13
Smssend	248
Stealer	14
Rufraud	2
Vidro	14
Vietsms	5
Tesbo	4
Adrd	8
DrDelux	5
Geinimi	23
Smforw	2
Nyleaker	21
Others	864

Appendix 6B: List of Features

Feature Name	Description
Geological Analysis Features	
Q[i]date	Percentage of the functions of an application for which creation time matched the range of the ith bin ($i = 0\ldots8$)
Q[i]size	Percentage of the functions of an application for which the length matched the range of the ith bin ($i = 0\ldots8$)
noAge	Percentage of functions with no coding time
min	Oldest functions coding time (in seconds from 1970) normalized by the average and standard deviation of the coding times of the functions in the reference database
max	Newest functions coding time (in seconds from 1970) normalized by the average and standard deviation of the coding times of all functions in the reference database
median	Median coding time of functions normalized by the average and standard deviation of the coding times of the functions in the reference database
mean	Average coding time of functions normalized by the average and standard deviation of the coding times of the functions in the reference database
stddev	Standard deviation of the functions coding times normalized by the average and standard deviation of the coding times of functions in the reference database
maxLen	The length (in opcodes) of the longest function normalized by the average and standard deviation of the lengths of the functions in the reference database
minLen	The length of the shortest function normalized by the average and standard deviation of the length of the functions in the reference database
meanLen	Average length of the functions in the application file normalized by the average and standard deviation of the lengths of the functions in the reference database
stddevLen	Standard deviation of the length of the functions in the application file normalized by the average and standard deviation of the length of the functions in the reference database

Continued

Feature Name	Description
medianLen	Median length of the functions in the application file normalized by the average and standard deviation of the lengths of the functions in the reference database
Graph Theory Analysis Features	
closeness full mean	Average closeness centrality of the vertices of the file in the reference graph
closeness full min	Minimum closeness centrality of the vertices of the file in the reference graph
closeness full max	Maximum closeness centrality of the vertices of the file in the reference graph
closeness full median	Median closeness centrality of the vertices of the file in the reference graph
closeness full kurtosis	Kurtosis of the closeness centrality of the vertices of the file in the reference graph
closeness full skew	Skewness of the closeness centrality of the vertices of the file in the reference graph
closeness full stddev	Standard deviation of the closeness centrality of the vertices of the file in the reference graph
closeness induced mean	Average closeness centrality of the vertices of the file in the induced graph
closeness induced min	Minimum closeness centrality of the vertices of the file in the induced graph
closeness induced max	Maximum closeness centrality of the vertices of the file in the induced graph
closeness induced median	Median closeness centrality of the vertices of the file in the induced graph
closeness induced kurtosis	Kurtosis of the closeness centrality of the vertices of the file in the induced graph
closeness induced skew	Skewness of the closeness centrality of the vertices of the file in the induced graph
closeness induced stddev	Standard deviation of the closeness centrality of the vertices of the file in the induced graph
pageRank full mean	Average page rank of the vertices of the file in the reference graph
pageRank full min	Minimum page rank of the vertices of the file in the reference graph
pageRank full max	Maximum page rank of the vertices of the file in the reference graph
pageRank full median	Median page rank of the vertices of the file in the reference graph

Continued

Feature Name	Description
pageRank full kurtosis	Kurtosis of the page rank of the vertices of the file in the reference graph
pageRank full skew	Skewness of the page rank of the vertices of the file in the reference graph
pageRank full stddev	Standard deviation of the page rank of the vertices of the file in the reference graph
pageRank induced mean	Average page rank of the vertices of the file in the induced graph
pageRank induced min	Minimum page rank of the vertices of the file in the induced graph
pageRank induced max	Maximum page rank of the vertices of the file in the induced graph
pageRank induced median	Median page rank of the vertices of the file in the induced graph
pageRank induced kurtosis	Kurtosis of the page rank of the vertices of the file in the induced graph
pageRank induced skew	Skewness of the page rank of the vertices of the file in the induced graph
pageRank induced stddev	Standard deviation of the page rank of the vertices of the file in the induced graph
eigenvector full mean	Average eigenvector centrality of the vertices of the file in the reference graph
eigenvector full min	Minimum eigenvector centrality of the vertices of the file in the reference graph
eigenvector full max	Maximum eigenvector centrality of the vertices of the file in the reference graph
eigenvector full median	Median eigenvector centrality of the vertices of the file in the reference graph
eigenvector full kurtosis	Kurtosis of the eigenvector centrality of the vertices of the file in the reference graph
eigenvector full skew	Skewness of the eigenvector centrality of the vertices of the file in the reference graph
eigenvector full stddev	Standard deviation of the eigenvector centrality of the vertices of the file in the reference graph
eigenvector induced mean	Average eigenvector centrality of the vertices of the file in the induced graph
eigenvector induced min	Minimum eigenvector centrality of the vertices of the file in the induced graph
eigenvector induced max	Maximum eigenvector centrality of the vertices of the file in the induced graph

Continued

Feature Name	Description
eigenvector induced median	Median eigenvector centrality of the vertices of the file in the induced graph
eigenvector induced kurtosis	Kurtosis of the eigenvector centrality of the vertices of the file in the induced graph
eigenvector induced skew	Skewness of the eigenvector centrality of the vertices of the file in the induced graph
eigenvector induced stddev	Standard deviation of the eigenvector centrality of the vertices of the file in the induced graph
degree full mean	Average degree of the vertices of the file in the reference graph
degree full min	Minimum degree of the vertices of the file in the reference graph
degree full max	Maximum degree of the vertices of the file in the reference graph
degree full median	Median degree of the vertices of the file in the reference graph
degree full kurtosis	Kurtosis of the degree of the vertices of the file in the reference graph
degree full skew	Skewness of the degree of the vertices of the file in the reference graph
degree full stddev	Standard deviation of the degree of the vertices of the file in the reference graph
degree induced mean	Average degree of the vertices of the file in the induced graph
degree induced min	Minimum degree of the vertices of the file in the induced graph
degree induced max	Maximum degree of the vertices of the file in the induced graph
degree induced median	Median degree of the vertices of the file in the induced graph
degree induced kurtosis	Kurtosis of the degree of the vertices of the file in the induced graph
degree induced skew	Skewness of the degree of the vertices of the file in the induced graph
degree induced stddev	Standard deviation of the degree of the vertices of the file in the induced graph
Textual Analysis Features	
Function [i] tfidf value	The *tf-idf* value of function i in the file ($i = 110,000$)

References

1. Trend Micro. Project Website, http://www.trendmicro.com/vinfo/us/security/research-and-analysis/predi ctions/2016
2. A. Mylonas, A. Kastania, and D. Gritzalis, Delegate the smartphone user? Security awareness in smartphone platforms, *Computers & Security*, 34: 47–66, 2013.
3. M. La Polla, F. Martinelli, and D. Sgandurra, A survey on security for mobile devices, *IEEE Communications Surveys & Tutorials*, 15(1): 446–471, 2013.
4. V. Rastogi, Y. Chen, and W. Enck, Appsplayground: Automatic security analysis of smartphone applications, in *Proceedings of the Third ACM Conference on Data and Application Security and Privacy*, ACM, San Antonio, TX, USA, pp. 209–220, 2013.
5. R. Fedler, J. Schütte, and M. Kulicke, On the effectiveness of malware protection on Android, *Fraunhofer AISEC*, 45: 2013.
6. S. Alam, Z. Qu, R. Riley, Y. Chen, and V. Rastogi, Droidnative: Semantic-based detection of Android native code malware, *arXiv preprint arXiv:1602.04693*, 2016.
7. J. Del Vecchio, F. Shen, K. M. Yee, B. Wang, S. Y. Ko, and L. Ziarek, String analysis of android applications (n), in *Automated Software Engineering (ASE), 2015 30th IEEE/ACM International Conference on*, IEEE, Lincoln, Nebraska, USA, pp. 680–685, 2015.
8. A. Ali-Gombe, I. Ahmed, G. G. Richard III, and V. Roussev, Opseq: Android malware fingerprinting, in *Proceedings of the 5th Program Protection and Reverse Engineering Workshop*, ACM, Los Angeles, California, USA, p. 7, 2015.
9. A. Bose, X. Hu, K. G. Shin, and T. Park, Behavioral detection of malware on mobile handsets, in *Proceedings of the 6th International Conference on Mobile Systems, Applications, and Services*, ACM, Breckenridge, Colorado, USA, pp. 225–238, 2008.
10. X. Xiao, X. Xiao, Y. Jiang, X. Liu, and R. Ye, Identifying Android malware with system call co-occurrence matrices, *Transactions on Emerging Telecommunications Technologies*, 2016.
11. P. Faruki, S. Bhandari, V. Laxmi, M. Gaur, and M. Conti, Droidanalyst: Synergic app framework for static and dynamic app analysis, in *Recent Advances in Computational Intelligence in Defense and Security*, Springer, pp. 519–552, 2016.
12. Z. Yuan, Y. Lu, and Y. Xue, Droid detector: Android malware characterization and detection using deep learningdroid detector: Android malware characterization and detection using deep learning, *Tsinghua Science and Technology*, 21(1): 116–125, 2016.
13. I. You and K. Yim, Malware obfuscation techniques: A brief survey, in *Broadband, Wireless Computing, Communication and Applications (BWCCA), 2010 International Conference on*, IEEE, pp. 297–300, 2010.
14. W. Wong and M. Stamp, Hunting for metamorphic engines, *Journal in Computer Virology*, 2(3): 211–229, 2006.
15. A. Aizawa, An information-theoretic perspective of tf–idf measures, *Information Processing & Management*, 39(1): 45–65, 2003.
16. A. Shabtai, L. Tenenboim-Chekina, D. Mimran, L. Rokach, B. Shapira, and Y. Elovici, Mobile malware detection through analysis of deviations in application network behavior, *Computers & Security*, 43: 1–18, 2014.
17. T.-E. Wei, C.-H. Mao, A. B. Jeng, H.-M. Lee, H.-T. Wang, and D.-J. Wu, Android malware detection via a latent network behavior analysis, in *2012 IEEE 11th International Conference on Trust, Security and Privacy in Computing and Communications*, IEEE, Liverpool, United Kingdom, pp. 1251–1258, 2012.

18. M. M. Saudi, M. Z. Abd Rahman, A. A. Mahmud, N. Basir, and Y. S. Yusoff, A new system call classification for Android mobile malware surveillance exploitation via sms message, in *Advanced Computer and Communication Engineering Technology*, Springer, pp. 103–112, 2016.
19. A. S. Shamili, C. Bauckhage, and T. Alpcan, Malware detection on mobile devices using distributed machine learning, in *Pattern Recognition (ICPR), 2010 20th International Conference on*, IEEE, Istanbul, Turkey, pp. 4348–4351, 2010.
20. D. Papamartzivanos, D. Damopoulos, and G. Kambourakis, A cloud-based architecture to crowdsource mobile app privacy leaks, in *Proceedings of the 18th Panhellenic Conference on Informatics*, ACM, Athens, Greece, pp. 1–6, 2014.
21. I. Burguera, U. Zurutuza, and S. Nadjm-Tehrani, Crowdroid: Behavior-based malware detection system for android, in *Proceedings of the 1st ACM Workshop on Security and Privacy in Smartphones and Mobile Devices*, ACM, Chicago, IL, USA, pp. 15–26, 2011.
22. A. Reina, A. Fattori, and L. Cavallaro, A system call-centric analysis and stimulation technique to automatically reconstruct android malware behaviors, in *Sixth European Workshop on Systems Security (EuroSec)*, Prague, Czech Republic, April, 2013.
23. M. Zhao, T. Zhang, F. Ge, and Z. Yuan, Robotdroid: A lightweight malware detection framework on smartphones, *Journal of Networks*, 7(4): 715–722, 2012.
24. Y. Zhang, M. Yang, B. Xu, Z. Yang, G. Gu, P. Ning, X. S. Wang, and B. Zang, Vetting undesirable behaviors in Android apps with permission use analysis, in *Proceedings of the 2013 ACM SIGSAC Conference on Computer & Communications Security*, ACM, Berlin, Germany, pp. 611–622, 2013.
25. K. Tam, S. J. Khan, A. Fattori, and L. Cavallaro, Copperdroid: Automatic reconstruction of Android malware behaviors, in *The 2015 Network and Distributed System Security (NDSS) Symposium*, San Diego, CA, February 8–11, 2015.
26. W. Zhou, Y. Zhou, X. Jiang, and P. Ning, Detecting repackaged smartphone applications in third-party Android marketplaces, in *Proceedings of the Second ACM Conference on Data and Application Security and Privacy*, ACM, San Antonio, TX, USA, pp. 317–326, 2012.
27. A. S. Yuksel, A. H. Zaim, and M. A. Aydin, A comprehensive analysis of Android security and proposed solutions, *International Journal of Computer Network and Information Security*, 6(12): 9, 2014.
28. B. A. Debelo, W. Pak, and Y.-J. Choi, Sandroid: Simplistic permission based Android malware detection and classification, in *2013 9th International Wireless Communications and Mobile Computing Conference (IWCMC)*, Sardinia, Italy, July 1–5, 2013.
29. D.-J. Wu, C.-H. Mao, T.-E. Wei, H.-M. Lee, and K.-P. Wu, Droidmat: Android malware detection through manifest and API calls tracing, in *Information Security (Asia JCIS), 2012 Seventh Asia Joint Conference on*, IEEE, Tokyo, Japan, pp. 62–69, 2012.
30. S. Y. Yerima, S. Sezer, and I. Muttik, Android malware detection using parallel machine learning classifiers, in *2014 Eighth International Conference on Next Generation Mobile Apps, Services and Technologies*, IEEE, Cardiff, Wales, UK, pp. 37–42, 2014.
31. H.-Y. Chuang and S.-D. Wang, Machine learning based hybrid behavior models for Android malware analysis, in *Software Quality, Reliability and Security (QRS), 2015 IEEE International Conference on*, IEEE, Prague, Czech Republic, pp. 201–206, 2015.
32. D. Arp, M. Spreitzenbarth, M. Hubner, H. Gascon, and K. Rieck, Drebin: Effective and explainable detection of Android malware in your pocket, in *Network and Distributed System Security (NDSS) Symposium*, San Diego, CA, February 23–26, 2014.

33. K. O. Elish, D. D. Yao, B. G. Ryder, and X. Jiang, A static assurance analysis of Android applications, technical report, TR-13-03, Department of Computer Science, Virginia Polytechnic Institute & State University, http://hdl.handle.net/10919/23302, 2013.
34. Z. Dong, H. Ye, Y. Wu, S. Cheng, and F. Jiang, Android apps: Static analysis based on permission classification, *ZTE Communications*, 11(1): 62–66, 2013.
35. L. X. Min and Q. H. Cao, Runtime-based behavior dynamic analysis system for Android malware detection, *Advanced Materials Research*, 756, 2220–2225, Trans Tech Publ, 2013.
36. E. Menahem, A. Shabtai, and A. Levhar, Poster: Detecting malware through temporal function-based features, in *Proceedings of the 2013 ACM SIGSAC Conference on Computer & Communications Security*, ACM, Berlin, Germany, pp. 1379–1382, 2013.
37. R. Tian, L. M. Batten, and S. C. Versteeg, Function length as a tool for malware classification, in *Malicious and Unwanted Software, 2008. MALWARE 2008. 3rd International Conference on*, IEEE, Alexandria, VA, USA, pp. 69–76, 2008.
38. R. Islam, R. Tian, L. Batten, and S. Versteeg, Classification of malware based on string and function feature selection, in *Cybercrime and Trustworthy Computing Workshop (CTC), 2010 Second*, IEEE, Ballarat, Victoria, Australia, pp. 9–17, 2010.
39. A. Shabtai, R. Moskovitch, C. Feher, S. Dolev, and Y. Elovici, Detecting unknown malicious code by applying classification techniques on opcode patterns, *Security Informatics*, 1(1): 1, 2012.
40. F. Sebastiani, Machine learning in automated text categorization, *ACM Computing Surveys (CSUR)*, 34(1): 1–47, 2002.
41. S. J. Pan and Q. Yang, A survey on transfer learning, *IEEE Transactions on Knowledge and Data Engineering*, 22(10): 1345–1359, 2010.

Chapter 7

Detecting Android Kernel Rootkits via JTAG Memory Introspection

Mordechai Guri, Yuri Poliak, Bracha Shapira, and Yuval Elovici

Contents

7.1 Introduction

Over the past few years, mobile devices have emerged as a preferred target for cyber criminals. This trend is fueled by the valuable personal and organizational information stored on those devices. Android is by far the most popular mobile operating system (OS); its numerous vulnerabilities, coupled with the ease of distributing malicious code through its popular app market, have made this OS a favorite target of attackers [1]. For example, the DroidDream attack [2] was distributed through legitimate applications on the Android market and infected about 50,000 mobile devices in the course of a few days. More recently, an Android "bootkit," that is, a rootkit that modifies the device's boot partition and boot script (codenamed "Old-boot") infected over 500,000 mobile devices within a period of 6 months in China alone [3]. In 2015, researchers have uncovered a rootkit that resides deep inside Android devices, while receiving commands from its operator across the internet [4]. In 2016, a rootkit-level backdoor was found preinstalled on 3 million Android phones, many of them used by people in the United States [5].

7.1.1 Kernel Rootkits

Mobile and desktop malware can operate in user or kernel space. User space malware can only modify and inject code into the memory areas allocated to apps and user processes. Kernel space malware can manipulate objects that reside in the entire memory area of the OS. Although sophisticated mandatory access control (MAC) mechanisms such as SElinux [6] are integrated into current versions of Android, malware developers still manage to run their code in the kernel [3,7,8]. Rootkits are kernel space malware that use illicitly granted exclusive permissions to hide their existence from detection systems, by manipulating the kernel's internal data structures [9]. A malicious code that has penetrated the memory of the kernel can neutralize any security tool running on the OS. For instance, if a process sends a request to the kernel asking for the list of files in a specific directory, there is no guarantee regarding the returned list's integrity. Consequently, in order to detect the presence of rootkits, a *trusted* snapshot of the kernel memory must be obtained [10].

7.1.2 Infection Vectors and Roots

Installing a rootkit on a smartphone requires the device to have root access (rooted). With regard to the adversarial attack model, malicious apps which gain root privileges in mobile phones (e.g., by exploiting OS vulnerabilities) are commonly found in the wild [11–18]. Research conducted in 2016 discovered more than 1163 apps which are capable of rooting Android [19]. In addition, 10 million Android phones infected by malicious auto-rooting apps were detected in 2016 [20].

7.1.3 Problem with Current Rootkit Detection Approach

Antivirus is the most popular tool utilized to cope with user space malware because it is an integral part of the security multilayered approach. Generally, antivirus tools scan the system's files and sometimes the memory for known signatures of malicious code. Despite the fact that those tools have been proven effective against user space malicious code on personal computers, the effectiveness of antivirus tools for mobile platforms is questionable, mainly due to their high battery consumption, mobile OS architecture, and low detection rate. In any case, while existing mobile antivirus applications may potentially detect user space malware, they cannot detect kernel space malware programs such as kernel mode rootkits [21]. Malicious code that has penetrated into the memory of the kernel can bypass any security measure that is receiving services from the kernel. Kernel level [22,23] and hypervisor level [7,24,25] antivirus solutions are considered less practical at this point. Kernel level solutions are vulnerable to attacks from kernel level rootkits since it shares the same execution level. Hypervisor level solutions consume high CPU resources and are vulnerable to kernel level code via runtime or bootloader vulnerabilities [8,26].

7.1.4 The Proposed System

In this chapter, we present *JoKER (JTAG observe Kernel)*, a system that utilizes the JTAG hardware interface of the mobile device in order to obtain a trusted snapshot of the device memory for the detection of kernel rootkits. The JTAG standard [27] was developed to assist with system testing and postmanufacture debugging of the circuit board. JTAG's connectors are installed on the printed circuit board (PCB) of modern mobile devices such as smartphones and tablets. Our detection system uses two of JTAG's important debugging features:

1. The ability to *halt* the system instantly by sending special instructions to the main processor.
2. The ability to *access* the content of the device's volatile memory (RAM) while it is being halted. The overall system does not run on the mobile device, and therefore it can securely read the kernel's memory areas in a trusted manner.

Once the kernel memory is extracted, it is passed through an array of programmed scripts. Each script reconstructs specific data structures in the kernel and analyzes them for traces of suspicious modifications. We provide a detailed description of the system architecture and its implementation. Our evaluation shows that JoKER can successfully detect maliciously modified objects located in the Android kernel in a trusted manner.

7.1.5 Method Limitation

Using the JTAG interface requires a physical connection to the JTAG port which is placed on the smartphone's main board. This approach may appear rather awkward compared to more common software-based methods. However, external memory acquisition capabilities (from outside the device) provide the advantage of trusted memory inspection. Accordingly, our proposed system aims at finding stealthy and sophisticated rootkits where other detection methods, running within the device, cannot be trusted.

7.1.6 Our Contribution

JTAG was mentioned as a general forensic tool for embedded devices and Android systems in prior work [28,29]. This chapter introduces several contributions and advantages over previous related work in the field.

■ First, we are the first to propose an automated system for the Android OS and ARM architecture focused on detecting kernel rootkits, utilizing JTAG-based memory forensics. Since it is external, hardware-based, and transparent to the malicious code, our method is trusted and hence cannot be subverted. We present the overall architecture and detailed working implementation of the detection system, both at the hardware and software level.
■ Second, we discuss five rootkit mechanisms for the Android kernel and show how they can be identified by our system.
■ Third, we introduce a new method for detecting hidden processes by analyzing the Android kernel cache mechanism.
■ Fourth, we show how to overcome several challenges that we encountered during our low-level examination of the system. Those challenges

include translating between physical and virtual memory addresses, along with resolving notorious kernel synchronization issues.

7.2 Related Work

While existing mobile AV apps may detect user space malware, they are generally ineffective for the detection of kernel level rootkits [7,30]. Tools such as the Linux Memory Extractor (LiME) [22] and DMD [23] are helpful for acquisition and analysis of volatile memory in Android devices. However, since these tools operate from within the OS, their use can be subverted by a rootkit and hence cannot be considered as trustworthy. Android kernel securing and hardening [6,31,32] have also been proposed for defending the kernel memory space against rootkits. Hypervisors [7,24,25] and the Trusted Platform Module (TPM) [33,34] have been researched as a potential trusted point of acquisition for kernel space memory. More recently, Sun et al. presented TrustDump, a hardware-assisted system for reliable memory acquisition on smartphones using ARM TrustZone support [10]. The mechanism employed by TrustDump runs in the TrustZone's secure domain to ensure a separation between the OS and the memory acquisition tool. However, as long as the security mechanism runs on the same physical device as the monitored OS, it can be compromised via runtime or bootloader vulnerabilities [8,26]. Other type of TrustZone integrated security mechanisms such as the Samsung Knox [35] found to be vulnerable to privilege escalation vulnerabilities within its Real-time Kernel Protection (RKP) component [36]. JTAG was discussed as a tool for forensic imaging of embedded systems [29] and more generally in the context of Android devices [28]. Table 7.1 shows the different hardware and software layers used for malware detection and memory acquisition in mobile devices.

7.3 System Design

The JoKER system consists of four components, as depicted in Figure 7.1: (A) the mobile device, (B) the JTAG controller, (C) the memory analyzer program which

Table 7.1 Different Hardware and Software Layers of Malware Detection and Memory Acquisition Mechanisms for Mobile Devices

Approach	Implementation	Run In Device	Bypass Examples
Kernel	[31,32]	Yes	[3,25,30]
Hypervisor	[24,37]	Yes	[8,26]
TrustZone	TrustDump [10]	Yes	[8,26]
JTAG	JoKER	No	–

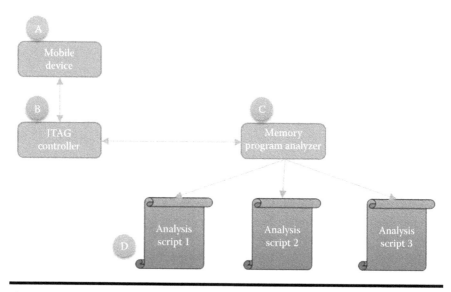

Figure 7.1 Schematic layout of the system's components.

extracts the kernel's raw memory from the device and manages its analysis, and (D) a set of scripts to analyze and detect rootkits in selected areas of the kernel. The mobile device is an Android device that is scanned for the presence of rootkits. This device should have a JTAG port with a compatible soldered connector so that it can be connected to a JTAG controller. The JTAG controller is the hardware component that can communicate with the CPU and the memory controller on the target device through the on-chip debugging (OCD) connectors. The memory extraction program receives the raw content of the device's RAM by communicating with the OCD. Finally, the scripts reconstructs and analyzes the kernel memory.

The detection process consists of three main phases: (1) halting the processor of the target device, (2) extracting the kernel's data structures from the RAM, and (3) applying a forensic analysis algorithm to find rootkit footprints in the extracted binaries. These steps are described in the following subsections.

7.3.1 System Halting

JTAG can halt the main processor of the mobile device by sending a halt command to the OCD [38]. We use this functionality at an early stage in the detection process to ensure that no code is executed on the device. This fact plays a major role in the detection mechanism's design, since the suspicion that the system is being monitored can prompt a running malware to mask its presence. Halting the processor by a single command ensures that a malware cannot prepare to mask itself before the system halt.

7.3.2 Extracting Kernel Memory

The second phase involves extraction of kernel-related memory areas for further analysis which will be conducted on a separate computer. Modern JTAG interfaces offer rich debugging functionality such as direct read and write access to the RAM and flash memory [38]. We have used JTAG's commands to extract raw memory from the RAM of the device as it is halted. The decision of which memory regions to extract is based on the specific analysis techniques used. To demonstrate the system, we used techniques adapted from studies related to Linux-based rootkits [39]. Rootkits attack various data structures on Linux systems, primarily the system call table, the exception vector table (EVT), and the kernel's processes list. We therefore focus on extracting the related memory regions for further analysis.

7.3.3 Reconstruction and Analysis

During the third and final phase, the detection system applies analysis algorithms to the extracted raw memory. This process involves scanning for suspicious modifications of memory regions. The scripts check the integrity of the system call table, the EVT, and the software interrupt handler. Since these objects should not be modified on a regular Android system, we validate their integrity by comparing them to a clean Android system.

Another script detects stealthy processes which are hidden from the kernel's processes list. Unlike the system call table, the EVT, and the software interrupt handler, the processes list is a *dynamic* kernel object which is changed frequently. Detecting hidden rootkit processes is challenging, since rootkits typically remove their entry from this list in order to evade detection. To that end, our system analyzes the kernel's cache which is responsible for maintaining pools of the OS's internal objects. We have applied cross-view methodology by comparing the objects in the kernel's processes list to a baseline that consists of active processes reconstructed from the protected cache pools. A difference between the two views indicates the presence of hidden processes. This method may reveal the presence of a rootkit and can also pinpoint the processes that the rootkit has tried to hide.

7.4 Implementation

We implemented the JoKER framework according to the design outlined in Section II. We used the RIFF Box JTAG controller [40] to communicate with a JTAG capable Android device. The testing described in this chapter was conducted on Samsung Galaxy (S2 and S4) mobile phones with a JTAG interface.

On the software side, the JTAG controller (Figure 7.1B) extracts relevant memory regions from the device (Figure 7.1A) via a set of PRACTICE scripts. PRACTICE [41] is a script language which operates on the Lauterbach TRACE32 microprocessor development tool and its related product line. These tools are aimed at providing easy programming access to OCD systems. TRACE32 supports communication with the JTAG interface, among other interfaces. The memory analyzer program (Figure 7.1C) receives the raw memory data and feeds it to a set of Python scripts (Figure 7.1D). Each script receives the memory contents as an array of bytes, performs its own forensic analysis, and returns the results. All logs are saved by the memory analyzer program, and the final results are presented to the user.

7.4.1 System Setup

Figure 7.2 presents the system setup as constructed and installed in our lab.

The setup follows the outline discussed earlier and consists of the RIFF Box JTAG controller (Figure 7.2C), which supports communication with a variety of mobile and embedded devices on the market. The JTAG controller is connected to the JTAG port on the target device (Figure 7.2A), through a flat cable with a connection to the JTAG port on the device (Figure 7.2B). The other end of the JTAG box is connected to a computer that hosts the controller program, through a USB cable (Figure 7.2D). The laptop computer also hosts the PRACTICE scripts that are

Figure 7.2 The JoKER system, as constructed and installed in the lab.

responsible for the memory extraction and the Python scripts that are responsible for the memory analysis (Figure 7.2E).

7.4.2 Memory Analysis

JoKER is a generic framework that can be enriched with a wide range of detection and analysis scripts. In order to test the system, we implemented five scripts, each targeting a different type of rootkit technique. The scripts include: (1) system call table integrity checks, (2) EVT integrity checks, (3) two types of software interrupt handler (SWI handler) integrity checks, and (4) revealing hidden processes by analyzing the kernel's cache. To the best of our knowledge, the former method is new and is introduced for the first time in this chapter. The analyzed kernel structures are presented in Table 7.2. For clarity, a flow of a system call in the Android kernel, along with the relevant tables, is outlined in Appendix 7A.

Prior to the system operation, the analysis scripts are initialized with the physical address of the objects within the kernel memory in the specific version of the examined Android. These parameters can be extracted from the kernel's symbol list located at /proc/kallsyms. Note that these parameters can be retrieved from any clean device having the same version of the kernel. We have developed a loadable kernel module (LKM) which is executed on a clean version of Android OS (downloaded from the official website) with an identical version of the kernel and reports the parameters' values back to the system.

7.4.2.1 Physical to Virtual Memory Translation

Since JTAG refers to memory in physical addresses, we had to translate between the virtual addresses (OS view) and the physical addresses (JTAG view). Since the

Table 7.2   List of Android Kernel Objects Analyzed during Our Implementation of JoKER

Structure	Description
System Call Table	A static structure which contains pointers to low-level system functions
EVT	A static structure which contains pointers to exceptions and interrupt handlers
SWI	A static structure which contains pointers to interrupt handlers
kmem_cache structure	A dynamic structure which contains information on the kernel's cached objects

input addresses are part of the kernel space, they can be calculated from the virtual address by subtracting a fixed offset. An exceptional case is the address of EVT, since on ARM-based architectures the virtual address of EVT must be 0x00000000 or 0xffff0000. To calculate the physical address of EVT, we used ARM's assembly instructions which translate the virtual address to a physical address by traversing the page tables in our LKMs. Note that on Android distributions that disable the LKM mechanism and omit the kernel's symbol list, it is still possible to extract the initializing parameters by using the Runtime Kernel Patch (RKP) strategy for accessing the kernel space memory as has been demonstrated in Reference 9.

7.4.3 Detection Scenarios

Some detection techniques involve checking the integrity of various structures of the kernel. This scenario is relevant when performing a "before and after" forensic examination, for example, when examining an application that may bring a malicious payload into the device. In such cases, the forensic analyst will have to examine the system at two points: before installing the application (a "clean" snapshot) and after installation.

7.4.4 Detection Flow

The flow chart in Figure 7.3 outlines the process of rootkit detection after the initializing parameters have been set, the JTAG controller has been connected to the target device, and the communication with the control software has been initialized. We assume that a clean snapshot of the kernel's memory has been taken previously from a device with a clean kernel version.

The main steps of the detection algorithm are as follows:

Steps (1–4): Halting the CPU of the target device and validating the integrity of the current system call table against a clean version of this table. In cases of inconsistency, a rootkit alert is triggered. *Steps (5–7)*: Validating the integrity of the current EVT against a clean version of EVT. In cases of inconsistency, a rootkit alert is triggered. *Steps (8–10)*: Validating the integrity of the SWI handler pointer against a clean version of the SWI handler and validating the integrity of the SWI handler code. In a case of inconsistency, a rootkit alert is triggered. *Step (11)*: Extracting the kernel's processes list, by parsing each task_struct node in the list starting at the INIT process. *Step (12)*: Reconstructing the list of task_struct that appears in the cache mechanism of the kernel. *Step (13)*: Comparison between the list of task_struct which was extracted from the kernel's list and those extracted from the cache. *Step (14)*: If the kernel's processes list and the cache differ by the task_structs, a hidden process is found, and a rootkit alert is triggered.

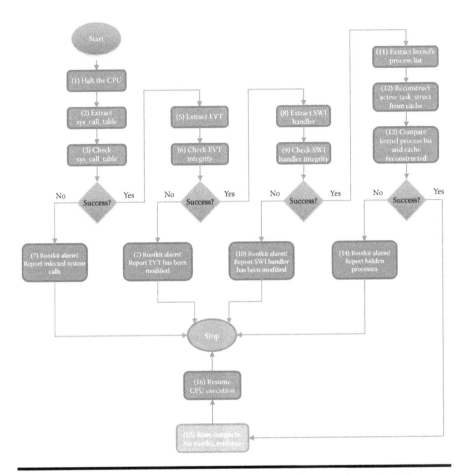

Figure 7.3   Outline of the detection flow in our implementation of JoKER.

7.5 Evaluation

We evaluate the detection system by testing it against five types of suspicious kernel modification code. To that end, we implemented five kernel modules which perform the malicious operations. The reason for using self-constructed rootkits rather than originals is due to the fact that samples of rootkits for current mobile phones have not been released to the research community (source or binary). Interestingly, although kernel rootkits have been widely researched in the context of desktop operating systems, there are no documented samples of rootkits for recent versions of Android. In addition, rootkits which target the desktop version of the Linux kernel cannot be installed on the Android kernel. This is due to differences in the kernel architecture between the OSs and the modified versions of LIBC in the Android OS. We evaluate the system with

Android kernels 2.6.35 and 3.4.0 installed on Samsung Galaxy S2 and Galaxy S4, respectively.

7.5.1 Kernel Rootkits

The rootkit mechanisms have been implemented in the form of LKMs. Each of the five rootkits (samples 1–5) exposes a different malicious functionality. Sample 1 modifies the address of four system calls in the system call table. Sample 2 modifies an indirect pointer of the SWI handler which is stored in the instruction at offset 0×8 in EVT. Sample 3 modifies the address of the SWI handler which is stored in EVT. Sample 4 modifies the offset of the system call table which is stored in the SWI handler routine. Sample 5 hides a process by removing it from the kernel's processes list. Note that the use of self-constructed rootkits as an evaluation method has been a part of previous studies in the field [24,30].

7.5.2 Syscall Table Hooking

The first rootkit was implemented as a kernel module (syscallTableHook.ko) which modifies the address of four system call addresses in the system call table: read(), write(), open(), and close(). We chose four basic system functions that can be used maliciously in order to intercept the file system, sensors, and network access operations. The experiment starts by executing a PRACTICE script to get a binary snapshot of the kernel's system call table before and after the execution of the rootkit. In Figure 7.4, we see the two snapshots of the system call table in a binary form of the hex editor viewer. As can be seen, four modified addresses in the table have been detected. The original addresses of the system calls are marked in blue, while the modified addresses are marked in red. In the next step, the script checks which system calls have been changed. This is achieved by parsing the header file (unistd.h) from the source tree of the Android kernel. This file contains the order and names of the system call functions in the table. Next, the Python script receives the two snapshots of the system call tables and the list of functions from the kernel and returns the names of functions that have been modified. The output of the system is shown in Figure 7.5.

7.5.3 Exception Vector Table Hooking

In the ARM architecture, each exception or interrupt is branched to the EVT. This table is a central component of the OS, and as such it is naturally the target of various hooking techniques. When a software interrupt happens in the system, the processor loads the instruction at offset 0×8 in EVT to the instruction register

00	01	02	03	04	05	06	07	08	09	0A	0B	0C	0D	0E	0F
000: CC	68	2D	C0	9C	9E	2C	C0	B0	6A	28	C0	54	55	36	C0
010: EC	52	36	C0	4C	2F	36	C0	60	2B	36	C0	9C	49	2E	C0
020: 7C	2F	36	C0	50	1C	37	C0	14	15	37	C0	C0	6A	28	C0
030: 2C	3A	36	C0	9C	49	2E	C0	64	19	37	C0	F4	38	36	C0
040: EC	93	2F	C0	9C	49	2E	C0	9C	49	2E	C0	24	43	36	C0
050: 7C	5D	2D	C0	9C	FF	37	C0	9C	49	2E	C0	0C	93	2F	C0
060: FC	8F	2F	C0	9C	49	2E	C0	48	1E	2D	C0	9C	49	2E	C0
070: 9C	49	2E	C0	B0	76	2D	C0	9C	49	2E	C0	9C	49	2E	C0

00	01	02	03	04	05	06	07	08	09	0A	0B	0C	0D	0E	0F
000: CC	68	2D	C0	9C	9E	2C	C0	B0	6A	28	C0	78	40	03	BF
010: 00	40	03	BF	28	40	03	BF	50	40	03	BF	9C	49	2E	C0
020: 7C	2F	36	C0	50	1C	37	C0	14	15	37	C0	C0	6A	28	C0
030: 2C	3A	36	C0	9C	49	2E	C0	64	19	37	C0	F4	38	36	C0
040: EC	93	2F	C0	9C	49	2E	C0	9C	49	2E	C0	24	43	36	C0
050: 7C	5D	2D	C0	9C	FF	37	C0	9C	49	2E	C0	0C	93	2F	C0
060: FC	8F	2F	C0	9C	49	2E	C0	48	1E	2D	C0	9C	49	2E	C0
070: 9C	49	2E	C0	B0	76	2D	C0	9C	49	2E	C0	9C	49	2E	C0

Figure 7.4 System-call table, before (upper) and after (lower) the rootkit operation.

```
The address of system call < read > has been changed
original address: c0365554
new address: bf034078

The address of system call < write > has been changed
original address: c03652ec
new address: bf034000

The address of system call < open > has been changed
original address: c0362f4c
new address: bf034028

The address of system call < close > has been changed
original address: c0362b60
new address: bf034050
```

Figure 7.5 Output of the analysis script that indicates which functions have been modified, along with their addresses.

for execution (Figure 7.6). In this case, the instruction that will execute is ldr pc, [pc, #1040]. This instruction loads the program counter with the address of the software interrupt handler address that resides in the offset 1040 (0 × 420) relative to the current program counter.

The second rootkit was implemented as a kernel module (HookBranchInstruction.ko) which modifies the EVT. Our implementation technique is similar to Reference 9, applying two types of modifications to the EVT. First, it copies the

```
[000] ffff0000: ef9f0000 [Reset]          ; svc 0x9f0000 branch code array
[004] ffff0004: ea0000dd [Undef]          ; b    0x380
[008] ffff0008: e59ff410 [SWI]            ; ldr  pc, [pc, #1040] ; 0x420
[00c] ffff000c: ea0000bb [Abort-perfetch] ; b    0x300
[010] ffff0010: ea00009a [Abort-data]     ; b    0x280
...
[420] ffff0420: c003df40 [vector_swi]
```

Figure 7.6   A snapshot of EVT in the kernel.

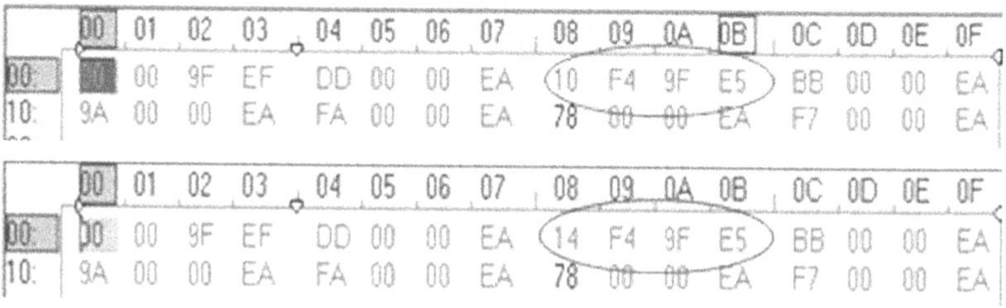

Figure 7.7   The EVT, before (upper) and after (lower) the rootkit's modification.

address of a new SWI handler to the memory at offset 0×424 in the EVT. Second, the rootkit changes the instruction at offset 0×8 to load the address at offset 0×424 (the new handler) to the table instead of the original address. This technique allows the attacker to hook the SWI handler and intercept interrupts and exceptions in the system. Our system extracts the kernel's memory, before and after the rootkit's installation, by using a PRACTICE script. A Python script reconstructs and compares the two views.

As can be seen in Figure 7.7, the instruction at offset 0×8 of the EVT has been changed from 0xe59ff410 to 0xe59ff414. This modification causes the processor to load the address that resides at offset 0×424 of the table instead of the address at offset 0×420. The difference between the two EVTs is identified and reported to the system as a rootkit alert.

7.5.4 Hooking the Address of the SWI Handler Routine

Another hooking approach is modifying the SWI handler routine, and in this case a rootkit injects the address of its own handler function. By intercepting all interrupts and exception calls, the rootkit can perform malicious operations in a hidden manner. In our example, the rootkit (exvHookSwiHandlerAddress.ko) copies the binary content of the original SWI handler to another address in the kernel space, modifies the binary code of the handler, and then inserts the address of the new handler at offset 0×420 of the EVT. Our system extracts the kernel's memory before and

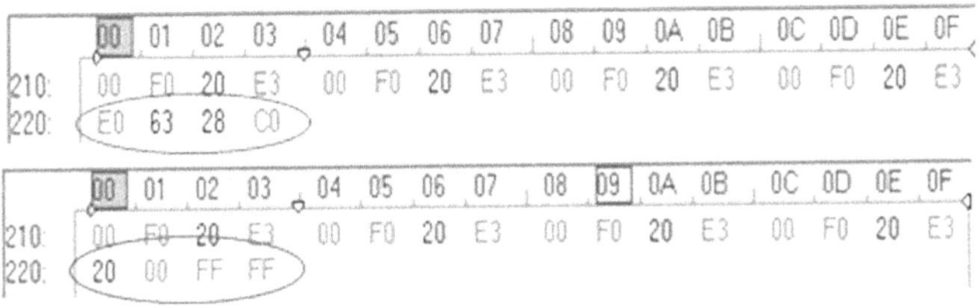

Figure 7.8   The address of the SWI handler, before (upper) and after (lower) the rootkit's modification.

after the rootkit's installation, by using a PRACTICE script. Then a Python script reconstructs and compares the two views.

As can be seen in Figure 7.8, the address of the SWI handler has been changed at offset 0 × 220 in the second part of the EVT which is offset 0 × 420 from the base of the table. This difference indicates that a malicious modification has occurred. The event is reported to the system as a rootkit alert.

7.5.5 Hooking the Code of the SWI Handler Routine

The last hooking technique involves hooking the binary code of the software interrupts routine itself.

Figure 7.9 shows the part of the SWI handler that locates the address of the system call table with an offset relative to the current program counter. The system call table itself is located after the code of the handler. By manipulating the marked instruction, a rootkit can direct any software interrupt to its own system call functions. Our implemented rootkit (hookSysCallTableAddressInSwiHandler.ko) iterates over the entries of the instruction which loads the system call table pointer. Next, the instruction is replaced with a new instruction—ldr r8, [pc, #offset], where #offset is the relative offset of our system call table.

```
000000c0 <vector_swi>:
    ...
    100: e1a096ad mov      r9, sp, lsr #13 ; get_thread_info tsk
    104: e1a09689 mov      r9, r9, lsl #13
    108: e28f8094 add      r8, pc, #148    ; load syscall table pointer
    10c: e599c000 ldr      ip, [r9]        ; check for syscall tracing
```

Figure 7.9   Part of the SWI hander routine code in the kernel memory.

	00	01	02	03	04	05	06	07	08	09	0A	0B	0C	0D	0E	0F
000:	48	D0	4D	E2	FF	1F	8D	E8	3C	80	8D	E2	00	60	48	E9
010:	00	80	4F	E1	3C	E0	8D	E5	40	80	8D	E5	44	00	8D	E5
020:	00	B0	A0	E3	20	00	18	E3	00	A0	A0	13	04	A0	1E	06
030:	A8	C0	9F	E5	00	C0	9C	E5	10	CF	01	EE	80	00	08	F1
040:	AD	96	A0	E1	89	96	A0	E1	98	80	8F	E2	00	C0	99	E5
050:	FF	A4	DA	E3	09	76	2A	12	84	80	9F	15	30	00	2D	E9

	00	01	02	03	04	05	06	07	08	09	0A	0B	0C	0D	0E	0F
000:	48	D0	4D	E2	FF	1F	8D	E8	3C	80	8D	E2	00	60	48	E9
010:	00	80	4F	E1	3C	E0	8D	E5	40	80	8D	E5	44	00	8D	E5
020:	00	B0	A0	E3	20	00	18	E3	00	A0	A0	13	04	A0	1E	05
030:	A8	C0	9F	E5	00	C0	9C	E5	10	CF	01	EE	80	00	08	F1
040:	AD	96	A0	E1	89	96	A0	E1	80	00	9F	E5	00	C0	99	E5
050:	FF	A4	DA	E3	09	76	2A	12	84	80	9F	15	30	00	2D	E9
060:	01	0C	1C	E3	08	00	00	1A	17	0E	57	E3	65	EF	4F	E2

Figure 7.10   SWI code in memory, before (upper) and after (lower) the rootkit operation.

Figure 7.10 depicts modifications of the SWI handler routine identified by our detection system. An NOP instruction (0xe320f000) has been changed to the address of the malicious system call table (0xc02864c8). The instruction that loads the address of the system call table into register r8 has been changed from add r8, pc, 0 × 98 to ldr r8, [pc, 0 × 80]. The PRACTICE script generates snapshots of the SWI handler, the Python script compares them, and a rootkit alert is triggered when a relevant modification is detected.

7.5.6 *Direct Kernel Object Manipulation*

Direct kernel object manipulation (DKOM) is a technique used by a rootkit in order to hide itself from the OS layer. By directly accessing the data structures in the kernel, a rootkit can hide resources such as processes and thread descriptors, network connections, and other objects in the memory. To examine the effectiveness of our system against DKOM, we implemented a rootkit (dkomRootkit.ko) which manipulates the linked list of the kernel's structures representing the list of processes and threads (task_structs). We executed a process on the device which simulates the malicious program (MalApp) that the rootkit intends to hide. The program itself is executed as a user-level process. Our rootkit scans the linked list of the kernel's task_structs, searching for a task with the name "MalApp" and removes it. Note that although the process is removed from the link list, it still exists in the scheduler's internal list; hence, its execution is not terminated. To detect the hidden process, we developed a new cross-view strategy which uses the kernel's cache pool. The kernel cache contains the cached version of the task_struct while it is in use, or shortly after termination for reuse. Rootkits typically do not interfere with the cache pool, as it

```
--------compare cache <-> tasks list (cross-view)---------
Task with pid:  3129 , name:  printer , file name:
section_task_struct#0x9  found in cache but not in tasks list
Number of tasks that appear in list but not in cache:  0
Number of tasks that appear in cache but not in list:  1
```

Figure 7.11   Results of the comparisons (cross view) between the kernel's processes list and the processes list reconstructed from the cache.

is an internally managed memory region. We used this fact to conduct an analysis of the cache pool and identify traces of hidden processes. Our script reconstructs the processes list from the kernel's processes list and the cache. The results of the comparison (Figure 7.11) show that all of the tasks appearing in the linked list also appeared in the cache, but there is a task_struct that appears in the cache that is not part of the linked list of process descriptors.

For the interested reader, we mention that in most Android distributions the cache mechanism does not have pointers to all of the slabs [42] that contain the task_structs. Therefore, we obtained the slab addresses in the following manner: We traversed each page frame number (PFN), translated it into a physical struct page address, and then checked whether the struct page represented a slab with task_struct objects. From each matching slab, we extracted all of the task_structs.

7.5.6.1 Kernel Consistency

When evaluating the cross-view detection approach, we noticed that when the list of the task_structs is extracted from the cache on the clean system, some of the structures might not appear in the kernel's processes list. Although rare, this behavior should be understood and eliminated when dealing with clean systems. We found that the reason for this exceptional behavior is the way that the JTAG box communicates with the device. When our system starts executing any of the PRACTICE scripts, the processor of the target device is instructed to halt immediately. The problem is that when halting occurs, the kernel of the device is, very briefly, in an unstable state. The cache mechanism reuses the task_struct objects. Thus, when a process ends its execution, the kernel should unlink it from its list of processes and only then mark it as an unused object in the cache. This process of object reuse is not an atomic operation, and halting the system's core takes place in the middle of the unlinking operation. This momentary unstable state causes some active processes to appear as if absent from the kernel's processes list. To distinguish between malicious processes (intentionally absent from the list) and "dummy" processes (absent because of the inconsistency), we analyzed the task_structs of these "dummy" processes. In so doing, we determined that the pid, comm, state, and flags fields in the task_structs can serve to indicate whether it is being halted. In some of these objects, the value of the pid was 0, but the name of the process (comm) was not "swapper." Obviously, such an object cannot represent a runnable process, as the only process in the system with a pid of 0 can be the swapper. Other active objects had a negative value in their state field. This field contains information about the runnable state of the process, and a negative value represents a nonrunnable state. The last indicator is the flags field which contains information about the state of the process. The value of this field is a bitwise OR of all of the characteristics that represent the state of the process at the moment. If the least significant bits equal 2, the process is in a shutting down mode. Since the kernel's nonconsistent task_structs can be filtered by the indicators listed above, we redesigned the detection system to filter these objects before comparing the task_structs in the cache and the linked list. We executed our redesigned detection mechanism on a typical clean system and validate that it does not issue false alarms as a side effect of the kernel cache behavior.

7.6 Conclusion

In this chapter, we present JoKER, a framework which utilizes the hardware's JTAG interface for trusted memory forensics. Our system demonstrates how kernel level rootkits in the Android OS can be detected in an automated manner by employing various memory forensic techniques. Unlike conventional methods, our method is trustworthy, since it is external, hardware-based, and undetectable by the malicious code within the device.

The JoKER framework extracts areas of the kernel's memory, reconstructs them for further analysis, and raises a rootkit alert when positive evidence for the presence of a rootkit is encountered. We present the overall layout of the framework, along with a detailed description of its implementation. Our system is evaluated under several attack patterns, demonstrating that it can successfully detect crafty kernel mode rootkits, whether persistent or nonpersistent. We implemented five types of rootkits, used to evaluate our system, and show how our system detects them. A new method is introduced for detecting hidden processes by analyzing the Android kernel cache data structure. We also discuss some technological challenges involved with our method, such as translation between physical and virtual memory addresses and resolving kernel synchronization issues. The detection system demonstrates the cross-view paradigm in which the inspected system is examined at multiple levels in order to expose contradicting traces suggesting the presence of a rootkit, and eliminate false alarms. Note that although JTAG's original purpose is system testing and verification, in this chapter we show that it can also be used for low-level malware detection. We believe that our current experimental system can serve as a platform or prototype for future research concerning trusted detection of mobile device rootkits and similar kernel-level malware.

Appendix 7A: System Call Flow in the Android OS

The flow of "read" system call in Android OS from the application level to the kernel level:

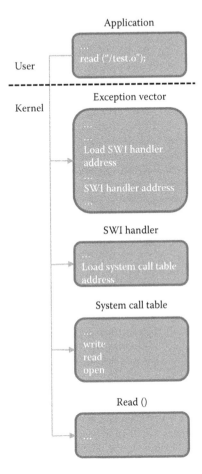

References

1. Kaspersky, *Bulletin, Kaspersky Security*, 12, 2013. Online, Available: http://securelist.com/analysis/kaspersky-security-bulletin/58265/kaspersky-security-bulletin-2013-overall-statistics-for-2013/. Accessed November 29, 2014.
2. F-Secure, Mobile threat report Q4, 2011, *F-Secure*, 12, 2011. Online, Available: https://www.f-secure.com/documents/996508/1030743/Mobile + Threat + Report + Q4 + 2011.pdf. Accessed November 29, 2014.
3. Z. Xiao, Q. Dong, H. Zhang, and X. Jiang, *Oldboot: The First Bootkit on Android*, Qihoo 360 Technology Co. Ltd, Online, Available: http://blogs.360.cn/360mobile/2014/01/17/oldboot-the-first-bootkit-on-android/. Accessed December 8, 2014, 17 1, 2014.

4. fireeye, Kemoge: Another Mobile Malicious Adware Infecting Over 20 Countries, Oct 2015. Online, Available: https://www.fireeye.com/blog/threat-research/2015/10/kemoge_another_mobi.html. Accessed 2016.

5. arstechnica, Powerful backdoor/rootkit found preinstalled on 3 million Android phones, *Arstechnica*, Nov 2016. Online, Available: http://arstechnica.com/security/2016/11/powerful-backdoorrootkit-found-preinstalled-on-3-million-android-phones/. Accessed December 2016.

6. S. Smalley and R. Craig, Security Enhanced (SE) Android: Bringing flexible MAC to Android, in *NDSS Symposium*, San Diego, CA, Vol. 310, pp. 20–38, 2013.

7. J. G. Suarez-Tangil, P. Tapiador, Peris-Lopez, and A. Ribagorda, Evolution, detection and analysis of malware for smart devices, *Communications Surveys & Tutorials*, 16: 961–987, 2013.

8. D. Rosenberg, QSEE TrustZone Kernel integer overflow, *BlackHat*, 1–4, 2014.

9. D-H. You and B-N. Noh, Android platform based linux kernel rootkit, in *Malicious and Unwanted Software (MALWARE)*, Fajaro, IEEE, 2011.

10. H. Sun, K. Sun, Y. Wang, J. Jing, and S. Jajodia, TrustDump: Reliable memory acquisition on smartphones, *ESORICS*, pp. 202–218, 2014.

11. Y. Zhang, Kemoge: Another Mobile Malicious Adware Infecting Over 20 Countries, FireEye, 7 Oct 2015. Online, Available: https://www.fireeye.com/blog/threat-research/2015/10/kemoge_another_mobi.html. Accessed January 17, 2016.

12. D. Goodin, Malicious apps in Google Play made unauthorized downloads, sought root, *Arstechnica*, 07: 01, 2016. Online, Available: http://arstechnica.com/security/2016/01/malicious-apps-in-google-play-made-unauthorized-downloads-sought-root/. Accessed January 17, 2016.

13. Q. Song, Spyware Android.Spywaller uses a legitimate firewall to thwart security software, *Symantec*, 28 Dec 2015, Online, Available: http://www.symantec.com/connect/blogs/spyware-androidspywaller-uses-legitimate-firewall-thwart-security-software. Accessed January 17, 2016.

14. C. Xiao, AppBuyer: New iOS Malware Steals Apple ID and Password to Buy Apps, Palo Alto Networks, 12 Sept 2014. Online, Available: http://researchcenter.paloaltonetworks.com/2014/09/appbuyer-new-ios-malware-steals-apple-id-password-buy-apps/. Accessed January 17, 2016.

15. M. Guri, Y. Poliak, B. Shapira, and Y. Elovici, JoKER: Trusted detection of kernel rootkits in android devices via JTAG interface, in *Trustcom/BigDataSE/ISPA, 2015 IEEE*, Helsinky, Vol. 1, pp. 65–73, 2015.

16. O. H. T. Forever, Rooting Tools for All Android Devices [Collection], ON HAX, 10 June 2015, Online, Available: http://onhax.net/rooting-tools-for-all-android-devices-collection-with-guide-is-here. Accessed February 11, 2016.

17. http://www.iroot.com/iroot-apk. Online, Available: http://www.iroot.com/.

18. Cecilia, 80% China's Mobile Users Rooted Smartphones in 2014, China Internet Watch, 10 Apr 2015. Online, Available: http://www.chinainternetwatch.com/12926/80-china-smartphone-users-rooted/. Accessed February 11, 2016.

19. Trend Micro; Jordan Pan, User Beware: Rooting Malware Found in Third Party App Stores, Feb 2016. Online, Available: http://blog.trendmicro.com/trendlabs-security-intelligence/files/2016/02/Appendix_User-Beware-Rooting-Malware-Found-in-Third-Party-App-Stores_a.pdf. Accessed February 11, 2016.

20. arstechnica, 10 million Android phones infected by all-powerful auto-rooting apps, Online, Available: http://arstechnica.com/security/2016/07/virulent-auto-rooting-malware-takes-control-of-10-million-android-devices/

21. M. Zheng, P. P. C. LeeJohn, and C. S. Lui, ADAM: An automatic and extensible platform to stress test Android anti-virus systems, in *DIMVA 2012: Detection of Intrusions and Malware, and Vulnerability Assessment*, International Conference on Detection of Intrusions and Malware, and Vulnerability Assessment. Springer, Berlin Heidelberg, pp. 82–101, 2012.

22. A. Heriyanto, Procedures and tools for acquisition and analysis of volatile memory on android smartphones, in *Proceedings of The 11th Australian Digital Forensics Conference*, SRI Security Research Institute, Edith Cowan University, Perth, Western Australia, 2013.

23. J. Sylve, A. Case, L. Marziale, and G. Richard, Acquisition and analysis of volatile memory from android devices, *Digital Investigation*, 8(3–4): 175–184, 2012.

24. A. Kunk, P. Bohman, and E. Shaw, VMM based rootkit detection on Android, in *University of Illinois at Urbana Champaign*, May 2010.

25. J. O. Bickford, R. O'Hare, A. Baliga, V. Ganapathy, and L. Iftode, Rootkits on smart phones: Attacks, implications and opportunities, in *Proceedings of the Eleventh Workshop on Mobile Computing Systems & Applications*, Santa Cruz, CA, 2010.

26. D. Rosenberg, Azimuth, *Azimuth Security*, 4: 8, 2013. Online, Available: http://blog.azimuthsecurity.com/2013/04/unlocking-motorola-bootloader.html. Accessed November 29, 2014.

27. Wikipedia, Joint_Test_Action_Group (JTAG), *Wikipedia*, Online, Available: http://en.wikipedia.org/wiki/Joint_Test_Action_Group. Accessed November 29, 2014.

28. Z. R. I. Jovanovic, *Android Forensics Techniques. International Academy of Design and Technology*, 2012.

29. I. Breeuwsma, Forensic imaging of embedded systems using JTAG (Boundary-scan), *Digital Investigation* 3(1): 2006.

30. R. C. Brodbeck, Covert Android Rootkit Detection: Evaluating Linux Kernel Level Rootkits on the Android Operating System, Master's thesis, IR Force Institute of Tech Wright-Patterson AFB OH Graduate School of Engineering and Management, Ohio, 2012.

31. M. Lange, L4Android: A generic operating system framework for secure smartphones, in *Proceedings of the 1st ACM Workshop on Security and Privacy in Smartphones and Mobile Devices*, Chicago, IL, pp. 39–50, 2011.

32. A. Shabtai, Y. Fledel, and Y. Elovici, Securing android-powered mobile devices using SELinux, *Security & Privacy, IEEE*, 8(3): 36–44, 2010.

33. https://www.trustedcomputinggroup.org/

34. M. Nauman, S. Khan, X. Zhang, and J-P. Seifert, Beyond Kernel-Level Integrity Measurement: Enabling Remote Attestation for the Android Platform, *Lecture Notes in Computer Science*, 6101, 1–15, 2010.

35. Online, Available: https://www.samsungknox.com/en

36. M. Burgess, Major security flaw in Samsung Knox could give hackers "full control" of your phone, Online, Available: http://www.wired.co.uk/article/samsung-knox-security-vulnerabilities

37. K. Barr, P. Bungale, S. Deasy, V. Gyuris, P. Hung, C. Newell, H. Tuch, and B. Zoppis, The VMware mobile virtualization platform: Is that a hypervisor in your pocket?, in *SIGOPS Operating System Reviews* 44(4): 124–135, 2010.

38. Corelis, JTAG Tutorial, Corelis, Online, Available: http://www.corelis.com/education/JTAG_Tutorial.htm. Accessed November 29, 2014.
39. T. Shields, Survey of rootkit technologies and their impact on digital forensic, *Personal Communication*, 11: 2008.
40. JTAG RIFF Box, Online, Available: http://www.jtagbox.com/
41. L. GmbH, PRACTICE Script Language Reference Guide, 2 2014. Online, Available: http://www2.lauterbach.com/pdf/practice_ref.pdf. Accessed November 29, 2014.
42. Wikipedia, Widipedia (Slab_allocation), Online, Available: http://en.wikipedia.org/wiki/Slab_allocation. Accessed November 29, 2014.

Chapter 8

Various Shades of Intrusion: An Analysis of Grayware, Adsware, Spyware, and Other Borderline Android Apps

Andrea Saracino and Fabio Martinelli

Contents

8.1 Introduction

Android is the most common operating system for mobile devices, such as smartphones and tablets. In the last few years, the number of applications (apps) available both in official (i.e., Google Play Store) and unofficial markets has risen exponentially. Among these apps, millions have access to the Internet, thus being able to download data on the hosting mobile device, and/or upload information extracted from it [1]. The Internet access feature is generally used to provide services to the user, spanning from instant messaging to video streaming. However, the Internet access can also be a vector for intrusion attempts, since it exposes the device to network access, allowing third-party apps to receive information and commands from outside the device, and to send data extracted from the device itself. This feature has been exploited to build specific malware threatening Android users' privacy, known as spyware [2]. Moreover, even genuine app developers found in the Internet access not only a way to provide services to their users but also to exploit monetization strategies based on (unsolicited) advertisement and private data extraction for user profiling. In fact, app developers currently have three main sources of revenues from their mobile apps, namely, *in-app purchases*, *advertisement*, and *user data collection*. While the in-app purchase is merely an addition to the normal price of the application for buying additional features and contents, which is triggered by the user themselves, the other two monetizing strategies are generally unwanted, although often (barely) tolerated by users. Advertisement consists of application banners, pop-ups, and videos to advertise commercial products, either of the developer or third parties. Advertisement contents are generally not preloaded in the app; instead, they are downloaded at runtime in background. Thus, advertisements not only worsen the user experience, by interrupting the normal app execution to show potentially unwanted contents, but also consume user data traffic and, indirectly, energy. This issue, which is accentuated in those countries where data plans are limited to few hundreds of megabytes [1], is also related to the third monetizing strategy, that is, collecting user data.

Several apps collect user data for offline or online processing. This processing is generally aimed at providing contextual advertisement, which should be in line with user preferences and current needs. Information that can be used to profile a user are worth money. The collection of such information is generally performed by including in the apps, components and services of advertisement providers, such as *Google Ads* and *Google Analytics*. These providers directly give money to the developer for both displayed advertisements and collected data, becoming in the long run, the main revenue for the majority of mobile app developers. While the users generally tolerate in-app advertisement, they are often not aware of the energy and traffic overhead generated by some apps, which may represent a considerable monetary cost noticeably surpassing in the long run the typical price of a mobile app [3].

In this chapter, we present an analysis of the different security threats that exploit data traffic as an attack vector, analyzing both the behavior of malicious and genuine apps, together with that set of apps that lies in the gray zone, that is, not classified as malicious, yet still presenting critical behaviors generally unwanted, which might also threaten the user's privacy. Afterward, we present *Data-Sluice*, a framework that allows a fine-grained control of incoming and outgoing data on a per-app basis, with the objective of detecting intrusions and other unwanted behaviors, and eventually preventing them. Hence, we present the analysis on a set of popular Android apps studying both the type of exchanged data and their amount. Next, the proposed framework has been exploited to enforce policies to stop or limit the traffic generated by apps. In particular, specific policies have been implemented to prevent apps (in particular spyware) from sending out privacy-sensitive data, or more generally, traffic unrelated to the desirable execution of a specific app. Then, reported results show how it is possible to consistently reduce or remove both the traffic overhead and the disclosure of privacy-sensitive information, which affects negatively the user experience. By removing, for example, undesired advertisement banners or annoying pop-ups, the user experience would be improved instead. The effectiveness of the proposed framework has been tested on a set of very popular applications showing critical security features presented by them. Hence, Data-Sluice has been used to render ineffective a set of malicious apps (spyware), preventing them from sending out privacy-sensitive user's information.

The contributions of this chapter are reported in the following:

■ A taxonomy of threats exploiting data traffic as an intrusion vector, spanning from genuine apps with massive network usage, to spyware, passing through adsware and grayware apps is presented.

■ A characterization of the amount and type of traffic generated by very popular Android apps is presented.

■ Data-Sluice, a framework to enforce a fine-grained control on data traffic generated by Android application, is described.

■ Finally, an application of Data-Sluice to successfully stop the malicious actions performed by a set of 197 spyware apps, belonging to nine different families, is discussed.

This chapter extends and enriches Reference 4, presenting (i) a taxonomy of data-related threats, spanning from web-based apps to spyware; (ii) a discussion and comparison of different strategies for detecting unwanted behaviors; (iii) a small set of additional experiments, including additional ad providers and apps; and (iv) a deeper and extended review of related work. The rest of the chapter is organized as follows. Section 8.2 reports background notions on Android native security mechanisms. Section 8.3 describes grayware, adsware, and spyware, characterizing their actions and the level of threat they bring. Section 8.4 reports a set of strategy for the intrusion detection of malicious exploitation of data traffic and briefly describes the Xposed Framework, a tool that is exploited by the proposed framework to detect intrusion and prevent it. Section 8.5 details on the Data-Sluice framework, presenting its components and their interaction with the Android system. Section 8.6 presents the experimental results on data traffic overhead reduction on a set of eight popular applications, reporting an analysis of the filtered traffic. Afterward, the experiment on a set of 197 malicious apps is reported. Finally, a performance analysis showing an improvement of the battery duration is presented. In Section 8.7, we survey some related work from research and industry. Finally, Section 8.8 presents some concluding remarks.

8.2 Android Security

The Android native security mechanisms are the Permission System and Application Sandboxing, which enforce, respectively, access control and isolation. Through the Permission System, every security-critical resource (e.g., camera, GPS, Bluetooth, and network), data, or operation is protected by means of a permission. If an application needs to perform a security-critical operation or access a security-critical resource, the developer must declare this intention in the app `AndroidManifest.xml` (manifest for short) file, asking the permission for each needed resource or operation. Permissions declared by the application are shown to users when installing the app, to decide if they want to consider the application as secure or not. If the application tries to perform a critical operation without asking the permission for it, the operation is denied by Android. The manifest file is bound to the application by means of a digital signature. The integrity check is performed at deploy time; thus the Android system ensures that if an application has not declared a specific permission, the protected resource or operation cannot be accessed. In the latest Android versions, users can dynamically revoke and regrant specific permissions to applications; however, this practice requires a level of knowledge and expertise greater than that of average users.

On the other hand, isolation is enforced through the synergy of two elements: the runtime environment and the underlying Linux kernel. In Android, every application runs in a virtual machine (VM); thus, each application has its own memory space, can act as if it is the only application running on the system, and is isolated from other apps. Moreover, each VM is registered as a separate user of the Linux kernel. This means that each installed app is considered as a user at the kernel level, able to run its own processes and with its own home folder. The home folder of each application stores application files on the device internal memory; thus, it is protected from unauthorized access by the Linux kernel itself. In fact, files stored in the home folder can be accessed only by the application itself. However, since the device internal memory is limited, the amount of data that can be stored in the home folder is limited and generally using the internal memory is a deprecated practice.

These two native mechanisms are able to ensure a good protection for Android devices and users; still, from 2011, several malicious developers started to develop malicious apps that leverage different strategies to damage the user or the device, avoiding the protection of the native mechanisms. These malicious apps (malware) target either the user money or the user privacy. In particular, malware attempting to steal private information are extremely common. In fact, smartphones and tablets currently store several privacy-sensitive information, such as text messages, social network accounts, contacts, International Mobile Equipment Identifier (IMEI), and International Mobile Subscriber Identifier (IMSI), which could be exploited to clone SIM cards and even credit card information. These pieces of data can be monetized at different levels, motivating attackers to create specific malicious apps known as spyware.

8.3 Taxonomy of Data-Related Threats

In order to have access to the network, Android apps must declare the INTERNET permission in the manifest file. As discussed, apps with the INTERNET permission can both upload and download data to and from the mobile device. In this section, we will present a taxonomy, based on the introduced level of threat, of apps that might endanger the device exploiting the data connection, discussing the motivations behind the dangerous or malicious behavior.

8.3.1 Web-Based Apps

Most of Android and, in general, mobile apps exploit the network to perform their normal functionalities and offer services to the users. Typical example of apps based on the Internet are social networking and instant messaging apps such as Facebook, Whatsapp, Telegram, Twitter, and Skype. Most of the traffic generated by these apps should be considered *desirable traffic*; hence, it is necessary to provide a specific service to the user. Still, it must be noticed that some apps can generate a very large amount of traffic in a short time. This might create an issue for users with monthly data traffic limit. Moreover, it is still possible to find that some part of the data traffic generated by these apps does not fall in the desirable traffic category, falling instead in one of the two categories described in the following.

8.3.2 Grayware

As a matter of fact, we have witnessed in the last 2 years a progressive reduction of the paid apps on Google Play Store. Paid apps should be the most straightforward monetization strategy for app developers, that is, when a user wants to install an app in its device, the user will pay a price, specified by the app developer to Google Play Store, which will give the money to the developer, keeping a certain percentage. However, the greatest majority of apps can be downloaded completely for free, including the most popular ones. It is hence possible to infer that app sale is not the most common and rewarding monetization strategy. Several apps have replaced the price of the app with the possibility to buy additional app features, after the app has been installed for free. This strategy, known as *in-app purchase*, allows to generate much higher revenue than the app price itself. This model is also used by apps requiring periodic subscriptions to provide premium services, such as the extremely popular music streaming app Spotify, which removes advertisement between songs only to premium accounts, on a monthly fee.

Another monetization strategy that is exploited by several apps is data collection and analytics. Android offers a set of APIs to developers, to include in-app services for data collection. These APIs can be used to read the user location, measure the interaction time with a specific app, recording search keyword patterns, visited websites, and other information that might be used to shape the habits and preferences of the user. This set of information is extremely valuable, but it causes an unavoidable privacy violation, which is not desirable for the user. Apps showing such a behavior fall in a controversial category known as grayware [5]. Such a controversy is related to the view of the service provider or app developer, which, giving the app free, considers the privacy loss and the correlated revenue as the effective price the user is paying for the app. However, grayware apps are borderline legal, due to their similarity with more dangerous apps that fall instead in the category of malicious. Additional reasons have to be found in the fact that the user is often unaware of the kind of information they are submitting, which are often misused to send unsolicited advertisement by third parties, and in the fact that the right to privacy advantages the user in legal settings, especially in those countries where opt-out is not a common marketing strategy.

8.3.3 Adsware

Advertisement is another direct source of income for app developers. As for analytics, Google provides a set of APIs to implement in-app advertisement, in the form of banners, pop-up, or status bar notifications, and even short advertisement videos. Contents to be shown are downloaded from different ad providers. A revenue is provided to both the API provider (i.e., Google) and to the developer each time the advertisement content is shown and/or the user clicks on the advertisement, generally to be redirected to the website of the advertised product. In-app advertisement is seen by most users as a kind of spam, which alters the app usage experience and generates unsolicited notifications, which might become insistent, especially if generated by several apps.

However, the major issue of adsware is the performance overhead and the consequent indirect costs that might easily surpass the financial price of paid apps. In fact, advertisement contents are provided at runtime by the ad providers; hence they are downloaded before being shown or directly streamed, as it happens for videos. The amount of data that can be generated by in-app advertisement is not negligible. Moreover, this cost can become consistent if ads are downloaded over a 3G/4G connection, which normally has monthly traffic limit, or if the user pays on a traffic base. Moreover, this overhead is also reflected on a reduced battery charge duration, caused by the increased radio activity. Indeed, as we have experimentally verified, several apps download advertisement contents even when the app is in the background, showing pop-up or notifications even when the user is not interacting with the app.

Apps that include advertisement and perform behaviors that are not desired by users are also known as *adware* and are also extensively discussed, since unsolicited advertisement is still considered illicit [6]. It is worth noting that there exist several tools to mitigate the effects of ads on user experience, known as *AdBlocker*. Moreover, it has been found that often advertisement notifications link to websites to download and install malicious software, thus infecting the device [7]. For these reasons, controlling the traffic generated by genuine apps including adsware is necessary to timely detect intrusion attempts.

8.3.4 Spyware

Spyware is a term that refers to a set of apps and more generally to malicious code specifically designed to extract privacy-sensitive information concerning the device and its user, sending it to the malware developer [2]. For this reason, spyware are considered as a malware class. Notwithstanding, it must be noticed that there exist spyware apps distributed even through official markets, generally with purposes that are not illegal, such as parental control. Hence, in the following, we consider as malicious spyware those apps that steal user private data in a covert manner. In the wild, there can be found several spyware families that generally follow two patterns to conceal their actions. The first and most common pattern consists of hiding the malicious code inside a genuine app, where the malicious behavior is performed in the background, while the genuine app code runs normally to not make the user suspicious. This pattern, common also to other malware classes, is called *trojanized app* and is very hard to detect, especially for spyware, which does not require suspicious permissions. The second pattern consists of designing stealthy apps, that is, apps without activities (GUI) that after being installed are not visible in the launcher. These apps can just be visualized by browsing the All app list of the Settings menu, where generally the malicious app further complicates the identification task using hideous names. Spyware can send out information related to geolocation, contact numbers, text messages, browser history, and even device or mobile subscription identifier such as IMEI and IMSI, needed to attempt cloning SIM cards.

8.4 Detection Strategies

The task of detecting apps maliciously exploiting the data connection is quite challenging. One of the basic approach used for the detection of suspicious software is the analysis of declared permissions. The rationale of the basic Android security mechanism can be exemplified by the permission *SEND_SMS*, used by apps to send text messages outside. This permission is exploited by a malware class known as SMS Trojan [8], which, despite being the most common, is easily identified due to the peculiar permission. In fact, the number of apps effectively needing to send text messages is quite limited and is unlikely that a gaming app should ask for such a permission. This analysis task on permissions has been further simplified by research frameworks, which automatically analyze permissions, to infer malicious patterns and communicating the risk to the user [9,10]. On the other hand, the threats analyzed in the former section mainly require, as discussed, the INTERNET permission. The same permission is required independently from the amount, type, or recipient/source of data, by any app that needs to interact with the network. Differently from SMS, almost any app has a valid rationale to ask for the INTERNET permission, making thus the permission analysis strategy almost ineffective.

Another strategy that proves to be more effective is the multilevel dynamic analysis, whose approach has been extended by Data-Sluice. The multilevel analysis consists of monitoring at run time features concerning the Android OS, the APIs invoked by the apps, and the user activity. An effective framework exploiting this approach is MADAM, presented in References 8 and 11, whose architecture is schematically depicted in Figure 8.1. As shown, the framework analyzes features from four different levels, namely, package level, user level, application level, and kernel level, with the features being mainly numerical. By monitoring features at different levels, it is possible to automatically correlate, by means of classifiers, patterns shaping normal, genuine behaviors and suspicious ones. The features being numerical, the main rationale is to identify intrusions by monitoring increases of the system and API calls not related to the increase of user activity. This approach is effective in detecting the actions of malware belonging to several classes, including SMS Trojan, Rootkit, and Ransomware. However, this approach that is based on machine learning techniques hardly spots the difference between the behavior of genuine apps using the Internet and the one of spyware apps. In fact, actions performed by spyware are on the behavioral side, identical to the operation of apps receiving data from the network or uploading data to the network. Moreover, the

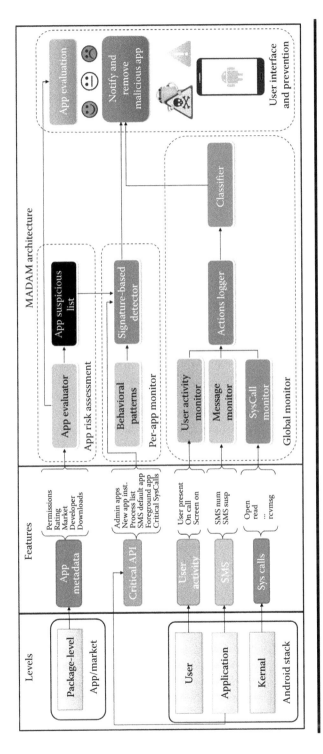

Figure 8.1   Architecture of MADAM, a multilevel intrusion detection system.

amount of data sent out by spyware is generally limited; hence, it does not cause consistent changes in the system and API calls that are easily spotted by the multilevel analysis.

Hence, a strategy that is more oriented on the semantic of invoked APIs is necessary to detect intrusion attempts of spyware and other data-related threats. To this end, it is necessary to analyze for any opened connection and http request, the data recipient or sender and, when possible, the payload of the request. To further complicate this task, several spyware encrypt the payload to avoid analysis such as data tainting [12] or blacklisting of malicious packets. However, it is still possible to exploit the blacklist and whitelist approaches on data source and sinks, to filter those packages considered as malicious or suspicious. Still, having such a control on the system APIs requires the modification of the original Android code. These modifications may pass through dynamic loading of kernel modules [11] or custom version of the Android OS [12,13], or by exploiting code inlining tools like the one described in the following subsection.

8.4.1 Xposed Framework

The Xposed Framework (or simply Xposed)* is an advanced custom developer tool designed to give a much greater control of the Android system and the running apps, compared to the one granted by the Android available APIs. Xposed comes in the form of an Android app called *Xposed Installer*, which will install an extended app process executable in the /system/bin folder. This executable will add to the classpath an additional jar file, which will be called at phone startup. This jar file redefines the address of specific API calls, forcing the system to invoke custom versions of these API calls specified in Xposed-compatible apps called as *modules*. Xposed modules allow developers to *hook* any method, either native of Android or defined in a third-party app exploiting a private method named hookMethodNative. Hence, once a method is hooked, Xposed allows to define the operations to be performed immediately *before* the method execution and the ones to be performed immediately *after*. In particular, it is possible to control and even change the actual parameters of the invoked method, up to completely prevent its execution. Xposed can be programmed to hook a specific method either globally, that is, controlling a method every time it is invoked, independently from the app that is invoking it, or on a package base, that is, controlling a specific method only when it is invoked by a specified app, identified uniquely by its package name. The Xposed Framework can be installed on any Android device; however, it requires the target device to be *rooted* (jailbroken).

---

* https://github.com/rovo89/XposedBridge/wiki/Development-tutorial

8.5 Data-Sluice Architecture

Data-Sluice comes as an app for Android devices, which implements the tools for selecting data traffic policies and to enforce them selectively on specific apps. Figure 8.2 depicts the architecture of Data-Sluice in its operational environment. The Data-Sluice core contains three main operative blocks: *Policy Storage*, *Analyzed*, and *Enforcer*. Policy Storage contains the policies on data traffic for each app. This component also offers a GUI for the user, which allows the user to select specific policies for each app. Figure 8.3 offers two screenshots of this GUI. The left screen shows the list of apps installed on the device, while the circle represents which of the policies, detailed in Section 8.6, is enforced. In the app, in particular, a blue circle means that no policy is enforced, a green circle corresponds to the *Log-Only* policy, a yellow circle corresponds to the *BlackList* policy, while a red circle corresponds to the *Always Block* policy. The right screenshot shows instead the interface to select the policy to be applied for each app, with the level option that specifies if hooked methods should be the high-level network APIs in Figure 8.2, or the `Socket.connect()` low-level method.

The Enforcer component hooks a set of methods used to open network connections and stream data in both directions. Hence, as soon as one of these methods is invoked by an app, the Enforcer passes the method call details to the Analyzer,

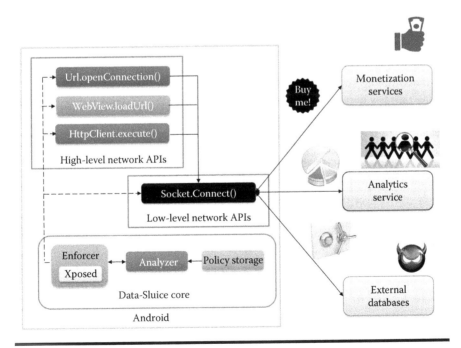

Figure 8.2   The architecture of Data-Sluice.

Figure 8.3   Two screenshots of the Data-Sluice GUI.

which will verify the policy defined for the caller. The Analyzer will then decide either to permit or deny the execution of the hooked method. The Enforcer exploits the Xposed Framework to hook methods, inserting a callback to the Analyzer inside the `BeforeHookedMethod()` function provided by Xposed. The task of monitoring the outgoing data traffic is particularly challenging, since different apps may exploit different source codes to handle network connections, and such a code is not known beforehand, since the app's source code is not generally made available by the developer. Thus, we have identified a set of methods needed to control data traffic by disassembling and inspecting popular free apps. The `DecompileAndroid`[*] online tool has been used to this end. Table 8.1 lists the Android native methods that are hooked by Data-Sluice to control data traffic. Controlling these four methods allows to control the data traffic at different levels, which, as shown in Figure 8.2, we consider as incoming or outgoing toward three possible sets of entities: (i) *monetization services*, which send to the user advertisement contents, such as page to be shown in pop-up and banners, or video advertising products, (ii) *analytics service*, which collect user information for profiling purposes, and (iii) *external databases*, mainly controlled by attackers to store private data stolen from a victim device. The high-level methods `Webview.loadUrl()` and `Url.openConnection()` are commonly used to download data, show banners and pop-up with advertisement either for other developer's product, or more likely, for third parties. The amount of data downloaded may vary from few kBs to several MBs, considering that a pop-up can also be used to show heavy advertisement videos. The `HttpClient` class is a

---

[*] www.decompileandroid.com

Table 8.1 Methods Hooked to Control Data Traffic

Class	Method	Description
Webview	loadUrl()	Used by pop-ups and banners to download and show advertisement
Url	openConnection()	Opens an http connection toward a specified URL
HttpClient	execute()	Used to download contents such as videos for advertisement
Socket	connect()	Kernel-level function used to open a connection and stream data

general Java class defined in the *Apache* Common Library,[*] which defines the methods used to open http connections. Some apps exploit this general method instead of the previous one more specific for Android, specifically those apps opening connections in background, without showing pop-up to the user. For this reason, the HttpClient.execute() function is particularly important, since it is exploited also to send out user information, both from legal apps and malicious ones. In particular, all the analyzed spyware samples have been found to call this method to transit device or user private data, such as IMEI/IMSI or the text message history. By hooking these three methods, it is possible to extract both the other end of the connection and the payload, which for these functions is typed. Thus, it is possible to specify policies specific for both these parameters, selectively blocking only the unwanted traffic. The Socket.connect() method is instead a low-level function, which by itself is invoked by all the previous three methods in order to open a connection and stream data on the socket. Hooking the socket ensures that all invoked connections are captured, even if they present usability issues, compared to the other three methods of hooking. In fact, analyzing the payload of a socket will result in reading an untyped stream of bytes, which are hard to parse and handle. For this reason, enforcement on this method is performed only when it is not possible to control data traffic for an app through the other three methods, that is, the app uses custom primitives not implemented in the Android native libraries.

Hence, Data-Sluice is implemented as an Xposed module, which performs selective hooking on four methods invoked by specific packages (apps) specified by the user through a GUI. Controlling these four methods, Data-Sluice is able to govern the opened connections of selected apps, which are handled by the implementation of policies that will be described in the following. Recall that, being an Xposed

---

[*] http://hc.apache.org

module, Data-Sluice requires the target device to be rooted. For this reason, Data-Sluice is not currently designed to be a solution for everyone, but as a research tool to analyze at runtime the data traffic generated by Android apps. Still, some functionalities could be added in subsequent Android releases, or by device manufacturer, to exploit the functionalities used to tackle spyware apps. After installation, the Data-Sluice app continuously runs in the background, starting automatically every time the device is booted.

8.6 Application

This section details the application of Data-Sluice, which through different security policies filters the unwanted traffic of Android apps. In particular, we discuss the effect of Data-Sluice on a set of eight popular apps on reducing the overhead traffic and removing the burden of unwanted advertisement banner and pop-up in an intensive usage context. The benefits on battery consumption is also presented. Finally, an application of Data-Sluice on a large set of malicious apps, belonging to the spyware behavioral class is presented. All experiments have been performed on a Samsung Galaxy S2 smartphone, equipped with the Android 4.2 Jelly Bean OS, equipped with a SIM card providing a 4 GB monthly data plan.

8.6.1 Blocking Unwanted Data

In the first set of experiments, Data-Sluice has been used to control the traffic generated by a set of popular Android apps and to limit or remove in-app advertisement. For this set of experiments, and for each application, the phone has been used for six periods of 6 h. In the first three periods, the smartphone has been used with Data-Sluice inactive, while in the second set, the framework has been used with different policies specific per application, as discussed in the following. Data-Sluice is, in fact, configurable with three possible policies for each controlled app:

- *Always Block*: Blocks all connections generated by the app.
- *BlackList*: This mode performs a selective traffic filtering, blocking traffic coming from or directed toward configurable addresses of advertising service providers, monetization services, or servers known to be used by spyware malicious apps. Table 8.2 reports the list of the main monetizing service blocked by Data-Sluice when the blacklist mode is active. The table also contains the percentage of apps on Google Play Store communicating with this service and the percentage on the overall installed apps on user devices, as reported in Reference 14.
- *Log-Only*: This mode does not block any connection, but logs all the events of `Webview.loadUrl`, `Url.openConnection()`, and `HttpClient.execute()` issued by a specific app. It also reports the parameters of each invoked method.

Table 8.2   Monetizing Service Stopped by Data-Sluice

Ad Network	% of Apps	% of Installs
Admob	43.85	50.53
Chartboost	4.16	11.32
AdColony	1.76	9.34
MoPub	2.13	9.21
InMobi	2.86	8.95
Unity Ads	1.69	8.10
MillennialMedia	2.50	6.78
TapJoy	1.25	5.74

It is worth noting that the policy is app-specific, that is, apps for which a policy is not specified will work normally.

Tests have been performed on a set of nine popular apps and are presented in Table 8.3. The second and third column of Table 8.3 reports the average amount of traffic generated in the three experiments of 6 h, respectively, with Data-Sluice not active and active. The last column specifies if the in-app advertisement has been successfully stopped from showing banners, pop-ups, or videos to the user. N/A means that the app is already ads free, but may generate traffic by sending information to analytics services. As shown, Data-Sluice is very effective in both reducing the data traffic generated by controlled Android apps and removing in-app advertisement. In the following, we detail the experimental results gathered from three relevant applications, that is, *Apus Launcher*, *Angry Birds*, and *Skype*.

8.6.1.1 Apus Launcher

Apus is an alternative launcher that substitutes the standard Android home screen and the organization of icons. The download number for this app from the Google Play Store official market ranges between 100 and 500 M. Being a launcher, the app is continuously active on the phone, generating a noticeable amount of traffic in the time interval of 6 h. By averaging the results of three periods, the app has generated an amount of 23.45 MB of traffic, with a deviation of few kBs for each period. Hence, in a day of usage, the Apus Launcher could generate up to 90 MB of traffic, which is 3.7 GB for a month of usage. Such an overhead is generally not acceptable from a launcher, especially considering that an average European monthly plan offers at most 2 GB of data.

Table 8.3 Application of Data-Sluice on a Set of Popular Apps

App Name	DS Off	DS On	Ads Stop
Angry Birds	37.38 MB	5.77 MB	Partially
Apus Launcher	23.45 MB	0	Yes
Aviary	5.84 MB	0	Yes
Crossy Roads	14.40 MB	0	N/A
Clean Master	5.36 MB	0	Yes
Candy Crush	14.2 MB	0	N/A
Flow Free	4.2 MB	0	Yes
King Calculator	4.1 MB	0	Yes
Skype	1.7 MB	1.8 MB	Yes

Considering the functionality of a launcher, it is possible to derive that the network interaction should be limited. For this app, the *Always Block* policy has been used, blocking all connections for Apus Launcher. As a result, the generated traffic in the 6 h period has dropped to zero for all the three experiments. Moreover, during the experiment, the user has not noticed any service degradation or launcher malfunctioning.

8.6.1.2 Angry Birds

Angry Birds is a very famous gaming app for Android whose download number from Google Play Store amounts to the range 100–500 M. Although the game is completely playable offline, the app performs some legit connections for in-game rewards and to record high scores on online charts. The game is distributed in two versions, a free one and another version that costs 0.99 euros per download. Both these versions generate a very large amount of traffic during usage. A thorough analysis performed through the *Log-Only* mode of Data-Sluice shows that the traffic is related to information sent toward data analytics services such as Google Analytics, and received by monetization services such as AdMob. Both versions show advertisement banners, reduced in the nonfree one. Moreover, at the end of each level, an advertisement video is shown. Videos provide the bigger impact on data traffic overhead and they are downloaded continuously during the game.

In the performed experiment, the user has played with the free version of Angry Birds for intervals of 30 min each hour. Hence, in each period of 6 h, Angry Birds has generated traffic equivalent to 3 h, amounting, in the average, to 37 MB with a limited deviation.

In the second set of experiments, Data-Sluice has been configured with the *BlackList* policy, to block traffic from monetization services. The final impact on the set of three experiments results in an average reduction of generated traffic to 5.77 MB, that is, 87% of reduction in generated traffic. Furthermore, the amount of shown advertisement has been noticeably reduced, thus improving the user experience. However, it has not been possible to completely stop the app from showing advertisement videos, though they have been consistently reduced. The reason is that the first video download is triggered using native C libraries, largely used in the Angry Bird app, instead of the standard Java APIs offered by Android. As a result, it is not possible to intercept the download event with one of the three Java methods hooked by Data-Sluice. Using the Always Block mode, also intercepting and blocking the `Socket.connect()` method, the generated traffic from Angry Birds is further reduced. Moreover, the app works correctly without crashing, since it believes that the phone is not connected to the network. However, the user will not receive the in-game reward, and nor will any other of the legit online functionality work.

8.6.1.3 Skype

Another interesting use case is the Skype app. Differently from the previously analyzed ones, this app is expected to have a noticeable network activity, being an application for chatting and VoIP-like phone calls. Skype comes as a free app but offers in-app purchase functionalities, with the possibility of adding credits to call landline phones and to have the premium version, which is advertisement-free. Data-Sluice has been used to successfully remove the advertisement presence also from the free version. To this end, Data-Sluice applies the *BlackList* policy on the Skype app. During experiments, Skype has been used mainly for chatting and to perform a single call lasting 1 min. The difference in produced data between the experiments with Data-Sluice on and off, respectively, is not appreciable.

8.6.1.4 Performance Analysis

Another set of experiments has been run in order to estimate the impact of Data-Sluice on battery consumption. Three experiments have been run in order to fully discharge the smartphone battery according to the following configuration: (i) the device screen always on, (ii) all the eight apps from the previous set of experiments installed on the phone, and (iii) the user alternating 30 min of active interaction with Angry Birds, Skype, and Apus Launcher to 30 min of idleness. The three experiments have been performed both with the Data-Sluice module on, keeping the same configuration of the experiments on data traffic, and Data-Sluice module off. Figure 8.4 represents the average battery discharge cycle for the two experiments.

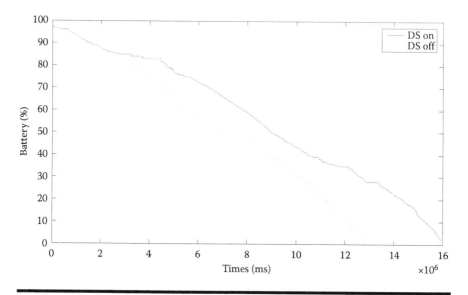

Figure 8.4   Battery depletion cycles with module on and off.

As shown, there is a noticeable difference in the two discharge cycles. Apart from the fluctuations following unplugging from power supply (as already noted in Reference 8), the battery discharge when Data-Sluice is off is faster than the case in which Data-Sluice is active. More precisely, the phone battery completely discharges in 3 h and 20 min when Data-Sluice is off, while the device can run for 4 h and 26 min when the Data-Sluice is active. Hence, the overall impact of Data-Sluice on battery performance can be considered positive, since, reducing the generated data traffic noticeably, the energy consumption is also reduced.

8.6.2 Spyware Prevention

After proving the effectiveness of Data-Sluice in minimizing the data traffic overhead and in removing in-app advertisement, a further set of experiments has been performed to show how Data-Sluice can also be used for protection against malicious apps. In particular, Data-Sluice focuses in tackling the action of a malware class known as spyware. Recall that spyware is a class of malware that extracts pieces of private data from the mobile device, such as IMEI and IMSI, user contacts, messages, or social network account credentials. This malware class is particularly subtle and difficult to detect even with behavior-based intrusion detection systems [8,11]. In fact, the malicious code runs in the background of apps, which seems legit, that is, they present themselves and they behave as useful apps and even popular apps. As already pointed out in Section 8.2, this technique to spread malware is known as trojanized or repackaged app [15] and has the perk of letting the malware pass

Table 8.4 Spyware Families Tested on Data-Sluice

Malware Family	Samples	Effective	Traffic
Geinimi	87	Yes	Yes
Kmin	52	Yes	No
Plankton	11	Yes	No
Gone60	9	Yes	No
SMSReplicator	4	Yes	No
SMSZombie	10	Yes	No
NickySpy	2	Yes	No
Trackplus	6	No	Yes
DroidDream	16	Yes	Yes

unnoticed to the user, who will use the app without being aware of the malicious behaviors happening in the background. Through app repackaging, it is possible to insert the same malicious code in different genuine apps. Hence, each trojanized app is considered a malware *sample*, while two or more samples carrying the same malicious code constitute a *family*. The most hard-to-detect spyware families are the ones using http connections to send the stolen information on servers controlled by the attacker.

Data-Sluice can be exploited to prevent spyware from sending information through the network. To this end, Data-Sluice has been configured with the blacklist mode to stop any connection from the analyzed apps directed toward any numerical IP address. Data-Sluice has been tested against the set of malicious applications reported in Table 8.4. The same table also contains the number of samples per family, and if Data-Sluice has been effective in preventing data from being sent out of the phone (third column). The last column specifies if data traffic has been generated by the app with Data-Sluice active in the blacklist mode. In order to measure the effectiveness of the proposed framework, the logging mode of Data-Sluice has also been added to notify the invocations of `HttpClient.execute()`. Hence, we consider Data-Sluice to be effective for all those apps where calls to this method have been reported and blocked. `Geinimi` [16] is a spyware that sends out several private information to an external server, after registering the infected device to the server with a challenge response protocol. Data-Sluice is effective in avoiding any malicious action from Geinimi, since it blocks the connection for this registration process to happen. Moreover, as shown in Table 8.4, some Geinimi-infected apps present, in the genuine part, network functionalities, which still work correctly when Data-Sluice is active. A similar behavioral pattern is shown by the DroidDream malware family. This malware managed to spread even through the official market and its main malicious action consists of sending IMEI and IMSI to a server controlled by an attacker. Such attempts are blocked by Data-Sluice. As shown in Table 8.4, the TrackPlus malware is not blocked by Data-Sluice in this experiment. The reason is that, differently from other tested spyware, TrackPlus does not sent data to a numeric IP address; it instead directs data toward a destination indicated by a domain name. This behavioral difference is mainly due to the fact that TrackPlus should not really be considered a malware, since it is a legal software installed on the phone of a person, to be used for parental control purposes. Still, it is worth noting that Data-Sluice could successfully block the connections from TrackPlus, adding the server domain to the blacklist or using the Always Block mode instead.

Data-Sluice can thus be used as an enforcement framework for security to control and eventually block connections toward suspicious addresses. The enforcement engine of Data-Sluice can be used to implement several different strategies, including whitelisting of allowed destination addresses for specific apps. Moreover, in order to automatically select which app should be controlled by Data-Sluice, the framework can be integrated with an app rating mechanism like the one proposed in Reference 10 or 17. These frameworks evaluate the potential security threat brought by a specific app, exploiting multicriteria algorithms applied on declared permissions and other reputation indexes. Choosing the correct app to control through Data-Sluice allows users to use the useful part of a trojanized app, without being the victim of the malicious behavior, or can be used for early detection of suspicious connections, which may trigger an app removal process.

8.7 Related Work

Some works concerning the control of data traffic for smartphones already exist in the literature. In Reference 18, an analysis on data usage patterns for 3G and 4G is presented. This work mainly focuses on analyzing the amount of generated traffic, in order to help the service provider to design strategic data plans. Differently from Data-Sluice, the analysis performed in Reference 18 is static, based on stored data logs collected in November 2010. The overhead caused by advertisement in free apps has also been debated in Reference 3. This chapter presents an analysis to show that, in the long run, a free app containing advertisement will imply a monetary cost higher than the nonfree version. The analysis focuses on energy consumed, without considering data traffic aspects. Data-Sluice, on the other hand, mainly focuses on data traffic, which adds another monetary cost to the user, confirming the results of Reference 3 for what concerns energy leakage, as discussed in Section 8.6.1.4 and shown in Figure 8.4.

TaintDroid [12] is a framework for enforcing security policies on Android applications. This framework performs hooking of security-critical methods by flashing a custom operative system on the device. TaintDroid is mainly focused on detecting and tackling privacy leakages. Compared to Data-Sluice, it has a different focus, and requires a more invasive modification of the target device. A similar framework to TaintDroid is TISSA [13], which instead focuses the action on app permissions. Through TISSA, it is possible to dynamically revoke permissions from apps, to stop unwanted actions. However, since permissions are coarse-grained [9], it is not possible to exploit TISSA to implement fine policies. For example, TISSA can prevent an app from generating overhead data traffic by removing the INTERNET permission. Data-Sluice instead is more fine-grained, allowing to implement policies to filter specific data traffic streams. A framework for the enforcement of security policies on mobile devices is presented in Reference 19. The proposed framework verifies that a contract, describing the expected behavior of an app, matches with specified security policies, by means of formal verification or effective enforcement. Differently from this work, Data-Sluice is more focused on controlling data traffic, implementing policies able to limit the unwanted behavior, without negatively affecting the user experience.

Apart from the work on Android, some intrusion detection systems have also been developed for iOS mobile devices. In Reference 20, the authors present a cloud-based security framework developed for Android devices, which emulates the behavior of a real device in a cloud environment. Such a framework, though effective in verifying the behavior of more applications at the same time, can be deceived if the app verifies if it is running on a real device or an emulator. In fact, several apps hide by purpose some of their functionalities when they detect that they are not running on a real device. Being device-based, Data-Sluice is not affected by deception mechanisms able to understand if the app is going to be monitored. In Reference 21, a host and network-based intrusion detection mechanism for Android devices is presented. The framework is lightweight and scalable, with an analysis mainly based on generated connections, which exploits clustering and correlation to detect suspicious behaviors. Data-Sluice, on the other hand, looks for and actively blocks connections known to be unwanted and/or malicious. The work in Reference 22 presents a framework for dynamic analysis of iOS apps in terms of method invocation and extraction of behavior to be compared with the malicious one, to determine if the app contains a malicious code. The same authors present in Reference 23 a framework for intrusion detection, again for iOS with two operative modes, respectively, on-device and in the cloud, providing thus a hybrid approach able to ensure real-time detection still outsourcing calculus complexity to an external service. In Reference 24, an anomaly-based IDS for iOS is defined, to detect malicious behaviors also based on classification. The work presents a comparison between different classifiers used to evaluate the behavior of end-users, monitoring features such as phone calls, text messages (SMS), and web browsing history. Differently from these frameworks, Data-Sluice does not consider classification aspects, being mainly based on heuristics. Classification ensures generality, which matches the typical requirements of IDSs. However, for a small set of specific behaviors, the heuristic approach is more effective.

The work presented in Reference 25 describes the phenomenon of covert channel communications, that is, those communications that happen without the user's explicit consent and that are not necessary to provide an explicit and desired service to the user. This work demonstrates that covert channel communications are extremely common in Android apps, which further motivates our work. All the connections tackled by Data-Sluice are considered as covert channel communications.

Given the exponential growth of available Android malware [26], several works have focused on tackling the actions of Android malicious apps. The work presented in Reference 8 presents a framework effective against about 2700 malicious app samples, exploiting machine learning and behavioral analysis. Still, the framework is less effective against spyware, which instead is the specific target of Data-Sluice. The framework presented in Reference 27 looks for specific API call patterns statically analyzing apps binaries to extract features. A similar approach based on static and/or offline analysis is also applied on the ProfileDroid framework, proposed in Reference 28, and in CopperDroid, presented in Reference 29. Data-Sluice, on the other hand, is able to block the malicious behavior at runtime. Though the analysis of Data-Sluice is limited to the spyware class, it brings two advantages with respect to static analysis: (i) it does not require the overhead brought by code or binary analysis and (ii) it allows the user to benefit from the trojanized app functionalities, removing the effect of the malicious code. A framework for enforcement of security policies in mobile devices, both deterministic and probabilistic, is the Security-By-Contract (SxC) [30,31]. The framework works in the hypothesis that an application is shipped with a contract describing the security-relevant actions performed by the app. Such a contract is matched with one or more security policies, having a dynamic enforcement applied on those apps whose contract does not match the specific policy. Data-Sluice has the advantage of not requiring contracts to be generated, whose procedure might require a noticeable amount of resources [32]; still, the enforced policies are less general.

Finally, some Android apps have been developed in the attempt to give to the user a greater control on generated data. *My Data Manager* [33] is an app that aims at controlling the amount of data generated by an Android app, in order to help the user save money. However, differently from Data-Sluice, the app does not exploit any mechanism to enforce data control; instead, it sends alerts to the device user, in order to spot apps that are generating noticeable amount of traffic. The Android app *Data On/Off* [34] exploits the Java Reflection to completely disconnect the data (3G/4G) network. Data-Sluice is much more fine-grained, performing data filtering, instead of shutting down a radio interface. However, differently from Data-Sluice, Data On/Off does not have any specific requirement to be installed on a device.

8.8 Conclusion

Exploiting the Internet capability of mobile devices may impose direct and indirect costs to the users, which might also include serious privacy violations. In this chapter, we have shown that there exists a large number of apps that misuse the INTERNET permission, with different levels of gravity to steal maliciously information from the user, profiling its habit with the objective of generating aimed advertisement, negatively altering the usage experience of both apps and Android itself, and/or generating high overhead that impose a considerable cost on the user in terms of wasted traffic and battery life. As a solution to detect and prevent these unwanted or malicious behaviors, we have proposed Data-Sluice, a framework for fine-grained traffic control, which performs a per-app control of incoming and outgoing data. The framework has been tested against a set of nine families of spyware and a set of popular apps showing unwanted behaviors typical of the grayware and adware categories discussed in this chapter.

Acknowledgments

This research has been partially supported by H2020 EU funded project NeCS GA #675320, H2020 EU funded project C3ISP GA #700294, and EIT Digital MCloudDaaS.

References

1. Global mobile statistics 2014 Part a: Mobile subscribers; handset market share; mobile operators, http://mobiforge.com/research-analysis/, 2014.
2. Y. Zhou and X. Jiang, Dissecting Android malware: Characterization and evolution, in *Proceedings of the 2012 IEEE Symposium on Security and Privacy*, SP'12, IEEE Computer Society, Washington, DC, USA, pp. 95–109, 2012.
3. A. Pathak, C. Hu, and M. Zhan, Where is the energy spent inside my app? Fine grained energy accounting on smartphones with eprof, in *Proceedings of EuroSys: European Conference on Computer Systems*, ACM, 2013.
4. A. Saracino, F. Martinelli, G. Alboreto, and G. Dini, Data-Sluice: Fine-grained traffic control for Android application, in *2016 IEEE Symposium on Computers and Communication (ISCC)*, IEEE, Messina, Italy, pp. 702–709, 2016.
5. Grayware: Casting a shadow over the mobile software marketplace, http://www.symantec.com/connect/blogs/grayware-casting-shadow-over-mobile-software-marketplace, 2014.
6. Forbes readers served malicious ads after asking them to disable adblocker, http://www.trendmicro.com/vinfo/us/security/news/cybercrime-and-digital-threats/forbes-readers-served-malicious-ads-after-asking-them-to-disable-adblocker, 2016.
7. When online ads attack, http://www.trendmicro.com/vinfo/us/security/news/cybercrime-and-digital-threats/malvertising-when-online-ads-attack, 2015.

8. A. Saracino, D. Sgandurra, G. Dini, and F. Martinelli, Madam: Effective and efficient behavior-based Android malware detection and prevention, *IEEE Transactions on Dependable and Secure Computing*, (99): 1–1, 2016.

9. A. Porter Felt, E. Ha, S. Egelman, A. Haney, E. Chin, and D. Wagner, Android permissions: User attention, comprehension, and behavior, in *Proceedings of the Eighth Symposium on Usable Privacy and Security*, SOUPS'12, ACM, New York, NY, USA, pp. 3:1–3:14, 2012.

10. G. Dini, F. Martinelli, I. Matteucci, M. Petrocchi, A. Saracino, and D. Sgandurra, Evaluating the trust of Android applications through an adaptive and distributed multi-criteria approach, in *12th IEEE International Conference on Trust, Security and Privacy in Computing and Communications, TrustCom 2013/11th IEEE*, Melbourne, Australia, July 16–18, 2013, IEEE Computer Society, 2013.

11. G. Dini, F. Martinelli, A. Saracino, and D. Sgandurra, MADAM: A multi-level anomaly detector for Android malware, in *Computer Network Security—6th International Conference on Mathematical Methods, Models and Architectures for Computer Network Security, MMM-ACNS 2012*, St. Petersburg, Russia, October 17–19, 2012. Proceedings, pp. 240–253, 2012.

12. W. Enck, P. Gilbert, B.-G. Chun, L. P. Cox, J. Jung, P. McDaniel, and A. N. Sheth, Taintdroid: An information flow tracking system for real-time privacy monitoring on smartphones, *Communications of the ACM*, **57**(3): 99–106, 2014.

13. Y. Zhou, X. Zhang, X. Jiang, and V. W. Freeh, Taming information-stealing smartphone applications (on Android), in *4th International Conference on Trust and Trustworthy Computing (TRUST 2011)*, Springer, Pittsburgh PA, USA, pp. 93–107, 2011.

14. Googleplay stats, http://www.appbrain.com/stats, 2016.

15. X. Jiang, Security alert: New DroidKungFu variants found in alternative Chinese Android markets, http://www.cs.ncsu.edu/faculty/jiang/DroidKungFu2/, 2012.

16. T. Wyatt and T. Strazzere, Geinimi trojan technical teardown, https://blog.lookout.com/media/Geinimi_Trojan_Teardown.pdf, 2011.

17. B. Wolfe, K. O. Elish, and D. Yao, Comprehensive behavior profiling for proactive Android malware detection, in S. S. M. Chow, J. Camenisch, L. C. K. Hui, and S. M. Yiu, editors, *Information Security, vol 8783 of Lecture Notes in Computer Science*, Springer International Publishing, Hong Kong, pp. 328–344, 2014.

18. X. He, P. P. C. Lee, L. Pan, C. He, and J. C. S. Lui, A panoramic view of 3G data/control-plane traffic: Mobile device perspective, *NETWORKING 2012: 11th International IFIP TC 6 Networking Conference*, Prague, Czech Republic, May 21–25, 2012, Proceedings, Part I, Springer, Berlin, pp. 318–330, 2012.

19. N. Dragoni, F. Martinelli, F. Massacci, P. Mori, C. Schaefer, T. Walter, and E. Vetillard, Security-by-contract (sxc) for software and services of mobile systems, *At Your Service-Service-Oriented Computing from an EU Perspective*, 429–455, 2008.

20. S. Zonouz, A. Houmansadr, R. Berthier, N. Borisov, and W. Sanders, Secloud: A cloud-based comprehensive and lightweight security solution for smartphones, *Computers and Security*, **37**: 215–227, 2013.

21. K. Ariyapala, H. Giang Do, H. Ngoc Anh, W. Keong Ng, and M. Conti, A host and network based intrusion detection for Android smartphones, in *2016 30th International Conference on Advanced Information Networking and Applications Workshops (WAINA)*, IEEE, Crans Montana, Switzerland, pp. 849–854, 2016.

22. D. Damopoulos, G. Kambourakis, S. Gritzalis, and S. O. Park, Exposing mobile malware from the inside (or what is your mobile app really doing?), *Peer-to-Peer Networking and Applications*, **7**(4): 687–697, 2014.

23. D. Damopoulos, G. Kambourakis, and G. Portokalidis, The best of both worlds: A framework for the synergistic operation of host and cloud anomaly-based ids for smartphones, in *Proceedings of the Seventh European Workshop on System Security*, EuroSec'14, ACM, New York, NY, USA, pp. 6:1–6:6, 2014.

24. D. Damopoulos, S. A. Menesidou, G. Kambourakis, M. Papadaki, N. Clarke, and S. Gritzalis, Evaluation of anomaly-based ids for mobile devices using machine learning classifiers, *Security and Communication Networks*, **5**(1): 3–14, 2012.

25. J. Rubin, M. Gordon, N. Nguyen, and M. Rinard, Covert communication in mobile applications, in *Proceedings of the 30th IEEE/ACM International Conference on Automated Software Engineering*, IEEE, Long Beach, CA, USA, 2015.

26. Kindsight Security Labs, Kindsight security labs malware report H1 2014, http://resources.alcatel-lucent.com/?cid = 180437, 2014.

27. Y. Aafer, W. Du, and H. Yin, Droidapiminer: Mining api-level features for robust malware detection in Android, in T. Zia, A. Zomaya, V. Varadharajan, and M. Mao, editors, *Security and Privacy in Communication Networks, Volume 127 of Lecture Notes of the Institute for Computer Sciences, Social Informatics and Telecommunications Engineering*, Springer International Publishing, Sydney, New South Wales, Australia, pp. 86–103, 2013.

28. X. Wei, L. Gomez, I. Neamtiu, and M. Faloutsos, Profiledroid: Multi-layer profiling of Android applications, in *Proceedings of the 18th Annual International Conference on Mobile Computing and Networking*, Mobicom'12, ACM, New York, NY, USA, pp. 137–148, 2012.

29. A. Reina, A. Fattori, and L. Cavallaro, A system call-centric analysis and stimulation technique to automatically reconstruct Android malware behaviors, *EuroSec*, ACM, April 2013.

30. G. Costa, N. Dragoni, A. Lazouski, F. Martinelli, F. Massacci, and I. Matteucci, Extending security-by-contract with quantitative trust on mobile devices, in *2010 International Conference on Complex, Intelligent and Software Intensive Systems (CISIS)*, IEEE, Krakow, Poland, pp. 872–877, February 2010.

31. G. Dini, F. Martinelli, I. Matteucci, A. Saracino, and D. Sgandurra, Introducing probabilities in contract-based approaches for mobile application security, in *Data Privacy Management and Autonomous Spontaneous Security—8th International Workshop, DPM 2013, and 6th International Workshop, SETOP 2013*, Egham, UK, September 12–13, 2013, Revised Selected Papers, pp. 284–299, 2013.

32. A. Aldini, F. Martinelli, A. Saracino, and D. Sgandurra, Detection of repackaged mobile applications through a collaborative approach, *Concurrency and Computation: Practice and Experience*, **27**(11): 2818–2838, 2015.

33. My Data Manager—data usage, https://play.google.com/store/apps/details?id = com.mobidia.android.mdm, 2016.

34. Data On/Off, https://play.google.com/store/apps/details?id = it.miabit.android.dataonoff, 2013.

Chapter 9

Data Leakage in Mobile Malware: The What, the Why, and the How

Corrado Aaron Visaggio, Gerardo Canfora,
Luigi Gentile, and Francesco Mercaldo

Contents

9.1 Introduction

According to the 2016 Internet Security Threat Report released by Symantec, the volume of new malware for mobile devices is dramatically growing: the number of Android variants increased by 40% in 2015, compared to 29% growth in the previous year.

The combination of rich sensors and ubiquitous connectivity makes smartphones the perfect candidates for privacy attacks. It is a consolidated habit of apps writers to write code for tracking users and leaking their personally identifiable information [1–4], while users are generally unaware and unable to block them [5,6]. In fact, the only real defense for sensitive data is the user that should deny those permissions that request the usage of sensitive data. However, Android architecture does not provide for a mechanism for signaling the user which app is using which data and if data are transmitted to a third party. By exploiting this mechanism and the scarce attention of users to this problem, many apps share sensitive user data with third parties [7], without warning or acknowledging the user about that. Dynamic loading makes the situation worse, as a fragment of code that steals sensitive information could be loaded and executed by an app after the scanning of an antimalware.

One of the main goals of the malware targeting mobile devices is to steal sensitive information, by sending it to a drop server or a remote controller. This kind of data is mined by cyber criminals for many reasons: identity theft, scams, phishing, and harassment.

Considering how easily and how frequently data are stolen by malware in Android devices, it is necessary to strengthen the mechanisms for protecting sensitive data. In order to design such mechanisms, it is necessary to understand in detail which are the methods used by malware for gathering sensitive data.

In this chapter, we analyze which are the processes and techniques used by malware for capturing sensitive data, which kinds of sensitive data are collected, which are the most common code patterns used, and where sensitive data are sent after having been gathered.

In order to realize this study, we analyzed the data leakage implemented by 4593 malware, belonging to 11 malware families, obtained by official datasets or repositories releasing malware samples. Furthermore, the study demonstrates how widespread among malwares are those functions related to data leakage.

Usually, two kinds of data leakage can be accomplished: one between applications and another consisting of the shipping of (sensitive) data exfiltrated from a target device to a third-party server or destination that is external to the target device. Our study examines only the latter case.

Evidence about data leakage is extracted by using three tools that are considered the state of the art for this kind of analysis: FlowDroid [1], Amandroid [5], and Epicc [8]. The three tools are able to collect complementary pieces of information: FlowDroid identifies the overall set of connections between the sources and the sinks involved in the data exfiltration; Amanandroid extends some features of FlowDroid, capturing the pattern of actions that lead to the data theft; and, finally, Epicc retrieves explicit and implicit intents that could be leveraged for extracting sensitive data.

The remainder of the chapter is organized as follows: Section 9.2 thoroughly analyzes related work, Section 9.3 introduces the tool chain used for the analysis, and Section 9.4 illustrates the results of the experiment. Finally, Section 9.5 draws the conclusions.

9.2 Related Work

The problem of detecting data leakage has been addressed in the literature from three main perspectives. The first deals with the code patterns that exfiltrate sensitive information and send it to a third party. The second is the data leakage between apps, which occurs when an app allows third-party apps to intercept its methods that handle sensitive data. The third perspective concerns the detection of malicious software that is able to steal sensitive data.

ScanDal [2] is a static analyzer performing both flow-sensitive and flow-insensitive analysis of an application for detecting privacy leaks in Android applications. To detect privacy leak, ScanDal collects information on where the value was created. When a value is created at an information source, ScanDal denotes the program counter of the source and sends it to the analysis. When a value is created from existing values, ScanDal denotes the union of all the sets of the program counters from existing values. By this, ScanDal can collect every value that could be created from information sources. If such values flow out through an information sink, ScanDal detects it and considers it as a privacy leak. The authors analyzed 90 popular applications from Android Market and detected privacy leaks in 11 applications. They also experimented eight known malicious applications from third-party markets and detected privacy leaks in all the eight applications.

TaintDroid [3] is an extension of the Android operating system that tracks the flow of privacy-sensitive data through third-party applications. TaintDroid assumes that downloaded, third-party applications are not trusted, and monitors—in real time—how these applications access and manipulate users' personal data. TaintDroid labels data from privacy-sensitive sources and transitively applies labels as sensitive data propagate through program variables, files, and interprocess messages. Using TaintDroid, the authors monitored the behavior of 30 popular third-party Android applications, finding 68 instances of potential misuse of users' private information across 20 applications.

IccTA [6] uses a static taint analysis technique to find privacy leaks, for example, paths from sensitive data, called sources, to statements sending the data outside the application or device, called sinks. A path may be within a single component or across multiple components. To verify their approach, the authors developed 22 apps containing intercomponent communication (ICC)-based privacy leaks. Another tool proposed by the scientific community is CHEX [4], which uses static analysis to detect component hijacking vulnerabilities in Android applications by tracking taints between sensitive sources and externally accessible interfaces. However, it is limited to at most one object, which leads to imprecision in practice.

PCLeaks [7] performs data flow analysis to detect potential component leaks, which detects both component hijacking vulnerabilities and component launch (or injection) vulnerabilities. ContentScope [9] is another tool that tackles potential component leaks. The authors exclusively focus on the open content provider interface of Android apps and study potential risks that may lead to passive privacy leakage and unintended manipulation of security-sensitive data.

Papamartzivanos and colleagues [10] provide a solution for real-time tracking of the privacy flow of a user. Furthermore, they develop a collaborative infrastructure for exchanging information related to apps' privacy exposure level, and a behavior-driven detection mechanism in an effort to take advantage of the crowdsourcing data to its maximum efficacy.

Amandroid [5] performs an ICC analysis to detect data leakage between apps, which has been developed concurrently with IccTA. Amandroid builds an inter-component data flow graph (ICDFG) and a data dependence graph (DDG) to perform ICC analysis. Amandroid provides a general framework to enable analysts to build a customized analysis on likely data leakage between Android apps.

FlowDroid [1] detects those apps that permit data leakage. It helps to reduce the leaks or false-positives. Novel on-demand algorithms allow FlowDroid to maintain efficiency despite its strong context and object sensitivity.

Epicc [8] identifies a specification for every ICC source and sink. This includes the location of the ICC entry point or exit point, the ICC intent action, data type and category, as well as the ICC intent key/value types and the target component name. Note that where ICC values are not fixed, Epicc infers all the possible ICC values, thereby building a complete specification of the possible ways ICC can be used. The specifications are recorded in a database as flows detected by matching compatible specifications.

ComDroid [11] is able to detect application communication vulnerabilities; differently from other approaches, it analyzes Dalvik bytecode. The tool is able to examine interapplication communication and present several classes of potential attacks on applications. Outgoing communication can put an application at risk of broadcast theft (including eavesdropping and denial of service), data theft, result modification, and activity and service hijacking. Incoming communication can put an application at risk of malicious activity and service launches and broadcast injection. The authors analyzed 20 applications and found 34 exploitable vulnerabilities; 12 out of the 20 applications have at least one vulnerability.

DidFail [12] leverages FlowDroid [1] and Epicc [8] to detect ICC leaks. Currently, it focuses on ICC leaks between activities through implicit intents. Thus, it will miss leaks involving explicit intents and components other than activities. Also, it does not handle some parameters for implicit intents (such as mimetype and data) and thus generates false links between components. The consequence of that is a higher false-positive rate.

SCanDroid [13] and SEFA [14] are two tools that perform ICC analysis. However, neither of them keeps the context between components and thus are less precise than IccTA by design. ComDroid [11] and Epicc [8] are two tools that tackle the ICC problem, but mainly focus on ICC vulnerabilities and do not taint data.

The authors of Reference 15 propose a framework including a set of criteria for evaluating security solutions for smartphones. Their study focuses on the assessment of security solutions, assessing the completeness and the quality of protection capabilities of these solutions. Our study analyzes how data leakage is performed on a wide dataset of Android malware.

Damopoulos et al. [16] developed an anomaly-based intrusion detection system tailored for malware detection and privacy-invasive software. The solution is supported by cloud-based technology for exploiting a greater computational power. This work focuses on the solution while our work focuses on the analysis of data leakage in Android malware.

The authors in Reference 17 propose a cloud-based smartphone-specific intrusion detection and response engine, which continuously performs an in-depth forensics analysis on the smartphone to detect any misbehavior. The solution is designed specifically for intrusion detection.

Andromaly [18] is a framework for detecting malware on Android mobile devices, through a host-based malware detection system that continuously monitors different events and features. This solution uses a classifier in order to determine whether an application is malicious or not. The focus of this chapter is not specifically on data leakage but on malware that can also steal sensitive data.

9.3 Tool Chain

To fully characterize the data leakage phenomenon in the Android environment, we adopt the tool chain depicted in Figure 9.1.

The tool chain employed in our analysis is composed of the following open-source software: FlowDroid [1], Amandroid [5], and Epicc [8].

First of all, we submit each application under analysis to FlowDroid tool in order to understand *what* is the data leakage performed in the device, that is, we analyze the kind of data that is stolen from the device and the communication channel used to send the data.

To discover the reason *why* the mobile application can send information to a third-party server, we employ Amandroid to deeply discover the implicit intents mechanism used by the application under analysis. Finally, to analyze *how* an application sends the stolen information outside the device, Epicc tool identifies the events that typically activate the data exfiltration in the Android environment.

In this section, we provide a high-level description of each of the three tools.

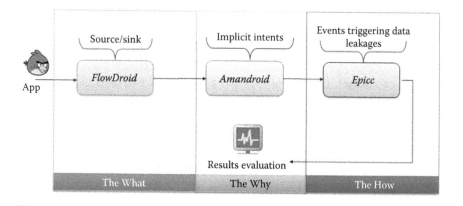

Figure 9.1 The tool chain used in the study. We submit the application under analysis to FlowDroid tool, in order to discover the data leakage patterns (the *what*). To understand *why* it is possible that user personal information is sent to third-parties servers, the implicit intents that mobile attackers are able to exploit are intercepted by Amandroid. Finally, by using the Epicc tool, we know *how* the malicious payloads are able to steal personal information.

9.3.1 FlowDroid

FlowDroid is a tool designed for the Android platform able to analyze contaminated paths within the application, that is, the paths that are able to produce a data exfiltration. This analyzer identifies the full set of the possible connections between the sources and the sinks crossed during a data exfiltration.

In addition, it has the ability to model the full life cycle of an Android application, with particular regard to the management of callback methods and internal interfaces.

Figure 9.2 shows the source code of an activity, which is basically the user interface of an Android application. Whenever the activity is called via the onRestart() method, the application reads a password entered using a text field (line 5). When the user clicks on one of the activity buttons, invoking the callback method sendMessage(), the user removes the password and sends it by SMS (line 24). Therefore, this flow between the password entered by the user and the sent SMS constitutes the data stream contaminated, causing the data loss (the lost data are represented by the password typed by the user). Android developers can explicitly define the callback method in the Java file of the activity or implicitly in the XML layout file (see the code snippet in Figure 9.2): FlowDroid analyzes the source code and processes accurately the associated metadata with the callback methods, that is, the XML layout files. In addition, the information is actually lost, only if the onRestart() method is called before the invocation of the sendMessage() method; the analysis

```
 1 public class LeakageApp extends Activity{
 2 private User user = null;
 3 protected void onRestart(){
 4    EditText usernameText =
        (EditText)findViewById(R.id.username);
 5    EditText passwordText =
        (EditText)findViewById(R.id.pwdString);
 6    String uname = usernameText.toString();
 7    String pwd = passwordText.toString();
 8    if(!uname.isEmpty() && !pwd.isEmpty())
 9       this.user = new User(uname, pwd);
10 }
11 //Callback method in xml file
12 public void sendMessage(View view){
13    if(user == null) return;
14    Password pwd = user.getpwd();
15    String pwdString = pwd.getPassword();
16    String obfPwd = "";
17    //must track primitives:
18    for(char c : pwdString.toCharArray())
19       obfPwd += c + "_"; //String concat.
20
21    String message = "User: " +
22       user.getName() + " | Pwd: " + obfPwd;
23    SmsManager sms = SmsManager.getDefault();
24    sms.sendTextMessage("+44 020 7321 0905",
25       null, message, null, null);
26 }
```

Figure 9.2   A code snippet explaining data leakage in Android.

performed by FlowDroid is able to understand this order, which is defined flow-sensitive.

Unlike Java programs, Android applications do not have the main method. Android applications can contain multiple components, each one characterized by its own life cycle: the information about the component that will be launched when the application is started is provided by the application's manifest file. For this reason, the static analysis requires more effort, as it has multiple entry points from which to start the analysis. In addition, callback methods have to be considered in the analysis, because they permit to record various information and to interact with the user interface and the activity. Also, the order in which these methods are invoked is important, because it cannot be determined *a priori* (because, as previously explained, an Android application does not have the main method). However, the callback methods can only be accessed when the corresponding component is running. To address this problem, FlowDroid bases its analysis on the use of the Interprocedural, Finite, Distributive, Subset framework (IFDS) [19], which emulates the life cycle of the components and the callback methods. The IFDS creates a call-graph to analyze the application components starting from the methods that characterize the life cycle (i.e., `onCreate()`, `onStop()`) implemented in the respective component classes. This graph is then used to scan the calls to the callback methods and gradually extended to include the discovered callbacks, until it reaches a termination point. Once the dummy main method is built, FlowDroid calculates a call-graph using this method as the entry point. This method, although expensive, provides a precise mapping between the components and the callback methods, reducing not only the probability of the detection of false-positives, but also the runtime analysis. This method will take account only of the life cycle of the components and their respective callback methods that can actually occur during the execution and that are defined in the XML configuration file of the application (i.e., the manifest file).

9.3.2 Amandroid

Amandroid is able to analyze the flow of data between the platform-specific Android components, by using a flow approach, object- and context-sensitive, performing a static analysis of applications. It extends some FlowDroid's features, emphasizing the capability to capture the dependencies between the control and the component data.

Furthermore, in order to manage the control between the components and the flow of data, the tool addresses the security problems that arise from the interactions among multiple components of the same application or between components of different applications through the construction of an interprocedural graph (ICFG). It is also able to acquire the data flow between the respective components crossed, building an intercomponent data graph (ICDG) used to obtain the DDG, that is, the graph of the data dependency.

Owing to the complexity related to the manipulation of the intercomponent, a static analyzer needs a model of the Android system to track the invocation of the component lifecycle's methods. The Amandroid model of the Android environment is inspired by FlowDroid [1], which uses a "dummy main" method to capture all the possible sequences of the methods' invocations as permitted by Android. The Amandroid model also extends the FlowDroid's model by capturing the control and data dependencies among components.

Amandroid decompresses the apk file of the app to analyze and retrieve the dex file, converting it into an IR format for subsequent analysis. The *dex2IR* translator is a modification of the original *dexdump* tool shipped with the Android platform tool set; the C++ source of the original *dexdump* is available in the Android build package, while the authors modified it so that it can also produce the app representation in the IR format.

The environment model is generated in this way in order to emulate the interaction of the Android system with the application; thus, it builds the graph of the data flow between the components of the whole application (IDFG). Finally, it includes the control flow graph that covers all the accessible components. By this way, it keeps track of all the created objects (even dynamic ones) that flow at any point of the program from the moment they are created, possibly modified, until their termination point. Amandroid creates an additional graph called DDG, which suggests explicit information in the flow.

The DDG and the IDFG can be applied to various security analysis: data leak detection, data injection detection, API's misuse detection, and so on.

The environment model employed by Amandroid extends in many points the FlowDroid one. As a matter of fact Amandroid, rather than using a model of the entire application (app level) as FlowDroid, uses an environment model for each component belonging to the application (component level).

Unlike the app-level model, the model used by Amandroid is more effective to capture the impact of the Android system on the control of data. The point is that each component has its own model that invokes the callback methods.

To obtain this environment model, Amandroid starts to collect the basic information from the content layout files in the resource folder, to be able to collect the callback methods, and then generates the E_c body containing the methods of the life cycle of the component C; finally, it collects the other callback methods in C, through an incremental reachability analysis, following the approach used by Flow-Droid. These operations are carried out before carrying out the analysis of the data flow (IDFG).

9.3.3 Epicc

Epicc aims at detecting the vulnerabilities within Android applications.

This approach is able to provide a high precision considering that it investigates how the components interact by solving the parameters of the ICC calls; more precisely, Epicc is able to retrieve explicit and implicit intents with the corresponding receiver; in this way, it is possible to deduce the possible sender–receiver couples (with the exchanged data) of the intents declared by an Android application.

Unlike FlowDroid and Amandroid, Epicc does not perform a taint analysis, that is, it does not analyze the flow of data between the components and consequently it is unable to find the paths in which the actual leakage of sensitive data occurs.

9.3.3.1 ICC Vulnerabilities

Android provides a communication system for exchanging data between the components of the same application, or with different applications, based on the intents.

These messages can be sent explicitly by specifying the name of the target component or they may be implicit by specifying the required action: in that case, the system will identify the recipient.

Android then determines those intents that must be delivered to the component by matching the name of the target component (in the case of explicit intent) or action, a category and additional data in the case of implicit intent.

Moreover, several applications can set the same intent filter and then be candidates to receive the same message; in that case, Android will show a window with the candidate applications and the user will choose the application to perform.

In addition, the operating system can also set priorities for all those applications that handle the same type of intent filter, thus making the implicit message to be sent to the application with the highest priority.

These types of implicit intent do not offer any guarantee that the message has been delivered to the appropriate recipient, and then the malicious app can safely intercept an implicit intent by simply declaring an intent filter for all the actions, categories, and data of that intent.

9.3.3.2 Epicc Analysis Model

Epicc is focused on the connection between the several components that constitute an Android application, and examines the component's communication in the same application and the component's communication among multiple applications.

For each input of an application A, Epicc returns the following results:

■ A list of the links formed by the A entry points that can be called by components of A or from external applications
■ A list of the links formed by the A exit points, useful to A in order to send intents to another component of A or to another application
■ A list of links between the components of A and a list with the links with A and the components of external applications

To better explain how Epicc works, we consider the code snippet shown in Figure 9.3 belonging to a banking application.

In Figure 9.3, an exit point is represented by the `startActivity()` method, from which we obtain the result for the *i* destination and for all the other destinations.

The Epicc analysis comprises the following steps:

1. Given an input application, Epicc decompiles it and extracts the manifest file and information about the packages, the permission, the intent filter, and a list of associated components.
2. Epicc performs a matching between the entry points and the exit points, obtaining the so-called ICC connections.

```
1  public void onClick(View v) {
2     Intent i = new Intent();
3     i.putExtra("Balance", this.mBalance);
4     if (this.mCondition) {
5        i.setClassName("a.b",
              "a.b.MyClass");
6     } else {
7        i.setAction("a.b.ACTION");
8        i.addCategory("a.b.CATEGORY");
9        i = modifyIntent(i);
10    }
11    startActivity(i); }
12
13 public Intent modifyIntent(Intent in) {
14    Intent intent = new Intent(in);
15    intent.setAction("a.b.NEW_ACTION");
16    intent.addCategory("a.b.NEW_CATEGORY");
17    return intent; }
```

Figure 9.3   An example of intent-based communication.

3. The obtained objects are stored into a DBMS.
4. It then proceeds with the string analysis, that is, the phase through which it is possible to identify, for example, the names of the components or the values of the arguments contained in the various library functions.
5. Through the values obtained, the tool invokes the interprocedural distributive environment (IDE) analysis, which is able to compute the values of intent used and the methods of the ICC calls. It also computes the values of the intent filters that receive intents through broadcast receivers dynamically registered.
6. The exit points are matched with the entry points previously computed.
7. The exit points are stored into the DBMS.
8. The values dynamically associated to the broadcast receivers are matched to the exit points.
9. The entry points are stored into the DBMS.

9.4 Experiment

In this section, we discuss the result of the experiment we performed to discover the patterns of data leakage in Android malware.

**Table 9.1 Malware Families
Involved in the Experiment with the
Corresponding Number of Samples**

Family	Number of Samples
FakeInstaller	925
Ransomware	683
DroidKungFu	667
Plankton	625
OpFake	613
GinMaster	339
BaseBridge	330
Kmin	147
Geinimi	92
Adrd	91
DroidDream	81

9.4.1 Dataset

The dataset includes 4593 real-world samples gathered from the Drebin project's dataset [20,21], and 672 samples of ransomware. The Drebin project's dataset [20,21] is a very well-known collection of malware used in many scientific works, which includes the most diffused Android families. Ransomware is a malware that impedes the access to smartphone resources and demands a payment for restoring the functionality and the resources. The ransomware real-world samples examined in the experiment were gathered from a freely available collection[*] for research purposes. The samples are labeled as ransomware, koler, locker, fbilocker, and scarepackage [22] and appeared from December 2014 to June 2015.

The malware dataset is also partitioned according to the *malware family*: each family contains samples that have in common several characteristics, such as payload installation, the kind of attack, and the events that trigger malicious payload [23].

We submitted the full dataset to confirm the maliciousness to the VirusTotal service.[†] VirusTotal provides a public API to submit and scan applications, to access finished scan reports of 57 different antimalwares.

Table 9.1 shows the number of malware samples with the family they belong to.

---

[*] http://ransom.mobi/
[†] https://www.virustotal.com

9.4.2 Patterns of Data Leakage

In this section, we present the results of the evaluation and we discuss the data leakage patterns extracted from the analyzed malware.

Table 9.2 shows the sources retrieved by FlowDroid, that is, the methods used by the malware to gather information, and provides how many times each method occurs in the samples of each malware family.

We explain in the following the kind of information retrieved by the involved methods:

- `getDeviceId`: It returns the unique device ID, for example, the IMEI for GSM and the MEID or ESN for CDMA phones.
- `getLongitude`: Gets the longitude, in degrees.
- `getLatitude`: Gets the latitude, in degrees.
- `getCountry`: It returns the country/region code for this locale, which should be an empty string.

Table 9.2 Cumulative Occurrences of the Source Methods Retrieved by FlowDroid on the Overall Dataset Ordered by Malware Family

	Adrd	*BaseBridge*	*DroidDream*	*DroidKungFu*	*FakeInstaller*	*Geinimi*	*GinMaster*	*Kmin*	*Opfake*	*Plankton*	*Ransomware*
getDeviceId	204			1299	330		11,251	340		1299	602
getLongitude	29	32	96	524		92	473			524	1
getLatitude	29	32	96	524		92	473			524	1
getCountry	6	26	24	187		14	91			187	6
getLastKnownLocation	13	9	3	269		71				269	1
getSubscriberId	189	7	4	946	48	72	10,691	422	1	946	1
getSimSerialNumber	89				4		10,604	72		661	
getInstalledPackages	5	7		98			109			98	
getInstalledApplications	7		36								
getLine1Number	58	2		825	465	2	10,615			825	48
getCid		2			1					1	
getLac		2			1					1	

- `getLastKnownLocation`: It returns a location indicating the data from the last known location fix obtained from the given provider.
- `getSubscriberId`: It returns the unique subscriber ID, for example, the IMSI for a GSM phone.
- `getSimSerialNumber`: It returns the serial number of the SIM, if applicable.
- `getInstalledPackages`: It returns a list of all the packages that are installed on the device.
- `getInstalledApplications`: It returns a list of all the application packages that are installed on the device.
- `getLine1Number`: It returns the phone number string for line 1, for example, the MSISDN for a GSM phone.
- `getCid`: This method returns the cell tower location. The cell site or cell tower is a cellular telephone site where antennae and electronic communications equipment are placed, usually on a radio mast, tower, or other high place, to create a cell (or adjacent cells) in a cellular network.
- `getLac`: It returns the location area code (LAC). The served area of a cellular radio network is usually divided into location areas. Location areas comprise of one or several radio cells. Each location area is given a unique number within the network, that is, the LAC. This code is used as a unique reference for the location of a mobile subscriber. This code is necessary to address the subscriber in the case of an incoming call.

As shown in Table 9.2, all the analyzed malware families retrieve the subscriber ID (i.e., the IMSI) using the `getSubscriberId` method. The methods usually employed by malware writers to gather the information about the localization are the `getLongitude` and `getLatitude` methods, while the `getCid` and `getLac` methods, which retrieve localization using gsm cell location and location area code, are less employed from malware writers, because compared to GPS precision, they are less accurate.

Table 9.3 shows the sink categories retrieved by FlowDroid, that is, the channel used to send the gathered information.

We explain in detail each sink category retrieved:

- *Log*: This category represents the Android API for creating logs and sending them outside the device. This category can be used to debug the application but also by malware writers to gather information. Error, warning, and info logs are always kept. Malware writers are typically interested by the info logs of the application.
- *HTTP*: This category represents the URL connection with support for HTTP-specific features. It is the preferred method to write personal information into a socket used by malware. The HTTP channel is usually used to communicate with a command and control server, to send the personal

Table 9.3 Sink Categories Retrieved by FlowDroid

	Adrd	*BaseBridge*	*DroidDream*	*DroidKungFu*	*FakeInstaller*	*Geinimi*	*GinMaster*	*Kmin*	*Opfake*	*Plankton*	*Ransomware*
Log	677	254	216	2410	170	270	4344	647	3	2410	143
HTTP	429	400	347	5134	882	685	5482	1876	1	5134	1056
SharedPreferences	211	1333	53	2716	1404	102	7039	867		2716	3319
File	108	426	93	311	77	3	270	72		311	304
Media	12				2	71					600
SMS	4				12	1387			72	61	12	13

information to attackers and/or to third-party servers, but also to download at runtime the malicious payload [24]. The classes that are involved in this category are: `DefaultHttpClient`, `BasicNameValuePair`, `URL`, `URLConnection`, `HttpClient`, `OutputStream`, and `Write`.

■ *SharedPreferences*: Android provides many ways for storing the data of an application. One of these is *SharedPreferences*. *SharedPreferences* permits to save and retrieve data in the form of a (key,value) pair. *SharedPreferences* are stored as a file in the file system of the device. They are, by default, stored within the app's data directory, and only the UID associated to the specific running application has the permission to access them. The class belonging to this category is `SharedPreferences.Editor` and is invoked by the following methods used to insert information in the SharedPreferences: `putFloat`, `putInt`, `putBoolean`, `putLong`, and `putString`.

■ *File*: This category represents the storage on a file. Android uses a file system that is similar to disk-based file systems on other platforms. All the Android devices have two file storage areas: "internal" and "external" storage. These names come from the early days of Android, when most devices offered built-in nonvolatile memory (internal storage), plus a removable storage medium such as a micro SD card (external storage). The main difference between internal and external storage is that the second one is world-readable, so files saved here may be read outside of the owner's control, while, by using the internal one, files saved are accessible only by the app itself.

■ *Media*: This category is referred to all the media generated by the application, for instance, the `onPictureTaken()` method from

`Camera.PictureCallback` class, which is called when an image is available after a picture is taken.

- *SMS*: SMSs are used by malware to send messages to premium rate numbers without the user's involvement. Malware also captures the user's banking information such as account number and password [23]. Malware also uses SMSs in order to communicate with the C&C server and/or to send SMSs with the malicious links to propagate the infection. The class involved in this category is SMSManager with the invocation of the following methods related to SMS sending: `sendTextMessage`, `sendMultipartTextMessage`, and `sendDataMessage`.

Table 9.4 shows the URLs retrieved by Amandroid with the corresponding number of implicit intents.

Amandroid found malicious URLs in Adrd, BaseBridge, DroidDream, Fake-Installer, Geinimi, GinMaster, and Kmin families. DroidDream is the family that makes an intensive use of malicious URLs with the corresponding implicit intents.

Table 9.4 URLs Retrieved by Amandroid with the Corresponding Number of Implicit Intents (#II) and the Family They Belong to

Family	URL	#II
Adrd	http://www.coolcode.org/android/Download_Service.apk htttp://www.10086apk.com	16 3
BaseBridge	http://www.androidlicenser.com/store_fronts/3/buy	10
DroidDream	http://pay.sztone.com/czwap/r.aspx http://market.android.com/search http://pay.sztone.com/billing/billing.aspx http://www.opda.com.cn http://www.kfkx.net/AndroidOptimizer/Weibo2.0.4-2464_ 0001.apk http://wp.me/pP0KO-f/ http://market.android.com/details http://www.google.com.hk/m/search	28 8 8 4 4 4 3 2
FakeInstaller	http://yandex.ru	97
Geinimi	http://www.dseffects.com/android/games/MonkeyJump2/ hi.php http://www.dseffects.com/android.php	6 9
GinMaster	http://market.android.com/searh http://www.netmite.com/android/andme_signed.apk http://www.amoneron.com/slugs/scoretable.php	31 16 11
Kmin	http://www.5j5l.com/ThemeDowner/91pandahome2.apk	534

The malicious payload, as explained in Reference 23, can be installed in different ways into a legitimate application:

■ *Repackaged*: With repackaging, the malicious payload is embedded into the application at installation time: the attacker decompiles a trusted application to obtain the source code, and then adds the malicious payload and recompiles the application.

■ *Update attack*: An apparently innocuous application is installed on the victim's device. The user is asked to update the application, which consists of downloading the malicious payload on the victim's device; thus the user has installed an app that does not exhibit any harmful behavior. With this technique, the malicious payload is not embedded into the application at installation time.

■ *Drive-by-download*: With this technique, the user is asked to download an add-on by clicking on a URL or by scanning a QR code: the add-on represents the malicious payload that will be embedded into the legitimate application.

■ *Rogueware*: It is a form of malicious software and Internet fraud that misleads the user into believing there is a malware on the device, and manipulates the user into paying money for a fake malware removal tool (that usually actually introduces malware into the computer). It is a form of scareware that manipulates users through fear.

Once installed, the malicious payload is triggered by a set of events, as demonstrated in Reference 23; the most used events able to activate malicious actions are

■ *BOOT*: Most of malware payloads is launched when the boot is completed (BOOT_COMPLETED event), activating a background service that does not require user interaction.

■ *SMS*: The SMS_RECEIVED event is transmitted to the system when a new SMS messaged is received. With this event, the malware has the ability to respond to specific incoming SMS messages to undertake malicious actions.

■ *NET*: The CONNECTIVITY_CHANGE event is transmitted when a change in the data connection occurs, for instance, when the connection switches from GPRS to HSDPA.

■ *BATT*: Within this malware feature, we group together a set of events related to battery consumption: ACTION_POWER_CONNECTED (i.e., the device is connected to the power), ACTION_POWER_DISCONNECTED (i.e., the device is disconnected from the power), BATTERY_LOW (i.e., low battery), BATTERY_OKAY (i.e., the battery is now okay after being low), and BATTERY_CHANGED_ACTION (broadcast containing the charging state, level, and other information about the battery).

■ *SYS*: With this malware feature, we refer to many system events: USER_PRESENT (useful to recognize when the phone has been unlocked or not), INPUT_METHOD_CHANGED (an input method has been changed), SIG_STR (listening to signal strength when the phone sleeps), and SIM_FULL (the SIM storage for SMS messages is full).

■ *USB*: The malware is activated when the device is plugged/unplugged, by using the USB cable: it uses the UMS_CONNECTED event in order to know when the device is plugged and the UMS_DISCONNECTED to know when the device is unplugged.

■ *PHONE*: The malware responds to the READ_PHONE_STATE event: it allows to access the phone state, including the phone number of the device, the current cellular network information, the status of any ongoing calls, and a list of any phone account registered on the device.

■ *PKG*: This category comprises the following events: BROADCAST_PACKAGE_REMOVED, which allows an application to broadcast a notification about the removal of an application package; DELETE_PACKAGES, which allows an application to delete packages; GET_PACKAGE_SIZE, which allows an application to find out the space used by any package; INSTALL_PACKAGES, which allows an application to install packages, PACKAGE_USAGE_STATS, which allows an application to collect usage statistics, and REQUEST_INSTALL_PACKAGES, which allows an application to request installing packages.

■ *CLOUD*: This category represents the set of Google Cloud Messaging (GCM) events, which is a free service that enables developers to send messages between a server and a client app. This includes downstream messages from a server to a client app, and upstream messages from a client app to a server.

Table 9.5 shows the events to activate the malicious payload retrieved by Epicc.

Epicc tool, as shown in Table 9.5, identifies the BOOT events as the events used by all the families in the dataset to trigger the malicious behavior. Events used by most of the malware families are also the SMS event and the NET event. FakeInstaller is the only family that used the GMC events in order to communicate with the attackers and/or third-party servers.

9.5 Conclusion

In this chapter, we examined the most common code patterns and mechanisms that Android malware is used to adopt to obtain sensitive information from mobile devices. The analysis spans over about 5000 real-world Android malwares, belonging to the most diffused families, including recent ransomware samples.

It emerged that all the families involved in the experiment contain malicious payload that are able to exfiltrate personal information. Ordinarily, the information stolen are the IMEI, the phone number, and the GPS coordinates, while the

Table 9.5 Events to Activate the Malicious Payloads Retrieved by Epicc

Family	Installation				Activation								
	Repackaging	*Update Attack*	*Drive-by-Download*	*Rogueware*	*BOOT*	*SMS*	*NET*	*PHONE*	*USB*	*PKG*	*BATT*	*SYS*	*CLOUD*
Adrd	✓	✓			✓	✓	✓			✓			
BaseBridge	✓	✓			✓	✓	✓		✓		✓	✓	
DroidDream	✓		✓		✓	✓		✓		✓			
DroidKungFu	✓				✓	✓	✓	✓	✓	✓	✓	✓	
FakeInstaller	✓				✓	✓	✓	✓				✓	✓
Geinimi	✓				✓	✓							
GinMaster	✓		✓		✓		✓			✓		✓	
Kmin		✓			✓	✓						✓	
OpFake	✓				✓	✓	✓	✓				✓	
Plankton	✓	✓			✓			✓	✓				
Ransomware		✓	✓	✓	✓	✓	✓	✓				✓	

most used channels to send the private data are the HTTP connection and the SharedPreferences.

References

1. S. Arzt, S. Rasthofer, C. Fritz, E. Bodden, A. Bartel, J. Klein, Y. Le Traon, D. Octeau, and P. McDaniel, Flowdroid: Precise context, flow, field, object-sensitive and lifecycle-aware taint analysis for Android apps, *ACM SIGPLAN Notices*, 49(6): 259–269, 2014.
2. J. Kim, Y. Yoon, K. Yi, J. Shin, and S. Center, Scandal: Static analyzer for detecting privacy leaks in Android applications, In H. Chen, L. Koved, and D. S. Wallach (eds.), *MoST 2012: Mobile Security Technologies 2012*, IEEE, Los Alamitos, CA, USA, pp. 1–10, May 2012.
3. W. Enck, P. Gilbert, S. Han, V. Tendulkar, B.-G. Chun, L. P. Cox, J. Jung, P. McDaniel, and A. N. Sheth, Taintdroid: An information-flow tracking system for realtime privacy monitoring on smartphones, *ACM Transactions on Computer Systems (TOCS)*, 32(2): 5, 2014.

4. L. Lu, Z. Li, Z. Wu, W. Lee, and G. Jiang, Chex: Statically vetting Android apps for component hijacking vulnerabilities, in *Proceedings of the 2012 ACM Conference on Computer and Communications Security*, ACM, New York, NY, USA, pp. 229–240, 2012.

5. F. Wei, S. Roy, X. Ou, and R. Robby, Amandroid: A precise and general inter-component data flow analysis framework for security vetting of Android apps, in *Proceedings of the 2014 ACM SIGSAC Conference on Computer and Communications Security*, ACM, New York, NY, USA, pp. 1329–1341, 2014.

6. L. Li, A. Bartel, T. F. Bissyandé, J. Klein, Y. Le Traon, S. Arzt, S. Rasthofer, E. Bodden, D. Octeau, and P. McDaniel, IccTA: Detecting inter-component privacy leaks in Android apps, in *Proceedings of the 37th International Conference on Software Engineering-Volume 1*, IEEE Press, Los Alamitos, CA, USA, pp. 280–291, 2015.

7. L. Li, A. Bartel, J. Klein, and Y. Le Traon, Automatically exploiting potential component leaks in Android applications, in *2014 IEEE 13th International Conference on Trust, Security and Privacy in Computing and Communications*, IEEE, Los Alamitos, CA, USA, pp. 388–397, 2014.

8. D. Octeau, P. McDaniel, S. Jha, A. Bartel, E. Bodden, J. Klein, and Y. Le Traon, Effective inter-component communication mapping in Android: An essential step towards holistic security analysis, in *Presented as Part of the 22nd USENIX Security Symposium (USENIX Security 13)*, pp. 543–558, 2013.

9. Y. Z. X. Jiang, Detecting passive content leaks and pollution in Android applications, in *Proceedings of the 20th Network and Distributed System Security Symposium (NDSS)*, 2013.

10. D. Papamartzivanos, D. Damopoulos, and G. Kambourakis, A cloud-based architecture to crowdsource mobile app privacy leaks, in *Proceedings of the 18th Panhellenic Conference on Informatics*, PCI'14, ACM, New York, NY, USA, pp. 591–596, 2014.

11. E. Chin, A. P. Felt, K. Greenwood, and D. Wagner, Analyzing inter-application communication in Android, in *Proceedings of the 9th International Conference on Mobile Systems, Applications, and Services*, ACM, New York, NY, USA, pp. 239–252, 2011.

12. W. Klieber, L. Flynn, A. Bhosale, L. Jia, and L. Bauer, Android taint flow analysis for app sets, in *Proceedings of the 3rd ACM SIGPLAN International Workshop on the State of the Art in Java Program Analysis*, ACM, New York, NY, USA, pp. 1–6, 2014.

13. A. P. Fuchs, A. Chaudhuri, and J. S. Foster, Scandroid: Automated security certification of Android. https://www.cs.umd.edu/~avik/papers/scandroidascaa.pdf

14. L. Wu, M. Grace, Y. Zhou, C. Wu, and X. Jiang, The impact of vendor customizations on Android security, in *Proceedings of the 2013 ACM SIGSAC Conference on Computer & Communications Security*, ACM, New York, NY, USA, pp. 623–634, 2013.

15. A. Shabtai, L. Tenenboim-Chekina, D. Mimran, L. Rokach, B. Shapira, and Y. Elovici, Mobile malware detection through analysis of deviations in application network behavior, *Computers & Security*, 43: 1–18, 2014.

16. D. Damopoulos, G. Kambourakis, and G. Portokalidis, The best of both worlds: A framework for the synergistic operation of host and cloud anomaly-based ids for smartphones, in *Proceedings of the Seventh European Workshop on System Security*, EuroSec'14, ACM, New York, NY, USA, pp. 6:1–6:6, 2014.

17. A. Houmansadr, S. A. Zonouz, and R. Berthier, A cloud-based intrusion detection and response system for mobile phones, in *Proceedings of the 2011 IEEE/IFIP 41st International Conference on Dependable Systems and Networks Workshops*, DSNW'11, IEEE Computer Society, Washington, DC, USA, pp. 31–32, 2011.

18. A. Shabtai, U. Kanonov, Y. Elovici, C. Glezer, and Y. Weiss, "Andromaly": A behavioral malware detection framework for Android devices, *Journal of Intelligent Information Systems*, 38(1): 161–190, 2012.
19. T. Reps, S. Horwitz, and M. Sagiv, Precise interprocedural dataflow analysis via graph reachability, in *Proceedings of the 22nd ACM SIGPLAN-SIGACT Symposium on Principles of Programming Languages*, ACM, pp. 49–61, 1995.
20. D. Arp, M. Spreitzenbarth, M. Huebner, H. Gascon, and K. Rieck, Drebin: Efficient and explainable detection of Android malware in your pocket, in *Proceedings of 21th Annual Network and Distributed System Security Symposium (NDSS)*, IEEE, Los Alamitos, CA, USA, 2014.
21. M. Spreitzenbarth, F. Echtler, T. Schreck, F. C. Freling, and J. Hoffmann, Mobilesandbox: Looking deeper into Android applications, in *28th International ACM Symposium on Applied Computing (SAC)*, ACM, New York, NY, USA, 2013.
22. N. Andronio, S. Zanero, and F. Maggi, Heldroid: Dissecting and detecting mobile ransomware, in *International Workshop on Recent Advances in Intrusion Detection*, Springer, Cham, Swiss, pp. 382–404, 2015.
23. Y. Zhou and X. Jiang, Dissecting Android malware: Characterization and evolution, in *2012 IEEE Symposium on Security and Privacy*, IEEE, Los Alamitos, CA, USA, pp. 95–109, 2012.
24. F. Mercaldo, V. Nardone, A. Santone, and C. A. Visaggio, Download malware? No, thanks. How formal methods can block update attacks, in *Formal Methods in Software Engineering (FormaliSE), 2016 IEEE/ACM 4th FME Workshop on*, IEEE, New York, NY, USA, 2016.

MOBILE NETWORK SECURITY AND INTRUSION DETECTION

Chapter 10

Analysis of Mobile Botnets Using a Hybrid Experimental Platform

Apostolos Malatras, Andrea Ciardulli,
Ignacio Sanchez, Laurent Beslay, Thierry Benoist,
and Yannis Soupionis

Contents

10.1 Introduction

The emergence of mobile botnets as noteworthy security risks is an undeniable fact that is attested by the large number of this kind of attacks that have been witnessed recently. Lately, the convergence of traditional forms of computing with mobile computing has spurred the development of botnets targeted at mobile devices. Such devices have increasingly higher computational capabilities as well as being constantly connected and inherently tied to the user account of the owner and thus to his/her personal data. These features make mobile devices a very attractive candidate for a series of security threats, mobile botnets being one of them. Similar to the case of traditional botnets, malicious attackers can exploit mobile networks for a wide variety of purposes ranging from pure lucrative gains such as carrying out banking physing campaigns to more specific purposes such as disrupting services in a particular region or impacting a specific collective by disclosing personal data. The particularities of the mobile ecosystem provide new opportunities for exploitation by malicious entities, such as the case where a mobile botnet is used to bypass two-factor authentication in banking attacks, that is, the MisoSMS mobile botnet.[*]

Mobile botnets draw their inspiration from traditional ones, additionally incorporating elements to adhere to and exploit the special nature of mobile devices and their platforms. They therefore take advantage of security vulnerabilities of mobile operating systems, that is, users are rarely updating their phones to the latest

---

[*] https://www.fireeye.com/blog/threat-research/2013/12/misosms.html

version since phones are phased out quite frequently and the same stands for security patches. Other vulnerabilities refer to the fragmentation of mobile operating systems and their app-based extensibility that exacerbate security concerns. The coordination and management of a mobile botnet is performed by the botmasters, who are responsible for infecting mobile devices and thus recruiting as large as possible a population of bots.

Botnets have different business models behind their operation. The botmaster in most cases rents out the services of the mobile botnet to other entities and in doing so benefits financially. The entities who rent out a mobile botnet can use it for their own purposes, exploiting the infected devices for either financial reasons or other motivations as mentioned before. It is these entities that define the type of attack that the botnet will be used for, for example, distributed denial-of-service (DDoS) attack. In some cases, the botmasters might utilize the mobile botnet for their purposes and accordingly define the types of attack that will be launched.

Botnets are assembled by infecting mobile devices, namely, bots, in order to be able to execute code on them that will allow the botmasters to remotely control them. To this end, utilization is made of the services of a command and control (C&C) server, which is a server that is controlled by the botmaster. According to the type of attack that will be performed by the botnet, the C&C server issues the appropriate commands to the infected bots. These entities coupled with the appropriate communication channels collectively form the notion of a mobile botnet. There exists a plethora of features that can be utilized to build a taxonomy of botnets ranging from the infection vector to the detection and take-down of the botnet, which we review in this chapter, also describing in detail the various possible architectures and typical examples of mobile botnets attacks.

The research topic regarding mobile botnets is steadily growing; nonetheless, a number of challenging issues still remain in particular when considering their detection and take-down operations. Many hindering factors contribute to these challenges. Mobile botnets are inherently highly distributed and dynamic, whereas the intrinsic features of mobile devices, for example, the use of side channels for communications (NFC, Bluetooth, embedded sensors, etc.), dynamic IP addressing, and malware masquerading as legitimate apps, further complicate the landscape. Systematic research efforts are required to address such challenging issues. The lack of tools and methodologies to assist in proper and repeatable experimentation is a major shortcoming that does not allow to effectively replicate mobile botnet deployments in the lab in order to analyze them. Currently, there do not exist many efforts in this direction, which would greatly benefit the research community by empowering the study of the behavior of mobile botnets and accordingly promote solutions to counter their adverse effects.

In this respect, the focus of our research reported in this chapter is on building and disseminating an experimental hybrid platform to perform tests and trials of mobile botnet research works. We thus aim to promote systematic research efforts in the realm of mobile botnets, support the corresponding research community, and allow for the development of reliable and applicable solutions to address related security concerns. The platform that we propose comprises of a hybrid infrastructure in that it enables experiments to be run on both actual mobile devices as well as emulated ones. Such a design on the one hand facilitates flexibility and scaling of tests and experiments, while on the other hand it supports the monitoring of realistic device behavior in regard to the effect of mobile botnets. Furthermore, the platform promotes the rapid reconfiguration of experimental settings in order to examine the effect that various network, device, or other type of parameters might have on the operation of a mobile botnet. The platform can also be used for training purposes, whereby security researchers can set up and simulate mobile botnets in the lab and experiment on their detection and take-down operations in a controlled environment.

The hybrid experimental platform makes use of the open-source Cuckoo[*] Sandbox in order to perform malware analysis on the mobile devices. We present in the following the implementation and configuration details, as well as a detailed description of the underlying infrastructure that has been built and set up. It needs to be noted that extensibility and scalability have been at the forefront of our requirements and these extend to the infrastructure as well as to the platform itself. Moreover, we also discuss the management interface that we have developed that allows researchers to define, run, monitor the progress, and control their experiments at a high level. The platform allows for monitoring of all devices, real or emulated ones, throughout all the phases of the mobile botnets' operation and the users are presented with all relevant results and collected data, for example, network traffic, API calls, and memory dumps. Lastly, we report on the validation of the hybrid experimental platform by illustrating a set of showcase examples of mobile botnets experiments as executed using our proposed platform.

The structure of this chapter is as follows. In Section 10.2, we describe in depth the notion of mobile botnets and their particularities and examine different features of their functionality, while in Section 10.3, we review related work in regard to the testing and experimentation of mobile botnets, and examine them in relation to our work. In Section 10.4, we discuss our proposed taxonomy of mobile botnets' features that can assist in classifying them and hence promote their methodical analysis. The design of the hybrid experimental platform is presented in Section 10.5 based on a set of functional requirements. We also justify our design choices and provide details on the hybrid aspect of the platform and how it can be effectively supported. Implementation and setup of the platform is the focus of Section 10.6, where the infrastructure is elaborated upon. In addition, we describe the management interface for the hybrid experimental platform and how it can be used to define, run, monitor, and analyze experiments. In Section 10.7, we illustrate two examples of

---

[*] https://www.cuckoosandbox.org/

typical mobile botnet experiments as tested on our platform. Elements such as the configuration of the infrastructure and the parameterization of the platform will be highlighted. The chapter concludes in Section 10.8 with a summary and pointers to future work that we plan to undertake, while additionally we aim at pinpointing open challenges and issues in order to encourage further research.

10.2 Mobile Botnets

Botnets are composed of a, usually large, number of compromised machines that aim to perform certain activities based on the requirements set by the owner of the botnet, namely, the botmaster [1]. Botnets are generally used for malicious purposes by the botmaster in order to disrupt the operation of services or extract lucrative gains from unsuspected users. The communication between the botmaster and the compromised machines usually takes place over the Internet or another type of network, hence the definition of botnets as networks of bots [2]. A mobile botnet refers to a botnet that is targeted at and exploits mobile devices.

The emergence, growing popularity, and dispersion of mobile botnets is in line with the convergence of traditional computing paradigm to the mobile, ubiquitous one and it is spurred by the ever-pervasive nature of smart mobile devices with high computational and communication capabilities [3–5]. Mobile botnets exploit security vulnerabilities that exist in mobile OS and their component-based architecture that builds on extensible systems with the use of applications that exacerbate security concerns. Accordingly, this established paradigm shift toward ubiquitous computing gives ground to the recent proliferation of mobile as well as hybrid botnets [6].

Mobile as well as traditional botnets consist of the following fundamental components [7]:

- *Botmaster*: The owner/initiator of the botnet, who is in charge of defining the functionality of the botnet and its action and is the recipient of any relevant lucrative operations. The botmaster is usually defining the high-level nature of the attack (the functionality of the botnet), whereas this is translated to actual commands by the C&C server. Recruiting campaigns for new bots are among the objectives of the botmaster [8].
- *C&C server/infrastructure*: It is responsible for sending commands for controlling and contacting the bots to retrieve information. It has knowledge of the bots that are under its control and can communicate with them. The connections to the bots does not need to be always open, but can be activated when new commands are issued to the bots. The C&C server executes the functionality defined by the botmaster in the botnet. The C&C infrastructure can be owned by the botmaster or abused by him/her, for example, IRC server or online social network used to distribute commands to the bots [9].

- *Bots*: End-user devices that have been infected by the botnet malware and are thus susceptible to receive commands and controls from the botmaster, via the C&C server. We can distinguish between:
  - *Servant bots*: Servants forward commands received by the C&C or other servants to client bots. The standard bot functionality, that is, execution of received commands and reporting back to the C&C, also applies to servants.
  - *Client bots*: Client bots are the leaves in the chain of control of a botnet. They are listening for commands from the C&C server and they report back to the C&C server or their delegated servant.
- *Communication channels*: They refer to the networking infrastructure available to the devices involved in a botnet. Accordingly, these can be exploited for a series of different functionalities, such as the initial infection of the botnet, its propagation, the commands issued by the botmaster and the C&C server, and the information collected and reported by the bots. Different communication channels impose their own constraints in regard to the various botnet architectures [10–12].

Mobile botnets can be used in a similar manner to traditional ones to enable DDoS attacks, disrupt service provisioning, and also to take financial advantage of end-users. Furthermore, the particularities of the mobile ecosystem open the door for more exotic and targeted types of attacks, giving rise to interesting use-case scenarios. These scenarios rely on two main features inherent to mobile botnets:

- *Contextualization*: A key distinguishing feature is the exploitation of the context of users, which is possible to infer and access, thanks to the rich sensor set available to mobile phones, for example, location and proximity.
- *Application based*: An application installed on a mobile device can access data (personal information) and send messages, thus facilitating its distribution through a worm-like scenario this is also possible on regular computers, but is less inherent to the application deployment itself.

The aforementioned features greatly facilitate the organization and deployment of targeted types of attacks. Specific sets of users could be targeted based on their application preferences and history, but also based on specific properties of the data collected by the sensors on their mobile devices, for example, all users residing in a certain city or building. For example, mobile botnet malware can be utilized to mount attacks on critical infrastructures (CIs) by forcing a backdoor into Supervisory Control And Data Acquisition (SCADA) systems through the mobile phones of employees who have access to the site of the CIs and hence are very likely to be connected to their networks [13]. Another possible use-case scenario for mobile botnets might involve localized social networks or large events where WiFi access is shared by many users and they are asked to download an application to access value-added services related to the social network or the event in question. The botmaster in this case can take advantage of the context and focus the attack on users with a particular profile.

Another interesting aspect of mobile botnet malware involves the use of sensors as side channels for communication, as suggested in Reference 14. In this case, since mobile platforms are equipped with sensors, attackers can exploit them to send commands or activate preconfigured commands on the infected mobile phones based on specific measurements. The advantage is the possibility to use this information outside of traditional data connectivity channels (3G or other) and contextualize the attack. Examples of triggering sources available to the botmaster include acceleration and specific path inside a tunnel, for example, or climbing stairs measured by the gyroscope, altitude (inside a building, outdoors) measured through GPS or accelerometer, an inaudible (for the human ear) sound pattern picked up by the microphone, any behavior like holding the phone near the face or voluntarily taking a self-image, reading a specific NFC message, etc.

10.3 Related Work

Research work on mobile botnets has witnessed a growth in the last few years [3–6, 15,16], attributed to the corresponding increase in related security incidents. To tackle such security threats, mobile botnet detection solutions that are considering the particularities of the mobile ecosystem have emerged, for example, honeypots such as the ones presented in References 17 and 18. Moreover, a plethora of research works exists on identifying, studying, and analyzing—either dynamically or statically—mobile malware [19]. Whereas such works provide essential information on certain aspects of the operation of mobile botnets, they do not consider their much more dynamic nature or the intricacies concerning the interactions between bots and C&C servers. Of particular interest is the work presented in Reference 20, which utilizes machine learning techniques and behavioral analysis to identify mobile botnet applications based on the communication patterns between bots and C&C server.

To date, conducting experiments in regard to mobile botnets has thus not been examined in a systematic manner by the research community. Different approaches have been exploited, such as simulations [5,21], ad hoc configurations [22], and collection of real-world statistics [23,24]. We argue that in order to have systematic research on mobile botnets, experiments should be performed systematically, allowing for them to be repeated in a contained environment. Such an approach is quite common for other fields, for example, widespread emulation platforms such as PlanetLab [25] and Emulab [26] or simulation environments such as OMNET++ [27], but has not to the best of our knowledge been studied in the specific context of mobile botnets. An initial attempt to design such an experimental testbed was proposed in Reference 23, but the authors did not proceed on testing and validating their high-level design. It is interesting to note that in Reference 23 the merits of adopting a hybrid approach with both real and emulated devices were highlighted.

Conducting large-scale experiments with mobile botnets is a main goal of our work and in this respect we align ourselves with works such as PlanetLab [25], Emulab [28], and DETERLab [29,30], which aim at supporting and facilitating repeatable, scalable, and verifiable experiments. The first deployment of our proto-type does not yet support scalability features, namely, large-scale experiments cannot be run due to constraints in the available infrastructure. In particular, the first deployment serves as a proof of concept for the validation of the desired function-ality; therefore, we did not prioritize scalability. Nonetheless, we aim at extending it using our prior work with the Experimental Platform for ICT Contingencies (EPIC) platform [31], which is developed in a JRC laboratory and uses the Emulab architecture and software. It will allow us to automatically and dynamically map physical components, for example, servers and switches, to a virtual topology. In other words, the Emulab software configures the physical topology in a way that it emulates the virtual topology of the implemented botnet as transparently as pos-sible. This way, we gain significant advantages in terms of repeatability, scalability, and controllability of our experiments. Moreover, we should state that emulation is particularly useful for security and resilience analysis [32] because in order to study those attributes, a researcher has to expose the system-under-test to high load and extreme conditions, under which software simulators fail to capture reality.

Our proposed hybrid experimental platform aims at providing a generic frame-work for mobile botnet research works to be tested and validated in a homogeneous manner. An important element of our approach involves the deployment of a mobile malware sandbox, namely, Cuckoo, to analyze at runtime the behavior of malware, while ensuring that it is tested in a contained environment. There have been quite a few proposals on mobile sandboxes, such as the work presented in Reference 33 or the AASandbox for Android applications [34], but we opted for Cuckoo due to its open-source nature and the fact that it is widely popular with the mobile malware community.

10.4 Mobile Botnets Taxonomy

Studying and analyzing mobile botnets requires more than the hybrid experimen-tal platform that is the major contribution of the research work presented in this chapter. From a theoretical perspective, establishing a taxonomy of mobile botnets is equally important in that it allows to conduct a comparative analysis of such botnets in order to demonstrate that mobile botnets present unexpected dimensions, which require the development of new strategies for detection and take-down actions. The in-depth taxonomy of mobile botnets that we present here—extending our initial work that was presented in Reference 35—supports systematic research efforts in a consistent manner, thus promoting methodological research and establishing a common foundation among researchers. Moreover, based on this analysis, we further describe possible botnet architectures for which the specificity of mobile botnets is again highlighted. We thus aim at instigating research on the protection against this relatively novel security risk, which increases nonetheless at an ever-growing rate.

10.4.1 Taxonomy

In line with the established paradigm shift that has witnessed the convergence of traditional, desktop computing systems with their mobile counterparts, botnet developers have also adapted their strategies to the mobile ecosystem. The development and evolution of botnets specifically targeted at mobile environments covers all possible aspects, from networking and propagation, to the intended use of the botnet and its impact. Mapping the current status of mobile botnets, identifying their key characteristics, and contrasting them with those of traditional ones, is the focus of the taxonomy of features of mobile botnets. By highlighting the distinct features and particularities of mobile botnets, we aim to instigate related research in a systematic manner.

In what follows, we list various aspects of botnets' functionalities and operations that should be considered in order to classify them in accordance to our taxonomy. Evidently, these aspects have a different notion when examining traditional botnets in comparison to the more recent mobile botnets. We aim to highlight the particularities, opportunities, and challenges that emerge when factoring in the mobile dimension. It is interesting to note that there also exist hybrid botnets, namely botnets, some parts of which are in the mobile network and others in fixed, commodity systems.

10.4.1.1 Network/Connectivity

When we refer to networking in the context of botnets, we refer to the network where the different components of botnets reside, and we can distinguish among wired, wireless (e.g., WiFi, cellular, NFC, Bluetooth), as well as hybrid wired/wireless ones. By considering network analysis as the mechanism to detect a botnet and its infection propagation, each of the existing networking standards is expected to exhibit a different behavior since they have diverse characteristics, that is, bandwidth, packet, signaling, jitter, delays, data propagation delays, physical medium, and link control [11]. Moreover, the architectural design of the different network standards further emphasizes the differences between them.

Traditional botnets: Traditional botnets are built on standard IP-based and mostly wired networks that do not impose any structural constraints on the design of the botnets since the topologies of such networks are inherently flat. Since wired connectivity has limited variability over time, it is quite common for the IP addresses of connected machines to remain stable over long periods of time. This observation assists both in detecting botnet behaviors and in their take-down, since it is relatively easy to track down the location and owner of an IP address of a wired connection by contacting the respective ISP [36].

Mobile botnets: WiFi networking is equivalent to wired networking from a functional point of view, so what applies to traditional botnets can be considered to characterize WiFi-connected devices. With the increasing data rates offered by WiFi standards, there is no particularity that affects traffic in a different manner than wired networks. Of interest are the various open WiFi hotspots that allow users to connect without authenticating themselves, which increases the difficulty in detecting malicious botnet activity. Contrary to wired networks, IP addresses change frequently when devices connect to WiFi networks, but it is still possible to detect the location of the IP address since the access points IP address is in most cases static.

Conversely, 3G cellular connections follow a strict hierarchical design, where there is less flexibility. The 3G network standards impose architectural constraints (centralized architecture) and the network infrastructure is owned by telecom operators and therefore it is out of the control of the end-users, including the botmaster and the owners of the compromised bots. Cellular networks are not free of charge; therefore, users might be able to detect strange behavior in their accounts, for example, the number of SMS messages or exceeding data plan, which could be attributed to malicious botnet behavior [5]. Up until recently, the data rates offered by cellular connections were limited and this led to a corresponding limited usefulness in terms of botnet propagation and operation. Moreover, IP communications over 3G networks can be easily tracked back to their owner, since IP addressing is managed by telco operators.

Other types of mobile botnets operate on short-range communication networks, for example, Bluetooth or NFC. This implies temporary close proximity of the different components of botnets, and therefore they do not refer to wide-scale infection scenarios. However, such networks are outside of the telco providers' control since they are formed on an ad hoc basis, thus making them extremely difficult to detect and counter.

10.4.1.2 Platform

Three types of platforms can be distinguished:

- ◼ The platform of the devices used by the botmaster and the C&C one
- ◼ The infection platform that might be a dedicated one or the one of the C&C
- ◼ The ones of the slaves that present potential security vulnerabilities that could be exploited by botmasters

The infection platform refers to that of the system used for the initial infection and compromise of a device and it is usually in the form of a website, a mail server, an IRC channel, etc. The platform used by the botmaster could be any sort of platform, for example, PC, server, or mobile, to contact the C&C server and issue commands or receive information from the bots. The C&C and the bots platforms are the platforms where the C&C has been installed and the compromised platforms respectively. Moreover, platforms have different operational and functional characteristics that could be used to detect botnets, where additionally the availability and implementation of various applications and networking protocols generally differs from platform to platform [3,4]. Such differences could also hinder the propagation of mobile botnets in particular, where interoperability is limited compared to traditional ones.

Traditional botnets: The most prominent platforms of traditional botnet slaves include commodity desktop OS, that is, Windows, MacOS, and Linux. The main difference observed in practice between these platforms is the user base, that is, the number of potential victims using each platform with Windows emerging as a fron-trunner. Windows also supports a larger variety of software and a security model that used to be weak on older versions (such as Windows XP that still represents around 15% of the market share despite it no longer being supported by Microsoft). For these reasons, Windows still remains the main target for installing botnet malware.

Mobile botnets: The most common mobile botnet platforms include the typical mobile OS, that is, Android, iOS, Windows Mobile, Symbian, etc. Again, the potential victim base does make a difference, and Android being today well ahead is the target of choice, and does indeed receive a lot of attention from botnet developers: depending on observers, 80%–95% of malware discovered on malware platforms are Android based [6]. There still exist some fundamental differences: it is always possible for an Android user to install software that does not come from official market places, whereas it is not often possible on iOS platforms, which definitely adds to the risk potential.

In terms of functionalities, the main evolution in the mobile world has been the advent of smartphones (and tablets), which share very similar capacities: touch screen, high-speed mobile Internet access, application download, two cameras, Bluetooth, and NFC. This opened up even more possibilities for botnet developers, such as the following scenarios [16]:

■ Reading SMS messages (e.g., used to receive one-time passwords for banking) or sending premium SMS.
■ Users might use their smartphone to access online bank accounts, removing most of the two-factor aspect if the phone is also used to receive one-time passwords.
■ Users might be less careful when using their smartphone and clicking on links, or validating alert boxes, because of the size of the screen.

10.4.1.3 Architecture

The architecture of botnet is essential to its analysis as well as to detect it and take it down. Clearly, centralized architectures will be more efficient in dispersing commands and collecting information (fewer number of exchanged messages will be necessary and therefore communication delay will be limited), whereas they are easier to take down due to the apparent single-point-of-failure. Hierarchical architectures are more scalable, but their take-down is easier compared to flat architectures where there is no clear dependency on a fixed organization scheme. Conversely, flat architectures do not scale well in terms of C&C [2]. Owing to their importance in regard to botnets' operation, we discuss mobile botnet architectures in more detail later in this section.

Traditional botnets: Owing to the flat topologies of the underlying networks, traditional botnets can exhibit all the aforementioned architectures. It is up to botmasters to enforce particular architectures according to their preferences and their requirements in terms of scalability, distribution of C&C commands, business model (e.g., they might wish to employ a hierarchical architecture in order to lease different parts of the tree hierarchy to different customers) or even strategy for detection avoidance, etc.

Mobile botnets: Mobile botnets have particularities in that the number of exchanged messages/data between botmaster and slaves should be reduced, since it can be subject to charges and affect the battery that could be noticeable by users. Moreover, the fact that 3G networks in particular are tightly controlled by telecom operators could deter the deployment of P2P-like architectures on their networks. Such constraints do not favor the deployment of complex architectures, such as hierarchical ones, especially when SMS is the medium of propagation. In addition, P2P architectures can only be realistically achieved when IP communication is enabled, since using SMS in multihop communications is not considered to be viable due to the great number of exchanged messages [15].

10.4.1.4 Propagation of Infection

Assuming that a botmaster wishes to release its botnet in the wild, the way it will be propagated depends on the botnet design itself. In one approach that is consistent with centralized architectures, the botnet (malware) could be pushed from one central location to all the slaves. Alternatively, the botnet malware could be propagated in a selective manner according to some context feature, for example, user, device, or network characteristics. Conversely, P2P propagation would see the botnet malware randomly or selectively distributed among the slaves with no central point to guide the process, similarly to flooding. Evidently, detection and take-down of the botnet is highly based on the type of propagation, while additionally the context information that might be used for selective infection is of high interest and could yield significant insight on the motivation and impact of the botnet. Different types of propagation include one-to-many, flooding, selective flooding, user-driven, and context-driven.

Traditional botnets: All aforementioned methods of propagation can be applied to traditional botnets. It is relatively more time-consuming and difficult to opt for user-driven and context-driven propagation due to the difficulties in acquiring relative information.

Mobile botnets: Mobile devices, with the plethora of onboard sensors and the collection of a large number of personal data over time, constitute an ideal environment for context- or user-driven propagation of the infection. Consistent with the architectures that are most applicable for mobile botnets, flooding for propagation of botnets is not a realistic option when SMS is the medium of propagation, with one-to-many being a preferred choice. Selective flooding could be supported by the use of Bluetooth or NFC (or the display of QR codes) to initiate infections in some specific locations or in an advanced cases by means of onboard sensors [14].

10.4.1.5 Infection Means

The means for infection and its corresponding payload aiming at initially infecting and compromising a device and thus making it a part of a botnet are numerous. The baseline for all is the installation of a program on the compromised device that will allow the botmaster to control it via the C&C.

Traditional botnets: Users are commonly tricked to download and execute a file on their personal computer, possibly by means of spam emails. Alternate means of infection include:

- Drive-by download/installation (exploit kit platforms)
- Firmware update
- USB or other removable media, as well as mobile phones that are plugged into devices for charging or sharing files
- Any document or file opened on a device that could exploit a vulnerability of the devices' installed software

Mobile botnets: Mobile devices nowadays are essentially equivalent to desktop computers in terms of processing power, memory, etc. The same stands for networking where all IP communication is the norm for mobile as well as traditional desktop computing devices. Therefore, it becomes clear that the means of botnet infection for mobile devices are expected to be the same as the ones for commodity computers. Mobile devices have additional channels of communications that are unique to them and have been already exploited by botmasters, namely, SMS and Bluetooth. Being aware of the decreasing rate of SMS exchanges, consideration will be given to the possibility to exploit SMS alternatives, for example, instant messaging, to infect mobile devices. Furthermore, one can also consider:

- WiFi session hijack (inject content in the communication to any website or service)

- Bluetooth or other proximity communication means
- Installation through a market or through another application (and through displayed advertising)
- SMS that could send a link to an application installation

10.4.1.6 Motivation/Impact

The motivation and the impact of the botnet is defined by the botmaster. The main points of interest for the botmaster refer to data theft (financial, personal, etc.), financial gain, service disruption (DoS) in terms of network service [37] or device operation, rooting of the device, and resource depletion.

Traditional botnets: In most traditional botnet cases, it is a combination of these points that is of interest to the botmaster. For example, extracting money from compromised bots by disrupting their services and promising to restore them to normality, that is, ransomware-related botnets, is an option for botmasters. Owing to the determinant role of personal data supporting online transactions or more generally fuelling the social life of individuals, more and more cases of botnets being used for data theft are becoming the norm.

Mobile botnets: Evidently, the same types of motivation that can be found in traditional botnets can also be considered for mobile botnets. Additionally, the ubiquitous nature of mobile devices that nowadays hold a huge amount of personal information, that is, financial, photos, contact lists, etc., has made them extremely attractive for malicious attacks that enable gaining access to such data. Moreover, the fact that premium SMS services and calls can be made by smartphones allows the attackers to gain financially from mobile botnets, an aspect that is not explicitly (or not anymore because of the disappearance of landline modems) available to commodity PCs [5]. Phones are also often attached to a market place account that could be authorized to make purchases with a preconfigured credit card number. Mobile platforms are often the tool used to receive or generate one-time passwords.

Classic attacks using botnets focused on DoS to either crash the device of the slave or use it to crash/attack a remote service. The same applies to mobile botnets with a couple of additional considerations. First, the notion of crashing the device of the slave can be expanded in the mobile realm to include the depletion of the limited resources, for example, battery, and also the rooting of the device. Second, the shared, all-listen nature of the wireless medium facilitates DoS attacks at this level since there is a sole communication channel that is shared by everyone. Furthermore, smartphones are equipped with a plethora of sensors that can yield significant information about their users, for example health, frequently visited locations, and application utilization. This extremely rich dataset, which at first might seem insignificant, offers the possibility to contextualize the attack and focus it on only the potentially most valuable targets or the ones with the most favorable profile for the attack.

10.4.1.7 Detection

The actual detection techniques used for traditional botnets do not differ significantly from the ones for mobile botnets.

Traditional botnets: Traditionally, standard/normal behavior of devices, applications, and networks are being monitored and then compared with their corresponding runtime behavior. Generally speaking, detection techniques are based on machine learning (device behavior, network analysis), application analysis (static, dynamic), use of honeypots, and *a posteriori* analysis (based on effect, e.g., financial, system outage).

Mobile botnets: In addition to the standard techniques used for traditional botnets, in mobile botnets, since applications are installed on smartphones and granted permissions to access local resources, the notion of application analysis should be extended to cover aspects such as whether the granted permissions are actually used [38]. Moreover, the notion of honeypots that is a key detection mechanism for botnets should be reconsidered to take into account the particularities of mobile environments in terms of architecture, network, application markets, etc. [39]. Some malware distribution methods will also specifically target mobile platforms as mentioned before, pushing for a need to adapt detection mechanisms.

10.4.1.8 Target

Targets of attacks of botnets are usually without discrimination all users whose device is subject to the vulnerability exploited by the botnet to infect and propagate itself. Alternatively, botnets could focus on targeted individuals based on specific criteria, for example, company or government employees or country-specific (based on network identity or location information when transmitted by default) [40].

Traditional botnets: Targeted attacks are mounted based on system features and not so much based on the user's characteristics. This is because it is not a common case for user's soft identities to be so tightly integrated with their device in traditional, commodity systems compared to their mobile devices. Therefore, targeted attacks could focus on systems with particular versions of software or OS running, specific IP ranges, and so forth.

Mobile botnets: Conversely, the inherent particularities of mobile phones as carriers of physical environment sensors enable more targeted selection of potential slaves. For example, devices under a specific geographic region could be targeted, or device owners with specific patterns of movements or even people that have the same set of electronic devices at home (magnetometer could be used for relevant identification). This grouping of device owners and consideration of physical world concepts could therefore be utilized to target selectively the distribution of the botnet, as well as the C&C communications, for example, by issuing different commands to different sets of slaves.

10.4.2 Functional Architectures

To fully assess the operation and functionality of mobile botnets, we describe here their architectures in further detail, presenting their functional components and the interactions between them. We thus aim to pinpoint the differences between traditional and mobile botnets in terms of the deployed architecture. The architecture of a botnet refers to its structure and to the communication and interactions between the botmaster, the C&C, and the bots (both servant and client ones). For simplicity, we consider a sole botmaster in all cases.

10.4.2.1 Centralized Architecture

In centralized architectures, there is a single C&C server issuing the commands and communicating with the botnets (Figure 10.1). Evidently, to take down the botnet, one needs merely to take down the C&C server. The bots hold the information necessary to communicate with the C&C server via a communication channel, for example, IRC, HTTP, or SMS for mobile botnets. This 1-to-N architecture is clearly very efficient in sending out commands due to the direct connections between the C&C and participating bots. Nonetheless, robustness and reliability of such botnets is minimal as they suffer from a single-point-of-failure weakness.

Mobile botnets using the centralized architecture are easier to detect than their traditional counterparts. The reason lies in the nature of the mobile environment and in particular mobile networks. The latter are inherently centralized, that is, all traffic from a mobile device goes through the base station or cell tower that the device is registered to. Moreover, the network of mobile operators is tightly monitored and controlled. Therefore, network flows with the destination or source

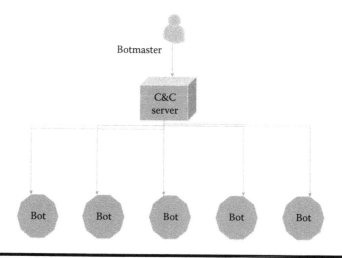

Figure 10.1　Centralized architecture of a botnet.

being constant and generating high amounts of traffic (both characteristics of centralized botnet traffic) will be easily detected by intrusion detection systems located in the GGSN (Gateway GPRS Support Node) gateway of 3G networks. IP-based intrusion detection systems can be effectively installed alongside the GGSN to filter IP traffic from and to mobile nodes, as proposed in Reference 17. Conversely, in traditional Internet-based botnets, the distribution of the infrastructure and its flat architectural style hinder such detection solutions.

Mobile botnets that do not use IP connections as communications channels but instead utilize still SMS will benefit from such an approach, since the direct connectivity limits the number of control messages that need to be exchanged between C&C server and bots. It is only a matter of scale as far as the C&C is concerned, whereas from the bots' perspective only one SMS needs to be sent out to the C&C server. Because the costs (or the billing traces) associated to SMS might be detected by users and therefore raise suspicion regarding the presence of a botnet malware, we argue that this architecture is favorable for mobile botnets.

10.4.2.2 Hierarchical Architecture

In hierarchical architectures, a tree-like structure is assumed by participating C&C and bots, with servants acting as C&C for client bots. At each level of the hierarchy, taking down one of the servants leads to the identification and possibly cleaning up of its corresponding client. To take down the entire botnet, one needs to take down the root of the tree structure, namely, the C&C. However, there have been cases reported, where taking down the C&C did not shatter the botnet, since the servant nodes were configured to rearrange themselves in a P2P-style architecture in the occurrence of such an event (e.g., TDL-4 botnet [41]). In general, hierarchical architectures are more scalable than centralized ones, since the C&C is relieved from the duty to keep track of all bots, delegating parts of this task to servants. Figure 10.2 illustrates a typical hierarchical architecture.

It is interesting to note that such an architecture allows the utilization of more than one communication channel (e.g., HTTP, SMS, IRC), since the different subtrees of the architecture are subject to the selection of the channel by their respective roots, that is, servants. While being more robust than the centralized architectures, nevertheless the robustness and reliability of hierarchical botnets are not really high since they suffer from multiple (of more limited scope) single-points-of-failure, as well as the main root of the tree.

Mobile botnets applying the hierarchical architecture paradigm have similar features to centralized ones. The nature of mobile networks facilitates the detection of such botnets, since the communication paths are relatively static and therefore monitoring of network flows can divulge possible botnet behaviors. When considering SMS-based mobile botnets, such an architecture increases the costs for the bots, that is, the infected end-user devices, with servants evidently incurring higher costs. This could lead to them being more easily detected by their users, upon examination of

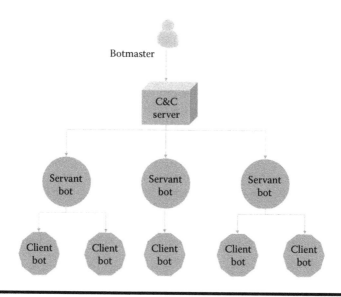

Figure 10.2 Hierarchical architecture of a botnet.

bills. Owing to the operation of mobile networks, this architecture scales only in terms of manageability and not energy efficiency and communication costs, which are actually increased. Instead of having one connection between C&C server and bot, a multihop connection exists. However, each hop does not involve direct communication as in traditional botnets; in mobile botnets, all communications have to pass through the base station and other elements of the mobile network infrastructure, thus increasing complexity and detectability. This particularity of mobile botnets (lack of direct communication) clearly distinguishes them from traditional botnets and can be potentially exploited to assist in their detection. For example, a base station would be the ideal location to place a honeypot.

Centralized and hybrid architectures are often associated with DGA (Domain Generation Algorithms) [42] when using an IP communication channel. This is a very common way for botmasters to issue commands to the bots via the C&C server and receiving information from them. In the mobile world, one could imagine a phone number generation algorithm but it would be very theoretical as it is difficult to control which phone numbers one will be using unless the botmaster has some sort of control over this functionality, which currently is the responsibility of telecom operators.

10.4.2.3 P2P Architecture

P2P botnet architectures are flat architectures with no hierarchical structure imposed (Figure 10.3). In this context, there is no dedicated C&C server. All infected devices

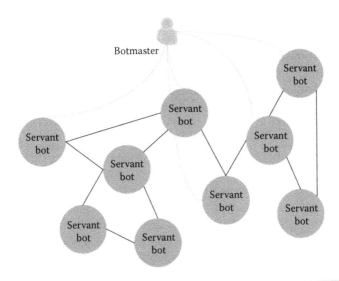

Figure 10.3 **P2P architecture of a botnet.**

are considered to be servant bots and commands are propagated through the botnet using a diffusion model that resembles flooding. Solid lines indicate direct communication paths between servants, whereas dotted lines indicate the issuing of commands from the botmaster. The latter can choose to initiate command propagation in the botnet exploiting different servants. It is important to note that direct communication paths exist in traditional botnets, whereas in mobile botnets, the communication is only conceptually direct and has to go through the base station or cellular tower. This type of architecture is robust and resilient to take-down operations, since all of the servants need to be taken down in order for the botnet to become inactive. The botmaster is not directly aware of the entirety of its botnet and cannot assess its size: this is a major drawback for the business model of botmasters who rent out the services of their bots to malicious entities.

Mobile P2P botnets are difficult to implement, owing to the lack of direct communication links. Whereas one can claim to have built such a botnet using SMS as a propagation channel, the nature and structure of mobile networks reduces this botnet to a centralized (or hierarchical) one. Moreover, mobile P2P botnets suffer from great costs in propagating commands, since a lot of messages need to be exchanged to ensure commands are disseminated and that information is reported back to the botmaster. Because of these particularities, they are subject to charges that could be noticeable by users and therefore lead to their detection. Detecting network flows to identify distinct patterns of mobile P2P botnets is much more difficult compared to hierarchical and centralized architectures. Since topologies are not static, network flow information will be dynamic as well, thus leading to highly efficient botnets in terms of low detectability. This comes at a high cost of exchanged messages as mentioned before, but also in terms of implementation, since connections to other servants need to be maintained and monitored and measures to compensate for communication loss need to be put in place. The P2P architecture could also present an advantage for mobile botnets for which all bots are not always connected or reachable. Updates and commands can be initiated to any of the available bots and later be further distributed to the others.

10.4.2.4 Hybrid Architecture

Hybrid architectures for botnets combine features of the aforementioned categories. A most common case is the deployment of a hybrid hierarchical–P2P architecture (as shown in Figure 10.4), where the servant nodes form a P2P architecture among themselves to increase robustness and resilience. Clearly, the combination of more than one architectural paradigms exploits the benefits of them, albeit at a higher implementation and communication cost. Owing to the diversity of mobile networks infrastructures and their increasing integration with traditional networks, such architectures could be considered for mobile botnets.

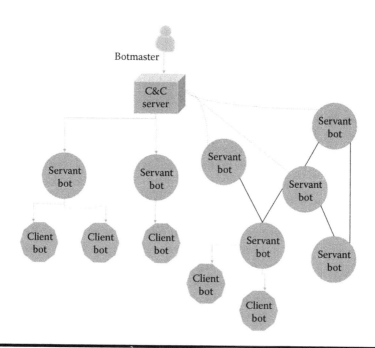

Figure 10.4   Hybrid architecture of a botnet.

10.5 Design of the Hybrid Experimental Platform for Mobile Botnets Research

To instigate homogeneous research efforts in the area of mobile botners, we propose a hybrid experimental platform that comprises both actual and emulated mobile devices. In what follows, we first describe the functional requirements of the proposed platform, namely, we define its desired features, as well as present a high-level design of the platform that will guide its implementation and eventual deployment. The design of the platform and the described architecture are based on our preliminary work that was presented in Reference 43. In that work, we discussed the architecture and desired requirements of the platform from a higher level point of view. Conversely, in this chapter, we have implemented the platform and have thus accordingly modified the architecture subject to limitations and modifications that the practical implementation has rendered necessary.

10.5.1 Functional Features

To elicit the design requirements for the hybrid experimental platform, it is necessary to first set out what its functionality will be, that is, what functional features it will support. In doing so, the requirements imposed on the design of the platform will be clarified, as well as its potential use cases. Accordingly, we can identify the following functional features for the proposed hybrid experimental platform for mobile botnets research:

- The platform shall support the observation of the operation of mobile botnets in hybrid configurations enabling the use of both emulated and actual mobile devices.
- Management communication traffic should be explicitly distinguished from the communication traffic that reflects botnet malware activity in the hybrid experimental platform.
- It shall support the parallel execution of multiple experiments (and therefore possibly multiple mobile botnets) subject to the availability of resources, that is, emulated and actual mobile systems. For the case of emulated systems, availability refers to computational and memory resources of the host machine that is used for emulation.
- It shall cater for the heterogeneity of mobile platforms and systems taking into account the well-established diversity of mobile phone OS, platforms, and device capabilities, as well as their networking capabilities and corresponding network protocols.
- The hybrid emulation platform shall be used to test various aspects of mobile botnets functionality, including operation, distribution, infection, detection, etc.

- The platform shall support the execution of experiments with configurable parameters such as the size of the botnet and the types of mobile systems used in order to allow for sensitivity analyses to be conducted.
- It shall operate in a user-friendly manner to facilitate researchers and to promote it as wide as possible distribution. One approach could involve the setting up of experiments regarding mobile botnets based on event-driven scenarios. Experiments can be as simple as launching a mobile botnet on a set of devices and monitor its evolution, but should also scale to more complex configurations such as ones involving network partitions, contextualization of mobile botnets infection using sensor data, etc.
- The platform shall support the collection of results and measurements regarding mobile botnets' operation, for example, infection rate, CPU and memory utilization, and number of exchanged messages.
- It shall provide a dedicated report concerning every experiment executed on the platform.
- The platform shall cater for the integration of realistic sensor data (in what concerns the emulated devices) in order to experiment with particular mobile botnets' settings.
- The platform shall support the remote configuration of both the emulated and real devices to be able to provide means to dynamically reconfigure them, enable/disable features, modify their operation, reconfigure the underlying network topology, etc.
- The platform shall support the execution of experiments in a contained environment where no access to the Internet will be provided, but also more realistic ones that will grant Internet access to the participating devices.
- It shall be secure to protect researchers against adverse effects of operating mobile botnets under experimental settings.
- It shall not be tied down to proprietary software solutions and products.

10.5.2 Design Requirements

To accommodate the aforementioned functional features, it is evident that the proposed experimental platform needs to be generic. Its design should allow for a great variety of experiments to be conducted, under different settings and with diverse objectives. To accomplish this, it is important to consider a modular architecture that can be easily extended to cater for different configurations and to adapt to experimental settings that have not been initially foreseen. Moreover, a major design requirement of the platform is its flexibility. It should allow for experiments considering diverse populations of bots, namely, with varying size. The setup of the experiments should cater for this diversity by supporting the definition of both simple and complex scenarios involving the infection, distribution, operation, and detection of mobile botnets.

Since mobile botnet malware pose a severe security risk, provisions should be put in place to hinder any adverse effects to the platform itself stemming from the testing of such malware. It is thus important to separate the management layer of the platform to the one dealing with botnet malware experimentation, both in terms of physical machines, but also at the communication/networking layer.

Furthermore, the platform should be easily extensible so as to be able to support prospective enhancements and modifications without significantly affecting its operation and design. A rigid architectural design would make it extremely cumbersome to modify and extend the platform. Since the platform needs to cater for a variety of experiments, all configuration settings should be modifiable dynamically and at runtime in order to gain from maximum flexibility. The notion of extensibility required of the hybrid experimental platform should extend to also address the need for heterogeneity support. In particular, the wide variety of OS, platforms, and architectures for mobile devices necessitates that the platform is capable of catering for this heterogeneity.

One additional requirement involves scalability. In order to realistically experiment with mobile botnets, it is crucial to be able to support large number of bots and possibly C&C servers. The reason behind this lies in the fact that the number of bots in a mobile botnet is usually rather high—although not as high as in traditional botnets [6]—and thus realistic research scenarios should consider large number of devices. To satisfy this requirement and taking into account the high costs involved in procuring and maintaining a large-scale testbed of mobile devices, the design should support the integration of emulated devices with comparable features. In this respect, the size of the mobile botnets supported by the experimental platform will only be limited by the available resources, that is, actual mobile devices in the infrastructure, as well as the memory constraints regarding the parallel execution of multiple emulators. Since mobile devices' emulators are typically consuming a large number of computational resources, the hosts on which these emulators operate might become overloaded. Accordingly, load balancing emerges as an additional design requirement to avoid adverse situations. It needs to be noted that support for scalability should also extend to the number of distinct botnets that can be concurrently tested on the platform.

The platform is meant to be used by researchers of mobile botnets in order to facilitate the systematic execution of experiments and to allow for reproducible results. This is a design requirement of paramount importance. In this line, definition and execution of experiments should be characterized by repeatability and the platform should support monitoring of the progress of the experiments so as to be able to produce results in a consistent manner.

Finally, the platform should not be based on proprietary solution, but instead utilize as much as possible open-source tools. The reason for this is dual. On the one hand, the support of the open-source community to extend the platform and to maintain its functionality is much desired. On the other hand, open access to the platform will increase its user base and will allow researchers to have a complete overview of their experiments, even behind the scenes, thus enabling them to modify them according to their individual needs.

10.5.3 Functional Architecture

We adopted a modular design for the experimental platform to adhere to the previously described requirements. The functional architecture is illustrated in Figure 10.5. We can distinguish two parts in the architecture, namely, the emulated one and the actual one. The former refers to the emulated devices that can be put at the disposal of users in order to conduct experiments, whereas the latter represents the actual real devices that belong to the infrastructure of the platform. These devices can connect to the Internet and interact with each other or even with the emulated ones by means of either cellular connections or wireless ones.

The emulated part of the architecture is the one that will support the requirement for scalability, since the number of actual devices is limited due to cost constraints. Moreover, emulated devices are more easily programmed and managed remotely compared to actual devices that exhibit certain limitations in this respect. To support remote configuration of real devices, they should be running under the development mode—as far as the Android platform is concerned—and be part of the same network as the hybrid platform. In some cases, for more advanced remote configurations to be possible, a remote access needs to be running on the real device.

A layer-based approach has been adopted in regard to the emulated part of the architecture. At the bottom layer, a series of hardware servers is used in order to deploy the emulated devices. The number of servers is directly proportional to the number of emulated devices that we envisage the platform to support, since the memory and computational capabilities required of each server to run the emulator are substantial. Servers can be added or removed at any time from the infrastructure layer, attributed to the modular design that we have undertaken. The only effect to the platform is in the number of emulated instances that will be supported subject to the modifications in the servers.

The virtualization layer is built on top of the infrastructure one and it refers to the instantiation and management of multiple virtual machines (VMs) that will be used to run the emulated mobile devices that will be part of the mobile botnet. There is no hard dependency, neither on the selected host OS for the servers nor for the virtualization software, that is, the software that will be used to launch and manage the instances of the VMs. The virtualization layer is remotely managed by the management interface provided to users of the proposed experimental platform. In this way, users can choose to dynamically enable or disable VMs according to the requirements of their experiments, as well as modify the configurations of currently running VMs.

The emulation layer involves the tools and systems needed to run emulated instances of mobile devices. It mainly consists of two elements, namely, the Android

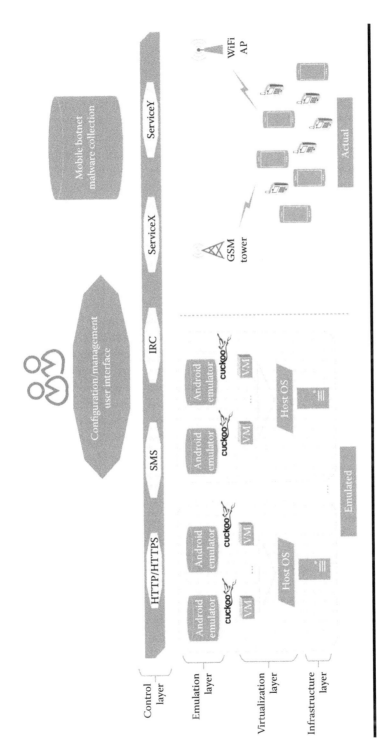

Figure 10.5 Functional architecture of hybrid experimental platform for mobile botnet research.

Emulator and the Cuckoo Sandbox,[*] which is an open-source malware analysis system. For each VM of the virtualization layer, there is exactly one instance of an Android Emulator and one Cuckoo Sandbox operating. The reason behind this decision is our desire to emulate as much as possible an actual mobile ecosystem and such we wish to avoid any issues that might occur from executing multiple instances of these elements on the same VM, for example, starvation of resources or competition for resources. In addition, this design alleviates any concerns regarding networking of the emulated devices, thus simplifying their remote management and configuration. Owing to the more open nature of Android, the platform will inherently support the Android Emulator but it is clear that the design is extensible enough to accommodate emulators of other mobile platforms as well, for example, iOS or Windows Phone.

The actual part of the hybrid experimental platform comprises a set of mobile devices that operate under standard settings. We favor the use of diverse types of mobile phones in order to cater as much as possible for heterogeneity, while additionally and for the same reason mobile devices have two modes of acquiring network connectivity. Similarly to the emulated part, actual mobile devices are operating on Android, with the possibility to extend to other types of OS being part of our future work and by no means limited by the platform capabilities. To support remote configuration and interaction of the devices, the developer mode has been enabled. Both cellular and wireless connectivity have been provisioned for these devices. The reason lies in the fact that not all mobile botnets simply require Internet connectivity—which can be easily supported by connecting all devices through a WiFi access point—but some of them operate on the basis of SMS exchange and take advantage of particularities of the cellular network. Accordingly, we have deployed a GSM base station to allow the platform to support such functionality and therefore cater for a wider variety of mobile botnets.

Both the emulated and the actual part of the hybrid platform are controlled and managed through a high-level configuration and management interface. The main functionalities of this interface consist of providing services to researchers to set up and execute mobile botnet-related experiments using the hybrid platform. In this respect, it enables the selection of emulated and actual devices that will take part in the experiment, its remote configuration, the definition of settings and other options to launch an experiment, and the collection of reports and results pertaining to the experiment. While services are provided to users, the requirement for flexibility drives our adoption of a modular design based on web services that promotes the modification and possible extension of the management interface. The latter communicates with the underlying layers of the hybrid experimental platform via the control layer, which builds on standard means of communications such as HTTP/HTTPS, SMS, and IRC to name a few. These communication methods

---

[*] https://www.cuckoosandbox.org/

refer to both the communication between the interface and the platform itself, as well as the communication between elements, namely, mobile devices, of the platform, for example, to emulate the functionalities of the C&C server, for example, issuing of commands to the active bots.

The hybrid experimental platform for mobile botnet research aims to serve as a focal tool to test and examine related mobile malware. In order to promote exchange of results and findings between researchers, as well as to assist in mapping the highly disperse field of mobile botnet malware, we consider establishing a mobile botnet malware collection (MBMC). This collection that will be constantly updated by the research community will on the one hand serve as a repository of mobile botnet malware and on the other hand will be the central collection of reports and experiments regarding these malware and therefore greatly promote and facilitate corresponding research works.

10.6 Implementation of the Hybrid Experimental Platform for Mobile Botnets Research

Based on the design requirements and the functional architecture that was presented in the previous section, we discuss in what follows details on the implementation of the hybrid experimental platform. We first give details on the hardware deployment in our laboratory settings and then describe the software elements of our proposed architecture. A major aspect of our work involves the networking capabilities of the experimental platform, which are detailed in terms of communication interfaces, that is, REST-based web services that we have introduced in order to support management of the platform. The section concludes with a presentation of the management/configuration interface that can be used to both configure and control the platform, as well as to design and launch experiments related to mobile botnets research.

10.6.1 Hardware Deployment

The deployment of hardware elements to support the operation of the hybrid experimental platform follows the overall design of the architecture and accordingly distinguishes between the control and the experiment plane, as depicted in Figure 10.6. The main reason behind this separation lies in the fact that we wish to hinder any possible interference between the two planes. A typical example of such interference involves increased network traffic in the control plane leading to unacceptable and unexpected levels of network delays or jitter in the experimental plane. Moreover, since the experiment plane is used to study mobile botnets, it is reasonable that it remains separated from the control plane due to the potentially harmful nature of the malware corresponding to mobile botnets' activities that could adversely affect the proper operation of the control plane. Control traffic and malware-related traffic

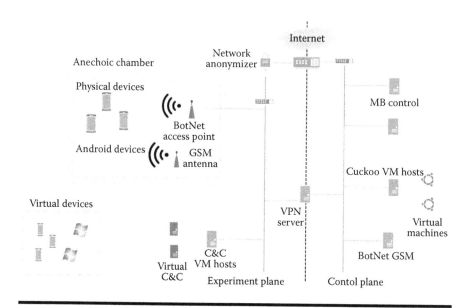

Figure 10.6 **Deployment of hardware elements of the infrastructure to support the operation of the hybrid experimental platform.**

must therefore always remain separated to promote the security and integrity of the platform.

The two planes are physically separated, but in order to enable their support at the network level as well, we introduce a VPN server that provides separate address spaces to them. The VPN server is located in a dedicated machine and provides IP addresses to both physical and virtual elements of the experimental platform. In terms of network infrastructure, the platform also comprises a dedicated gateway for Internet access. As an additional measure of security and privacy, the experiment plane also makes use of the services of a Tor gateway that supports anonymization. To promote the extensibility of the platform, we installed the Tor gateway on a dedicated Raspberry PI 2, which is noted as Network Anonymizer in Figure 10.6. The motivation behind supporting anonymized Internet access for the experiment plane is the fact that malicious malware activity that will be tested in the platform should not optimally be traced back to the researchers' environment to avoid possible retaliation from the botmasters.

The experiment plane consists of both physical and virtual devices that are effectively utilized for the testing and experimentation of mobile botnet malware. The physical devices are a set of mobile phones, which in the current state of the platform are only Android ones. These devices acquire network connectivity either via a dedicated WiFi access point or via a USRP (Universal Software Radio Peripheral) transceiver that we have set up in the lab to support cellular, for example, 3G, connectivity. Both these modes of networking infrastructure are provided, since various mobile botnet malware are specifically targeting the particularities of the network connection used by a mobile device and we therefore wished to proactively address all possible needs. The set of physical devices can be easily extended and more platforms can be added to the platform to increase its scalability. Virtual devices can either be emulated Android devices or emulated Windows/Ubuntu platform to cater for hybrid mobile botnets. While conceptually they reside in the experiment plane, from an operational point of view, they are executed on the Cuckoo VM Hosts that are located in the control plane as discussed later.

The C&C infrastructure also resides in the experimental plane and because many mobile botnets are equipped with more than one C&C server, we are using a dedicated host (C&C VM Host) to virtualize several C&C server according to the needs of the experiments. It is interesting to note that for the experiments envisaged with the hybrid platform, the C&C server can either be part of the platform (assuming that we have access to its source code and can therefore replicate its operation) or alternatively the physical and virtual devices could also connect to an actual, operational C&C server in the wild. Evidently, in terms of security, the latter option is of much higher risk.

All the elements required to manage the hybrid experimental platform can be found in the control plane. In this respect, in the control plane, there exists a dedicated machine to manage the hybrid experimental platform, namely, the Mobile Botnet Control (MB Control). Moreover, a set of host machines, that is, Cuckoo VM Hosts, are also part of the control plane. Since scalability and extensibility is a paramount requirement, this set of hosts can be extended to reflect an increase in the population of emulated devices that the platform can support. Whereas the virtual devices that are operating on the Cuckoo VM Hosts are conceptually part of the experiment plane, the fact that we wish to manage them—instantiate, turn off, configuration, etc.—using the MB Control requires these machines to be physically part of the control plane, since in our architecture, the two planes are part of different network spaces and can only communicate via the VPN server. In the same line, there is also a dedicated machine, that is, Botnet GSM, that is responsible for the management traffic referring to the physical devices that are connected to the USRP GSM Antenna.

10.6.2 Software Architecture

The requirements placed on the platform have spurred our decision to select open-source software products and solutions for the implementation of the platform. In particular, the VPN server is executing on an Ubuntu machine and it is running OpenVPN[*] and a standard DHCP server that provides addressing to the different elements of the platform. OpenVPN has been configured to be the only

---

[*] https://openvpn.net/index.php/open-source/overview.html

point where traffic is permitted to cross from the control plane to the experiment plane.

In the experiment plane, the C&C VM Host consists of a typical instantiation of the LAMP model and accordingly it has a Ubuntu OS, an Apache HTTP server, MySQL relational database management system, and PHP installed. Moreover, in order to support virtualization of multiple C&C servers, the VirtualBox[*] virtualization software is utilized. The Network Anonymizer element is a Raspberry PI 2 that operates on Raspbian[†] and has Tor installed.

In the control plane, the MB Control host is also based on the LAMP model and additionally has GitLab installed to manage the repository of the code relating to the implementation of the hybrid experimentation platform. The Botnet GSM host is responsible for interacting with the USRP GSM Antenna and thus uses OpenBTS[‡] on a Ubuntu system. Lastly, the Cuckoo VM Hosts are executing VirtualBox to instantiate a series of Ubuntu VMs, each of which is equipped with the standard Android Emulator[§] and the Cuckoo Sandbox for malware analysis. Cuckoo is a core element of our platform, since its functionalities provide us with advanced analysis of malware activity on the emulated platforms, ranging from networking to memory analysis and tracing API calls. We need to underline here that the physical devices of the platform are analyzed in an *a posteriori* manner, namely, after the execution of an experiment, each one of the devices is manually analyzed to get access to memory dumps and API calls. As far as network traffic dumps are concerned, they are being monitored at runtime using network capture tools such as *tcpdump*.[¶]

10.6.3 Communication Interfaces

To manage the platform and facilitate the setting up and execution of experiments, we have devised a series of communication interfaces, which are implemented as REST-based web services. They allow to interact with the distinct elements of the platform, configure parameter settings, and instantiate experiments. Resource representation is based on JSON, benefiting from its simplicity and without sacrificing expressiveness. The full set of communication interfaces that we have implemented allows for detailed interaction with the hybrid platform and for controlling its supported functionalities. We describe in what follows typical examples of such interfaces reflecting the main use cases.

When a new Cuckoo host needs to register to the experimental platform, and thus promote its scalability by giving more resources for experiments, a call to the *mbcontrol/register-host* web service is performed as seen in Figure 10.7 with a JSON

---

[*] https://www.virtualbox.org/
[†] http://raspbian.org/
[‡] http://openbts.org/
[§] https://developer.android.com/studio/run/emulator.html
[¶] http://www.tcpdump.org/

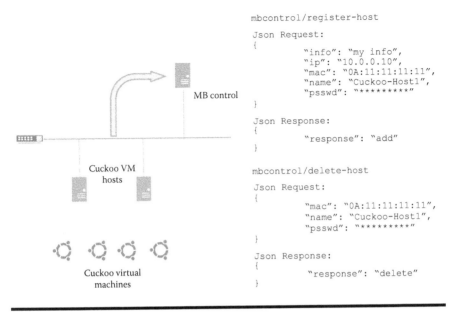

```
mbcontrol/register-host

Json Request:
{
        "info": "my info",
        "ip": "10.0.0.10",
        "mac": "0A:11:11:11:11",
        "name": "Cuckoo-Host1",
        "psswd": "*********"
}

Json Response:
{
        "response": "add"
}

mbcontrol/delete-host

Json Request:
{
        "mac": "0A:11:11:11:11",
        "name": "Cuckoo-Host1",
        "psswd": "*********"
}

Json Response:
{
        "response": "delete"
}
```

MB control

Cuckoo VM hosts

Cuckoo virtual machines

Figure 10.7 Registering or deleting a Cuckoo host to the pool of hosts that are made available to the hybrid experimental platform.

request that contains its IP and MAC address, as well as its name and a chosen password that can be later used for authentication. Subject to a successful completion of the process, the response is a simple confirmation.

Conversely, to delete a host from the platform's pool of available hosts, a call needs to be made to the *mbcontrol/delete-host* web service as shown in Figure 10.7. The JSON request attached to this call needs to include the MAC and the name of the host to be deleted, as well as the password that was used to originally register the host for security reasons. Once again, subject to a successful removal of the record for the host, the JSON response entails a simple confirmation of the outcome.

The MB Control initiates experiments and keeps track of their progress. It is therefore a prerequisite that it is aware of the Cuckoo virtual machines that are at its disposal, in order to be able to plan the execution of an experiment. To acquire this information (Figure 10.8), the MB Control calls the *x.x.x.x/get-cuckoos* web service on every Cuckoo VM Host (*x.x.x.x* denotes the IP address of these machines) that has been registered as illustrated before. The JSON response that is returned to the MB Control includes a listing of available Cuckoo virtual machines (IP address and status) that can be used for experimentation purposes.

At a higher level of granularity in terms of controlling the experiments, it is possible for the MB Control to directly interact with the Cuckoo virtual machines and instruct them on various aspects of their operation. The Cuckoo Sandbox offers a

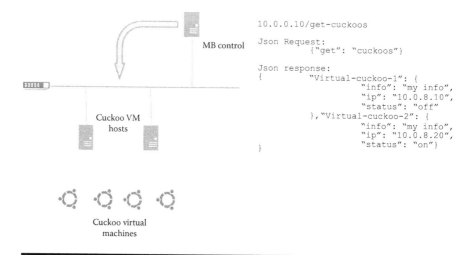

Figure 10.8 List of all available instances of Android Emulator and Cuckoo running on a specific Cuckoo VM Host.

rich REST web services API[*] that can be used through the web services that our platform supports in order to allow for its advanced (re-)configuration. An example of how a task could be rescheduled is shown in Figure 10.9, the only required information being the IP of the Cuckoo virtual machine and the standard port used by the MB Control to interact with it.

10.6.4 Management/Configuration Web Interface

The hybrid experimental platform aims at instigating further research on mobile botnets by providing an established tool to conduct experiments. To facilitate its use by researchers and developers alike, we have implemented a prototype management interface that abstracts from the underlying complexity of the web services that were previously described. By means of the graphical management interface, users of the platform can discover the capabilities of the platform, namely, how many devices it supports and how many are active, while at the same time they can initiate new experiments with mobile botnets malware, monitor their progress, and obtain reports with the analysis of the corresponding malware. The management interface that can be seen in Figure 10.10 is at a prototype stage and it is foreseen that it will be significantly enhanced to improve its usability and to augment the set of functionalities that it supports.

---

[*] http://docs.cuckoosandbox.org/en/latest/usage/api/

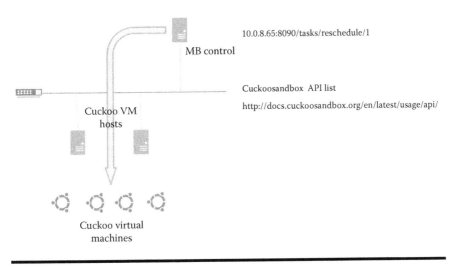

Figure 10.9   Managing active Cuckoo instances, for example, rescheduling tasks, using the Cuckoo Sandbox API.

10.7 Validation

In this section, we describe the set of experiments we have performed to validate the platform illustrating its ability to support the dynamic analysis of actual mobile botnets.

The validation of the platform involves checking that malicious applications to be analyzed can be deployed into several available nodes, virtual or physical, that compose the platform (or a subset of them if the experimenter would like to run the experiment at a smaller scale). Each node can be configured to have specific software and hardware settings allowing the experimenter to test the dynamics of the botnet in a heterogeneous environment. During the running of the experiment, the platform analyzes dynamically the execution of a botnet at the level of the network and each of the mobile nodes, physical or virtual, providing a full report at the end of the experiment consolidating the monitoring data gathered from the nodes. Among other things, this report includes memory dumps, network communications, accessed files, and screenshots.

Since the purpose of the platform is to support research on mobile botnets threats to gain a better understanding of how they operate and contribute to the design of effective strategies to detect and prevent them, the experiments described below make use of actual mobile botnet malware samples that have been found in the wild. These experiments using actual malware serve as a first validation of the prototype, complementing the set of tests performed during the design and development phase that used synthetic samples (i.e., programs specifically developed by us to simulate a certain functionality expected to be found in mobile botnet malware),

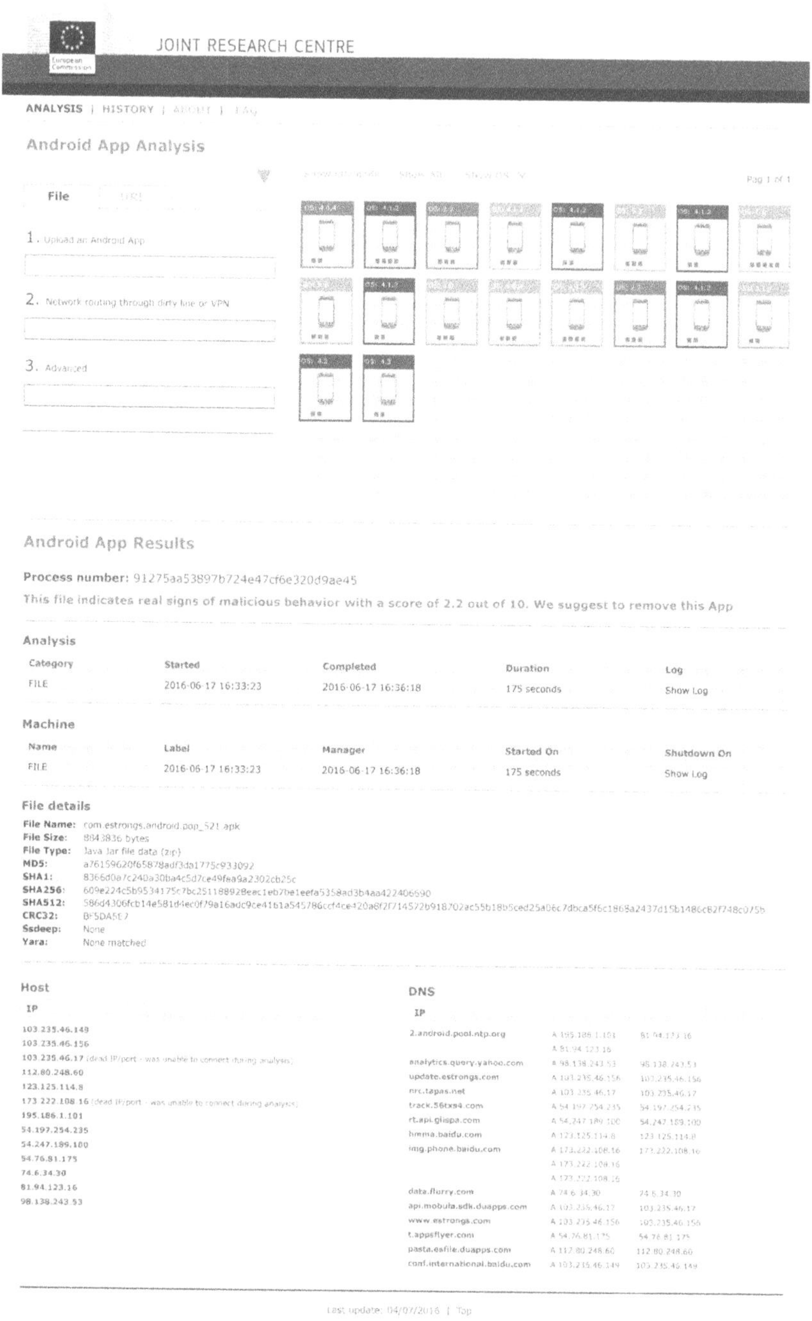

Figure 10.10 **Platform control interface to submit samples and manage nodes (upper part) and to review results (lower part).**

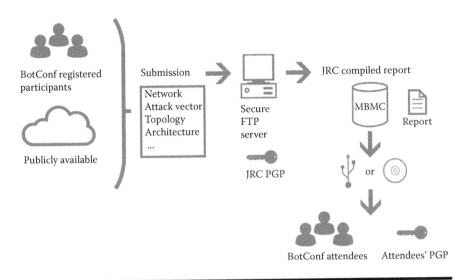

Figure 10.11   Mobile botnet malware collection initiative workflow.

and testing that the overall functionality of the platform works as expected and can lead to useful results in the analysis of mobile botnet malware. Indeed, the use of actual malware samples ensures that the number of assumptions we make on this type of emerging threats are minimized. For example, specific countermeasures can be considered to mitigate the risk of malware using techniques to detect sandboxes and prevent the debugging of code.

The collection of mobile botnet malware samples to support this type of research is an ongoing collaborative effort undertaken by industry and research community. In Botconf (Botnet Fighting Conference) 2015,[*] we presented our MBMC initiative [44] aimed to collect and categorize mobile botnet malware found in the wild, along with source code and C&C components whenever possible, building a database of malware samples. To achieve this goal, in the months prior to the conference, we set up an online submission system allowing registered participants to securely submit mobile malware samples. Submissions were encrypted with a PGP/GPG public key specifically generated for this purpose. During the conference, registered participants were given a DVD copy of the collected database encrypted using the PGP/GPG public key that was provided in the registration procedure, as depicted in Figure 10.11. Our research team continued to update this database and used malware samples from it to test the prototype presented in this chapter, as described in the next sections.

---

[*] https://www.botconf.eu/botconf-2015/

10.7.1 Analysis of Dendroid

In this first experiment, our aim is to test the capabilities of our prototype to support the execution and analysis of an actual mobile botnet. In order to do so, we decided to use a botnet found in the wild known as Dendroid [45]. The usage of an actual malware sample found in the wild helps us to realistically test the capabilities of our platform, not only in being able to dynamically monitor the botnet but also to ensure that its execution works properly in the virtual environment setup for that purpose and that this environment does not interfere with the functionalities of the malware. Dendroid botnets are composed of a C&C server and of a malicious application running on mobile devices making them remotely controllable by the attacker unbeknown to the legitimate owners. The following are some of the actions Dendroid C&C can command:

- ■ Send SMS, and monitor or block SMS received by the target.
- ■ Spying the victim taking pictures or recording video/audio.
- ■ Download pictures taken by the phone.
- ■ Download web browser history and bookmarks.
- ■ Retrieve user authentication credentials for the several accounts configured in the phone.
- ■ Record phone calls.

In order to test the capabilities of our platform to host and dynamically analyze this type of botnet, we have set up the following experiment. In a self-contained environment, as depicted in Figure 10.6 of Section 10.6.1, we deployed both the C&C of the malware and a set of infected phones. To test the several features provided by our prototype, we have set up a total of five virtual phones emulated by the platform and two physical phones located inside the anechoic chamber. As described previously, in Section 10.6.1 and depicted in Figure 10.6, all physical devices involved in the experiment are connected to our GSM base station and to our WiFi network effectively connecting all the infected devices to the experimental plane network. The possibility to grant access to GSM and WiFi networks to nodes, as possible means of communication at the disposal of the malware, offers a greater degree of flexibility and contributes to avoid potential anti-debugging mechanisms. This helps to detect cover channels of communication between the C&C and the infected nodes (e.g., in those cases where SMSs are used as a means of communication).

The C&C application of Dendroid is written in PHP and it is designed to run on web servers capable of running PHP applications. For our experiment, we deployed the C&C on an Apache web server over GNU/Linux. The malware sample to be deployed over the mobile nodes during the experiment was configured with the specific username, password, and URL of the C&C previously deployed. For this experiment, the experimental plane network was isolated from the Internet and the entire experiment was run in a self-contained network environment. To prepare

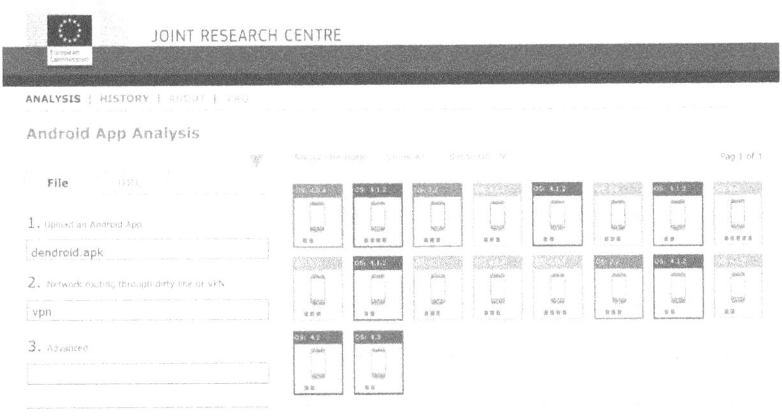

Figure 10.12   Submitting dendroid.apk for analysis using the platform webapplication.

the infection, we started the execution of the C&C server and deployed the infected APK to the set of virtual and physical phones using the management interface of the prototype, as seen in Figure 10.12.

Using this configuration, we run several experiments with different durations from 2 min up to a total of 2 h. In all the cases, the botnet runs successfully over the platform and the botmaster,[*] simulated by one of the members of our team, was able to execute commands over the infected physical and virtual mobile devices. After the finalization of each experiment, the platform provided detailed information about all network communication of the nodes, as well as traces of the execution of the malware in each node. From the logs of the network communications between the mobile phone and the C&C, as shown in Figure 10.13, it was possible to observe that the malware sends sensitive information in clear text, such as username and password. Communication between C&C and nodes, including the commands to the bots, as displayed in Figure 10.14, and their responses, could also be observed.

---

[*] The first time we deployed the malware we discovered it did not start on the emulated devices. Statically analyzing the malware at a deeper level, we discovered some anti-debugging code detecting the presence of a virtual device by checking the id of the system. In order to make the malware work also on emulated devices, we simply removed these lines of code and rebuilt the APK. Future versions of our platform will offer the functionality to mimic the id of actual devices in order to workaround this anti-debugging technique without the need of modifying the code of the malware.

```
⌃10.0.          → 10.0.
00000000: 4745 5420 2f67 6574 2d66 756e 6374 696f   GET./get-functio
00000010: 6e73 2e70 6870 3f55 4944 3d65 3565 3537   ns.php?UID=e5eS7
00000020: 6331 6666 3361 6635 3261 2650 6173 7377   c1ff3af52a&Passw
00000030: 6f72 643d                5454 502f 312e   ord=      HTTP/1.
00000040: 310d 0a55 7365 722d 4167 656e 743a 2044   1..User-Agent:.D
00000050: 616c 7669 6b2f 312e 362e 3020 284c 696e   alvik/1.6.0.(Lin
00000060: 7578 3b20 553b 2041 6e64 726f 6964 2034   ux;.U;.Android.4
00000070: 2e31 2e32 3b20 7364 6b20 4275 696c 642f   .1.2;.sdk.Build/
00000080: 4d41 5354 4552 290d 0a48 6f73 743a 2031   MASTER)..Host:.1
00000090:                     436f 6e6e 6563        .Connec
000000a0: 7469 6f6e 3a20 4b65 6570 2d41 6c69 7665   tion:.Keep-Alive
000000b0: 0d0a 4163 6365 7074 2d45 6e63 6f64 696e   ..Accept-Encodin
000000c0: 673a 2067 7a69 700d 0a0d 0a              g:.gzip....
```

```
⌄10.0         → 10.0
```

```
⌄10.0.        → 10.0.
00000300: ...                                       HTTP/1.1.200.OK
...
```

```
⌃10.0.        → 10.0
```

```
⌄10.0.        → 10.0.
```

```
⌃10.0.        → 10.0.
```

```
⌃10.0         → 10.0.
00000000: 4745 5420 2f67 6574 2d66 756e 6374 696f   GET./get-functio
00000010: 6e73 2e70 6870 3f55 4944 3d65 3565 3537   ns.php?UID=e5eS7
00000020: 6331 6666 3361 6635 3261 2650 6173 7377   c1ff3af52a&Passw
00000030: 6f72 643d           5454 502f 312e        ord=   .HTTP/1.
00000040: 310d 0a55 7365 722d 4167 656e 743a 2044   1..User-Agent:.D
00000050: 616c 7669 6b2f 312e 362e 3020 284c 696e   alvik/1.6.0.(Lin
00000060: 7578 3b20 553b 2041 6e64 726f 6964 2034   ux;.U;.Android.4
```

Figure 10.13 Sample of network trace showing information sent to Dendroid's C&C.

The results of the experiments show that the platform is able to support the execution of an actual mobile botnet found in the wild without affecting its functionalities. The monitoring capabilities of the platform also worked properly, providing traces of the execution of the malware in each node, as well as consolidated network traffic where the communications between the components of the botnet could be observed. This set of dynamic data can support subsequent analysis of the behavior of the botnet and help to better understand the techniques employed by this type of threat.

History Of: All Bots   [All Bots]   [Auto Scroll: On]   [View Awaiting Commands]

8de6676bc8085afa: [2016_03_18_15:33:48] - Opened Dialog: ALERT : ???
8de6676bc8085afa: [2016_03_18_15:45:34] - Opened Dialog: TEST : VPN
8de6676bc8085afa: [2016_03_18_15:51:26] - Opened Dialog: TEST 1 : VPN1
43e84edb99335579: [2016_03_22_15:54:01] - Screen On Complete
43e84edb99335579: [2016_03_22_15:54:13] - Screen On Complete
43e84edb99335579: [2016_04_01_13:58:02] - Taking Photo
43e84edb99335579: [2016_04_01_13:58:07] - Take Photo Complete
43e84edb99335579: [2016_04_01_13:58:24] - Screen On Complete
43e84edb99335579: [2016_04_01_13:58:46] - Taking Photo
43e84edb99335579: [2016_04_01_13:58:51] - Take Photo Complete
43e84edb99335579: [2016_04_01_13:58:58] - Screen On Complete
43e84edb99335579: [2016_04_01_14:00:15] - Taking Photo
43e84edb99335579: [2016_04_01_14:00:20] - Take Photo Complete
858c2c52848162d1: [2016_04_19_11:14:48] - Screen On Complete
43e84edb99335579: [2016_04_19_11:15:36] - Screen On Complete
43e84edb99335579: [2016_04_19_11:16:30] - Call Initiated: 32321321
43e84edb99335579: [2016_04_19_11:16:30] - Calling: 32321321
43e84edb99335579: [2016_04_19_11:16:42] - Webpage Opened: http://www-google-com
43e84edb99335579: [2016_04_19_11:17:24] - Webpage Opened: http://www-google-com

Figure 10.14   Commands sent by the C&C.

10.7.2 *Dropper and Payload*

In order to further analyze the use case of the infection vector, we have designed the following experiment that uses a malware sample recently found in the wild. This particular malware sample arrived in the form of an SMS from an unknown sender containing a link to download an application. A quick check using available online malware analysis services revealed no malicious results [46].

Given that in this case the C&C application was not available to us, we designed our experiments to provide Internet connectivity to the experimental plane network following the architecture previously described in Section 10.6.1. We set up our experiment to use five virtual phones and two physical phones, as shown in Figure 10.15. The experiment was left running for a period of 1 h.

At the end of the experiment, the logs provided by the platform reported some suspicious activities performed by the application, providing a strong indication of possible malicious activity. First of all, the sample triggered the installation of what appeared to be a well-known file explorer application. However, the network report of the platform revealed the download of another application, as shown in Figure 10.16.

Further analysis performed over that application extracted from the execution of the experiment revealed that it was a second-stage malware. We used the same experiment setup to analyze this second application and from the network logs it clearly appeared, as shown in Figure 10.17, that the infected phones sent personal information of the phone to an unknown Internet web service. A few days after the execution of this experiment, antiviruses started to detect this piece of malware categorizing it as a "Dropper" given its role as the first-stage malware downloading from the network additional malware code.

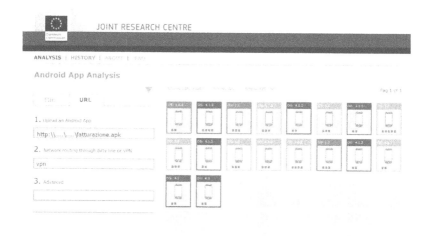

Figure 10.15 Submitting fatturazione.apk directly from the scam link.

An interesting finding analyzing this second-stage malware in a subsequent experiment was that it used a peculiar anti-debugging technique. The malware detects the presence of a virtual unattended environment by analyzing the user tapping on the screen. Fortunately for us, the latest version of the Cuckoo Sandbox engine that we used in the core of our platform was able to emulate random movements to counteract this anti-debugging technique.

10.8 Conclusions

In this work, we have presented our prototype of hybrid mobile botnet analysis platform. Unlike other existing sandboxes that aim to analyze standalone mobile malware samples, our platform is designed to allow researchers to go beyond standard static and dynamic analysis and experiment with the entire botnet ecosystem as a whole, including mobile nodes and C&C infrastructure. In doing so, it becomes possible to analyze not only the interaction between infected phones and the respective C&C but also the behavior of the entire botnet under several contexts. As an example, a researcher could design an experiment to determine how a particular mobile botnet behaves in the context of a population of infected phones interconnected by a GSM/3G network versus the same botnet with a population intercommunicated through a WiFi network. The ability of launching fully customized experiments with variable number of phones, operating systems, applications, communication networks, and type of sensors opens the door to an entire new range of experiments that could lead to a better understanding of this type of threats and the design of more effective methods to counteract them.

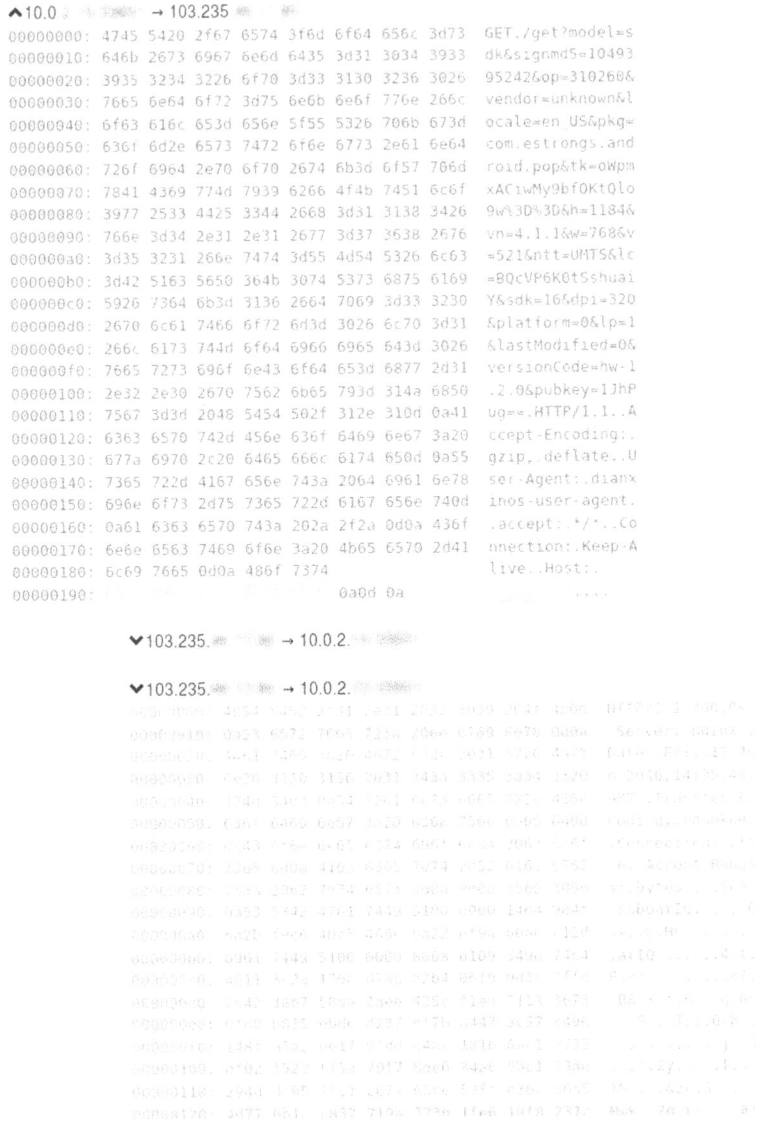

```
∧10.0.           → 103.235.
00000000:  4745 5420 2f67 6574 3f6d 6f64 656c 3d73   GET./get?model=s
00000010:  646b 2673 6967 6e6d 6435 3d31 3034 3933   dk&signmd5=10493
00000020:  3935 3234 3226 6f70 3d33 3130 3236 3026   95242&op=310268&
00000030:  7665 6e64 6f72 3d75 6e6b 6e6f 776e 266c   vendor=unknown&l
00000040:  6f63 616c 653d 656e 5f55 5326 706b 673d   ocale=en_US&pkg=
00000050:  636f 6d2e 6573 7472 6f6e 6773 2e61 6e64   com.estrongs.and
00000060:  726f 6964 2e70 6f70 2674 6b3d 6f57 706d   roid.pop&tk=oWpm
00000070:  7841 4369 774d 7939 6266 4f4b 7451 6c6f   xACiwMy9bfOKtQlo
00000080:  3977 2533 4425 3344 2668 3d31 3138 3426   9w%3D%3D&h=1184&
00000090:  766e 3d34 2e31 2e31 2677 3d37 3638 2676   vn=4.1.1&w=768&v
000000a0:  3d35 3231 266e 7474 3d55 4d54 5326 6c63   =521&ntt=UMTS&lc
000000b0:  3d42 5163 5650 364b 3074 5373 6875 6169   =BQcVP6K0tSshuai
000000c0:  5926 7364 6b3d 3136 2664 7069 3d33 3230   Y&sdk=16&dpi=320
000000d0:  2670 6c61 7466 6f72 6d3d 3026 6c70 3d31   &platform=0&lp=1
000000e0:  266c 6173 744d 6f64 6966 6965 643d 3026   &lastModified=0&
000000f0:  7665 7273 696f 6e43 6f64 653d 6877 2d31   versionCode=hw-1
00000100:  2e32 2e30 2670 7562 6b65 793d 314a 6850   .2.0&pubkey=1JhP
00000110:  7567 3d3d 2048 5454 502f 312e 310d 0a41   ug==.HTTP/1.1..A
00000120:  6363 6570 742d 456e 636f 6469 6e67 3a20   ccept-Encoding:.
00000130:  677a 6970 2c20 6465 666c 6174 650d 0a55   gzip,.deflate..U
00000140:  7365 722d 4167 656e 743a 2064 6961 6e78   ser-Agent:.dianx
00000150:  696e 6f73 2d75 7365 722d 6167 656e 740d   inos-user-agent.
00000160:  0a61 6363 6570 743a 202a 2f2a 0d0a 436f   .accept:.*/*..Co
00000170:  6e6e 6563 7469 6f6e 3a20 4b65 6570 2d41   nnection:.Keep-A
00000180:  6c69 7665 0d0a 486f 7374   live..Host:.
00000190:                       0a0d 0a           ....
```

∨103.235.   → 10.0.2.

∨103.235.   → 10.0.2.

Figure 10.16   Network traces showing the download of the payload.

The architecture that we have described in this chapter is designed to maximize the flexibility of the platform and its ability to monitor the activity of the malware while providing a high level of security. The separation between the control plane and the experimental plane provides an extra layer of security effectively isolating the activity of the malware to the experimental network. The possibility for the platform

```
▲10.0.        —  103.235.
00000000: 4745 5420 2f61 6475 6e69 6f6e 2f73 6c6f   GET./adunion/slo
00000010: 742f 6765 7453 7263 5072 696f 3f68 3d31   t/getSrcPrio?h=1
00000020: 3138 3426 773d 3736 3826 6d6f 6465 6c3d   184&w=768&model=
00000030: 7364 6b26 7665 6e64 6f72 3d75 6e6b 6e6f   sdk&vendor=unkno
00000040: 776e 2673 646b 3d31 3626 6470 693d 3332   wn&sdk=16&dpi=32
00000050: 3026 7376 3d31 2e30 2e32 2673 766e 3d53   0&sv=1.0.2&svn=S
00000060: 454c 462d 312e 302e 3226 706b 673d 636f   ELF-1.0.2&pkg=co
00000070: 6d2e 6573 7472 6f6e 6773 2e61 6e64 726f   m.estrongs.andro
00000080: 6964 2e70 6f70 2676 3d35 3231 2676 6e3d   id.pop&v=521&vn=
00000090: 342e 312e 3126 746b 3d6f 5770 6d78 4143   4.1.1&tk=oWpmxAC
000000a0: 6977 4d79 3962 664f 4b74 516c 6f39 7725   iwMy9bfOKtQlo9w%
000000b0: 3344 2533 4426 6f70 3d33 3130 3236 3026   3D%3D&op=310260&
000000c0: 6c6f 6361 6c65 3d65 6e5f 5553 266e 7474   locale=en_US&ntt
000000d0: 3d55 4d54 5326 6c73 3d65 3635 6532 6238   =UMTS&ls=e65e2b8
000000e0: 3363 3762 3137 3532 6630 3064 3733 3333   3c7b1752f00d7333
000000f0: 3566 3061 3464 3233 3826 6169 643d 6535   5f0a4d238&aid=e5
00000100: 6535 3763 3166 6633 6166 3532 6126 7369   e57c1ff3af52a&si
00000110: 643d 3130 3333 3625 3243 3130 3033 3125   d=10336%2C10031%
00000120: 3243 3130 3834 3526 7265 733d 3130 3830   2C10845&res=1080
00000130: 2a34 3630 2532 4332 3434 2a32 3434 2532   *460%2C244*244%2
00000140: 4331 3730 2a31 3730 2532 4331 3038 2a31   C170*170%2C108*1
00000150: 3038 2048 5454 502f 312e 310d 0a49 662d   08.HTTP/1.1..If-
00000160: 4d6f 6469 6669 6564 2d53 696e 6365 3a20   Modified-Since:.
00000170: 5468 752c 2030 3120 4a61 6e20 3139 3730   Thu,.01.Jan.1970
00000180: 2030 303a 3030 3a30 3020 474d 540d 0a41   .00:00:00.GMT..A
00000190: 6363 6570 742d 456e 636f 6469 6e67 3a20   ccept-Encoding:.
000001a0: 677a 6970 0d0a 486f 7374   gzip..Host:.
000001b0:
000001c0:                6f6e 6e65 6374 696f   ..Connectio
000001d0: 6e3a 204b 6565 702d 416c 6976 650d 0a55   n:.Keep-Alive..U
000001e0: 7365 722d 4167 656e 743a 204d 6f7a 696c   ser-Agent:.Mozil
000001f0: 6c61 2f35 2e30 2028 5831 313b 2055 3b20   la/5.0.(X11;.U;.
00000200: 4c69 6e75 7820 7838 365f 3634 3b20 656e   Linux.x86_64;.en
00000210: 2d55 533b 2072 763a 312e 392e 322e 3138   -US;.rv:1.9.2.18
00000220: 2920 4765 636b 6f2f 3230 3131 3036 3238   ).Gecko/20110628
00000230: 2055 6275 6e74 752f 3130 2e30 3420 286c   .Ubuntu/10.04.(l
00000240: 7563 6964 2920 4669 7265 666f 782f 332e   ucid).Firefox/3.
00000250: 362e 3138 0d0a 0d0a   6.18....
```

Figure 10.17   Network traces showing information sent by the infected node.

to route Internet traffic through the TOR network allows the design of experiments that require the botnet under analysis to access external components such as C&C infrastructure in a safer way preserving the anonymity of the researcher.

One key feature of our platform is its hybrid dimension allowing the usage of both emulated and physical mobile devices. This important feature that drove the design of our architecture aims to strike a balance between the scalability of the platform, by means of the flexibility provided by the usage of emulated devices, and the resilience against sandbox detection techniques commonly employed by malware found in the wild. Indeed, as illustrated by the experiments described in our work, mobile malware often use anti-debugging and sandbox detection techniques to detect the presence of a sandbox and abort the execution of the malicious code.

The usage of a mix of physical and mobile devices in the experiments supported by the platform allows the detection of such behavior and provides the researcher with means to customize the settings of the experiment, in terms of number and type of devices, to better fit the specific purposes of the analysis.

The prototype we have developed is a work in progress but it has already provided promising results demonstrating its ability to analyze actual mobile malware and providing meaningful insights into the dynamics of mobile botnets. The excellent open-source Cuckoo Sandbox used by the platform has proven to be the right choice to support the execution and monitoring of the virtual mobile nodes. Future versions of Cuckoo Sandbox, which is under active development and quickly approaching release 2.0, will be integrated into our platform to benefit from future improvements and new features, such as new anti-debugging evasion techniques.

Support for the monitoring of sensor activities, such as NFC, accelerometer, and Bluetooth, is planned to be added in future versions of our platform in order to support the analysis of mobile botnets that might make some special use of them under specific circumstances. Future work will test the scalability of the platform running experiments with a large number of nodes and consider the integration with the EPIC platform currently hosted at the JRC. The integration with EPIC [47] can boost the scalability of the solution and efficiently recreate realistic network topologies and conditions (e.g., delay and loss characteristics of wide area network [WAN] links) of the Internet infrastructure to support more realistic experiments.

References

1. R. A. Rodríguez-Gómez, G. Maciá-Fernández, and P. García-Teodoro, Survey and taxonomy of botnet research through life-cycle, *ACM Comput. Surv.*, **45**(4): 45:1–45:33, 2013.
2. H. R. Zeidanloo and A. A. Manaf, Botnet command and control mechanisms, in *Computer and Electrical Engineering, 2009. ICCEE'09. Second International Conference on*, vol. **1**, pp. 564–568, Dec 2009.
3. C. Mulliner and J. P. Seifert, Rise of the ibots: Owning a telco network, in *Malicious and Unwanted Software (MALWARE), 2010 5th International Conference on*, Nancy, France, pp. 71–80, Oct 2010.
4. C. Xiang, F. Binxing, Y. Lihua, L. Xiaoyi, and Z. Tianning, Andbot: Towards advanced mobile botnets, in *Proceedings of the 4th USENIX Conference on Large-Scale Exploits and Emergent Threats*, LEET'11, USENIX Association, Berkeley, CA, USA, pp. 11–11, 2011.
5. Y. Zeng, K. G. Shin, and X. Hu, Design of sms commanded-and-controlled and P2P-structured mobile botnets, in *Proceedings of the Fifth ACM Conference on Security and Privacy in Wireless and Mobile Networks*, WISEC'12, ACM, New York, NY, USA, pp. 137–148, 2012.
6. R. Nigam, A timeline of mobile botnets, in *Proceedings of the 4th Edition of the Botnet Fighting Conference—Botconf 2014*, 2014.

7. S. S. C. Silva, R. M. P. Silva, R. C. G. Pinto, and R. M. Salles, Botnets: A survey, *Comput. Netw.*, **57**(2): 378–403, 2013.

8. B. Stone-Gross, T. Holz, G. Stringhini, and G. Vigna, The underground economy of spam: A botmaster's perspective of coordinating large-scale spam campaigns, in *Proceedings of the 4th USENIX Conference on Large-Scale Exploits and Emergent Threats*, LEET'11, USENIX Association, Berkeley, CA, USA, pp. 4–4, 2011.

9. Z. Zhang, R. Ando, and Y. Kadobayashi, Hardening botnet by a rational botmaster, in M. Yung, P. Liu, and D. Lin, editors, *Information Security and Cryptology*, Springer-Verlag, Berlin, Heidelberg, pp. 348–369, 2009.

10. W. T. Strayer, D. Lapsely, R. Walsh, and C. Livadas, Botnet detection based on network behavior, in *Botnet Detection: Countering the Largest Security Threat*, Springer US, Boston, MA, pp. 1–24, 2008.

11. I. Vural and H. Venter, Mobile botnet detection using network forensics, in *Proceedings of the Third Future Internet Conference on Future Internet*, FIS'10, Springer-Verlag, Berlin, Heidelberg, pp. 57–67, 2010.

12. D. Zhao, I. Traore, B. Sayed, W. Lu, S. Saad, A. Ghorbani, and D. Garant, Botnet detection based on traffic behavior analysis and flow intervals, *Comput. Secur.*, **39**: 2–16, 2013.

13. S.-H. Seo, K. Yim, and I. You, Mobile malware threats and defenses for homeland security, in *Multidisciplinary Research and Practice for Information Systems: IFIP WG 8.4, 8.9/TC 5 International Cross-Domain Conference and Workshop on Availability, Reliability, and Security, CD-ARES 2012, Prague, Czech Republic, August 20–24, 2012. Proceedings*, Springer, Berlin, Heidelberg, pp. 516–524, 2012.

14. R. Hasan, N. Saxena, T. Haleviz, S. Zawoad, and D. Rinehart, Sensing-enabled channels for hard-to-detect command and control of mobile devices, in *Proceedings of the 8th ACM SIGSAC Symposium on Information, Computer and Communications Security*, ASIA CCS'13, ACM, New York, NY, USA, pp. 469–480, 2013.

15. J. Hua and K. Sakurai, A sms-based mobile botnet using flooding algorithm, in *Proceedings of the 5th IFIP WG 11.2 International Conference on Information Security Theory and Practice: Security and Privacy of Mobile Devices in Wireless Communication*, WISTP'11, Springer-Verlag, Berlin, Heidelberg, pp. 264–279, 2011.

16. A. Karim, S. A. A. Shah, and R. Salleh, Mobile Botnet attacks: A thematic taxonomy, in *New Perspectives in Information Systems and Technologies*, Springer International Publishing, Cham, vol. **2**, pp. 153–164, 2014.

17. E. Gelenbe, G. Görbil, D. Tzovaras, S. Liebergeld, D. Garcia, M. Baltatu, and G. Lyberopoulos, NEMESYS: Enhanced network security for seamless service provisioning in the smart mobile ecosystem, in *Information Sciences and Systems 2013: Proceedings of the 28th International Symposium on Computer and Information Sciences*, Springer International Publishing, Cham, pp. 369–378, 2013.

18. M. Wählisch, S. Trapp, C. Keil, J. Schönfelder, T. C. Schmidt, and J. Schiller, First insights from a mobile honeypot, in *Proceedings of the ACM SIGCOMM 2012 Conference on Applications, Technologies, Architectures, and Protocols for Computer Communication*, SIGCOMM'12, ACM, New York, NY, USA, pp. 305–306, 2012.

19. Sufatrio, D. J. J. Tan, T.-W. Chua, and V. L. L. Thing, Securing Android: A survey, taxonomy, and challenges, *ACM Comput. Surv.*, **47**(4): 58:1–58:45, 2015.

20. A. Karim, R. Salleh, and M. K. Khan, Smartbot: A behavioral analysis framework augmented with machine learning to identify mobile botnet applications, *PLoS ONE*, **11**(3): 1–35, 2016.

21. M. R. Faghani and U. T. Nguyen, Socellbot: A new botnet design to infect smartphones via online social networking, in *Electrical Computer Engineering (CCECE), 2012 25th IEEE Canadian Conference on*, Montreal, QC, Canada, pp. 1–5, 2012.

22. M. Anagnostopoulos, G. Kambourakis, and S. Gritzalis, New facets of mobile botnet: Architecture and evaluation, *Int. J. Inf. Secur.*, **15**(5): 455–473, 2016.

23. M. Eslahi, M. R. Rostami, H. Hashim, N. M. Tahir, and M. V. Naseri, A data collection approach for mobile botnet analysis and detection, in *2014 IEEE Symposium on Wireless Technology and Applications (ISWTA)*, Kota Kinabalu, Malaysia, pp. 199–204, Sep 2014.

24. H. Pieterse and M. S. Olivier, Android botnets on the rise: Trends and characteristics, in *2012 Information Security for South Africa*, Johannesburg, Gauteng, South Africa, pp. 1–5, Aug 2012.

25. B. Chun, D. Culler, T. Roscoe, A. Bavier, L. Peterson, M. Wawrzoniak, and M. Bowman, Planetlab: An overlay testbed for broad-coverage services, *SIGCOMM Comput. Commun. Rev.*, **33**(3): 3–12, 2003.

26. B. White, J. Lepreau, L. Stoller, R. Ricci, S. Guruprasad, M. Newbold, M. Hibler, C. Barb, and A. Joglekar, An integrated experimental environment for distributed systems and networks, in *Proceedings of the Fifth Symposium on Operating Systems Design and Implementation*, USENIX Association, Boston, MA, pp. 255–270, 2002.

27. A. Varga, The OMNeT++ discrete event simulation system, in *Proceedings of the European Simulation Multiconference (ESM 2001)*, vol. **9**, SCS Prague, Czech Republic, pp. 65, sn, 2001.

28. B. White, J. Lepreau, L. Stoller, R. Ricci, S. Guruprasad, M. Newbold, M. Hibler, C. Barb, and A. Joglekar, An integrated experimental environment for distributed systems and networks, in *Proceedings of the 5th Symposium on Operating Systems Design and Implementation*, ACM, NY, pp. 255–270, 2002.

29. T. Benzel, The science of cyber security experimentation: The DETER project, in *Proceedings of the 27th Annual Computer Security Applications Conference*, ACSAC'11, ACM, New York, NY, USA, pp. 137–148, 2011.

30. J. Mirkovic, T. V. Benzel, T. Faber, R. Braden, J. T. Wroclawski, and S. Schwab, The DETER project: Advancing the science of cyber security experimentation and test, in *Technologies for Homeland Security (HST), 2010 IEEE International Conference on*, pp. 1–7, Nov 2010.

31. C. Siaterlis, B. Genge, and M. Hohenadel, Epic: A testbed for scientifically rigorous cyber-physical security experimentation, *IEEE Trans. Emerg. Topics Comput.*, **1**(2): 319–330, 2013.

32. J. Mirkovic, A. Hussain, S. Fahmy, P. Reiher, and R. K. Thomas, Accurately measuring denial of service in simulation and testbed experiments, *IEEE Trans. Depend. Secure Comput.*, **6**(2): 81–95, 2009.

33. M. Spreitzenbarth, F. Freiling, F. Echtler, T. Schreck, and J. Hoffmann, Mobile-sandbox: Having a deeper look into Android applications, in *Proceedings of the 28th Annual ACM Symposium on Applied Computing*, SAC '13, ACM, New York, NY, USA, pp. 1808–1815, 2013.

34. T. Blsing, L. Batyuk, A. D. Schmidt, S. A. Camtepe, and S. Albayrak, An Android application sandbox system for suspicious software detection, in *Malicious and Unwanted Software (MALWARE), 2010 5th International Conference on*, Nancy, France, pp. 55–62, Oct 2010.

35. A. Malatras, E. Freyssinet, and L. Beslay, Mobile botnets taxonomy and challenges, in *Intelligence and Security Informatics Conference (EISIC), 2015 European*, Nancy, France, pp. 149–152, Sep 2015.

36. A. Karasaridis, B. Rexroad, and D. Hoeflin, Wide-scale botnet detection and characterization, in *Proceedings of the First Conference on First Workshop on Hot Topics in Understanding Botnets, HotBots'07*, USENIX Association, Berkeley, CA, USA, pp. 7–7, 2007.

37. P. Traynor, M. Lin, M. Ongtang, V. Rao, T. Jaeger, P. McDaniel, and T. La Porta, On cellular botnets: Measuring the impact of malicious devices on a cellular network core, in *Proceedings of the 16th ACM Conference on Computer and Communications Security*, CCS'09, ACM, New York, NY, USA, pp. 223–234, 2009.

38. A. Egners, U. Meyer, and B. Marschollek, Messing with Android's permission model, in *2012 IEEE 11th International Conference on Trust, Security and Privacy in Computing and Communications*, Liverpool, UK, pp. 505–514, 2012.

39. C. Mulliner, S. Liebergeld, and M. Lange, Honeydroid—Creating a smartphone honeypot, in *Poster Session of the IEEE Symposium on Security and Privacy (May 2011)*, IEEE, 2011.

40. J. Oberheide and F. Jahanian, When mobile is harder than fixed (and vice versa): Demystifying security challenges in mobile environments, in *Proceedings of the Eleventh Workshop on Mobile Computing Systems and Applications*, HotMobile'10, ACM, New York, NY, USA, pp. 43–48, 2010.

41. S. Greengard, The war against botnets, *Commun. ACM*, **55**(2): 16–18, 2012.

42. M. Antonakakis, R. Perdisci, Y. Nadji, N. Vasiloglou, S. Abu-Nimeh, W. Lee, and D. Dagon, From throw-away traffic to bots, Detecting the rise of DGA-based malware, in *Proceedings of the 21st USENIX Conference on Security Symposium*, Security'12, USENIX Association, Berkeley, CA, USA, pp. 24–24, 2012.

43. A. Malatras and L. Beslay, Building a hybrid experimental platform for mobile botnet research, *Le Journal de la cybercriminalité & des Investigations numériques*, **1**(1): 29–39, 2016.

44. Mobile Botnet Malware Collection Website, https://www.botconf.eu/mobilebot/, Online; 2015.

45. VirusTotal Dendroid Analysis, https://www.virustotal.com/en-gb/file/8336f 74515b252233cffd8b5b50dab0be40c13f31c0f641445216424c9be/analysis/, Online; 02 October 2014.

46. VirusTotal Dropper Analysis, https://www.virustotal.com/en-gb/file/5303e12c5704c 4726cf2723e604509af2d8ba6c4b4140a00b50d4381f272ff26/analysis/1471616814/, Online; 07 July 2016.

47. C. Siaterlis, A. P. Garcia, and B. Genge, On the use of Emulab testbeds for scientifically rigorous experiments, *IEEE Commun. Surveys Tuts.*, **15**(2): 929–942, Second Quarter 2013.

48. G. Bontempi and Y. Le Borgne, An adaptive modular approach to the mining of sensor network data, in *Proceedings of the Workshop on Data Mining in Sensor Networks, SIAM SDM*, SIAM Press, Newport Beach, CA, pp. 3–9, 2005.

49. K. I. Diamantaras and S. Y. Kung, *Principal Component Neural Networks: Theory and Applications*, John Wiley & Sons Inc., New York, NY, USA, 1996.

50. A. Hyvarinen, J. Karhunen, and E. Oja, *Independent Component Analysis*, Wiley, New York, 2001.

51. M. Ilyas, I. Mahgoub, and L. Kelly, *Handbook of Sensor Networks: Compact Wireless and Wired Sensing Systems*, CRC Press, Boca Raton, FL, USA, 2004.

52. A. Jain and E. Y. Chang, Adaptive sampling for sensor networks, in *Proceedings of the 1st International Workshop on Data Management for Sensor Networks: In Conjunction with VLDB 2004*, ACM, NY, pp. 10–16, 2004.

53. I. T. Jolliffe, *Principal Component Analysis*, Springer-Verlag, NY, 2002.

54. S. Madden, M. J. Franklin, J. M. Hellerstein, and W. Hong, TAG: A Tiny AGgregation Service for Ad-Hoc Sensor Networks, in *Proceedings of the 5th ACM Symposium on Operating System Design and Implementation (OSDI)*, ACM Press, Boston, MA, vol. **36**, pp. 131–146, 2002.

55. S. R. Madden, M. J. Franklin, J. M. Hellerstein, and W. Hong, TinyDB: An acquisitional query processing system for sensor networks, *ACM Trans. Database Syst. (TODS)*, **30**(1): 122–173, 2005.

56. K. V. Mardia, J. T. Kent, J. M. Bibby, Multivariate Analysis, Academic Press, New York, 1979.

57. Y. Yao and J. Gehrke, The cougar approach to in-network query processing in sensor networks, *ACM SIGMOD Record*, **31**(3): 9–18, 2002.

Chapter 11

Applying Low-Cost Software Radio for Experimental Analysis of LTE Security, Protocol Exploits, and Location Leaks

---

Roger Piqueras Jover

Contents

The Long-Term Evolution (LTE) is the newest cellular communications standard globally deployed. Regardless of previous generations, with the coexistence of different technologies for mobile access, all operators are globally converging to LTE for next-generation mobile communications. With a fully redesigned PHYsical layer, built upon Orthogonal Frequency Division Multiple Access (OFDMA), LTE networks provide orders of magnitude higher rates and lower traffic latencies, combined into a strong resiliency to multipath fading and overall improved efficiency in terms of bits per second per unit of bandwidth. This highly improved Radio Access Network (RAN) is architected over the Enhanced Packet Core (EPC) network to provide connectivity to all types of mobile devices [1].

LTE cellular networks deliver advanced services for billions of users, beyond traditional voice and short message communications, as the cornerstone of today's digital and connected society. Moreover, mobile networks are one of the main enablers for the emergence of Machine to Machine (M2M) systems, with LTE expected to play a key role in the Internet of Things (IoT) revolution [2]. Consequently, M2M and the IoT are often analyzed as key elements within the LTE security ecosystem [3].

Given the widespread usage, with a subscription count in the billions, securing the connectivity of mobile devices is of extreme importance. The first generation of mobile networks (1G) lacked the support for encryption and the legacy 2G networks lacked mutual authentication and implemented an outdated encryption algorithm [4]. The wide availability of open-source implementations of the GSM protocol stack has resulted in many security research projects unveiling several exploits possible on the GSM insecure radio link.

Specific efforts were thus made to ensure confidentiality and authentication in mobile networks, resulting in much stronger cryptographic algorithms and mutual authentication being explicitly implemented in both 3G and LTE. Based on this, LTE is generally considered secure given this mutual authentication and strong encryption scheme. As such, confidentiality and authentication are wrongly assumed to be sufficiently guaranteed. LTE mobile networks are still vulnerable to protocol exploits, location leaks, and rogue base stations.

Based on the analysis of real LTE traffic captures obtained from live production networks in the areas of New York City and Honolulu, this chapter discusses the insecurity rationale behind LTE protocol exploits and rogue base stations. Despite the strong cryptographic protection of user traffic and mutual authentication of LTE, a very large number of control plane (signaling) messages are exchanged over an LTE radio link in the clear regularly. Before the authentication and encryption steps of a connection are executed, a mobile device engages in a substantial conversation with any LTE base station (real or rogue) that advertises itself with the correct broadcast information. This broadcast information is sent in the clear and can be easily sniffed [5], allowing an adversary to easily configure and set up a rogue access point. Real samples of LTE broadcast traffic are captured and analyzed to exemplify this and discuss potential ways this information could be leveraged with a malicious intent.

Motivated by this LTE protocol insecurity rationale, this chapter explores different areas of mobile network security with low-cost software radio tools. Based on an open-source implementation of the LTE stack, openLTE [6], a series of exploits are demonstrated and implemented. Discussion is provided regarding the implementation of low-cost IMSI catchers as well as exploits that allow blocking mobile devices, which were publicly introduced for the first time in Reference 7 and shortly after also discussed in Reference 8. Moreover, based on the analysis of real LTE traffic captures, a new location information leak in the LTE protocol is introduced, which allows potentially tracking LTE devices as they move. Potential mitigations and security discussions are provided for these exploits as well.

This chapter extends the results presented in Reference 8 and summarizes the author's research in LTE mobile security and protocol exploits research over the last few years.

11.1 LTE Mobile Networks

LTE mobile networks, as illustrated in Figure 11.1, split their architecture into two main sections: the Radio Access Network (RAN) and the core network, known as the Evolved Packet Core (EPC) [1]. The RAN of an LTE network comprises of the mobile terminals, known as User Equipment (UE), and eNodeBs, or LTE base stations. The evolution of mobile networks toward LTE highly specializes and isolates the functionality of the RAN. In current mobile deployments, the LTE RAN is able to, independently from the EPC, assign radio resources to UEs, manage their radio resource utilization, implement access control, and, leveraging the X2 interface between eNodeBs, manage mobility and handoffs.

The EPC, in turn, is in charge of establishing and managing the point-to-point connectivity between UEs and the Internet. In order to do so, the EPC leverages a series of nodes. The Serving Gateway (S-GW) and the PDN (Packet Data Network) Gateway (P-GW) are the two routing anchors for user traffic connectivity to the PDN. Once a connection is established, user data flows from the UE to the eNodeB, and is then routed over the S-GW and P-GW toward the PDN. In parallel to the routing functionality of both gateways, the Mobility Management Entity (MME) handles logistics of the bearer establishment and release, mobility management, and other network functionalities, such as authentication and access control. In order to provide security for user traffic and execute mutual authentication, the MME communicates with the Home Subscriber Server (HSS), which stores the authentication parameters, secret keys, and user account details of all the

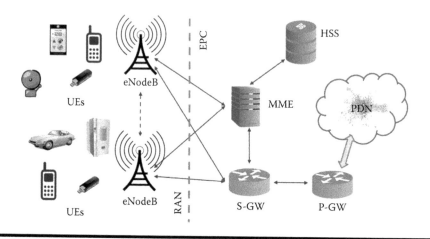

Figure 11.1   LTE network architecture.

UEs. The HSS is also leveraged upon incoming connections for mobile devices, in order to address paging messages.

Any mobile device or UE attempting to access the network must follow a series of steps illustrated in Figure 11.2. The process is initiated by the cell selection procedure, which involves the detection and decoding of the Primary Synchronization Signal (PSS) and the Secondary Synchronization Signal (SSS). Then, the Physical Broadcast Channel (PBCH) is decoded to extract the most basic system

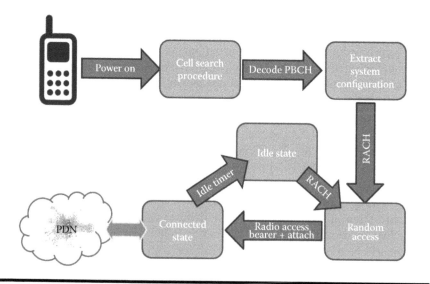

Figure 11.2   LTE cell selection and connection.

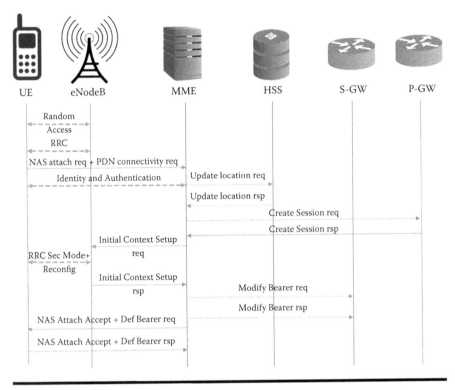

Figure 11.3 **LTE NAS attach procedure.**

configuration in the Master Information Block (MIB), such as the system band-width, which allows other channels in the cell to be configured and operated. The remaining details of the configuration of the cell are extracted from the System Information Blocks (SIB), which are unencrypted and can be eavesdropped by a passive radio sniffer. Then, the UE initiates an actual connection with the network by means of a random access procedure and establishes, via the NAS (Non-Access Stratum) Attach process, an end-to-end bearer in order to send and receive user traffic.

Figure 11.3 illustrates the connection process through which a mobile device attaches to the network and a point-to-point IP bearer is set up to provide data connectivity. The NAS Attach procedure contains a number of steps, involving all the elements in the EPC. The random access procedure assigns radio resources to the UE so it can set up a Radio Resource Control (RRC) connection with the eNodeB. The next step is to execute the identity/authentication procedure between the UE and the MME, which in turn leverages the HSS to configure security attributes and encryption. Finally, the point-to-point circuit through the SGW and PGW is set up, and the RRC connection is reconfigured based on the type of QoS (Quality of Service) requested. At this point, the UE is in the *connected* RRC state.

Another LTE procedure relevant in protocol security experimentation is paging. This is the process that initiates a mobile terminated connection and triggers, in the case of an incoming communication, the idle-to-connected state transition. At any time there is an incoming communication addressed to a given UE, the network must identify the cell in which the user is located. If this specific cell location was known at all times, a UE would be required to update its location with the network each time it moved to a new cell. This would be a costly operation that would result in great amounts of location update control packets within the network. In order to reduce the load of location update signaling, the location is only known with a much larger granularity. The network is at all times aware of the last Tracking Area (TA) where the UE was located. Upon receiving incoming traffic for a given UE that is in idle state, the EPC triggers the broadcast of a paging message over each cell within the TA where that UE is known to be [19]. As a result, the mobile terminal replies to this paging message, indicating this way its exact location in terms of cell or sector. This triggers the establishment of a bearer following a procedure similar to the one described above.

11.1.1 Mobile Network Identifiers

A number of identifiers are used in the operation of an LTE mobile network. The most important ones, in the context of this chapter, are the following:

- *IMSI (International Mobile Subscriber Identity)*: This is a unique identifier for the SIM card in a mobile device. The IMSI is a secret identifier that should be kept private and not transmitted in the clear as it can be used to track devices and other types of exploits [9].
- *TMSI (Temporary Mobile Subscriber Identifier)*: This is the identifier used to uniquely address a given device instead of the IMSI. Once the device connects to the network for the first time, a TMSI is derived and used thereafter. The TMSI is also refreshed periodically, though not as much as it should [4].
- *MSISDN (Mobile Subscriber ISDN Number)*: This is the id that identifies the user and owner of a mobile device, that is, the user's phone number. It is mapped to the TMSI in a similar way a url is mapped to an IP address.
- *IMEI (International Mobile Equipment Identifier)*: It uniquely identifies the hardware (i.e., smartphone) used to connect to the network. It can be thought as the serial number of a mobile device. The IMEI should ideally also be kept secret as, based on its value, one can easily know the type of mobile device the user has (make and model) and well as, in the case of an embedded M2M device, the software version it is running.

11.2 LTE Security Analysis Tools

The LTE security exploration work summarized in this chapter was possible thanks to a number of recent open-source implementations of the LTE protocol. Open-source implementations of the GSM protocol have been available for years, resulting in outstanding research work and substantial improvement in the understanding of GSM security. Some argue that open-source cellular implementations provide tools and resources to radio hackers to attack mobile networks. However, many argue that open source provided the tools to brilliant security researchers to find numerous flaws in the GSM protocol, improving the security of mobile communications overall.

Over the last few years, a number of open-source projects have been developed, providing the right tools for sophisticated LTE security research. Running on off-the-shelf software radio platforms, these open-source libraries provide, in some cases, the functionality of software-based eNodeB. With not too complex modifications of the code, they can easily be turned into LTE protocol analyzers, Stingrays, and rogue base stations, as it will be discussed throughout this chapter.

The main LTE open-source implementations being actively developed can be summarized as follows[*]:

- *openLTE [6]*: Currently, the most advanced open-source implementation of the LTE stack. It provides a fully functional LTE connection, including the features of the LTE packet core network. With proper configuration, it can operate NAS protocols and provide access to the Internet for mobile devices. It implements the HSS functionality on a text file storing IMSI–key pairs. It is the cornerstone of most current security research focused on LTE protocols.
- *srsLTE [10]*: Partial implementation of the LTE stack that provides full access to PHY layer features and metrics and full access to decoded broadcast messages. It provides advanced LTE network scanning functions along with several other tools. The srsLTE project recently introduced srsUE, an implementation of the UE stack that allows to emulate a mobile device against an eNodeB, which could be leveraged in security experimentation against the LTE infrastructure and the eNodeB.
- *gr-LTE [11]*: Open-source LTE implementation based on gnuradio-companion, which makes it ideal for beginners to start familiarizing with the LTE protocol and signal processing steps. It mostly implements the PHY layer.

Most open-source implementations can be run using standard off-the-shelf software radio, such as the USRP [12]. This tool allows both passive and active

---

[*] Note that this is not an exhaustive list, but a summary of the open-source tools that have been tested within the scope of the results in this project.

experimentation, as it provides transmit and receive features. Passive traffic capture and eNodeB broadcast information can also be performed with much simpler platforms, such as RTL-SDR radios [13]. Note that, to abide with regulations, active experimentation (i.e., UE and eNodeB emulation) must be carried out in an RF-shielded environment.

The experimentation summarized in this chapter has been done with a Core i7 Ubuntu machine operating a USRP B210 software radio platform. All the protocol exploits described in this chapter have been implemented by means of a customized version of openLTE, adding certain new features such as the collection of the IMSI of devices that attempt to establish a connection. All active experimentation was performed inside a Faraday cage using standard smartphones. One of the versions of eNodeB scanner was implemented using the same computing equipment and an RTL-SDR radio.

11.3 LTE Traffic Captures

Most traffic captures displayed in this chapter were obtained with a Sanjole Wavejudge LTE sniffer and protocol analyzer [14]. This tool provides the means to capture and analyze traffic at the PHY layer as well as real-time capture and decoding of MAC, RRC, NAS, and other layers. The majority of traffic capture analysis was performed with the Wavejudge software provided by the same vendor.

The LTE passive traffic captures were obtained from production networks in both the Honolulu and New York City areas, providing a real snapshot of operational LTE mobile networks in an urban environment. The traffic captures for active experimentation, that is, IMSI catcher and device blocking, were obtained within a Faraday cage with a modified version of openLTE acting as both the eNodeB and traffic sniffer, and the author's smartphone as the communicating endpoint. Note that all types of active radio experimentation must always be performed inside a Faraday cage in order to comply with U.S. federal regulations.

All user traffic, despite being fully encrypted and un-decodable, was filtered out and only control plane traffic is captured and analyzed. All traffic captures included in this chapter are edited to conceal the author's smartphone IMSI, cell ids, mobile network operator identifiers, and other sensitive data.

11.4 LTE Security and Protocol Exploits

This section discusses the rationale behind LTE protocol exploits based on the analysis of real LTE control plane traffic captures. Based on this, a series of security aspects and protocol exploits in the context of LTE are introduced, from leveraging the information extracted from MIB and SIB messages for rogue eNodeB configuration to denial-of-service (DoS) threats to temporarily block mobile devices and location leaks.

11.4.1 MIB and SIB Message Eavesdropping

The MIB and SIB messages are broadcasted and mapped on the LTE frame over radio resources known *a priori*. Moreover, these messages are transmitted with no encryption. Therefore, any passive sniffer is able to decode them. This simplifies the initial access procedure for the UEs but could be potentially leveraged by an attacker to craft sophisticated jamming attacks, optimize the configuration of a rogue base station, or tune other types of sophisticated attacks [15].

Figure 11.4 provides an example of the contents of an MIB and SIB1 message broadcasted by a commercial eNodeB in the area of New York City. From the information extracted from these messages, an adversary can learn the mobile operator

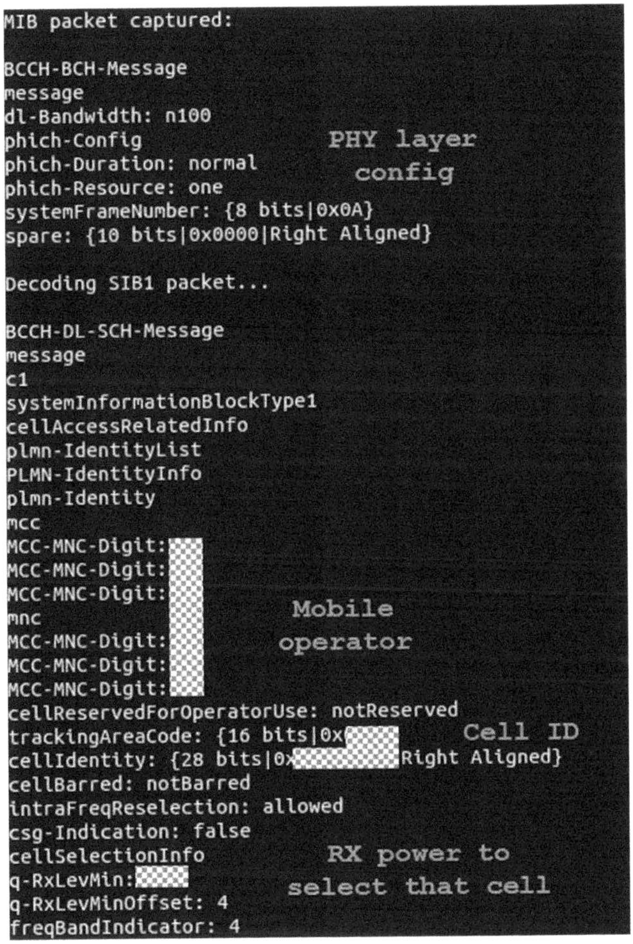

Figure 11.4   Real capture of MIB and SIB1 LTE broadcast messages.

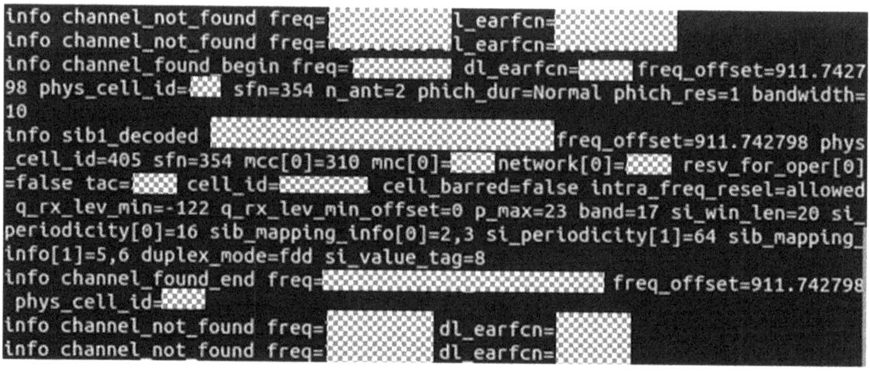

```
info channel_not_found freq=          l_earfcn=
info channel_not_found freq=          l_earfcn=
info channel_found_begin freq=          dl_earfcn=     freq_offset=911.7427
98 phys_cell_id=    sfn=354 n_ant=2 phich_dur=Normal phich_res=1 bandwidth=
10
info sib1_decoded                                      freq_offset=911.742798 phys
_cell_id=405 sfn=354 mcc[0]=310 mnc[0]=    network[0]=     resv_for_oper[0]
=false tac=    cell_id=          cell_barred=false intra_freq_resel=allowed
q_rx_lev_min=-122 q_rx_lev_min_offset=0 p_max=23 band=17 si_win_len=20 si_
periodicity[0]=16 sib_mapping_info[0]=2,3 si_periodicity[1]=64 sib_mapping_
info[1]=5,6 duplex_mode=fdd si_value_tag=8
info channel_found_end freq=                          freq_offset=911.742798
phys_cell_id=
info channel_not_found freq=          dl_earfcn=
info channel_not_found freq=          dl_earfcn=
```

Figure 11.5 Customized LTE cell scanner based on openLTE decoding and storing eNodeB MIB and SIB packets.

that operates that cell, the tracking area code, received power threshold to trigger a handoff to an adjacent cell, and a series of configuration parameters that could be leveraged to configure a rogue base station. One of the most useful pieces of information, from a protocol exploit point of view, is the list of high-priority frequencies. A rogue eNodeB configured to operate on one of the high-priority frequencies in an area will trigger most UEs to connect to it. Therefore, the impact of a rogue eNodeB can be optimized by leveraging configuration parameters in the SIB messages that can be easily eavesdropped with low-cost tools.

Moreover, an attacker can also extract the mapping of important control channels on the PHY layer from the SIB messages, such as the Random Access Channel (RACH) configuration. This can be leveraged to configure a smart jammer [5].

OpenLTE provides a traffic log feature that can be executed to passively scan broadcast information from nearby eNodeBs. A low-cost alternative can be implemented with an RTL-SDR radio running the LTE Cell Scanner tool [13,16] or the scanner function of the srsLTE project. The analysis of MIB and SIB traffic in this project has been carried out with a modified version of openLTE, which automatically detects new cells, decodes MIB and SIB messages, and stores the decoded information in a database file. Figure 11.5 provides a snapshot of the cell scanner while successfully detecting a cell in midtown Manhattan and decoding the information on both MIB and SIB1 messages.

11.4.2 LTE Insecurity Rationale

The strong encryption and mutual authentication algorithms implemented in LTE networks often lead to the misunderstanding that rogue base stations are not possible in LTE. Similarly, threats such as IMSI catchers, also known as Stingrays, rogue access points, and the like are generally assumed to exploit exclusively

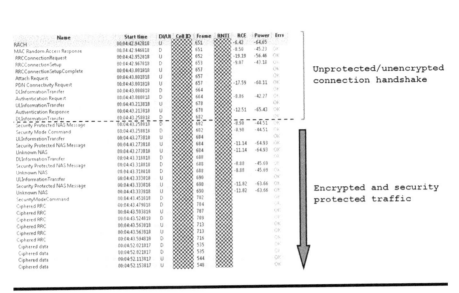

Name	Start time	Dl/Ul	Cell ID	Frame	RNTI	RCE	Power	Errs
RACH	00:04:42.942818	U		651		-6.42	-64.65	
MAC Random Access Response	00:04:42.946818	D		651		-8.50	-45.23	OK
RRCConnectionRequest	00:04:42.952018	U		652		-19.19	-56.46	OK
RRCConnectionSetup	00:04:42.967018	D		653		-9.07	-43.18	OK
RRCConnectionSetupComplete	00:04:43.001818	U		657				OK
Attach Request	00:04:43.001818	U		657				OK
PDN Connectivity Request	00:04:43.001818	U		657		-17.59	-60.11	OK
DLInformationTransfer	00:04:43.080818	D		664				OK
Authentication Request	00:04:43.080818	D		664		-8.86	-42.27	OK
ULInformationTransfer	00:04:43.213818	U		678				OK
Authentication Response	00:04:43.213818	U		678		-12.51	-65.43	OK
DLInformationTransfer	00:04:43.258918	D		682				OK
Security Protected NAS Message	00:04:43.258918	D		682		-8.98	-44.51	OK
Security Mode Command	00:04:43.258918	D		682		-8.98	-44.51	OK
ULInformationTransfer	00:04:43.273818	U		684				OK
Security Protected NAS Message	00:04:43.273818	U		684		-11.14	-64.93	OK
Unknown NAS	00:04:43.273818	U		684		-11.14	-64.93	OK
DLInformationTransfer	00:04:43.318818	D		688				OK
Security Protected NAS Message	00:04:43.318818	D		688		-8.88	-45.69	OK
Unknown NAS	00:04:43.318818	D		688		-8.88	-45.69	OK
ULInformationTransfer	00:04:43.333818	U		690				OK
Security Protected NAS Message	00:04:43.333818	U		690		-11.82	-63.66	OK
Unknown NAS	00:04:43.333818	U		690		-11.82	-63.66	OK
SecurityModeCommand	00:04:43.451818	D		702				OK
Ciphered RRC	00:04:43.479818	D		704				OK
Ciphered RRC	00:04:43.593818	U		707				OK
Ciphered RRC	00:04:43.524819	D		709				OK
Ciphered RRC	00:04:43.563818	U		713				OK
Ciphered RRC	00:04:43.563818	U		713				OK
Ciphered RRC	00:04:43.594818	D		716				OK
Ciphered data	00:04:52.021817	D		535				OK
Ciphered data	00:04:52.021817	D		535				OK
Ciphered data	00:04:52.113817	U		544				OK
Ciphered data	00:04:52.153817	U		548				OK

Unprotected/unencrypted
connection handshake

Encrypted and security
protected traffic

Figure 11.6   Initial connection to an LTE network sniffed from a real production network in the area of Honolulu, Hawaii.

well-understood vulnerabilities in latency GSM networks. However, any LTE mobile device will exchange a very large number of unencrypted and unprotected messages with any LTE base station, malicious or not, if it advertises itself with the right broadcast information. And this broadcast information can be eavesdropped by means of easily available tools, as discussed in Section 11.4.1. In order to optimize a malicious LTE access point, this can be configured with one of the high-priority frequencies, which can also be extracted from unprotected broadcast messages.

Figure 11.6 plots the actual message exchange between a mobile device and an eNodeB in order to establish a connection. The traffic was captured in a controlled environment within a Faraday cage with a single mobile device attaching to the cell. As highlighted in the figure, despite the fact that encryption is triggered upon mutual authentication, there is a large number of messages exchanged prior to the authentication step, with all this messages being sent in the clear and without integrity protection. In other words, up to the Attach Request message that a UE sends to the eNodeB, a rogue base station can impersonate a commercial eNodeB and the UE has no way to verify its legitimacy.

Aside from all the messages up to the Attach Request, there is a long list of other packets transmitted in the clear that can be eavesdropped and spoofed.

■ *UE measurement reports*: These reports often contain a list of nearby towers and the received power from each one of them, which can be potentially used to pinpoint the UE's location. In some rare cases, these contain explicitly the GPS coordinates of the device itself [17].

■ *Handover trigger messages*: These messages indicate a given UE that a handover procedure is to be executed, resulting in the connection being handed over from the current cell to a new cell. As discussed later, these messages can be leveraged to track a target device as it hands from tower to tower.

■ *Paging messages*: These messages, used by the network to locate devices in order to deliver calls and incoming connections, can be used to map a phone number to the internal id used by the network for each user.

■ *Connection reconfiguration messages*: Under certain circumstances, the network indicates the UE that a vertical handover is to be executed, that is, the connection is to be moved from the LTE radio access network to either 3G or 2G.

The remainder of this section discusses how to leverage these unprotected messages to build a Stingray, temporarily block devices, and potentially track users.

11.4.3 LTE IMSI Catcher

Although the IMSI should always be kept private and never transmitted over the air, it is intuitive that it will be required to communicate it at least once. The very first time a mobile device is switched on and attempts to attach to the network, it only has one possible unique identifier to be used in order to authenticate with the network: the IMSI. Once the device attaches to an LTE network, a TMSI is derived and can be used thereafter to keep the IMSI secret [1].

As a result, regular operation of an LTE network requires that a mobile device discloses in the clear its IMSI in specific circumstances, for example, in the event that the network has never derived a TMSI for that user before or in the rare event that the TMSI has been lost. Note that the IMSI is transmitted in a message before the authentication and encryption steps of the NAS attach process. This is precisely what an IMSI catcher exploits. Figure 11.7 presents a real capture of an *Attach Request* message in which a mobile device is disclosing its IMSI. This was a controlled experiment with the author's smartphone in an isolated environment.

An IMSI catcher [18], commonly known as Stingray, is an active radio device that impersonates, commonly, a GSM base station. In its most basic functionality, the IMSI catcher receives connection/attach request messages from all mobile devices in its vicinity. These attach messages are forced to disclose the SIM's IMSI, thus allowing the IMSI catcher to retrieve the IMSI for all devices in its vicinity. Stingrays, which have been widely discussed in the generalist media recently, are known for being often used by law enforcement to track and locate suspects [19].

More advanced Stingrays actually complete the network attach process, fully impersonating a real base station. At that point, they can effectively act as a Man in the Middle (MitM) for the device's connection as long as it forwards the traffic and calls into and from the real mobile network.

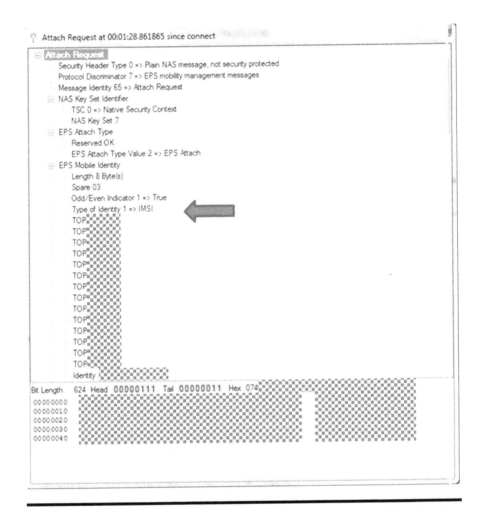

Figure 11.7 **IMSI of the author's smartphone being transmitted in the clear triggered by a software radio IMSI catcher.**

Although it is well understood that a MitM Stingray is only possible in GSM, there is a general assumption that LTE IMSI catchers are not possible or, at best, highly complex and expensive [20]. However, a fully LTE-based IMSI catcher is possible, very simple and very cheap to implement without requiring to jam the LTE and 3G bands to downgrade the service to GSM. In the context of this project, a fully operational IMSI catcher was implemented on a USRP B210 running a modified version of openLTE. A script was added to collect and store the IMSI being disclosed by those devices within the radius of coverage of the IMSI catcher. The total budget to build the IMSI catcher was under $2000, including the radio, antennas, GPS clock for the radio, etc. All experiments were carried out in a Faraday cage and only three IMSIs were ever collected, namely, the author's smartphone IMSI and the IMSI of two LTE USB dongles. Figure 11.7 contains the capture of an *Attach Request* message with the IMSI of the author's phone, which was captured by means of the software radio IMSI catcher.

11.4.4 Temporary Blocking Mobile Devices

The basic implementation of an LTE-based IMSI catcher described in Section 11.4.3 replies with an *Attach Reject* message to the *Attach Request* message, allowing the device to rapidly reconnect with the legitimate network. As with other preauthentication messages, the *Attach Reject* message is sent in the clear, which can be exploited by an attacker to temporarily block a mobile device.

3GPP defines a series of *cause codes* that the network utilizes to indicate mobile devices the reason why, for example, a connection is not allowed. These codes are defined as *EMM causes* in the LTE specifications [21].

Some of the EMM causes indicate the device that it is not allowed to connect to the network (i.e., *PLMN not allowed* EMM cause), which is a way a network provider can block a customer who engages in, for example, mobile fraud, and spam. Given that at the *Attach Request* stage in the connection there is no authentication or encryption established yet, any mobile device implicitly trusts the EMM cause codes regardless of the legitimacy of the base station. In other words, when the *Attach Reject* message is received, the base station is not authenticated yet but the UE has to obey the *Attach Reject* message.

If a rogue base station replies to an incoming connection with an *Attach Reject* message, it can fool the mobile device to believe that it is not allowed to connect to that given network. As a result, the device will stop attempting to connect to any base station of the provider that was being spoofed by the rogue eNodeB. Figure 11.8 illustrates this exploit by which a rogue base station effectively prevents any

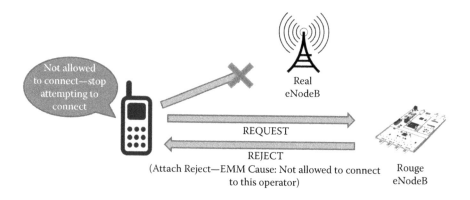

Figure 11.8   Mobile device temporary block by a rogue LTE base station.

mobile device in its radio communication range from connecting to the network, resulting in a DoS.

It is important to note that this is just a temporary DoS threat. By simply rebooting the device or toggling airplane mode, the device is capable again to connect to the network. Given that toggling airplane mode is commonly the first thing a user does when the cellular connection appears to be stalled, the impact of this DoS threat on smartphones would be minor. The impact could be more severe, though, in the case of an embedded device servicing an application in the context of the IoT. In the case of an embedded sensor deployed in the field with mobile network connectivity, it might be expensive and time-consuming to send a technician to reboot all the sensors being affected.

LTE mobile devices implement a timer (*T3245*), which is started when an *Attach Reject* message blocks the device from further attempting to connect. Upon expiration of this timer, the mobile device is allowed again to attempt to communicate and attach with the blocked network. According to the standards, this timer is configured to a value between 24 and 48 h [21]. Therefore, even in the context of an embedded sensor, the DoS would only be sustained for 1–2 days. In some applications, the impact of 24 h of connectivity loss could be very high, especially in critical and security applications.

This same attack can also be carried over by means of replying with a reject message to the *Traffic Area Update* (TAU) message. By means of configuring a rogue base station with a different Tracking Area value than the surrounding legitimate eNodeBs, one can trigger *TAU* messages from the devices that attempt to connect with the rogue base station.

This LTE protocol exploit, which was first introduced by the authors of Reference 7 and later also discussed in Reference 8, was implemented by means of a modified version of openLTE running on a USRP B210. All the experiments were carried in a controlled environment resulting in the blocking of the author's smartphone and two LTE USB dongles. No experimentation was carried over to determine the value of the timer T3245 because the author could not go about without phone for 24 h. Excellent further analysis and results of this threat were recently presented in Reference 17 in an outstanding paper by the authors.

11.4.5 Soft Downgrade to GSM

Similarly to the DoS threat discussed in Section 11.4.4, an attacker can trigger a soft downgrade of the connection to GSM, known for being highly insecure [4]. Exploiting the same *TAU Reject* and *Attach Reject* messages, a rogue base station can indicate a victim mobile device that it is not allowed to access 3G and LTE services on that given operator (*EPS services not allowed* EMM cause code). The target mobile device will then only attempt to connect to GSM base stations, as described in Figure 11.9.

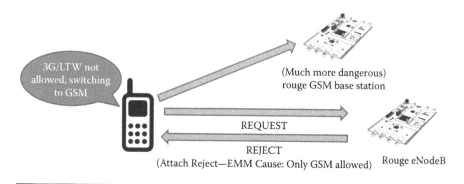

Figure 11.9   Mobile device soft downgrading to GSM by rogue LTE base station.

An attacker could combine this with a rogue GSM base station, which would open the doors to a full MitM threat, fully eavesdropping all mobile network traffic. This would allow the attacker to listen to phone calls, read text messages, and a long list of other known GSM threats [22]. Moreover, by configuring the rogue base station to target a specific user, one could leverage this technology for a spearphishing threat aiming to a specific device. As the device would not loose connectivity, the user might not realize it is connected through GSM unless it noticed the small icon on top of the screen. And even that is common in areas with spotty 3G and 4G coverage.

11.4.6 LTE Device Tracking with C-RNTI

This subsection introduces a previously unknown LTE exploit that could potentially allow a passive adversary (i.e., only sniffing capabilities) to locate and track devices and users as they move. This location leak threat was first disclosed by the author in Reference 8 and has already been discussed and analyzed with both GSMA and 3GPP.

The Cell Random Network Temporary Identifier (C-RNTI, RNTI for short from here on) is a PHY layer identifier unique per device within a given cell. In other words, there are no two devices within a cell with the same RNTI. But there could be two devices in adjacent cells with the same RNTI.

This 16-bit identifier is assigned to each mobile device during the random access procedure [1]. The eNodeB responds to the device's preamble with a MAC Random Access Response (MAC RAR) message that indicates, in the clear, the RNTI to be used by the device in that cell, as shown in Figure 11.10.

Passive analysis of real LTE traffic indicates that the RNTI is included in the header (unencrypted PHY layer encapsulation) of every single packet, regardless of whether it is signaling or user traffic. This allows, as shown in Figure 11.11, a passive

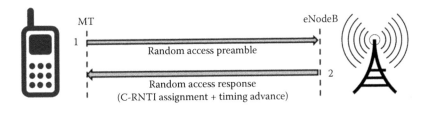

Figure 11.10   RNTI assignment in the RACH LTE procedure.

observer to easily map traffic, regardless of its encryption, to an individual device or user.

Mapping of a TMSI or MSISDN to the RNTI is trivial. By means of silent text messages or other mobile terminated traffic, an attacker can identify the current RNTI of a given device. Once this PHY layer id is known, a passive eavesdropper can know, for example, how long a given user stays at a given location. Assuming that an average smartphone will keep connecting to the network to receive email, SMSs, Whatsapp messages, check for pull notifications, and the like, there is going

Name	Start time	DI/UI	Cell ID	Frame	RNTI	UE Identity	Length	Errs
RACH	00:02:26.830866	U		988				
MAC Random Access Response	00:02:26.834868	D		989	8			OK
RRCConnectionRequest	00:02:26.840866	U		989	19841			OK
RRCConnectionSetup	00:02:26.853868	D		991	19841		24	OK
Ciphered data	00:02:26.855868	D		991	19681		1280	OK
Ciphered data	00:02:26.856868	D		991	19681		1280	OK
Ciphered data	00:02:26.857868	D		991	19681		1280	OK
Ciphered data	00:02:26.858868	D		991	19681		1280	OK
Unknown Data	00:02:26.871868	D		992	12381		52	1
Unknown Data	00:02:26.871868	D		992	12381		109	1
RRCConnectionSetupComplete	00:02:26.874866	U		993	19841			OK
Service Request	00:02:26.874866	U		993	19841			OK
Ciphered data	00:02:26.894868	D		995	19681		1280	OK
Ciphered data	00:02:26.895868	D		995	19681		1280	OK
Ciphered data	00:02:26.900868	D		995	19681		1280	OK
Ciphered data	00:02:26.901868	D		995	19681		1280	OK
Ciphered data	00:02:26.902868	D		995	19681		1280	OK
SecurityModeCommand	00:02:26.909868	D		996	19841			OK
Ciphered data	00:02:26.931868	D		998	19681		1280	OK
Ciphered data	00:02:26.932868	D		998	19681		1280	OK
SecurityModeComplete	00:02:26.932866	U		998	19841		32	OK
Ciphered data	00:02:26.933868	D		999	19681		1280	OK
Ciphered data	00:02:26.934868	D		999	19681		1280	OK
Ciphered data	00:02:26.952868	D		1000	19681		1280	OK
Ciphered data	00:02:26.953868	D		1001	19681		1280	OK
Ciphered data	00:02:26.954868	D		1001	19681		1280	OK
Ciphered data	00:02:26.955868	D		1001	19681		1280	OK
RRCConnectionReconfiguration	00:02:26.957868	D		1001	19841		34	OK
RRCConnectionReconfigurationC...	00:02:26.972866	U		1002	19841		32	OK
IP Data (IPv4 UDP)	00:02:26.972866	U		1002	19841		70	OK
Ciphered data	00:02:26.974868	D		1003	19681		1280	OK
Ciphered data	00:02:26.975868	D		1003	19681		404	OK
MAC Random Access Response	00:02:26.984868	D		1004	8			OK
RRCConnectionSetup	00:02:27.003868	D		1006	19843		24	OK

Figure 11.11   Identifying the traffic of a given device with the RNTI.

to be frequent and periodic data traffic originated or terminated at the device, allowing an eavesdropper to know the user is still there, while perhaps a partner of the eavesdropper is robbing the user's apartment. Moreover, assuming that the control plane traffic load for a mobile device is much lower than the actual user data load, the RNTI could also be leveraged to estimate the UL and DL data traffic load of a given device. This could potentially allow an adversary to identify the connectivity hotspot of an ad hoc LTE-based network, such as the ones being considered for both first responders [23] and tactical scenarios [24].

Examination of the 3GPP standards [25] indicate that the RNTI is defined as a unique id for identifying the RRC connection and scheduling dedicated to a particular UE. Although initially assigned as a temporary id, it is promoted to a static value after connection establishment or reestablishment. There is no explicit indications in the standards regarding the requirements for the RNTI being refreshed periodically. Observations of real LTE traffic from the major operators in the United States indicate that often the RNTI remains static for long period of times. The author's smartphone was observed to maintain the same RNTI for over 4 h.

Further analysis of LTE traffic uncovered a potential way a passive eavesdropper could track devices during mobility handover events. In LTE, handovers are network-triggered. Based on periodic measurement reports from the UEs, the eNodeBs decide whether it is necessary to hand the connection to an adjacent cell. As such, the handover is triggered by the source eNodeB through the *RRC Connection Reconfiguration* message. In this packet, the source eNodeB indicates the UE what is the destination eNodeB and provides some parameters necessary for the UE to connect to the new tower. Upon reception of this message, the UE performs a random access procedure with the destination eNodeB, which assigns it a new RNTI. At this point, the handover is complete and the UE completes the RRC connection with the destination eNodeB.

During the investigation of LTE security and protocol exploits, it was discovered that, in some cases, the RNTI that was initially assigned to the UE by the destination eNodeB was always quickly updated shortly after via an *RRC Connection Reconfiguration* message from the source eNodeB. Further inspection of the captures highlighted that, in the original *RRC Connection Reconfiguration* message sent from the source eNodeB to initiate the handover process, a new RNTI is explicitly provided by the eNodeB in a *Mobility Control Info* container [25]. This RNTI being explicitly provided matches the RNTI that is assigned to the UE at the destination cell via the *RRC Connection Reconfiguration* message.

Based on observations of real LTE traffic, the message that triggers the handover process appears to be sent in the clear. As a result, a passive eavesdropper can potentially track a given device in a cell and, upon a handover event, follow the connection to the destination cell (indicated in the message that triggers the handover) and intercept the RNTI that is assigned to the device in this new cell. This location and handover information leak was observed to not occur always. Discussions with the GSMA security team indicated that the *RRC Connection*

Reconfiguration message that triggers the handover should not be transmitted in the clear.

Figure 11.12 presents an example of such device tracking during a handover process. In the figure, a UE with RNTI 99 is connected to cell id 60. At some point, it receives an *RRC Connection Reconfiguration* message triggering the connection to be handed over to cell id 50 and explicitly indicates the UE that its new RNTI at the destination cell will be 10848 (0x2A60). The UE performs a random access procedure and is assigned an RNTI 112 via the MAC RAR message. However, shortly after the handover, the destination eNB (cell id 50) sends an *RRC Connection Reconfiguration* message that assigns the final RNTI at the destination cell, 10848. Note that, as part of the handover process, the UE still receives a couple of RRC messages from the source eNB (cell id 60) when it has already connected to the destination eNB and these are addressed to the old RNTI 99.

Although the standards do not explicitly indicate a need to refresh the RNTI periodically, this would be a strong mitigation against the threat of RNTI-based location tracking. For example, a new RNTI could be assigned each time a device transitions from idle to connected state. Also, it should be enforced that the handover trigger message is always encrypted as this message occurs after authentication and encryption set up. Nevertheless, it is important to note that these solutions might not be sufficient. An LTE analysis tool introduced in Reference 26 leverages the RNTI in order to map PHY layer measurements to a given device. In the event of an RNTI change, the authors devised an algorithm that was able to automatically map each device to its new RNTI. Keeping a fingerprint of signal measurements for each RNTI, the authors were able to track users as their RNTI changed with a precision of 98.4%.

11.5 Related Work

The wide availability of open-source tools for GSM experimentation has fueled a large array of very interesting security research on legacy networks. From the first demonstrations of GSM traffic interception and eavesdropping [4] to DoS threats against the GSM air interface [27], a number of GSM exploits have been reported in the literature and security conferences. Privacy and location leaks have also been deeply investigated in the context of legacy GSM networks [28].

LTE security research has been increasingly predominant over the last couple of years, mainly in network availability-related projects. For example, there has been interesting studies aiming to quantify and investigate the impact of large spikes of traffic load originated from M2M systems against the LTE infrastructure [29]. Also in the context of M2M, several studies have focused on the control plane signaling impact of the IoT against the LTE mobile core [30,31].

Applied LTE security research and protocol exploit experimentation have been close to nonexistent over the last few years. However, the recent availability of

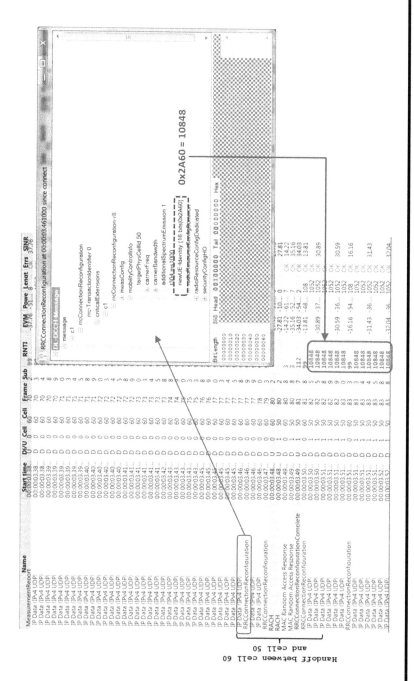

Figure 11.12   RNTI device tracking during a handover process.

open-source tools for LTE experimentation have provided the means for very interesting security research work, for example, some recent studies aimed at analyzing and evaluating sophisticated jamming threats against LTE networks [5,32,33]. Also leveraging LTE open-source tools, the authors of Reference 17 were the first to publicly disclose the implementation and analysis of the device blocking and soft downgrade to GSM exploits, which were also implemented in this manuscript. The same authors are responsible for some other excellent mobile protocol exploit experimentation, such as a study on intercepting phone calls and text messages in GSM networks [34] and mobile phone baseband fuzzing [35].

11.6 Conclusions

This chapter summarizes the experimentation and results of analyzing the security of next-generation LTE networks with low-cost software radio tools. Based on tools built upon the openLTE implementation of the LTE stack, the rationale behind a number of LTE protocol exploits is defined. Despite mutual authentication and strong encryption, there is still a large number of packets being exchanged, in the clear, between a UE and an eNodeB prior to these security functions being executed. A mobile device implicitly trusts these messages as long as they originate from an eNodeB broadcasting the right information, which opens the doors to a series of LTE rogue base station threats.

Basic software radio tools can be used to scan the LTE broadcast channels so that a rogue access point can be correctly configured. Once mobile devices attempt to attach to this malicious eNodeB, different exploits are possible by leveraging the unencrypted preauthentication messages. This chapter provides details on the implementation of an LTE-based IMSI catcher using of-the-shelf hardware with a budget under $2000.

Moreover, we investigate and implement DoS threats that block mobile devices and silently downgrade them to an insecure GSM connection. Such results are achievable by leveraging the EMM cause codes within the *Attach Reject* messages the rogue eNodeB transmits. Finally, a previously unknown LTE location leak is introduced and analyzed. By means of monitoring the RNTI PHY layer id, one could potentially track and follow a mobile user as it hands from tower to tower.

The growing number of LTE open-source implementations is lowering the bar for software radio experimentation and analysis of mobile protocols. Thanks to the efforts of the open-source community, such tools are fueling sophisticated research studies aimed at improving the security of mobile networks.

Acknowledgments

The author would like to extend his most sincere gratitude to Sanjole for providing some of the LTE traffic captures used in this chapter.

References

1. S. Sesia, M. Baker, and I. Toufik. *LTE, The UMTS Long Term Evolution: From Theory to Practice*, Wiley, New York, NY, 2009.
2. D. Lewis, Closing in on the future with 4G LTE and M2M. Verizon Wireless News Center, September 2012, http://goo.gl/ZVf7Pd
3. A. Prasad, 3GPP SAE-LTE security, in *NIKSUN WWSMC*, Princeton, NJ, July 2011.
4. K. Nohl and S. Munaut, *Wideband GSM sniffing*, in *27th Chaos Communication Congress*, Berlin, Germany, June 2010, http://goo.gl/wT5tz
5. M. Lichtman, R. Piqueras Jover, M. Labib, R. Rao, V. Marojevic, and J. H. Reed, *LTE/LTE-A jamming, spoofing and sniffing: Threat assessment and mitigation, Communications Magazine, IEEE*, 54(4): 2016.
6. OpenLTE—An open source 3GPP LTE implementation, http://openlte.sourceforge. net/
7. A. Shaik, R. Borgaonkar, N. Asokan, V. Niemi, and J.-P. Seifert, LTE and IMSI catcher myths, in *Proc. of BlackHat Europe*, Amsterdam, Netherlands, 2015.
8. R. P. Jover, LTE security and protocol exploits, *Shmoocon 2016*, Washington, DC, January 2016.
9. T. Engel, SS7: Locate. Track. Manipulate, in *FTP: http://events.ccc.de/congress/2014/ Fahrplan/system/attachments/2553/original/31c3-ss7-locate-track-manipulate.pdf*, 2014.
10. I. Gomez-Miguelez, A. Garcia-Saavedra, P. D. Sutton, P. Serrano, C. Cano, and D. J. Leith, srsLTE: An open-source platform for LTE evolution and experimentation. arXiv preprint arXiv:1602.04629, 2016.
11. gr-LTE—Gnuradio LTE cellular receiver, https://github.com/kit-cel/gr-lte
12. Ettus Research, USRP, http://www.ettus.com/
13. RTL-SDR, http://www.rtl-sdr.com/
14. Sanjole, WaveJudge 4900A LTE analyzer, http://goo.gl/ZG6CCX
15. R. P. Jover, J. Lackey, and A. Raghavan, Enhancing the security of LTE networks against jamming attacks, *EURASIP Journal on Information Security*, 2014(1): 7, 2014.
16. Evrytania LTE tools—LTE cell scanner, http://www.evrytania.com/lte-tools
17. A. Shaik, R. Borgaonkar, N. Asokan, V. Niemi, and J.-P. Seifert, Practical attacks against privacy and availability in 4G/LTE mobile communication systems, in *Proceedings of the 23rd Annual Network and Distributed System Security Symposium (NDSS 2016)*, Internet Society, San Diego, CA, 2016.
18. D. Strobel, IMSI catcher. Chair for Communication Security, Ruhr-Universität Bochum, p. 14, 2007.
19. The StingRays tale, *The Economist*, January 2016, http://goo.gl/wqwL5e
20. Insider Surveillance, Rayzone: Piranha LTE IMSI Catcher. Technical Report, June 2015, https://goo.gl/O2tO2o
21. Universal Mobile Telecommunications System (UMTS)—LTE. Non-Access-Stratum (NAS) protocol for Evolved Packet System (EPS)—Stage 3. 3GPP TS 24.301. v9.11.0, 2013.
22. D. Bailey and N. DePetrillo, The Carmen Sandiego project, in *Proc. of BlackHat (Las Vegas, NV, USA, 2010)*, 2010.
23. Nationwide Public Safety Broadband Network. U.S. Department of Homeland Security: Office of Emergency Communications, http://goo.gl/AoF41, June 2012.
24. A. Thompson, Army examines feasibility of integrating 4G LTE with tactical network. The Official Homepage of the United States Army, http://goo.gl/F60YNA, 2012.

25. LTE; Evolved Universal Terrestrial Radio Access (E-UTRA). Long Term Evolution (LTE) physical layer; Overall description. 3GPP TS 36.300. v8.11.0, 2009.

26. S. Kumar, E. Hamed, D. Katabi, and L. E. Li, LTE radio analytics made easy and accessible, *ACM SIGCOMM Computer Communication Review*, 44: 211–222, ACM, 2014.

27. D. Spaar, A practical DoS attack to the GSM network, *DeepSec*, http://tinyurl.com/7vtdoj5, Vienna, Austria, 2009.

28. D. F. Kune, J. Koelndorfer, N. Hopper, and Y. Kim, Location leaks on the GSM air interface, in *ISOC NDSS (Feb 2012)*, 2012.

29. C. Ide, B. Dusza, M. Putzke, C. Muller, and C. Wietfeld, Influence of M2M communication on the physical resource utilization of LTE, in *Wireless Telecommunications Symposium (WTS), 2012*, IEEE, New York, NY, pp. 1–6, 2012.

30. M. Jaber, N. Kouzayha, Z. Dawy, and A. Kayssi, On cellular network planning and operation with M2M signalling and security considerations, in *Communications Workshops (ICC), 2014 IEEE International Conference on*, IEEE, New York, NY, pp. 429–434, 2014.

31. J. Jermyn, R. P. Jover, I. Murynets, M. Istomin, and S. Stolfo, Scalability of Machine to Machine systems and the Internet of Things on LTE mobile networks, in *World of Wireless, Mobile and Multimedia Networks (WoWMoM), 2015 IEEE 16th International Symposium on a*, IEEE, New York, NY, pp. 1–9, 2015.

32. T. C. Clancy, M. Norton, and M. Lichtman, Security challenges with LTE-advanced systems and military spectrum, in *Military Communications Conference, MILCOM 2013–2013 IEEE*, IEEE, New York, NY, pp. 375–381, 2013.

33. M. Lichtman, J. H Reed, T. C. Clancy, and M. Norton, Vulnerability of LTE to hostile interference, in *Global Conference on Signal and Information Processing (GlobalSIP), 2013 IEEE*, IEEE, New York, NY, pp. 285–288, 2013.

34. N. Golde, K. Redon, and J.-P. Seifert, Let me answer that for you: Exploiting broadcast information in cellular networks, in *Proceedings of the 22nd USENIX Conference on Security*, USENIX Association, New York, NY, pp. 33–48, 2013.

35. C. Mulliner, N. Golde, and J.-P. Seifert, SMS of death: From analyzing to attacking mobile phones on a large scale, in *USENIX Security Symposium*, San Francisco, CA, 2011.

Chapter 12

A Comprehensive SMS-Based Intrusion Detection Framework

Abdullah J. Alzahrani and Ali A. Ghorbani

Contents

12.1 Introduction

At the core of the open and free access to third-party stores for mobile applications, Android empowered mobile devices to face serious security issues. While different from the Apple App Store, Google's Play Store had been a hub for mobile applications that could contain viruses and malware; thus, resulting in privacy invasion and worse, infecting the devices, leading to a failure to function.

The continuous development of botnets and other malware reveals the dynamic growth of its quantity [1]. One of the most serious threats to Internet security is the proliferation of botnets. In particular, mobile botnets have become a greater trend following the growth of traditional botnets in the Internet world. Mobile botnet is a network of compromised smartphones that share the same command and control (C&C) infrastructure and are controlled by a bot master to perform a variety of malicious attacks [2]. The increasing prevalence of mobile botnet attacks is driven by factors including the connectivity of the mobile phone that make communication with a C&C server easier, as well as mobile devices being lucrative attack platforms for attackers [3]. There could be various types or forms of botnet attacks within the mobile platform, including unwanted sending of emails and SMS/MMS, information theft using spyware, privacy issues, and many others.

SMS-based mobile botnet has become an evident trend in the field of cybercrime forensics, considering that SMS, unlike Internet access, has always been a primary service provided by all mobile devices, including smartphones and even tablets [4]. In 2013, SMS-based mobile botnet became evident as cybercriminals were pointed out as the developer of the then recently discovered Android malware that masqueraded itself as a Google application in order to steal messages from mobile phone users, particularly Android smartphone users [4]. One of the main components of a botnet is the C&C channel, which is used by attackers to carry out C&C communication. With the availability of SMS on smartphones, SMS messages are used to transfer C&C commands, send SMS spam, send premium-rate SMS messages without user knowledge, and distribute malware as propagation vectors.

In this chapter, we propose an SMS-based botnet detection formwork that uses multiagent technology based on observations of SMS messages and Android smartphone features. The proposed detection framework is based on a multilayer model that consists of three modules and JADE agents. We have developed an intelligent and proactive framework that scans incoming and outgoing text messages, monitors Android resources, and observes user usage that includes user connectivity time to block the attacks in order to prevent damage caused by botnet attacks. The framework creates a user profile that is used to perform behavioral profiling analysis in order to identity malicious SMS and cut the C&C channel. We developed an adaptive hybrid model of SMS botnet detectors by using a combination of signature-based and anomaly-based algorithms. This framework includes a defense module that was employed to generate signatures of malicious SMS messages, to update phone number blacklists (PNBLs), to analyze malicious applications, and to send feedback to Android smartphones so that the user can take action.

12.2 Related Works

A lot of studies have provided some attestation regarding the feasibility of an SMS-based mobile botnet and its potential damage to smartphone or mobile phone users. The research has proven that SMS messages can be used to transfer C&C instructions, distribute SMS spam, launch denial-of-service (DoS) attacks to send premium-rate SMS messages without user permission, and propagate malware via URLs sent within SMS messages. One prominent piece of research, about mobile botnets that use SMS as a propagation vector, was conducted by Hua et al. [5]. They proposed an Android-based mobile botnet, which utilizes the SMS as its platform to work. In their proposed botnet, Hua et al. [5] made use of an Internet server in order to create and establish the botnets' topologies, as well as to control the infected phone or nodes to send an SMS to its other neighboring nodes.

In light of this, the detection of mobile botnets is also a major problem because they are also hard to detect [6]. A review of the literature has identified various strategies and detection mechanisms to address the challenges of detecting mobile botnets. On detection of malware propagation vectors through SMS and MMS messages, Wang et al. [7] propose a malware detection system for mobile devices. Their system has the ability to analyze and predict malware propagation. One probable countermeasure that would aid in the detection of mobile botnets is outlined in the study by Zhen et al. [8]. They proposed an SMS commanded-and-controlled and P2P-structured mobile botnet. The principle behind the propagation of the botnet involves the involvement of users, as well as the exploitation of any mobile vulnerability. Here, a mobile botnet makes use of an SMS command and control channel, acting as a communication mode between the mobile bots and the botmaster. To facilitate reception of commands through SMS messages, each of the mobile bots has an 8-byte pass-code. More so, the SMS messages are designed as spam messages in order to avoid detecting the commands being transmitted through SMS messages. To deliver and propagate the command-contained SMS messages, the botmaster will exploit the SMS services that are being provided by the Internet to send the message. Based on the findings, their proposed mobile botnet using a modified structured P2P typology could only make use of SMS as its C2C channel.

Another research paper about the SMS botnet was a remarkable study by Mulliner et al. [9], wherein the researchers studied three forms of botnets that could work on Apple iPhone devices—P2P, SMS-based, and SMS-HTTP botnets. Taking into consideration the fact in the SMS-based botnet, every single command is generated and transmitted through SMS, and a tree of nodes (phones) is then created in order to limit the number of messages being exchanged. Nguyen et al. [10] propose methods to detect SMS C&C botnets from infected Android devices. Their approach is based on Android radio logs, whereby logs are required to be read and the radio activities associated with SMS-based botnet activities identified.

To complement these research efforts, several study groups have been evaluating and proposing mobile IDS frameworks. Alzahrani et al. [11] present an overview of the research in the field of intrusion detection techniques for the Android platform and explore the deficiencies of the existing experimentation practices. In another study, Damopoulos et al. [12] evaluate anomaly-based IDS for mobile devices using different types of machine learning classifiers to detect the misuse of mobile device based on user behavioral profiles. In another study, Damopoulos et al. [13] propose a framework and its design for a mobile IDS architecture that is used on the host and the cloud defense approaches. Also, the authors include a proof-of-concept implementation for IDS and state-of-the-art mobile hardware.

12.3 SMS-Based Intrusion Detection Framework

The SMS-based mobile botnet detection framework consists of two main systems: a multiagent system and an intrusion detection system (IDS). The architecture of the proposed detection framework consists of a multiagent system and three different modules: an SMS signature-based detection module, an SMS anomaly-based detection module, and an SMS defense module. The framework incorporates two components: Android mobile devices and a service provider that offers services. Figure 12.1 shows our complete framework design that functions as a comprehensive SMS-based IDS, and illustrates the interaction between agents and other modules.

12.3.1 Multiagent System

Although having the theoretical basis for an intelligent mechanism is important, more important still are the contributions that have been made by researchers in the artificial intelligence (AI) community, to evolve agent technology from theory to practice. Several multiagent platforms have been developed for smartphones. A multiagent system can be successfully applied for intrusion detection [1]. Therefore, this is the type of agent system we use. The main advantages of using a multiagent system are proactivity, autonomy, and self-awareness. The framework aims to secure the Android platform by developing a software layer that employs JADE middleware to create an Android user profile. In this section, we present a multiagent system that has been developed using the JADE platform, as shown in Figure 12.1. The basic idea is that agents have been developed to monitor Android platform devices and spot any unusual activity and abuse to the SMS service that may be caused

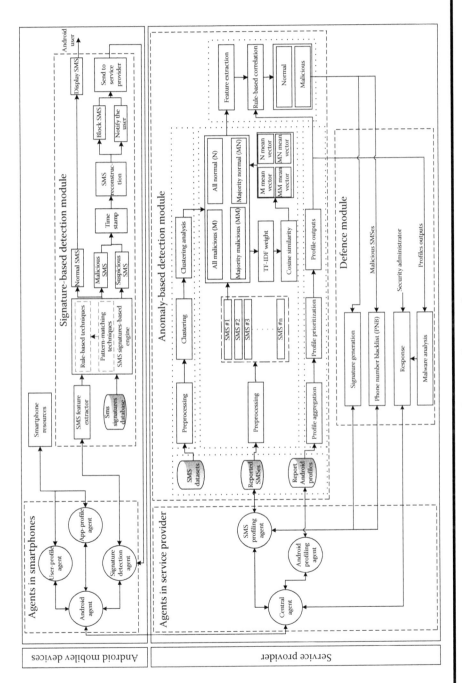

Figure 12.1   Overview of the proposed SMS botnet detection framework.

by malicious applications. This abnormal behavior will be reported to the service provider where more behavior analysis will be conducted.

Developing agent-based applications with JADE requires a runtime environment, a library of classes, and a suite of graphical tools. The JADE platform is a set of containers with a main container. Each agent has its own unique name that allows other agents to communicate with each other regardless of their location. In the proposed framework, the agents act as transducers that serve as an interface between external/legacy systems and other agents in the framework [14]. Since the multiagent system has to interact with the Android platform and other modules, the transducer approach is the most suitable for the SMS botnet detection framework.

Three agents operate on the service provider side, as shown in Figure 12.1. These agents have heavy roles to play, to maintain the list of subscriber agents, observe suspicious SMS messages and Android profiles to spot any abnormal behavior, find any correlation between the reported Android profiles, and perform actions. These three major agents in the service provider, which perform the majority of the activities concerning the Android user profiling, are the central agent, the Android profiling agent, and the SMS profiling agent. Based on the results of profiling analysis, these agents provide service and offer further analysis to achieve a high detection rate and make intelligent decisions, in order to detect SMS botnet activities. Table 12.1 illustrates all agent types with their relevant functionalities.

In Android device, each agent has a set of roles designed to allow it to achieve its goals. In order to access all the services, an Android mobile user must register with a service provider on the server. The agents are then used to detect SMS botnets, observe smartphone behavior and resources, and from that information create an Android user profile. As shown in Figure 12.1, there are four agents that reside in each mobile device, including an Android agent, a signature detection agent, an app-profile agent, and a user-profile agent. Table 12.2 illustrates all agent types with their relevant functionalities.

As shown in Figure 12.2, initially, Android agents interact with local agents to make sure they are running. A signature detection agent acquires SMS signatures and stays active to receive any signature update. In addition, this agent scans current text messages and labels all the SMS. If there is a malicious SMS, it is deleted; when a suspicious SMS is found, it is sent to the server. Also, this agent monitors incoming and outgoing text messages. If any malicious SMS or suspicious SMS is detected, this agent requests current profiles from app-profile and user-profile agents to be sent with the detected text messages. The app-profile agent creates an Android profile that includes all the features that need to be analyzed in order to investigate the suspicious SMS and to spot any abnormal activities. Also, this agent responds to any request coming from the detection agent and then sends the requested information to the Android profiling agent on the server. The user-profile agent keeps track of the user connectivity time and builds a user profile, as well as responding to any commands from the detection agent. Both profiling agents maintain a local copy of the constructed profiles and interact with the Android profile agent on the server.

Table 12.1 Central Service Agents and Their Responsibilities

Agent Type	Responsibilities
Central agent	1. Respond to smartphone devices and add them to the subscriber list. 2. Update, block, and delete smartphone agents as appropriate. 3. Manage the interaction communication between local agents. 4. Update the signatures database. 5. Send commands to start new agents or perform an action on Android devices if certain conditions are met. 6. Forward the Android profile, suspicious SMS, and SMS logs to Android profiling provider and SMS profiling provider.
SMS profiling agent	1. Handle the received suspicious SMS and then send it to the detection module. 2. Maintain an updated signature for each SMS detection agent. 3. Handle SMS logs and request an update within a specific time. 4. Interact with the detection module.
Android profiling agent	1. Maintain the profile database for all subscribing smartphones. 2. Update the profile changes when messages are received from other agents. 3. Respond to detection module requests. 4. Request more information from monitoring and human behavior agents if needed.

12.3.2 SMS Signature-Based Detection Module

As shown in Figure 12.1, the first step to effectively spot malicious SMS is to extract SMS features that have the potential to distinguish the behavior of SMS text messages. All the selected features have three significant characteristics: (1) they have been shown to be effective in distinguishing between the types of SMS messages, whether normal or malicious; (2) they can be used in real time, and impose no delay on the detection module; and (3) they keep our detection approach simple and fast.

Signature-based detection needs to be provided with up-to-date signatures of known botnets and malicious SMS. In order to develop an effective signature-based detection approach to combat malicious malware, we extract sender phone numbers and SMS content from our dataset; then, from the SMS content, we extract embedded URLs, commands, phone numbers, and phishing words as signatures for

Table 12.2   Android Smartphone Agents and Their Responsibilities

Agent Type	Responsibilities
Android agent	1. Smartphone user must have subscribed to the central service provider. 2. Read smartphone status. 3. Obtain the agent identification from the central service provider to establish the interaction. 4. Respond to requests from the central agent. 5. Manage the interaction communication between local agents. 6. Send data to the Android profiling agent. 7. Unsubscribe from the central service provider. 8. Notify the user when a new threat is detected.
Signature detection agent	1. Register with the SMS profiling service in the central server. 2. Obtain the SMS signature update. 3. Read incoming and outgoing SMS. 4. Receive the result from the SMS signature-based detection module. 5. Monitor SMS logs. 6. If SMS is normal, deliver it to the SMS application. 7. If SMS is malicious, delete the SMS and notify the user. 8. If SMS is suspicious, send a copy of the suspicious SMS to the SMS profiling agent.
App-profile agent	1. Register with the Android profiling service in the central server. 2. Report any access to browser or other apps when the SMS application tries to access. 3. Check the WiFi status and Internet access. 4. Monitor smartphone status, including battery usage, apps that are running, memory usage, etc. 5. Spot any setting changes.
User-profile agent	1. Register with Android profiling service in the central server. 2. Observe user connectivity time. 3. Maintain the whitelist and blacklist. 4. Report daily usage of Android mobile device.

our approach. The signature-based detection engine initially compares the selected features of a given SMS message (FromPhone#, ToPhone#, and Content) with provided signatures, and, if there is a match, the SMS is blocked. However, if the selected features do not match any of the signatures, the algorithm goes deeper and

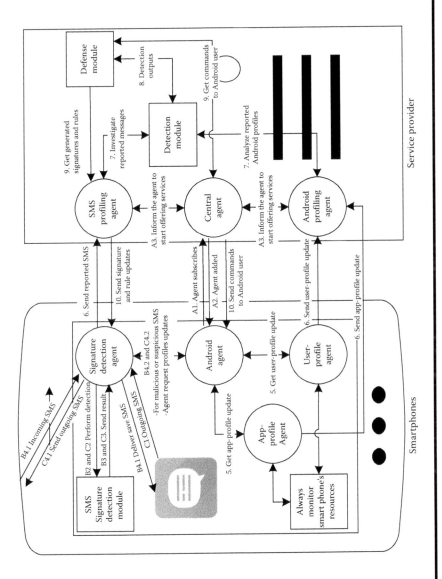

Figure 12.2   Architecture of the proposed multiagent system.

analyzes the body of the SMS. We extracted URLs, phone numbers, and commands by finding token strings in each SMS text body and matching them against the defined signatures.

The SMS messages are designed as spam message in order to avoid detecting the commands being transmitted through SMS messages. As some attackers use obfuscation techniques to avoid detection, we define three variable patterns to match obfuscated URLs, phone numbers, and commands using regular expression. If the SMS text has a URL, command, or phone number, the algorithm matches it against provided signatures. If there is a match, the SMS is blocked, but if there is no match, more evidence is sought to classify the SMS by applying rule-based techniques. A set of rules is then applied for unknown SMS. If the SMS matches a rule, it is classified as suspicious. Otherwise, it is considered normal. Finally, the output from the SMS signature-based detection algorithm labels the SMS as normal, suspicious, or malicious. If the SMS message is malicious, it will be deleted; however, if the SMS message is suspicious, it will be shown to the user with some information related to the detected SMS. If the SMS message is normal, a message to this effect will be displayed for the user.

12.3.3 SMS Anomaly-Based Detection Module

This module is where SMS collection, anomaly detection, and behavior-profiling analysis are conducted. Once the detection module receives the reported SMS from the Android device, it performs anomaly detection through specifically created and manipulated algorithms. Once SMS messages are deemed malicious, content analysis is performed in order to identify the type of attack. All the Android mobile profiles that contain the same SMS messages are grouped together based on their similarity.

12.3.3.1 SMS and Profiles Collection

The SMS and profiles collection is where input data are stored. There are three types of input sources, namely, labeled SMS datasets, reported text messages, and reported Android profiles. The SMS profiles collection is responsible for collecting, combining, storing, retrieving, and managing these data, to allow for more robust detection.

12.3.3.2 SMS Clustering

Unsupervised learning has benefited from significant efforts in pattern recognition and machine learning. Take, for instance, clustering. The main idea of clustering is to take a set of data and group its contents based on their similarities. The cluster method does not require class labeling of the data. There are several types of clustering methods, including hierarchical clustering, density-based clustering, and partition-based methods. One of the clustering approaches, the partition-based method, comprises two different algorithms, the K-means algorithm and the X-means algorithm. The unsupervised clustering algorithm that is used for SMS botnet detection is X-means clustering.

The dataset that is used for the cluster is a combination of malicious and legitimate SMS messages. As an output of the clustering, a number of clusters will have different kinds of messages. We analyze the result of clusters and group them into four class labels. The first class is called *All Malicious*, which consists of only malicious SMS; the second is *Majority Malicious*, in which the majority of the text messages are malicious messages; the third is *All Normal*, which contains all legitimate text messages; and the fourth is *Majority Normal*, where the majority of text messages are legitimate.

In the case of a cluster where 50% of the SMS messages are legitimate SMS messages and 50% of the SMS messages are malicious SMS messages, we analyze the content of the SMS messages. If the SMS messages have embedded URL, phone number, or command, the cluster will be grouped as *Majority Malicious*. Otherwise, the cluster will be grouped as *Majority Normal*. At this point, we can label all SMS messages that belong to the *All Malicious* class as malicious text messages but *Majority Malicious*, *All Normal*, and *Majority Normal* will require further analysis.

12.3.3.3 SMS Classification

In this stage, when new suspicious SMS messages are reported to the detection module, we begin by preprocessing each one. After that, we take each text message and add it to all the class labels. For each class label, we calculate the TF-IDF (Term Frequency–Inverse Document Frequency) weight, then apply the cosine similarity method to measure the similarity of the text message to each group by calculating the mean of each group. We find the minimum mean vector among the four class labels, and assign the text message to that class. After that, we remove the SMS message from other class labels, and update the class labels. We consider that all the SMS messages in All Malicious and Majority Malicious classes are malicious SMS, and we check the majority malicious class to look for any misclassification. If no misclassification is found, we then give a reason why the message is labeled as malicious. In order to verify the SMS messages in All Normal and Majority Normal classes, a further analysis is required, with additional information, to make a decision about the reported messages. The four class labels will then be sent to SMS correlation components.

12.3.3.4 Android Profiling Analysis

In order to provide accurate detection of SMS botnets, certain detection techniques are commonly used in IDSs. One of the techniques that contributes to detection is a behavior-based analysis technique that is employed to find proof of compromise rather than any specific attack. Android profiling analysis is used to analyze the reported profiles from Android smartphone devices, to detect outgoing SMS sent without user knowledge, and to perform further investigation of reported SMS to find the similarity between the reported profiles. The definition of the Android profiles here is the combination of the SMS profile, app profile, and user profile for each subscribed smartphone. To build an Android profile, we extracted the features that are related to SMS botnet behaviors. The profiles are collected in smartphones and then these profiles are reported to the service provider by the app-profile and user-profile agents.

A higher level of management is required for analyzing Android profiles before correlating them with SMS messages in the four class labels. The first phase of profile analysis is to aggregate the reported profiles based on the selected features. The aggregation of profiles takes into account the similarity between particular profile features. The similarity between values of each feature (e.g., Android_ID, FromPhone#, ToPhone#, sending_time, received_time, URLs, Command, Content, Phones#, Phishing Words, contact_list, dangerous permissions, services, connectivity time) has been well defined based on the characteristics of each feature. What alert aggregation is looking for is any deviation that can be recognized as abnormal behavior. This abnormal behavior is referred to as malicious activity.

The next phase of profile analysis is to prioritize each profile based on the following two features: dangerous permissions and user connectivity time. The objective is that, by means of the profile priority rank, an administrator can choose a high-risk profile as the selected profile for further correlation and analysis. If the profile has dangerous permissions and connectivity time, it will be considered a high-risk profile; otherwise, it will be considered low risk. The profile outputs will be stored in the abnormal profile table (APT).

From the perspective of Android malware, malware can request more permissions than actually legitimate applications. Additionally, it can often request permissions that have risks related to user privacy and device security, such as collecting user data, collecting device information, accessing Internet, or sending and deleting SMS. We have analyzed seven well-known SMS-based botnet families (MisoSMS–Zitmo–NickySpy–TigerBot–Sandroid–PletorWroba) to study the distribution of requested permissions for each family. We found 20 dangerous permissions that abuse the SMS message service by means of SMS mobile botnets, all of which were missed by the Android security. The list of these dangerous permissions is described in Table 12.3.

The last phase of profile analysis is to produce the APT. Figure 12.3 contains an illustration of an APT. APTs maintain records of all reported Android profiles. In the detection module, the SMS profiling and Android profiling agents decide about an SMS message and its profile on receipt. The APT divides profiles into two categories: (1) normal profiles, which consist of Android profiles with no indication

Table 12.3   List of Dangerous Permissions

Number	Dangerous Permissions	Number	Dangerous Permissions
1	INTERNET	11	WRITE_EXTERNAL_STORAGE
2	SEND_SMS	12	ACCESS_WIFI_STATE
3	RECEIVE_SMS	13	CHANGE_WIFI_STATE
4	READ_SMS	14	MOUNT_UNMOUNT_FILESYSTEMS
5	WRITE_SMS	15	ADD_SYSTEM_SERVICE
6	RECEIVE_BOOT_COMPLETED	16	CALL_PHONE
7	READ_PHONE_STATE	17	READ_CALL_LOG
8	READ_CONTACTS	18	WRITE_CALL_LOG
9	WAKE_LOCK	19	BROADCAST_STICKY
10	ACCESS_NETWORK_STATE	20	RECEIVE_MMS

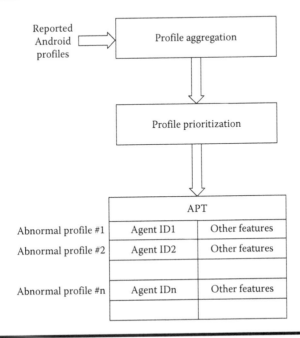

Figure 12.3   Profile analysis diagram and abnormal profile table.

of suspicious activity, and (2) abnormal profiles, which are recorded in the APT of the profile analysis, and require a suitable response.

12.3.3.5 SMS Correlation

In this approach, alerts are basically a set of logical facts about how Android platforms and SMS botnets work. In order to correlate two alerts directly using rule-based correlation, one predicate in the consequences condition of the first alert should be connected with one predicate in the prerequisites condition of the second alert. We have defined a set of correlation rules that can be applied against each SMS message in each class label, along with its reported profile to label SMS messages as either malicious or normal. Any match of the rule is reported, and the SMS messages labeled as malicious. The set of rules may define an attack scenario involving some unusual SMS content. In this case, the algorithm first checks whether or not the sender phone number of the SMS message is in the user contact list, and also applies permission-based methods to check for dangerous permissions. Additionally, we check the percentage of reported text messages that have the same features. These rules are predefined, and they label the SMS messages as either "malicious" or "features need to be checked." We consider all SMS messages that pass the rule-based methods to be normal SMS messages.

For each SMS, the feature extraction technique is used to extract six features from SMS messages. These features are: (1) "Sender_num," which refers to the sender phone number of the SMS author; (2) "Has_URL," which refers to whether an SMS message contains an embedded URL; (3) "Has_num," which refers to whether an SMS message contains an embedded number in it; (4) "Has_command," which refers to whether an SMS message contains an embedded command; (5) "Content," which represents SMS content; and (6) "Outgoing_SMS," which refers to the type of outgoing SMS message. First, we apply feature extraction to each SMS message. The results of the feature extraction are called alerts. In the second step, we will correlate each SMS message to its profile outputs. If the message has an alert, we apply the correlation rules. If any match of the rules is found, the SMS message will either be labeled as malicious or will require further analysis by an administrator. However, if there is no match, the algorithm will apply the next correlation rules.

12.3.4 SMS Defense Module

Typically, an SMS defense module begins by gaining insight into unknown SMS botnets and then generates signatures and rules. The defense module described in this chapter attempts to protect Android smartphones by introducing a proactive approach to generate signatures and rules. The defense module consists of four components, namely, signature generation, PNBL, malicious application analysis, and response action.

Each component describes one of four approaches that can be used to respond to an attack [15]. The first approach is to identify the malicious correspondent's phone number and block it. The second approach identifies the misbehavior of apps: it detects common features of the malicious applications and prevents those apps from running in the smartphones. The third approach is to identify the similar characteristics of malicious SMS messages and to group them by their common features. These common features include FromPhone#, ToPhone#, URLs, commands, phone# in SMS, size of SMS, time, and the app names, for example. After getting the result from the detection module, the administrator can interact with an Android user using the fourth approach, that is, by sending feedback to the user. The feedback will explain what the user should know and what actions the user should take.

12.3.4.1 Signature Generation

To ensure acceptable rates of false-positives and false-negatives during the signature detection process, we consider many exploits, and frequently update the signatures. The central agent sends the signature updates to all Android mobile devices. The first line of defense against SMS botnet activities is the signature detection module. For all messages labeled malicious, signatures will be created based on selected features.

Signatures are generated using the signature generation algorithm (see Algorithm 12.1). This module receives the malicious SMS messages reported by the detection module. At first, we compare the new SMS messages with the existing malicious SMS messages. If an SMS message already has a signature, the algorithm will attempt to match the message's other features. If there is any match, the SMS message will be ignored. If there is no match, the algorithm will generate a signature of the following features: FromPhone#, ToPhone#, URLs, Command, Phones#, Content, and Phishing Words. It will then repeat the same process until it has generated a signature for each malicious SMS message. The signature updates will be sent to all subscribed Android mobile devices.

12.3.4.2 Phone Number Blacklist

Blocking malicious SMS is the primary defense against SMS botnet attacks. Clearly, SMS-based attacks would be defendable by filtering if there were regularities in one or more of the attributes of the malicious SMS on Android smartphones. A PNBL contains a list of phone numbers that the SMS botnet detection app should block and should not accept any SMS text messages from. A PNBL can be queried with the signature detection module and allows an efficient way to perform lookups. As

---

Algorithm 12.1 Signature Generation Algorithm

Inputs:

$ms \leftarrow \{ms_1, \ldots, ms_n\}$ (*malicious SMSes*)

$fs \leftarrow fs_0, fs_1, fs_2, fs_3, fs_4, fs_5\}$ (*features signature*)

Outputs:

$SMSs \leftarrow SMS_{s1}, \ldots, SMS_{sn}\}$ (*sms signatures*)

Method

   for each $ms_i \in ms$ **do**

      if $fs^5_{i=0} \mathrel{!}= smss$ **then**

         $SMS_s \leftarrow$ *signature generation(ms)*

      else

         remove ms

      end if

   end for

send a copy of smss to android smartphones

---

an example, when detection results report that a set of malicious SMS messages having the same phone number (a common feature of malicious traffic) is initiating harm (sending SMS spam, commands, etc.), we would generate a signature of the phone number and then send it to the signature detection module in Android smartphones, so that the module could perform a signature scan and block SMS text messages from this phone number.

12.3.4.3 *Malicious Applications Analysis*

Malicious apps are the primary means by which SMS botnets, receiving commands through the SMS service, perform attacks. Analyzing reported apps and extracting their features is therefore a strong method of defense against SMS botnets. The profiling analysis step is done in the detection module, and the outputs are shown to a security administrator, who can perform static and dynamic analysis using common tools.

The profile outputs represent the degree of risk presented by an installed application (low, medium, or high), as gauged by a specific set of security rules. For example, the use of permissions is not as dangerous in some apps as it is in others. The profile outputs can include a normal feature of an attacked smartphone and can be part of a totally legitimate profile. However, a malicious app can also exploit this feature. Research experiments show that it may not be always possible to confirm the intent of using permission to recognize an attack. Nevertheless, security administrators are able to use this technique to understand the functionality of the malicious app, and to confirm features and characteristics of the malware.

12.3.4.4 Response

The results of the detection module determine the degree of threat or severity of an attack against the Android smartphone. Although identifying malicious SMS messages will help to block SMS botnets by taking down SMS bots and cutting the C&C channel, it also requires the Android user to cooperate by removing the malicious application.

The security administrator is able to send a request to users, asking them to perform an action, for the protection of their smartphones. We have developed an SMS botnet detection app that runs agents, performs signature detection, and provides an interface to allow the administrator to communicate and interact with the Android user. In the Android platform, users themselves have to uninstall the apps based on the information provided. The administrator provides an extensive explanation about the malicious app, including information about its publisher, and other apps from the same publisher. Also, the administrator indicates what dangerous permissions the app used, and notifies the user that the app is sending out SMS messages without the user's knowledge.

12.4 Evaluation and Discussion

To provide a comprehensive evaluation of the proposed framework, we collected a large set of SMS botnet samples that has seven botnet families. These families describe the behavior of SMS botnets. The SMS botnet dataset collected botnet samples from the Android Genome Malware project [16], malware security blog [17], and samples provided by a well-known anti-malware vendor.

12.4.1 Evaluation Methodology

In order to evaluate the proposed framework, we tested our implementation using the Android platform and ran real-time behavior monitoring for the collected features. We carried out different experiments to evaluate the overall framework that included a signature-based detection module, an anomaly-based detection module, and a multiagent system that was composed of data collection agents and service provider agents. The experiment consists of three parts. The first part evaluates the ability to detect malicious SMS messages on smartphones and reports the results to the service provider in order to perform an anomaly-based detection. The second part assesses the detection approach by monitoring Android features and creates Android profiles that are sent by an app-profile agent. In the third part, we study the scenario where malicious applications try to send out SMS messages at specific time and send to a premium-rate phone number by mimicking human behavior.

1. *Malicious SMS Botnet Detection*: A variety of SMS malware used to send premium-rate SMS without user knowledge was analyzed. These SMS were

Table 12.4   Overview of the Extracted C&Cs and URLs

Botnet Family	Total Samples	C&C	URLs
DroidDram	336	14	500
Geinimi	266	4	157
MisoSMS	101		195
NickiSpy	203	16	139
NotCompatible	76	5	5
PjApps	213	7	189
Pletor	85	3	12
Rootsmart	33	12	9
TigerBot	97	10	26
Wroba	94	10	36
Zitmo	81	23	29
Other malware family	1054	275	
Total	2639	379	1297

extracted, along with the Tophone#, and added to our dataset. In addition, we created a dataset that has malicious URLs, commands, and phishing words. We used Android mobile botnet dataset that were collected using different sources to extract the URLs by analyzing APK files. We also gathered Android botnet URLs from the UNB ISCX Android Botnet Dataset [18]. The botnet commands were extracted from the dataset samples by analyzing APK and from other sources [3]. The summary of extracted C&C and URLs are illustrated in Table 12.4.

In this experiment, we used two datasets. The first dataset is a labeled dataset called the British English SMS dataset [19] that has 425 malicious text messages. The second dataset is the NUS (National University of Singapore) dataset [20] that has over 55,000 unlabeled text messages to evaluate the proposed framework. In the first experiment using the British English SMS, we loaded all SMS messages and ran our application prototype, then reported the results. In the second experiment, we randomized the NUS dataset that has a total of 55,835 text messages. We divided the 55,196 SMS messages into 11 sets, each set having approximately 5000 SMS messages, as shown in

Table 12.5   NUS Experiments Datasets

Number of Sets	Number of SMS Messages	Number of Sets	Number of SMS Messages
Set #1	5003	Set #6	5024
Set #2	5002	Set #7	5000
Set #3	5008	Set #8	4999
Set #4	5000	Set #9	5002
Set #5	5000	Set #10	4987
Set #11	5171		

Table 12.5. We used 11 Android emulators to load each set to an emulator and ran our SMS botnet detection application prototype.

2. *Android Profile*: Creating Android profiles requires monitoring Android features that malicious apps use to initialize an attack. To test this approach, the app-profile agents obtained the list of installed applications and running applications. The app-profile agent also monitors the granted permissions and tracks any permissions related to SMS permissions. Table 12.6 shows

Table 12.6   Monitored Permissions

Permission Name	Permission Description
RECEIVE_SMS	To monitor incoming SMS
SEND_SMS	To send out SMS
READ_SMS	To read current SMS
WRITE_SMS	To write to SMS Content Provider
BROADCAST_SMS	To broadcast an SMS notification
INTERNET	To obtain full access to the Internet
ACCESS_NETWORK_STATE	To access ConnectivityManager
CHANGE_NETWORK_STATE	To change network state
ACCESS_WIFI_STATE	To access WifiManager
CHANGE_WIFI_STATE	To change Wi-Fi connectivity state
WAKE_LOCK	To monitor if the process is a wake

the permissions that the agent keeps, observes, and monitors. This includes accessing the state of the network and Internet permissions. In addition, this agent observes services that are running as an important element to determine which service is an application component that can carry different types of operations that run in the background.

3. *User Profile*: The user-profile agent is responsible for observing human behavior. Attackers monitor user usage of the Android device in order to launch an attack during specific times (e.g., when the user is sleeping or inactive). Based on this point, we decided to observe the user behavior, including user connectivity time, to monitor the time the phone is in wake or sleep modes. For instance, any SMS sent out while the smartphone is in sleep mode is flagged as malicious SMS and the information regarding the application that sends the SMS is recorded by the user-profile agent. We evaluate the capability of these agents by using the test applications and determine if the agent is monitoring whether the SMS is sent from the mobile device while it is in the wake mode and at a specific period of time.

In order to determine the efficiency of the proposed framework, we repackaged a real-world malware sample known as AndroidOS/Fakeplayer.A [21]. This malware pretends to be a movie player and shows messages in Russian. It sends SMS messages and contains the string "798657" to Russian SMS short code numbers (3353 and 3354) that may charge the user without their knowledge. We changed the short code number from 3353 and 3354 to "1-555-521-5562." We also developed an application called "DroidDreamTest" that can send out SMS messages at certain times. This application has similar silent patterns as DroidDream that only operate from 11 pm to 8 am [22]. We were able to monitor a test application and its behavior relating to SMS being sent out at specific times when the device was in the sleep mode and send out SMS without the user's permission. If the same SMS message is reported by more than one Android agent and it has the SMS botnet characteristics, it is flagged as malicious.

12.4.2 Experimental Results and Discussion

In this section, we discuss the results that have been obtained by the SMS signature-based detection module and the SMS anomaly-based detection module; we also discuss the performance of the JADE agents on smartphones.

1. *SMS Signature-Based Detection Results*: In smartphones, the signature detection agents obtained the results of the signature detection and then requested other agents to report the profiles to the service provider agents with malicious and suspicious SMS.

   For the first experiment, we illustrate the results of the signature detection on the British English SMS [19], as shown in Table 12.7. The signature

Table 12.7 Proposed Framework Experimental Results

Types	SMS Content	Phones#	URLs	Commands
Malicious	170	417	56	0

detection agent reported 425 malicious SMS to SMS profiling agent in the service provider server. The signature detection module detected 179 SMS messages that have corresponding content signatures and 417 phone number are malicious, which match the phone number signatures. The signature detection module spotted 56 malicious URLs and this dataset does not have any botnet commands. The detection accuracy on British English SMS dataset is 100% with zero false alarm since we have signature.

In the second experiment, the summary of NUS dataset signature detection results are shown in Table 12.8. The signature detection agent sends 3115 suspicious SMS text messages and 165 malicious SMS messages to the SMS profiling agent and sends commands to app-profile and user-profile agents requesting the current profiles be sent to the Android profiling agent. 139 of the SMS messages contained C&C botnet commands that have corresponding command signatures and 26 malicious SMS messages have malicious URLs.

2. *SMS Anomaly-Based Detection Results*: We performed the evaluation using various datasets. For the experiments, we chose two public datasets: (1) IIIT-D SMS Spam Dataset [23], a labeled dataset that has 1000 spam SMS messages and 1000 normal SMS messages; and (2) SMS Spam Collection Dataset [24], a labeled dataset that has 747 spam SMS and 4827 normal SMS.

Table 12.8 Proposed Framework Experimental Results

Types	Features	Number of SMS	Total	Percentage (%)
Malicious	SMS body	0		
	Phones#	3		
	URLs	23	165	0.5
	Commands	139		
	FromPhone#	0		
	ToPhone#	0		
Suspicious	Phones#	869		
	URLs	144	3081	5.5
	Commands	2182		
Normal			51,721	94

Table 12.9   Experimental Result of the Anomaly-Based Detection Module

Detection Metric	First Experiment	Second Experiment
Accuracy	0.933	0.968
Precision	0.943	0.986
Recall (TPR)	0.925	0.978
FNR	0.077	0.022
TNR	0.923	0.909
FPR	0.057	0.091
F-measure	0.934	0.982

In the first experiment, we evaluated the proposed module on the IIIT-D SMS Spam Dataset that has 2000 SMS messages. Table 12.9 shows the overall detection performance of this module on IIIT-D SMS Spam Dataset. By analyzing the result, the overall performance of the proposed system is improved significantly and it achieves more than 93% accuracy for all types of attacks.

In the second experiment, we used 5574 evaluated text messages [24]. Table 12.9 shows the results of Recall, Precision, and F-measure of the detection module. The detection module achieved a 96.84% rate of accuracy, with 97.76% high recall and 98.58% precision.

In intrusion detection, normal data usually outnumber intrusion [25]. For instance, with 95% normal and 5% attack, the accuracy metric is misrepresentative due to the probability that a system will rank a randomly chosen positive instance higher than a randomly chosen negative one. In this case, a system always classifies all data as normal with a high accuracy (97% in our example).

We reported the results of the experiment that had different percentages of malicious SMS messages. We built 10 different datasets from SMS Spam Collection Dataset [24] that had the following percentage of malicious SMS message 5%, 10%, 20%, 30%, 40%, 50%, 60%, 70%, 80%, and 90%. For example, in the first dataset, 95% of the SMS messages are normal and 5% of the messages are malicious. Figure 12.4 shows the capability of the proposed module. By changing the percentage of the malicious SMS messages, the detection module achieved 95.68% precision and 87.54% recall when 20% of the SMS messages are malicious. Although when malicious SMS messages exceed 60% of the dataset, the persistent results are increased and the recall result are decreased.

After evaluating the performance of this module, we had a set of reported suspicious text messages that had not yet been classified or labeled as normal

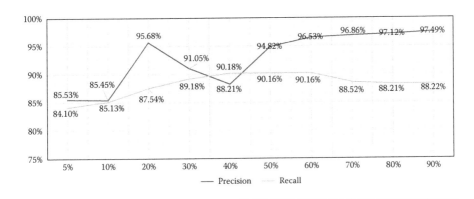

Figure 12.4 Second experiment: precision and recall.

or malicious. We started by combining all the datasets and we removed duplicated ones from the datasets. Although SMS spam messages are characterized by obfuscation, we kept many of the nonidentical messages that might still be close matches. We randomized spam SMS messages and normal SMS messages and chose 500 normal SMS messages and 500 spam SMS messages. In the first step, we clustered 1000 SMS messages using X-mean clustering technique and then applied our clustering analysis method. The results of SMS clustering algorithm are described in Table 12.10. In the second step, we applied an SMS classification algorithm by taking 165 reported malicious SMS messages that were received from signature detection agents and classified all malicious SMS to malicious class labels, which help to classify suspicious SMS messages. The SMS classification algorithm also classified the 3081 reported suspicious SMS messages to one of four class labels. The result of the classification are given in Table 12.10. For the NUS dataset, 2891 of the SMS messages are classified as normal and 95 SMS messages are classified as majority normal. 56 SMS messages are labeled as malicious and 39 SMS messages are labeled as majority malicious. In the third step, analyze the reported profiles by applying the Android profiles analysis algorithm that produces the APT. In the fourth step, employ the SMS correlation algorithm that applied correlation rules to label the instances in each one of four class labels. Table 12.11 shows the results of the detection on the NUS dataset. 941 of the SMS messages are labeled as malicious and 2152 of the SMS messages are labeled as normal.

3. *JADE Agents Evaluation Results*: To ensure the efficiency of the proposed framework, the agents achieve the following tasks: (1) the Android agent obtains signature updates from the central agent; (2) the signature detection agent performs signature detection on existing SMS messages and gets the results back; and (3) the results forward the SMS profiling agent, and the

Table 12.10 SMS Clustering and Classification Results

Type of Data	M	MM	N	MN	Total
Dataset SMSes	338	107	153	402	1000
Malicious SMSes	165	0	0	0	165
Suspicious SMSes	56	39	2891	95	3081

Table 12.11 Detection Module Results for NUS Dataset

Label	Total	Phones#	URLs	Commands	Phishing Words
Malicious	941	818	136	281	144
Normal	2152				

app-profile and user-profile agents get the current profiles and forward them to the Android profiling agent.

In the first task, Figures 12.5 and 12.6 show the CPU and memory usages to obtain signatures update at the first time with over 6600 signatures. From these charts related to signature updates, we can observe that the prototype application tends to consume more CPU power and more memory while communicating with the central agent to obtain signature update the first time. It took around 383 s to get all the signature records.

To address this shortcoming, we decided to pack the signature updates with an application prototype and frequently update the signature when a new threat has been discovered.

In the second and third tasks, Figures 12.7 and 12.8 are a summary of CPU and memory usage of the signature detection algorithm for 200

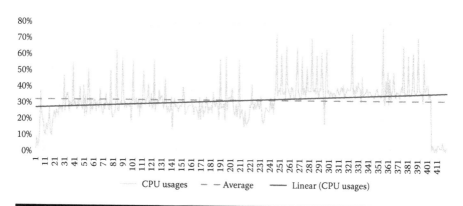

Figure 12.5 CPU usages for obtaining signatures update.

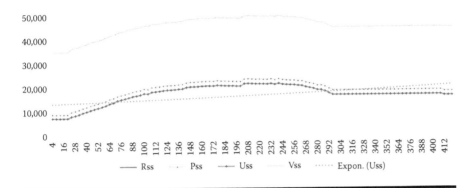

Figure 12.6   Memory usages for obtaining signatures update.

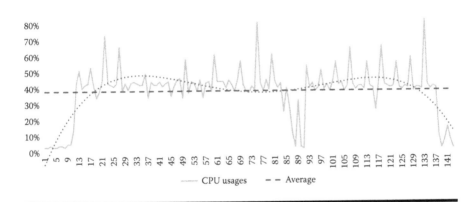

Figure 12.7   CPU usages for signature detection performance.

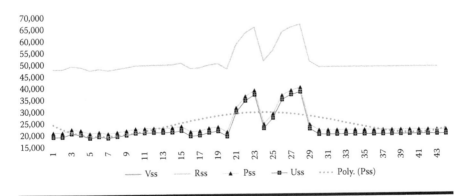

Figure 12.8   Memory usages for signature detection performance.

SMS massages and the reported results of three suspicious SMS messages with related profiles. To perform signature detection for 200 SMS messages required less than 60 s with 38% average of CPU power and 23,911 K of Pss usages and 22,062 K of Uss of memory usage.

12.5 Conclusions and Future Work

The framework covers four main components that work intelligently together to provide full protection and mitigation against SMS botnet activities. Based on the results of the SMS signature-based detection module, the JADE agents reported the profiles that are required to be used with suspicious SMS message to identify if it is malicious or not. These rules can be updated over time by finding new information about new botnet behavior, and includes specific characteristics of malware that has a unique behavior in the defense module. JADE agents depend on human knowledge and the set of rules that are programmed into them to make intelligent decisions autonomously. They continuously observe the smartphones, perceptively analyze and hold the characteristics of the abnormal behavior, and autonomously respond to it.

The work performed in this chapter provides a basis for future research of IDSs based on a multiagent system in mobile devices. One area of future work is applying a broader range of features for intrusion detection. These features need to be calculated in real time to enable the detector to keep up with the large number of reported SMS messages and their profiles. Another interesting area that can be investigated in the future is to extend the framework. If an agent can make decisions on the fly about suspicious SMS messages, these decisions are based on other agents' findings if they report the same SMS message with its characteristics.

References

1. O. Savenko, S. Lysenko, and A. Kryschuk, Multi-agent based approach of botnet detection in computer systems, in *Computer Networks*, A. Kwiecien, P. Gaj, and P. Stera (eds.) CN2012. CCIS vol. 291, Springer, Heidelberg, pp. 171–180, 2012.
2. G. Geng, G. Xu, M. Zhang, Y. Guo, G. Yang, and C. Wei, The design of sms based heterogeneous mobile botnet, *Journal of Computers*, 7: 235–243, 2012.
3. R. Nigam, *A Timeline Of Mobile Botnets*, Virus Bulletin. Available at: https://www.virusbtn.com/virusbulletin/archive/2015/03/vb201503-mobile-botnets. Accessed on March 12, 2016.
4. K. Hamandi, I. H. Elhajj, A. Chehab, and A. Kayssi, Android sms botnet: A new perspective, in *Proceedings of the 10th ACM International Symposium on Mobility Management and Wireless Access*, ACM, Paphos, Cyprus, pp. 125–130, 2012.
5. J. Hua and K. Sakurai, A sms-based mobile botnet using flooding algorithm, in *Proceedings of the 5th IFIP WG 11.2 International Conference on Information Security Theory and Practice: Security and Privacy of Mobile Devices in Wireless Communication*, Springer-Verlag, Heraklion, Crete, Greece, pp. 264–279, 2011.

6. D. Geer, Malicious bots threaten network security, *Journal of Computer*, 38: 18–20, 2005.

7. W. Wang, I. Murynets, J. Bickford, C. V. Wart, and G. Xu, What you see predicts what you get—Lightweight agent-based malware detection, *Security and Communication Networks*, 6(1): 33–48, 2013.

8. Y. Zeng, K. G. Shin, and X. Hu, Design of sms commanded-and-controlled and p2p-structured mobile botnets, in *Proceedings of the Fifth ACM Conference on Security and Privacy in Wireless and Mobile Networks*, ACM, Tucson, AZ, USA, pp. 137–148, 2012.

9. C. Mulliner and J.-P. Seifert, Rise of the ibots: Owning a telco network, in *Proceedings of the 5th International Conference on Malicious and Unwanted Software*, IEEE, Nancy, France, pp. 71–80, 2010.

10. A. Nguyen and L. Pan, Detecting sms-based control commands in a botnet from infected android devices, in *Proceedings of the 3rd Applications and Technologies in Information Security Workshop*, IEEE, Deakin University, Melbourne, Victoria, Australia, pp. 23–27, 2012.

11. A. J. Alzahrani, N. Stakhanova, H. Gonzalez, and A. A. Ghorbani, Characterizing evaluation practices of intrusion detection methods for smartphones, *Journal of Cyber Security and Mobility*, 3: 89–132, 2014.

12. D. Damopoulos, S. A. Menesidou, G. Kambourakis, M. Papadaki, N. Clarke, and S. Gritzalis, Evaluation of anomaly-based ids for mobile devices using machine learning classifiers, *Security and Communication Networks*, 5(1): 3–14, 2012.

13. D. Damopoulos, G. Kambourakis, and G. Portokalidis, The best of both worlds: A framework for the synergistic operation of host and cloud anomaly-based ids for smartphones, in *Proceedings of the Seventh European Workshop on System Security*, ACM, Amsterdam, Netherlands, p. 6, 2014.

14. M. Nikraz, G. Caire, and P. A. Bahri, *A Methodology for the Analysis and Design of Multi-agent Systems Using JADE*. Available at: http://jade.tilab.com/doc/tutorials/JADE_methodology_website_version.pdf. Accessed on May 19, 2013.

15. S. Stankovic and D. Simic, Defense strategies against modern botnets, *International Journal of Computer Science and Information Security*, 2: 1–7, 2009.

16. Y. Zhou and X. Jiang, Dissecting android malware: Characterization and evolution, in *Proceedings of IEEE Symposium on Security and Privacy*, IEEE, San Francisco, California, USA, pp. 95–109, 2012.

17. Contagiominidump, *Mobile Malware Mini Dump*. Available at: http://contagiominidump.blogspot.ca/. Accessed on December 22, 2015.

18. A. F. Abdul Kadir, N. Stakhanova, and A. A. Ghorbani, Android botnets: What urls are telling us, in *Proceedings of 9th International Conference on Network and System Security*, NSS, New York, NY, USA, pp. 78–91, 2015.

19. M. T. Nuruzzaman, C. Lee, and D. Choi, Independent and personal sms spam filtering, in *Proceedings of the 11th International Conference on Computer and Information Technology*, IEEE, Paphos, Cyprus, pp. 429–435, 2011.

20. T. Chen and M.-Y. Kan, Creating a live, public short message service corpus: The NUS sms corpus, Language Resources and Evaluation, 47: 299–335, 2013.

21. M. Mallen, *Trojan: AndroidOS/Fakeplayer.A*. Available at: https://www.microsoft.com/security/portal/threat/encyclopedia/entry.aspx?Name = Trojan:AndroidOS/Fakeplayer.A. Accessed on September 22, 2015.

22. T. Strazzere, *Do Androids Dream?* Available at: https://blog.lookout.com/blog/2011/03/06/do-androids-dream. Accessed on January 09 2015.

23. K. Yadav, P. Kumaraguru, A. Goyal, A. Gupta, and V. Naik, Smsassassin: Crowd-sourcing driven mobile-based system for sms spam filtering, in *Proceedings of the 12th Workshop on Mobile Computing Systems and Applications*, ACM, Phoenix, AZ, USA, pp. 1–6, 2011.

24. T. A. Almeida, J. M. G. Hidalgo, and T. P. Silva, Towards sms spam filtering: Results under a new dataset, *International Journal of Information Security Science*, 2: 1–18, 2013.

25. A. Ghorbani, W. Lu, and M. Tavallaee. *Network Intrusion Detection and Prevention: Concepts and Techniques*. Springer Science & Business Media, Philadelphia, New York, 2010.

INTRUSION DETECTION IN DYNAMIC AND SELF-ORGANIZING NETWORKS

Chapter 13

Intrusion Detection System in Self-Organizing Networks: A Survey

Razan Abdulhammed, Miad Faezipour, and
Khaled Elleithy

Contents

13.1 Introduction

An intrusion detection system (IDS) can be defined as the process of detecting and identifying the nonpermitted access to the services, activities, system information, and resources of network systems [1]. This concept was first described in the early 1980s [2].

Presently, there is a need to distinguish between the main challenges and constraints that face the design and implementation of an IDS for both traditional information communication systems (TICS) and non-TICS such as the cyber physical system (CPS), mobile ad-hoc network (MANET), vehicular ad-hoc network (VANET), and wireless sensor network (WSN). Furthermore, there is a need to reveal a comparison among the function of an IDS in these environments. This chapter tries to fill a part of this need and open the door for other researchers for further investigation (see Figure 13.1).

In general, and from a system architecture perspective, the core functions of an IDS consist of three essential elements [3]:

- Collecting data concerning an adversary
- Analyzing the data
- Responding to the analysis

Examples of adversary data collection may include system logging information from a local host or multi-hosts, user activity on operating systems and the recording

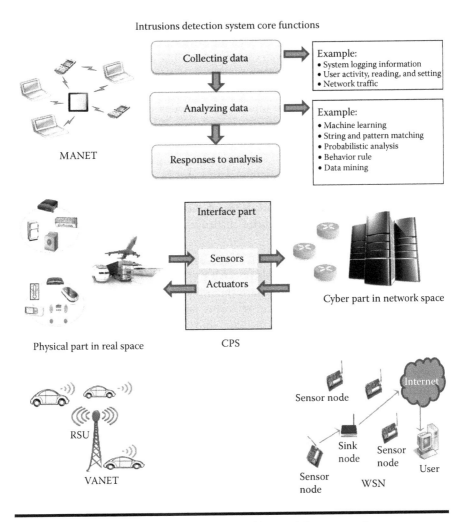

Figure 13.1 Intrusions detection systems self-organizing network.

traffic received and sent on network interfaces, or it can be of both types. An instance of analysis may include statistical methods such as the Markova process marker, multivariate, unvigilant, time series, and operational statistical moment [4]. Another example of analysis is machine learning which includes techniques such as Bayesian network, genetic algorithm, neural network, fuzzy logic, and outliner detection. Other examples of analysis are string and pattern matching, probabilistic analysis, and data mining which include frequent pattern mining, classification, clustering, and mining data streams. Examples of responses may include spreading an alarm, updating routing tables, and closing a session. The first two core functions have been treated numerously in the literature. In contrast, the third function was less treated [3]. Table 13.1 shows the most common commercial IDS tools.

Through the years, IDSs have evolved in handling diversified kinds of threats, security attacks, and new variabilities as well as handling older threats that have not disappeared or have evolved. This chapter describes the major categories of IDSs and provides some advice or information aimed at resolving the issues on how to select the right categories for the self-organizing networks (SON). SON is an automation technology designed to make the planning, configuration, management and optimization of mobile radio access networks simpler and faster. This chapter presents a brief introduction on intrusions detection in SON.

The major motivation and objective of this chapter is to provide a recent summary about self-organizing network IDSs of the literature as well as to provide a comparative analysis on the security challenges that are faced in the design of intrusion detection protocols.

The chapter takes a glance at the previous related work with the intent to compare this survey with other accessible research work. Early in the research into such systems, researchers tried to provide a taxonomy of IDSs. A survey by the authors of cited work [5] provided a strong systematic taxonomy of approaches with special emphasis on the detection principles, while others [6,7] provided a slightly perfunctory knowledge taxonomy of approaches.

Correspondingly, Axelsson [8] introduced a taxonomy that consists of a two classification headings detection principle and the operational aspects. Their survey grouped the systems in terms of the increasing difficulty of the problems that the system attempted to address. Furthermore, Sobh [9] introduced a structural taxonomy for IDS that categorized the system elements according to its assumptions and components. Also, Butun et al. [10] introduced a state-of-the-art survey with an extended taxonomy for IDSs in WSN, as well as taking a glance at the systems proposed for MANETs and its applicability to WSNs. A classification of the IDSs for MANETs is provided by Mandala et al. [11] and Anantvalee and Wu [12]. Moreover, Mitchell and Chen [13] argue that there are two broad categories of IDSs in the CPS. Furthermore, Erritali and El Ouahidi [14] presented and classified recent techniques of IDS in VANET environments according to the detection techniques and architecture. Table 13.2 presents a comparison of our survey with existing survey articles.

This chapter aims to develop a taxonomy that provides a common base that might be applicable to all networking environments under investigation and to not further aggravate the survey by using unnecessarily complex taxonomies. A limitation of this survey is that it does not address IDSs in cloud computing environments. Readers are directed to cite work Modi et al. [15] for a survey of IDS in cloud computing environments. It is important to note that the IDSs that are investigated in this survey are related to the techniques that are common among traditional and nontraditional networking environments, and they have been used by researchers. To illustrate further, cluster-based detection techniques are commonly used in WSN

Table 13.1　Most Common Commercial IDS Tools

Name	Year	Network	Authors	Audit Material	Detection Approach	Platform
EMERALD	1997	WiredLAN	SRI International	Network	Hybrid	Windows
Bro	2015	WiredLAN	V. Paxson	Network	Signature	Windows
Suricata	2002	WiredLAN	OISFoundation	Network	Signature	Windows
Kismet	2016	Wireless LAN	Mike Kershaw	Network	Signature	Windows, Linux, and MAC
SNORT	2000	WiredLAN	SRI International	Network	Hybrid	Windows
AirMagnet	2016	Wireless LAN	AirMagnet	Network	Anomaly	Windows, Linux, and MAC

Table 13.2 A Comparison of Our Survey with Existing Survey Articles

Reference No.	[8,9,15]	[10]	[11]	[14]	[12]	[13]	Our Survey
Network Covered	TICS	WSN	MANET	VANET	MANET	CPS	WSN, CPS, MANET, and VANET
Comparison Covered	None	None	MANET and WSN	None	None	CPS and TICS	Among WSN, CPS, MANET, and VANET

environments rather than other environments. Furthermore, agent-based detection techniques are more suitable for MANET environments rather than other environments. In addition, reputation-based techniques and voting-based techniques are more applicable to WSN environments, so the survey did not include this type of IDS. Moreover, a watchdog-based IDS is common in VANET environments rather than other environments. This survey does not include this type of IDS. Also, this survey does not comprise or contain as part or as a whole the system of methods and the notions that are applied to secure the IDSs. Readers who are interested in that topic may refer to Reference 16.

The rest of this chapter is organized as follows: In Section 13.2, an introduction to the IDS classification tree with all detailed information for each class and subclass is presented. Section 13.3 introduces the common metrics that are used by developers to evaluate the implemented IDS, which is followed by a discussion of the well-known intrusion detection datasets and benchmarks in Section 13.4. Then, in Section 13.5, a brief review of the security issues, challenges, and constraints in the targeted systems is provided. In Section 13.6, a codified compressed brief review of security attacks in SON is provided. Section 13.7 and its subsections present the sample of the surveyed papers related to IDS in SON. Section 13.8 presents a comparison among CPS, WSN, MANET, VANET, and TICS, which is the term that this study uses to refer to all types of networking systems except CPS, MANET, VANET, and WSN. Section 13.9 presents discussion, analysis, and critical challenges. Finally, concluding remarks are provided in Section 13.10.

13.2 IDS Classification Tree Hierarchy

Moving up in this section, a hierarchical classification tree is developed to organize the sample of the existing IDSs. This classification is especially for the

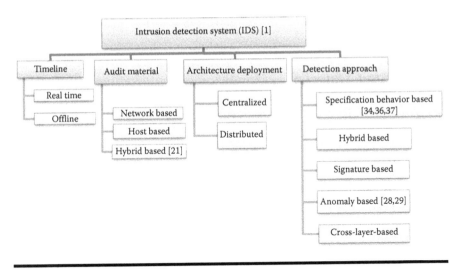

Figure 13.2 Intrusion detection system classification.

"self-organizing network" IDS and is chosen carefully to reflect the extensive groups of research work under investigation. Also, these classification criteria are important for this type of IDS as it represents a solid scientific base that groups and binds these IDSs together. Figure 13.2 shows the hierarchical classification tree of IDS based on four classification criteria:

1. *Timeline*: A measure that identifies "When" the analyzing process takes place to detect the intrusions.
2. *Detection approach*: A standard that identifies "How" the malicious activities will be interfered by IDS for identifying the intrusions.
3. *Architecture deployment*: A criterion that defines "How," the manner in which an IDS is deployed.
4. *Audit Material*: A criterion that determines "How" the information collection process is accomplished for data analysis.

In the following subsections, a discussion for each classification criterion is presented.

13.2.1 Timeline

According to the timeline criterion, IDS can be grouped into real-time IDS and offline IDS. In a real-time approach, the analyzing process takes place while the sessions are ongoing, and the IDS immediately sets on its alarm to indicate that an attack is detected. In an offline IDS, the analyzing process takes place after the information has been already collected. This type of IDS is easy to implement compared to the real-time one. However, the real-time IDS is useful for understanding the attacker's behavior.

13.2.2 Audit Material

The analyzed data (audit material) can be collected by different schemes [17]. These include: host-based, network-based, and hybrid-based approaches.

13.2.3 Architecture Deployment

The architecture deployment can be either centralized or distributed. In a centralized architecture, the deployment of the IDS function is actualized by means of a central station, while in the distributed architecture the deployment is actualized by means of distributed agents that perform the analysis by themselves, either individually or in a cooperative manner.

13.2.4 Detection Approach

Detection approaches are specification behavior-based, hybrid-based, signature-based, anomaly-based, and cross-layer-based approaches. These approaches have their own pros and cons which make them suitable for a specific type of networking environment. Next, detailed discussion about each classification approach is given.

13.2.4.1 Anomaly-Based Detection Technique

This technique was proposed in 1987 [18]. The fundamental concept behind this technique is to define the behavior of the network and/or system, and then this predefined behavior is compared with the normal behavior. The result will be either to accept it or it will trigger the alarm management system for further investigation. The function of the anomaly-based IDS is performed in two phases, which are the (1) training phase and (2) detection phase. A normal profile of the network traffic (network behavior) and/or system information logs are generated throughout the training phase. In the detection phase, the actual traffic is matched to the current normal profile that searches for any deviations. The network administrator and the network security experts prepare the accepted network behavior profiles. The constructed profiles are based on users logging information, servers logging information, and network connection features such as protocol type, flags, and so on. The applicability of this approach is defined by the attribute's nature as well as the features of the targeted system under investigation.

On the basis of processing methods, the anomaly-based detection (ABD) can be categorized into (1) data mining-based, (2) statistical-based, (3) traffic analysis-based, (4) probabilistic-based, and (5) machine learning based systems, as shown in Figure 13.3.

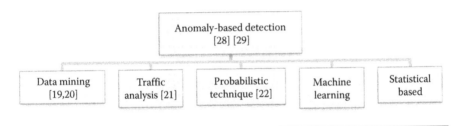

Figure 13.3   Classification of anomaly-based detection.

13.2.4.2 Signature-Based Detection Technique

This technique, which is a pattern-based detection [19] approach, implements an intruder profiling mechanism that looks for runtime features that correspond to a predetermined unique feature of misbehavior actions or attacks. These techniques are recognized not only by their low false positive rates (FPRs), but also by their ability to detect nonzero day attacks with high accuracy. On the other hand, it cannot pinpoint zero-day attacks or adjusted attacks. The main research challenge in signature-based detection (SBD) is not only to create a powerful attack dictionary, but also to add new attack patterns. Compared to the ABD technique, this technique requires more computation and resources. Signature-based detection can be classified further based on the processing approach into (1) pattern matching, (2) security rule specification, (3) state-based, and (4) data mining (see Figure 13.4).

13.2.4.3 Cross-Layer-Based Detection Technique

The basic concept behind the operation of this detection technique is to exchange certain parameters and information between the protocols that suit the targeted attack detection. The result is to combine two or more layers of information of the transport control protocol/Internet protocol suite [20] to detect multilayer security attacks. In a cross-layer-based design, more input information increases the complexity of the design. The most challenging issue is how to choose the minimum cross-layer information without affecting performance and efficiency. Furthermore,

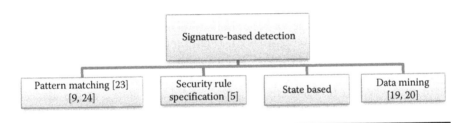

Figure 13.4   Classification of signature-based detection.

it is useful for designing new protocols and mechanisms. The conventional intrusion detection in wired and wireless networks focus on considering single-layer misbehavior features in the network layer.

13.2.4.4 Hybrid-Based Detection Technique

In 2007, a study by Kai et al. [21] proposed the design principles and evaluation results of the hybrid intrusion detection system (HIDS). The basic concept in this method is to use a combination of two or more previously mentioned methods. Many researchers use this technique to implement an IDS model which helps in enhancing the system's detection ability to disclose novel as well as known security attacks.

13.2.4.5 Specification-Based Detection Technique

In 2003, specification-based detection methodology was introduced by the authors of cited work [22], which provided the capability to detect nonzero day attacks as well as zero-day attacks, while exhibiting a low FPR In this method, the detection process is accomplished through developing the behavioral specification of legitimate system behaviors manually. This approach is used as a basis both for detecting attacks and characterizing the legitimate system behavior. The intruder activity can be identified by observing the normal system and/or users' behavior of the targeted system under investigation. It can recognize endeavors to exploit new and unexpected vulnerabilities as well as recognizes misuse of privileges kinds of attacks which do not really take advantage of any security vulnerability [23]. Therefore, this is not the best method to detect an insider attack when an unusual (but legitimate) program or an attacker is encountered. One of the advantages of this method is that it can avoid false positives since the specification can capture all legitimate behavior. However, developing an error-free, complete, and detailed specification for the system is a challenging task. The approaches that are used with this method can be statistics, neural networks, expert systems, computer immunology, state machines or extended finite state automata, and user intention identification.

13.3 Intrusions Detection Performance Metrics

Intrusions detection system performance can be evaluated through a set of metrics. These metrics include (1) false negative rate (FNR) or failure to report, representing the total number of times when the IDS incorrectly identifies a misbehaved node as a well-behaved node, (2) true positive rate (TPR), representing the total number of times when the IDS correctly identifies a misbehaved node as an intruder, (3) FPR or false alarm, representing the total number of times when the IDS incorrectly identifies a well-behaved node as a misbehaved node, (4) accuracy, (5) detection rate, estimated by calculating the number of detected intrusions over the total number of involved intrusions, and (6) receiver operating characteristic (ROC) curve for detection rate's sensitivity against false positive probability. It is noticeable that some researchers rarely try to enhance the performance evaluation metrics process by introducing new metrics to evaluate and analyze the IDS performance. One such metric is the detection latency [24], which tries to measure the time that is required to discover an intruder since it descended into the system. It should be noted that this metric is not identical in the case of the insider attackers and the case of outsider attackers. This metric is helpful in the CPS IDS since most of the CPSs are life-critical applications embedded within an infrastructure. In WSN environments where power consumption and energy saving are important factors, some researchers [25] estimated the time for a random number of sensors to run out of their energy. Others [26] measured the efficiency of the packet sampling operation represented by the percentage of analyzed packets that the IDS recognized as intended to do harm (malicious). In MANET and mobile device environments, the authors of cited work [27] used equal error rate (EER) to calculate the performance of their IDS. This is represented by the rate when the rejected error equals the accepted error.

Throughout the timeline of this survey, we noticed variant attempts to establish new brighter performance metrics. Nonetheless, a majority of them lack applicability as well as generality [28]. To sum up, there are no metrics, especially for specific types of IDSs. So in this survey, we decided to present a high-level definition of performance metrics for IDSs that are well known and well established.

13.4 Intrusion Detection Datasets

In this section, we will summarize popular benchmark datasets that have been used for designing IDSs by developers. Generally, in research, datasets fall into several categories [29] such as "baseline data," "simulations," "traffic generation," and "live network." The types of datasets and the pros and cons of attributes for the previously mentioned dataset categories are further explained in Reference 29. There are multiple datasets that are used by researchers to assist in the evaluation of IDSs and intrusion prevention systems (IPS). For example, Stolfo et al. [30] constructed the KDDCup99 from captured data in the DARPA'98 IDS [31]. KDDCup99 is another dataset that has duplicate redundant records. To overcome KDDCup 99 dataset problems, Tavallaee et al. [32] suggested the NSL-KDD [33] dataset. The NSL-KDD dataset comprises chosen records of KDDCup 99. NSL-KDD suffers from some problems such as the dependency on synthetic data in order to estimate real system performance. Moreover, the ISCX 2012 dataset [34,35], which was collected in 2010, served as a replacement for KDDCup99. The ISCX-UNB 2012 dataset has network packet filtering (NPF) attributes. One problem about the ISCX-UNB 2012 dataset is that it is labeled. Another dataset is the ITOC 2009 [36]. This dataset has added NPF and audit logging data but is devoid of labels. Other replacements to KDDCup99 that are similar to KDDCup99 exist, such as KDD 99. These datasets are very large and consist of many components. Another example of the baseline data that were designed for the purpose of comparison among algorithms includes LINCOLN LABs (2000) and DARPA (2000). The latter one is recognized as a standard, although it is a deficient dataset for these experiments [29]. The U.S. government introduced the dataset using a simulated background network traffic and added the pattern, attribute, and features of the known attacks to the simulated background network traffic. Researchers have extensively criticized the appropriateness of the dataset for research purposes [29,34]. Researchers consider DARPA an extremely outdated dataset that has not been verified or investigated in order to see how compatible it is to real network traffic [29,37,38]. Furthermore, the DARPA dataset is unable to meet the new trend of attacks such as Illusion attacks and Bogus Information attacks [39]. Moreover, it is fully inappropriate for research in wireless local area networks (WLAN), since the data have been gathered over a local area network (LAN) [28].

Recently, researchers from the University of Twente, Sperotto and Van developed a new dataset called the UT dataset in the form of Net Flow [40]. A honeypot host at the University was used to collect the flow of information in the network traffic during 2008. Notable datasets include: MAWI [39], NSA Data Capture [41], and the Internet Storm Center [42]. A CRAWDED [43] is a community resource for archiving wireless data at Dartmouth University. The website includes collected datasets such as cu/rssi [44] that incorporates information such as received signal strength indication (RSSI). Another dataset is the Utah-CIR dataset [45] that blends inter-arrival time packets information for variant type of wireless devices such as iPads, iPhones, IP-Cameras, and so on. In addition, the hope/nh_amd dataset represents RFID tracking data [46]. Additionally, the AWID family of datasets [47] can act as a reliable testbed for intrusion detection experiments in wireless networks [48]. Kolias et al. provide an extensive evaluation of AWID using variant machine learning algorithms [48]. Moreover, WSN-ID [49] is a dataset for WSN that has been collected using an NS2 simulator to detect denial of service (DoS) attacks in WSN implementing the LEACH protocol. Certainly, the potential absence of an efficient intrusion benchmark or dataset is a critical issue in academic research. A few developers established their own benchmark. There are two difficulties with this approach. The first issue is that one should know according to what specifications these data will be labeled as either anomalies or normal. The second one is that the datasets must be updated frequently over time. The update must include the instances of new technologies, applications, and users because these represent normal traffic. Also, the update must include the pattern of new attacks which used new techniques or exploit system vulnerability in an innovative method. The update process will assist not only in keeping research pertinent, but also in training the learning system for intrusion detection more efficiently and in a reiterated manner as technology and also cyber-attacks emerge and evolve [50]. A list of the most frequently used datasets are shown in Table 13.3. Choosing a specific dataset to assess the IDS performance is a process

Table 13.3 Most Common IDS Datasets

Collecting Criteria	Type of Dataset	Year	Network	Size	Abbreviation
Network traffic	Benchmark	1998	TICS	5 GB	DARPA98
NPF	Real Life	2009	TICS	12 GB	ITOC 2009
Network traffic	Benchmark	1999	TICS	4898431 Train 311029 Test	KDD99
Full packet payloads in pcap format	Real Life		TICS	80 GB	UNB ISCX 2012
Selected record of KDDCup 99	Benchmark	2009	TICS	125973 Test 22544 Train	NSL-KDD
SNORT dataset	Real Life	1999	TICS		SNORT
Labeled WSN data repository collected from a simple single-hop and multi-hop WSN deployment using TelosB motes	Synthetic	2010	WSN	1 M byte	LWSNDR
Collected from Network Simulator 2 and then processed to produce 23 features	Synthetic	2016	WSN	224796 Train 149865 Test	WSN-DS
AWID family of datasets	Benchmark	2015	WSN	150 GB	AWID
Honeypot host at the University was used to collect the flow of information in the network traffic	Real Life	2008	TICS	10 GB	UI

that depends on both the IDS problems and targeted security requirements. This study recommends choosing a dataset that is close or identical to real-time network traffic.

13.5 Distinguished Characteristics and Security Challenges in SON

Even though IDSs have evolved rapidly in the past few decades, major important issues remain. For example, detection systems should be more effective, detecting a wider range of attacks with fewer false positives. In addition, intrusion detection must be able to accommodate modern networks' increased size, speed, and dynamics. The well-known commercial IDS tools are primarily focused on the traditional network-based IDS environments and these security mechanisms may not be effective in SON such as VANET, MANET, WSN, and CPS. Several constraints and challenges are facing these environments and the wireless communication environment represents a common factor that has an influence on these environments. In the following paragraphs, we present distinguished characteristics and security challenges in SON.

13.5.1 Distinguished Characteristics and Security Challenges in CPSs

CPSs are characterized as geographically large-scale distribution systems. Usually, these systems are federated, heterogeneous, and life-critical systems. CPSs cooperate with the physical world in a trustworthy, reliable, safe, secure, efficient, and real-time manner. PSs incorporate not only cyber components such as traditional networking components, but also physical components such as sensors, actuators, and feedback control units. Examples of CPSs are smart grids, pervasive health care systems, unmanned aircraft systems, critical-infrastructure control systems such as electric power systems, and water treatment systems [51]. The common functions among all these systems are acquisition and control. The most significant features of CPSs are that these systems have multiple control loops, strict timing requirements, predictable network traffic, legacy components, and possibly wireless network segments [13,52]. In this study, we glance at the threat models for CPSs, which could be identified by the attack specific goal such as to disturb operations as well as to cause loss of data [13,52]. Generally, the communication protocols employed in the CPS network layer include Ethernet, Dial-Up Modem, MODBUS, which is a serial communication protocol that is designed for use with Modicon programmable logic controllers [53], RS-232, WiDom [54], IEEE 802.15.4 [55], and TCP-IP. Furthermore, cyber-attacks that might work against the sensors in WSNs will be extant in the actuator/sensor layer of CPS. Researchers and developers have nowadays shown increased interest in CPSs. Han et al. [56] showed in his study the challenges and techniques related to IDS in CPSs. Furthermore, Mitchell et al. [57] presented an excellent modeling, analysis, as well as the mechanisms of counter defense for security attacks related to CPS. The authors developed a stochastic petri nets analytical model to capture the dynamics between adversary behavior that include a surveilling attacker and destructive attacker and defense mechanism of a modernized electrical grid as an example of a CPS. Further, the model considers different types of failures that can happen to a CPS such as attrition failure, pervasion failure, and exfiltration failure. The authors apply optimal detection interval, data leak rate control, and redundancy as a countermeasure mechanism. The IDS has to perform an intrusion detection audit on a target node in every optimal detection interval. The results revealed the optimal design condition for the proposed design parameters. Thus, the proposed mechanisms are efficient and effective in securing a CPS and it is recommended to apply the same mechanism to secure other CPSs such as medical cypher physical and industrial control system.

Software patching and frequent updates are not well suited for CPSs since these are critical infrastructure and shutting down the system can lead to serious situations or may cause either customer dissatisfaction or financial loss. However, these are well suited for TICS. Running the TICS offline or shutting it down for a new or periodic update in the system may need a simple planning, while this update may require advanced planning since taking the system offline would require months of advanced planning. In addition, CPSs are autonomous decision-making systems that require making decisions in real-time manner. Thus, real-time requirements are other challenges in CPSs. This is while availability requirements are one of the challenges that affect TICS. Furthermore, speed of detection (detection latency) is another challenge and a key research problem for CPS IDSs since a DoS type of attack is devastating in CPS environments, especially for utility and health care applications. Furthermore, scalability, geographic dispersion, and federation are common issues in CPS environments [3]. Nonetheless, CPSs present relatively simpler network dynamics than TICS. To illustrate, servers may change rarely, topology is fixed, communication patterns are regular, user population is stable, and the number of protocols that govern communication is limited [58]. However, WSNs share the networked operation and low capability characteristics with CPSs [59,60].

13.5.2 Distinguished Characteristics and Security Challenges in a MANET

A MANET is characterized by its open medium and the wide distribution of nodes that operate as a host as well as a router. In contrast to WSNs, MANETs do not require a base station [61]. A MANET does not have any fixed topology. The nodes are mobile (they can connect or disconnect from the network at any time). Furthermore, a MANET has low deployment costs and it shares certain properties with WSN such as being battery and power constrained. The nodes in a MANET are either a laptop or a cell phone. These devices are required to send packets on to a further destination, and as a MANET grows, the huge forwarding process will affect their limited power and processing capabilities, since they occupy a considerable amount of processing power [62].There are several vulnerabilities that are related to the nature of a MANET's environment such as the fact that the links are wireless, the topology is dynamic, the resources are limited, the nature or the routing algorithms are cooperative, and finally, a network perimeter is absent. As a result, routing attacks are the most effective attacks in MANETs [63]. Moreover, MANETs have additional vulnerabilities that are shared with wired networks like DoS, eavesdropping, and spoofing. This is not an exhaustive list, and it only mentions the major attacks studied in the literature. A MANET's environments are exposed to some or all of the same challenges that WSN environments are exposed to such as eavesdropping, highly applicable to being hacked, and being a distributed environment in product of the lack of infrastructure. However, MANET nodes usually have a bigger battery and more power compared to WSN nodes. In addition, they have better computational capacity than WSN nodes, since the majority of these nodes are laptops, which could have a microprocessor with a maximum speed of 3.5 GHz. Finally, the nodes density in MANET environments is lower than its counterpart in WSN environments. Thus, developers should take into consideration all these challenges and constraints when developing, building, and adapting an IDS in these environments.

13.5.3 Distinguished Characteristics and Security Challenges in VANET

VANETs provide communication among close-by vehicles and roadside equipment. VANET is considered a special type of MANET. Nevertheless, it still differs significantly from MANET. VANET is characterized by its high node mobility in an organized fashion, volatile topology change, transient nature of participants, and no persistent communication links. VANET environments are attractive for attackers since vehicles in VANETs manage vital and sensitive information. As being a special case of MANET, all the vulnerabilities of MANET may be considered true in VANETs. Security attacks of VANETs is the current area of research where many activities are being observed. A study published in 2012 by Faezipour et al. [64] represents a good illustration as well as ongoing research of the security, privacy, and secure communication of intelligent VANETs. Furthermore, a study by the cited work [65] provided a summary of the all existing security problems in VANETs.

Unlike WSNs and MANETs, the nodes in VANETs are powered by huge batteries, effective for highly computational tasks. Thus, complex cryptographic calculations are applicable [66]. As a result, power-efficient protocols are not vital for VANET. The architecture, standard requirements, solutions, and challenges of VANETs are discussed and analyzed considerably by Karagiannis et al. [67]. VANETs are considered as one of the ad hoc networks. However, they differ from other ad hoc networks as well as TICSs. A VANET environment exhibits challenges and constraints that would affect a VANET IDS. For example, the nodes have the ability to move or be moved freely and easily with a frequently predefined pattern and predication and these play an important role both in the network protocol design and the IDS protocol design. Furthermore, the movement among the vehicles is at a high speed which would affect the topology as well as the connectivity of the nodes in VANETs. As a result, VANETs are categorized as having highly dynamic topologies and frequently disconnected network environments [68]. Furthermore, nodes have an adequate energy and computing power in terms of processing and storage. The latter distinguishes between VANETs and WSNs. Another partial issue of VANETs, especially in some applications like large transport vehicles (truck), is that the node has some hard delay constraints since these large vehicles often drive on automatic highway road systems. When a break event happens, the message should be transferred and arrived in a specific time to avoid crash [69]. All of these challenges and constraints as well as availability requirements are factors that influence the design and the applicability of the intrusion detection techniques.

13.5.4 *Distinguished Characteristics and Security Challenges in WSN*

WSNs can be categorized as having resources constrained, such as limited power, limited memory processing capability, less memory, short communication range, self-organization, and multi hop routing characteristics. Moreover, WSNs lack a physical line of defense to monitor the information flow such as gateways or switches. The security of WSNs is challenging, especially for applications where data confidentiality is of prime importance. In WSN environments, there are vulnerabilities of multiple layers.

SON environments can be categorized as being distributed networking environments that are missing the basic physical and organizational structure and facilities, such as a gateway, router, switch, base station, and so on, that the TICS includes. These are important elements for supporting networking operations like communication, routing, encryption, and real-time analysis. Another critical problem is that the nodes are highly exposed to being vulnerable to physical tampering or hijacking, which compromises network operation since these hacked nodes could supply misleading routing information to other network infrastructure (sinkhole, wormhole, and black hole attacks). Furthermore, the broadcasting nature in WSN environments exposes nodes to eavesdropping, which would reveal important information to an adversary and/or to jamming/interfering, and these can cause a DoS attack in the environment [10]. In addition, the WSN node's power supply is limited, since the majority of the WSN's nodes are powered by a 2 AA sized battery such as MICA [70] and MICAz [71]. This can affect their lifetime as well as their applicability to implement specific intrusion detection techniques. Besides that, the node intensity in WSN environments is highly distributed. Nonetheless, the short lifetime of sensor nodes in WSN environments because of the fast energy decay as well as lack of physical security may reduce nodes intensity over time. Finally, WSN's environments are missing centralized trusted authorities that can be trusted by the network. Thus, decisions should be made in a collaborative manner, which may expose the environment to trust authority management issues.

To sum up, the common distinguished characteristic among these types of networks is the wireless communication environment, which presents security threats and vulnerability such as DoS, eavesdropping, spoofing, and authentication, in addition to other variant security challenges that are faced in SON. Table 13.4 introduces a summary and comparison of the distinguished characteristics that present security challenges for the types of networks under review.

13.6 Security Attacks in SON

SON are vulnerable to different types of attacks. In this chapter, we explore some of the common existing attacks against SON. These attacks can work against a specific individual open system interconnection (OSI) layer. For example, malicious code and repudiation are types of attack on the application layer. Session hijacking and SYN flooding are examples of attacks on the transport layer. Flooding, blackhole, greyhole, wormhole, link spoofing, link withholding, byzantine, replay, and location disclosure are examples of attacks on the network layer. MAC malicious behavior and selfish behavior are examples of attacks on the data link layer. Finally, interference, jamming, eavesdropping, tampering, capturing, and hijacking are examples of attacks on the physical layer. Physical layer attacks are hardware oriented. The attacker needs to use an external hardware to mount such attacks. Furthermore, it is more difficult to mount such an attack if the topology of the network is constantly changing. For instance, in the case of MANETs and VANETs, the attacker must remain at the area near or surrounding the moving vehicle or node continuously to disrupt their communication, which is difficult. On the contrary, such attacks might be of value in stationary topology networks such as WSNs and CPSs.

On the basis of the attack interaction nature, attacks can be classified as passive or active. In a passive attack, a malicious node monitors the network to find out information. Eavesdropping, disclosure, and traffic analysis are examples of passive attacks. This type of attack is trivial in case of highly dynamic environments such as VANETs and MANETs since the topology itself changes rapidly and very frequently. In an active attack, an unauthorized node tries to modify the state of the network. Fabrication, dropping, timing attacks, and modification are examples of active attacks. Figure 13.5 presents a classification of attacks based on the attack interaction nature.

Routing attacks are significant and present a challenge in corporative routing environments such as MANET. Examples of these attacks include blackhole,

Intrusion Detection System in Self-Organizing Networks ■ **357**

Table 13.4 Summary and Comparisons of Distinguished Characteristics of SONs

Distinguished Characteristics	TICS	MANET	VANET	CPS	WSN
Network environment	Wired	Wireless	Wireless	Wireless	Wireless
Dynamic topology	Not applicable	High	High	Not applicable	High in dynamic WSN
Node distribution	Low	Low	Low	High	High
Cooperativeness of routing protocol	Moderate	High	Moderate	Moderate	Moderate
Real-time Requirements	Low	Low	Critical	Critical	Low
Infrastructure (lack of a clear line of defense)	Available	Lack	Available	Available	Lack
Software patching and frequent updates	Inconsiderable and has no influence on the network	Inconsiderable and has no influence on the network	Inconsiderable and has no influence on the network	Critical and has an influence on the network	Inconsiderable and has no influence on the network

(Continued)

Table 13.4 (Continued) Summary and Comparisons of Distinguished Characteristics of SONs

Distinguished Characteristics	TICS	MANET	VANET	CPS	WSN
Availability requirement	Critical and necessary	Required	Required	Necessary	Required
Resource constrained/limited resources	Very low	Moderate	Low	High	High
Scalability	Low	Moderate	Moderate	High	High
Location awareness	Not applicable	Required	Necessary	Required	Required in tracking applications
Tradeoff between authentication and privacy	Not Required	Not Required	Necessary	Necessary	Not Required
Computational/ processing ability	High	High	High	Limited	Limited
Bandwidth	High	Limited	High	Limited	Limited

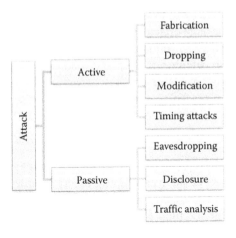

Figure 13.5   Attacks classification.

link with holding, link spoofing, reply, flooding, wormhole, and colluding miserly attacks. In the following section, we explain briefly these attacks.

In a blackhole attack, a malicious node either discards packets instead of forwarding them, or claims that it has an optimum route by sending a fake routing information, which makes good nodes route packets through the malicious node [63].

In a link withholding attack, a malicious node turns a blind eye to the requirement of advertising the link of certain group of nodes, which can cause a link loss to these nodes [17]. In link spoofing attacks, a malicious node advertises forged links with other adjacent or nonadjacent nodes in the network.

With high mobility nodes such as in MANET environments, the current topology may change rapidly, signifying replay attacks. In this attack, a malicious node records the legitimate control message of another node and retransmits it at a later time. This can cause an outdated update of the routing information.

In a flooding attack, a malicious node intends to exhaust the network resources such as bandwidth and node resources such as computational and battery power. This is accomplished through flooding the network traffic or disrupting the routing operation to cause severe degradation in the network performance or to bring down a network service. Moreover, in wormhole attacks, a malicious node collects traffic at one point in the network and selectively forwards it to another point in the network, and then replays it into the network from this final point [72].

Another type of attack that is relevant to the corporative routing algorithm is the gray hole attack. In this type of attack, a malicious node that is designated to act as a router shall forward and handle a subset of packets, while leaving others. A different version of this attack is the selective forwarding attack. In this type of attack, a malicious node behaves as a normal node most of the time but selectively drops sensitive packets periodically. This type of attack is more powerful than the gray hole attack and it is more difficult to disclose since the node behaves as a normal node frequently. Moreover, in colluding miserly attacks, multiple malicious nodes are engaged together in collusion to drop and modify routing packets, which can cause disrupting routing operations.

In session hijacking, a malicious node exploits a valid networking session to gain unauthorized access to information or services. This is an example of transport layer attacks.

In a jelly fish attack, a malicious node gets hold of forwarding packets through starting to delay and/or dropping packets for a specific amount of time before forwarding normally. In a repudiation attack, a legitimate node denies that it performed certain actions or transactions. This type of attack is difficult to prove without an adequate mechanism for auditing. The repudiation can be referred to as the denial of participation in all or part of the communication [63].

In a sybil attack, a malicious node in the network claims multiple identities. This is an active type of attack and can lower the performance and consumes more resources.

A new threat in VANETs is represented by illusion attacks. In illusion attacks, a malicious vehicle creates virtual traffic events through broadcasting traffic warning messages based on recent road conditions to produce illusions to vehicles in their neighborhood [73]. Furthermore, in location disclosure attacks, the sole target of the attacker is to gather information only related to the identity of nodes in the network which can later be used in one way or the other. This attack is of more relevance in VANETs and WSNs with military applications wherein a driver's location privacy needs to be protected always. Other new security threats in VANET environments are the bugs information attacks. In this type of attack, a malicious node disseminates false information in the network in order to affect the decisions of other drivers [74].

A desynchronization attack is an active attack and it is of a value in CPS and WSN environments. In this attack, a malicious node aims to disrupt the clock phase of multiple nodes.

An example of an attack in the application layer is the Microsoft security bulletins. In this attack, an adversary from a remote location exploits system availabilities to gain user right as the current user access. This type of attack is significant in CPS environments that are tied to consumer utilities such as electrical power system and smart grid.

Furthermore, a new security threat in the CPS is the slander or ballot stuffing. In this attack, a node falsely reports unfavorable evaluations of good nodes or favorable evaluations of bad nodes [3].

Additionally, a software-based attack is another type of attack at the application layer. In this type of attack, an adversary tries to make partial or minor changes to the software code in the memory or exploit known vulnerabilities in the software code. Such a type of attack can bring harm to both CPS as well as WSN environments.

Other examples of application layer attacks are the malicious codes attacks. In this attack, a malicious program can spread itself through the network with the intent to slow down or damage the computer system and network. Examples of this attack include: virus, worm, spyware, and Trojan horse, which are application layer attacks.

Moreover, in data exfiltration attacks, a malicious node copies, transfers, or retrieves data from the system without authorization, making sensitive information available outside of the compromised system [3]. This type of attack can be seen in CPSs and WSNs of military and medical applications.

Other new threats are the deluge (programming) attacks. This type of attack is an active attack that has an effect on the application layer. In this attack, the attacker tries to reprogram the targeted node for certain purposes. A different version of this attack is the command injection attack. In this attack, a malicious node executes an arbitrary command on the host operating system through a vulnerable application. This type of attack can harm CPS environments. Another type of attack is the impersonation attack. In this type of attack, a malicious node successfully claims the identity of one of the legitimate nodes in the network and/or communication protocol.

Another type of active attack is the fabrication attack. In the fabrication type of attack, a malicious node fabricates its own packets or data to cause confusion in the network operations.

Furthermore, multihop wireless networks such as MANETs and VANETs face many security challenges caused by the infrastructure networks, but the corporative nature of routing algorithms also brings new threats of routing attacks. Among these is the wormhole attack. This attack remains the most serious one since it can be organized and carried out by an outsider attacker.

Finally, a survey of the state-of-the-art MANET's security attacks with an emphasis on examining and analyzing routing attacks and its countermeasures was presented in References 17 and 75. In the same manner, Al-Kahtani [76] introduced a survey on security attacks in VANETs. The survey included a list of diverse attacks as well as the defense approaches against these attacks and considers possible future security attacks. Concerning WSN environments, Padmavathi and Shanmugapriya [77] present a classification mechanism for a wide variety of security attacks in WSN and the challenges that are faced in these environments. Furthermore, many studies available in the literature are explained in detail such as security attacks in WSN [77], MANET [17,75], VANET [76,65], and CPS [57] environments.

To sum up, the literature on security attacks and threats in traditional and self-organizing networking environments shows a variety of approaches that have been studied, investigated, analyzed, and counter measured by researches through decades. Existing security schemes for TICS cannot be applied directly to SON, which make these networks much more vulnerable to security attacks. Different solutions for these types of attacks are available in the literature. Yet these solutions are still not perfect with respect to the tradeoffs between effectiveness and efficiency. The future researches should focus on improving the effectiveness of the security schemes as well as on minimizing the cost to make them suitable for self-organizing network environments that share the wireless environment, resource constraints, and the dynamic topology with the exclusion of CPS and WSN environments that have a fixed topology. Furthermore, each proposed solution can work only with a specific attack and is still vulnerable to unexpected attacks. Therefore, self-organizing network researchers should furthermore focus on both exploring and preventing all possible attacks to make self-organizing network environments a secure and reliable environment. In this survey, we provide a codified compressed brief review of layered-based security attacks and threats in the examined environments.

13.7 Intrusions Detection System in SON

With the rapid expansion of traditional and nontraditional networking environments, the security of these environments has become very important. Every day, new kinds of attacks are being faced by these environments. Many methods have been proposed for the development of IDSs using different techniques. In this section, we describe the design and implementation of several IDSs that are related to SON.

13.7.1 Cyber Physical System

To date, there are few studies that have investigated CPS IDSs, and the present state of design is in the infancy stage. Few reported CPS IDSs are available in the literature. This study reviews the design as well as the implementation of a sample of IDSs with the intent to examine recent research in CPS IDSs. The findings are summarized in Table 13.4. Although Table 13.4 does not contain all the available work in the literature, it provides a representative sample of the most relevant work.

A study Linda et al. [78] implements an IDS using neural network modeling (IDS-NNM) for smart utility applications such as power management systems. The model used error-back propagation as well as Levenberg–Marquardt with window-based feature extraction. Several features were used in building the system such as IP address count, maximum packets per protocol count, and so on. The authors experimentally recorded five datasets, each containing 20,000 captured packets. Metasploit, Nessus, and Nmap utilities were used to construct an attack model with a high degree of complexity for generating 100,000 intrusions. This investigation used network traffic for collecting adversary data, machine learning for analysis, and did not address responding to the analysis.

In a similar manner, the work presented in Reference 79 studied an SBD IDS for smart utility applications. No dataset was included in the study; however, some water level readings as the audits sensor, valve settings as the actuator data, and the GOOSE messages arrival rate as closed control loop timing were used in the design. The attack model has not been discussed and the investigation considered dealing only with ABB System 800xA devices as the legacy hardware. This study used system activities for collecting adversary data, behavior rule-based analysis, and did not address responding to the analysis.

Regarding health care applications, Mitchell and Chen [80] suggest an SBD IDS implemented in a distributed architecture deployment for a pervasive health monitoring system application. The environment consisted of different medical devices such as patient controlled analgesia (PCA), vital sign monitor (VSM), and cardiac device (CD). The impact of attacker behavior was investigated to realize its efficacy on the medical cyber physical system (MCPS) IDS. The threat model focused on defeating inside attackers and subtle manipulation attacks. The authors profile different behavior specification such as pacemaker frequency, analgesic request, oxygen saturation, heart pulse, temperature, respiration, analgesic infusion rate, blood pressure, and cardiac device mode. The investigation considers dealing with the Welch Allyn Connex 6000 as the VSM legacy hardware that fits into the proposed model. The threat model in this system is sophisticated; it comprised a mechanism to deal with insider attacks and subtle manipulation and exfiltration types of attacks. The authors stated that their solution produced a nearly 100% detection rate for the proposed deployment configuration. This result emerged because safety is a key factor in MCPS. This study used user reading, behavior rules in analysis, and did not address responding to the analysis.

Likewise, Asfaw et al. [81] examined a distributed anomaly-based IDS design for a CPS. The deployment architecture is composed of two components: a mobile device and a server. The components are communicating with each other using a communication channel that was established over a wireless link. The function of the mobile device is to collect and forward the data to the centralized audit server. The audit logs of the information enclose both the medical record access and the location of the information. An association rule mining method using the Apriori algorithm was used to build the class rule generator (CBA-RG). Then these generated rules were used as the basic classification rules (CBA-CB) to identify whether the user requests were either normal or anomalies. The dataset was constructed using a single user observed recording that consisted of 20 normal records with a noise-free assumption (free of misbehavior), which considers a somewhat small sample. In addition, the authors did not report FNR or FPR. This work only considers exfiltration attacks. This study used system logging information, data mining for analysis, and did not address responding to the analysis.

In the light of CPSs, a specification-based IDS for an unmanned aircraft system (UAS) application was proposed in Reference 82. The proposed IDS works toward securing the infrastructure embedded in the system such as the sensors or the actuators. The threat model considers the compromised unmanned aerial vehicle (UAV) in a UAS. The proposed IDS investigated five malicious security behaviors to model the system. Four of these threats were concerned with the attacker who violated the integrity of the system. These include: degrading a UAV's lastingness to waste its power, increasing UAVs' vulnerability to activate its countermeasures needlessly, directing a UAV's weapon opposite to cooperative subjects, and finally capturing the UAV. The last one is concerned about the attacker who violated the system's privacy to surreptitiously withdraw the mission data. The threat model in this system is sophisticated since it considers different types of attack as well as both types of insider and outsider attackers.

Yang et al. [83] presented a signature-based IDS for supervisory control and data acquisition network (SCADA) which utilized IEC 60870-5-104 protocols. The system used a deep packet inspection (DPI) method that included both misuse detection-based and model-based approaches. Likewise, an inclusive and validated set of SNORT IDS rules was established to model the IEC 60870-5-104 system. The proposed implementation could precisely detect a few known, doubting, or malicious attacks as well as the sources of the attacks. A SNORT-based experimental process has been used to validate the established rules. The implemented IDS introduced some latency; however, it did not affect the availability and the timing requirement for the operation of SCADA data, as stated by the authors. This study used rules for analysis, and did not address responding to the analysis. However, the work considered legacy hardware components that distinguishes a CPS from other systems.

Equally important, the authors of cited work [84] investigated a hybrid intrusion detection approach for an electric power smart grid system. The proposed approach incorporated both the power information as well as the sensor placement to detect CONSUMER attacks [84]. The work used a placing algorithm to insert the intelligent grid sensor (GPS) on the lines as well as the feeder of the distribution network to supply effective observability of the network as well as to enhance the detection performance. This investigation used system and user information to collect adversary data, and the threat model is sophisticated as it compromises CONSUMER attacks.

Ghaeini and Tippenhauer [85] developed a scalable distributed hierarchical monitoring intrusion detection system (HMIDS) framework for a CPS. HMIDS framework incorporated Bro [86] open source IDS with added support for Ethernet/IP and control an information protocol (CIP) [87] to record industrial system control network traffic. The authors profile three types of attackers, namely cyber-criminal, insider, and strong attacker. The designated HMIDS works toward detecting both specific SCADA networks such as Stop CPU, Crash CPU, Crash ETHERNET and reboot Ethernet, and general attacks such as IP scanning, SYN flooding, and APR poisoning. HMIDS was validated in a realistic industrial control system (ICS) in the SWaT plant that emulates a water treatment system. The framework was able to detect all type of attacks under test with a 100% detection rate. However, the framework observed a false positive as a result of the IDS flagging all traffic through an identified ARP spoofed as attack, even when the content was nonmalicious.

13.7.2 MANET

The IDSs for MANETs have been very well investigated and the following is a sample of the literature. In the first place, Thamilarasu et al. [88] approached a decentralized architecture using a cross-layer detection system for MANETs. This threat model focuses on detecting jamming attacks. To detect the attack, an incorporated mechanism that differentiates actual network failures from a malicious jamming attack has been used. The detecting process is done in both the physical layer (PHY) as well as the media access control (MAC) layers through performing a channel monitoring on the MAC layer. The scheme is carried out in two phases. The attack detection is performed in the first phase and when the attack in this phase is confirmed, the second phase detection is triggered, which is implemented for obtaining network congestion using a cross-layer design technique. A simulated approach has been used to validate and examine the effectiveness of the designed IDS model. The results showed that this design performs well in terms of detection accuracy and FPRs.

Correspondingly, cited work [89] studied routing attacks detection in MANET using various IDS approaches. The study examined the likelihood of CLB IDS in MANET to overcome the complication associated with such networks. A decentralized IDS-based architecture and the optimum link state routing (OLSR) protocol were used. The study was able to detect sinking and spoofing attacks using a linear discriminant analysis (LDA) classification technique [90]. A comparison between cross- and single-layer schemes has been done, and the cross-layer-based scheme outperformed the single-layer-based scheme. The systems used data mining for analysis, and did not report responses to analysis.

Equally, cited work [91] proposed to design a real-time NIDS tool for wireless networks. The designed tool was based on the "Ad-hoc On-Demand Distance Vector routing protocol (AODV)" [21]. The presented IDS was developed in a distributed manner to detect spoofing attacks as well as packet dropping attacks in MANETs.

Moreover, authors of cited work [92] proposed an IDS for MANETs. The proposed system used and compared different classification algorithms with respect to classification error as well as weighted error. The GloMoSim Library was used to build a simulated $850*850$ m^2 MANET environment of 15 nodes each with a 250 m radio propagation range and 2 Mbps channel capacity as well as an ad-hoc on demand distance vector routing protocol with a predefined mobility pattern. The cited work used network traffic for collecting data concerning adversary and machine learning for classification. The work did not report responses to analysis. The threat model combined flooding, forging, packet drooping, and black hole attacks. The article reported FPR, FNR, and detection rate.

The author of cited work [93] proposed a distributed deployment of an anomaly-based IDS for MANETs. The proposed system uses a neuro-fuzzy classifier in binary form to detect the behavior of current activities with an emphasis on packet drooping attacks. The classifier used a variant set of features such as the number of data packet forwards, the total number of broken links, and the average number of hop counts. The performance of the proposed IDS as well as visualization of the attacks scenarios was evaluated using the Qualnet simulator along with the MATLAB toolbox. The simulated results showed that the proposed approach can detect the zero-day attacks and nonzero day attacks with high values of true positive and low FPRs with respect to other approaches such as the Sugeno fuzzy inference-based approach.

Santhi [94] proposed an IDS that uses an enhanced adaptive acknowledgement with digital signature algorithm namely (EAACK-DSA) to detect and isolate the malicious nodes in MANET environments. The fundamental idea behind EAACK-DSA is to digitally sign all the acknowledgment packets before transmission. The simulated result showed that the proposed approach was able to resolve false misbehavior report, receiver collision, and packet dropping attacks of the watchdog scheme. However, in a resource constrained environment, such an approach may add a burden to the network.

13.7.3 Vehicle Area Network

In the first place, cited work [95] established a modular cross-layer IDS to detect wormhole attacks in VANETs. The audit data were collected through different modules and in different layers. Two decision modules were used for the detection purpose. One is local in each node in the targeted VANET, while the other is a central decision module. An application module that uses the knowledge from the application layer evaluated every warning message received in the application layer. This was done through the knowledge provided from a specified application and in combination with the sensor data that were supplied through the context information module. The simulated results showed that the established modular design could effectively detect wormhole attacks through lower values for both TP and TN rates, and the threat model was sophisticated.

Alheeti et al. [96] proposed a hybrid IDS for semi- as well as automatic self-driving vehicles. An artificial neural networks (ANNs) algorithm has been utilized to disclose DoS attacks. Profiles for both malicious and normal behaviors have been generated to represent real-world traffic based on Manhattan mobility models, and using the "Simulation of Urban Mobility Model (SUMO)" and "MObilty VEhicles (MOVE)." The NS-2 simulator was used to create a VANET environment of 30 vehicles and six road side units (RSUs) with one malicious vehicle. The experiments showed that the system was able to identify anomalies efficiently and effectively with a TPR of 98.06%, a TNR of 87.75%, an FNR of 1.93%, and finally an FPR of 12.248%. This system used system and user logging information for collecting data concerning adversary and data mining for analysis. However, they did not address intrusion responses. Furthermore, the threat model was not sophisticated.

In Reference 66, a novel rule-based IDS that monitors and examines data sent in VANETs using plausibility checks was introduced in order to identify two types of attackers: the constrained and the unbounded. The proposed system consisted of three core components. These included the message revocation scheme, the plausibility checks, and the adaptive warning levels. The threat model was not sophisticated; it considers the constrained and unbounded type of attacks.

Yoon et al. [97] proposed an ABD IDS using statistical measures for a real-time embedded system. The proposed framework aims to protect the real-time embedded system from malicious entities such as W32.Stuxnet and Duqu worms. The basic idea behind this work was to statistically analyze and observe the execution profiles of the real-time system to find inherent properties. The results showed that the proposed system effectively detected on the fly the malicious code execution, keeping the physical system safe. This investigation used statistical analysis, and system and program activities for collecting data concerning the adversary. Nonetheless, the work did not report responses to analysis.

Beigi-Mohammadi et al. [98] proposed an IDS that aims to detect wormhole attacks in smart grid neighborhood area networks (NAN). The system computed the estimated hop count between the collector and the smart meters. This was elaborated in "Maple" and mingled with OPNET. Three real geographical regions have been used to model the proposed NAN and evaluate the IDS performance, which include rural areas, suburban areas, and urban areas. The proposed IDS performed well in terms of the detection rates. The highest overall detection rate of 96.9% was achieved for the urban NAN, while FPR and FNR were 4.8% and 3.45%, respectively. This investigation used the analytical approach for analysis, and the threat model was not sophisticated, though the system addressed legacy hardware components.

Sedjelmaci and Senouci [99] designed and carried out a hybrid-based detection technique in an accurate as well as lightweight manner. The AECFV's threats model points to keep the VANET safe against hazardous attacks such as black holes, selective forwarding, wormhole, and resource exhaustion, as well as Sybil attacks. The architecture of AECFV consisted of three level components: cluster members level, cluster headers level, and RSUs level. In the cluster members level, and in each node, there is one local intrusion detection system (LIDS). The LIDS used a combination of rules to identify malicious vehicles. At the second level, which is the cluster header level, there is one global intrusion detection system (GIDS) in each cluster head. The GIDS monitors the behavior of its cluster members to evaluate the trustworthiness of the monitored vehicle. The vehicle's trust level is computed at the RSU level by means of a global decision system. The system categorizes vehicles into a shortlist to place them according to their trust level. At the initial deployment of the network and before the formation of the cluster, all the nodes act as an LIDS. After cluster formation, only a few member nodes of the cluster initiate their LIDSs while other nodes deactivate their LIDSs in order to decrease the overhead in the network. Cluster-headers (CHs) are selected depending on both the mobility of the node and the trust level of the vehicle. The authors developed a new reputation mechanism to evaluate the level of trustworthiness of vehicles with respect to the information they provide and their behaviors. The work was tested using NS-3.17. The simulation area covered 3000×3000 m^2, two parallel highways (2×3 lanes) and an urban scenario. The simulation results showed that as the malicious nodes were set to the 45% percentile of the total number of the nodes, AECFV was able to exhibit faster attack detection, low FPRs, high detection rates, as well as less communication overhead comparing to other detection framework solutions such as VWCA [100], IDFV [101], and T-CLAIDS [102].

By the same token, the authors of cited work [102] proposed a detection framework that used an ABD technique called T-CLAIDS to identify the malicious vehicles in VANET. The normal behavior of the vehicle was modeled using a learning automata and the Markov chain model (MCM). The procedure of combining these two approaches helped in identifying attacks and producing a high detection accuracy rate, which was proven throughout the simulation experiment. However, linking together both algorithms in a VANET can produce a communication overhead and high computation cost, as a consequence of increasing the total count of vehicles. In addition, the work omitted to mention the threat model and attacks that the system could detect.

13.7.4 Wireless Sensor Network

WSNs and IDS are becoming key research topics and what follows presents a sample of the literature.

Hortos [103] designed a cross-layer-based detection (CLBD) IDS for WSN that initially sets the security attributes to detect different intruder attacks. For the purpose of the study, data were collected based on specific metrics such as end-to-end QoS, route availability, reliability, and energy usage during the network operation. A group of mobile software agents that can behave like a distributed ant colony among the nodes has been used to execute a pattern recognition algorithm. The pattern recognition algorithm uses a statistical-based method while moving among the layers to perform its function, and was applied on the gathered information during the network operation. The applied algorithm provided the best network global performance to help reduce the communication overhead, latency in network response, and fault tolerance. The simulation results showed that the designated CLBD IDS can successfully identify different types of attacks such as black hole, flooding, Sybil as well as distributed DoS attacks.

Likewise, the cited work introduced [104] embedding genetic algorithms CLBD designed with anti-phase synchronization based on the work presented in Reference 105 to manage the transmissions order in both MAC and data-link layers in WSN environments. The proposed approach has been established using an ant colony optimization (ACO) algorithm. Moreover, the proposed approach exploited the information of application, network, MAC, and PHY layers to build a trust model using a quantized data reputation algorithm, to reference the quantized information. The authors implemented a two-stage algorithm to disclose and recognize attack types in order to overcome the problems of both resource restrictions and huge sensor data loads in WSNs. A bio-inspired neural-network (NN) algorithm with cross-layer design at the first stage was used, while at the second stage a reputational version of the support vector machine (SVM) with K-nearest neighbor (KNN) was used. The authors ran a Monte Carlo simulation for a WSN of 16 sensor nodes with randomly placed mobile agents in the nodes. Simulation results showed that the proposed combination of algorithms proceeded well in detecting and identifying the black hole, routing request flooding RREQ, and routing request disruption RREQ attacks.

In addition, authors of cited work [106] used a cross-layer intercommunication to detect various types of WSN multilayer attacks. The cross-layer intrusion detection agent (CLIDA) can recognize intruders as they communicate with other network nodes. The system checks routing tables at the network layer to assure that the intended node is registered in the routing path as one of the neighbors, while the network cross layer of MAC as well as PHY Layer information was used to detect potential intruders. Furthermore, the intruder node authenticity has been examined through the packet's received signal strength indicator (RSSI). An action such as flagging a neighbor or dropping a packet can be taken when the intrusion is detected. This work implemented a topology that partitioned the network into multiple clusters, and the node (within a cluster) that has the maximum energy reservation is chosen to become the cluster head. A CLIDA agent consist of two segments. The first segment, which is the interaction interface, is responsible for facilitating the contact between the layers and specified application with the CLIDA agent. The second segment, which is the cross-layer data module, is responsible for providing the data and maintaining it up and through the cross-layer interaction interface.

On the other hand, the authors of cited work [107] proposed an anomaly-based wireless IDS that uses a data mining clustering and classification technique on the gathered data. The data were gathered from the wireless packets for the purpose of real-time detection of various security attacks such as DOS, WEP key cracking, MAC spoofing, and war driving. The architecture of the established system consisted of three parts: a wireless access point (AP), an oracle database, and network chemistry radio frequency (RF) sensors to capture the audit data features. The audit features include MAC address, count number of errors, and sent and received packets, in addition to channel signal rate, packet size, number of retries, and the receiving time. The WIDCA algorithm stands for the wireless intrusion detection clustering algorithm that uses the local sparsity coefficient (LSC) outlier with a density-based detection algorithm as the main mechanism to cluster and identify outliers or abnormal wireless connection records. The authors developed the system using Java language under Windows platform, tested it along with one other system, and then evaluated the efficiency of both systems. In addition, the designated system performance has been evaluated in an actual network environment through a predefined intrusion that has been carefully designed. The results showed that WIDCA outperformed SNORT-Wireless and Online version of the K-Means algorithm.

By the same token, authors of cited work [108] introduced a novel cross-layer-based IDS/IPS technique to detect misdirection attacks in WSNs. The proposed work aims to identify and stop the misdirection attack. The mobile nodes localized in the same communication range form a cluster, and a chosen cluster head based on both fairness and efficiency is selected to perform the detection operation of the intruder node in that cluster. The authors verified the proposed algorithm in OPNET through simulating a WSN under misdirection attack. The simulated scenario consisted of 14 sensor nodes, two routers, and one coordinator under the tree topology. The simulation showed that this technique was very efficient in detecting and preventing misdirection attacks, with considerably increased throughput, although some delays were introduced in the system.

Comparatively, in cited work [109], a hybrid-based IDS approach that applied stream flow and state transition analysis was implemented to disclose a sync-flood attack in WSNs. The fundamental idea behind the mechanism was to monitor the operation of the 3-Way Handshake of TCP to recognize the attack patterns. The authors suggested implementing the proposed approach in the NS-2 simulation package to check its effectiveness in securing sensor networks.

Regarding WSN, Alajmi [110] proposed an anomaly detection approach to detect selective forwarding attacks in WSN. It maintains the safety of data transmission between a source node and base station while detecting selective forwarding attacks in an attempt to provide a reliable, energy efficient, and scalable approach. The proposed approach incorporated a MAC pool IDS, a rule-based processing algorithm, and an anomaly detection algorithm. The MAC IDS authenticates the incoming traffic to check if a node is legitimate or malicious, while the rule-based processing algorithm checks the traffic against a list of rules. Finally, the anomaly detection algorithm can identify unknown attacks as a false positive, send an alert, and reject the traffic. However, the set of behavior rules specified may add an extra burden to the network nodes as well as consume their power.

Kolias et al. [111] proposed TermID, which is a distributed network IDS approach that is suitable for wireless networks. TermID uses classification rule induction and swarm intelligence principles in an attempt to achieve efficient model training, without exchanging sensitive data. This system includes: two operational units, the monitor nodes and the central node. The monitor nodes transform the input examples from their local dataset to intermediate summaries while the central node performs reduce operations on the global dataset and runs the main body of the rule construction process. TermID used the Aegean Wireless Intrusion Dataset version 2 (AWIDv2) that is manually broken down and distributed to each node. One of the limitations of TermID is that it supports nominal valued datasets. A summary list of the findings is offered in Table 13.5. In spite of our best efforts,

Table 13.5   Comparative Classification of the Presented IDSs

Ref.	Year	Algorithm	Attacker	Architecture	Attack	Detection Approach	Dataset	Network	TimeLine	Audit Material	Audit Features
[78]	2009	IDSNNM	Outsider	Centralized	Known attacks	Specification behavior	Data recorded from real network of an existing critical infrastructure.	CPS	Offline	Network	IP address count, flag code count, zero length packet count
[79]	2009	None	Outsider	Centralized	Related to generic object-oriented substation events	Specification behavior	Simulation, no dataset	CPS	Offline	Network	Water-level readings, valve settings, Arrival rate of GOOSE messages, GOOSE meta data
[82]	2012	None	Insider	Distributed	Subtle manipulation attacks	Specification behavior	Reading from medical devices	CPS	Offline	Host	Pacemaker frequency, Analgesic request, Oxygen saturation, Pulse, Temperature, Respiration, Analgesic Infusion rate, Blood pressure

(Continued)

Table 13.5 (Continued) Comparative Classification of the Presented IDSs

Ref.	Year	Algorithm	Attacker	Architecture	Attack	Detection Approach	Dataset	Network	TimeLine	Audit Material	Audit Features
[81]	2010	CBA-RG	Outsider	Distributed	Exfiltration	Anomaly	Experimental recording consisting of 20 actual records obtained from a single user source	CPS	Real time	Host	User ID, time,location, type, network address, patient requests of medical record
[82]	2009	None	Insider	Distributed	Integrity and privacy of the CPS attacks	Specification behavior	Simulation, no data set	CPS	Offline	Host	Not stated
[93]	2016		Insider	Distributed	Routing attacks	Anomaly	Data set constructed using Qualnet and MATLAB	MANET	Offline	Host	hop counts, RREQ, route
[111]	2016	TermID	Insider	Distributed	Evil twin, ARP,Beacon	Anomaly	AWIDv2	WSN	Offline	Network	Packets
[97]	2013	Secure Core	Insider	Centralized	W32.Stuxnet and Duqu worms	Anomaly	A profile for the execution code	TCIS	Real time	Host	Execution time, memory space

(Continued)

Table 13.5 (Continued) Comparative Classification of the Presented IDSs

Ref.	Year	Algorithm	Attacker	Architecture	Attack	Detection Approach	Dataset	Network	TimeLine	Audit Material	Audit Features
[98]	2014	None	Insider	Distributed	Wormhole attacks	Hybrid	Simulation	TCIS	OPNET	Hybrid	Hop count metric, location information of the smart meters, RREQ packets
[99]	2015	AECFV	Insider	Distributed	Selective forwarding, black hole, packet duplication, resource exhaustion, wormhole as well as Sybil attacks	Hybrid	Real-world traffic using SUMO	VANET	NS-3.17	Host	PDR PSR, MDR, and SSI
[104]	2012	None	Insider	Distributed	Black hole, routing request disruption attacks	Cross layer	Data from different layers of the same cluster	WSN	Monte Carlo simulation in open-source code	Host	Routing table changes, node location transmission speed Frequency transmission code

(Continued)

Table 13.5 (Continued)   Comparative Classification of the Presented IDSs

Ref.	Year	Algorithm	Attacker	Architecture	Attack	Detection Approach	Dataset	Network	TimeLine	Audit Material	Audit Features
[106]	2012	CLIDA	Insider	Distributed	Spoofed routing information, cloning nodes, Sinkhole attack, DOS attack	Cross layer	Data from different layers of the same Chains cluster	WSN	Ns-2	Host	RTS, RSSI of received packet
[108]	2013	None	Insider	Distributed	Misdirection Attack	Cross layer	No data	WSN	Simulation using OPNET	Host	Time stamp with the entry of each sent packet, sequence number
[85]	2016	HMIDS	Both	Distributed	IP scanning, SYN flooding, and APR poisoning	Anomaly	Real traffic	CPS	SWat Team	Network	Packet payloads
[102]	2014	T-CLAIDS	Outsider	Distributed	Not defined	Anomaly	Trace file using SUMO	VANET	NS-3.17	Host	Density, mobility direction of motion of the vehicle

(Continued)

Intrusion Detection System in Self-Organizing Networks ■ 375

Table 13.5 (Continued) Comparative Classification of the Presented IDSs

Ref.	Year	Algorithm	Attacker	Architecture	Attack	Detection Approach	Dataset	Network	TimeLine	Audit Material	Audit Features
[109]	2012	None	Outsider	Distributed	Sync flood attack	Hybrid	Simulation	WSN	Ns-2	Host	Port, flags, TCP three-way handshake
[95]	2012	None	Insider	Centralized	Wormhole attacks	Cross layer	Data from different layers of the same node	VANET	Simulation	Network	Audit data from the modules with extra information from other nonnetwork devices, such as GPS, radar, and sensors
[96]	2015	None	Insider	Centralized	DoS attacks	Hybrid	Real-world traffic using SUMO	VANET	Ns-2	Host	Delay, packet ID, payload size, payload type, source MAC, destination MAC
[94]	2016	EAACK-DSA	Insider	Distributed	Routing attacks	Anomaly	No data set	MANET	NS-2	Host	Acknowledge packets
[66]	2014	REST-NET	Insider	Distributed	Fake message	Signature	Beacon Data	VANET	Simulation	Host	Temperature, light, speed
[110]	2015	None	Insider	distributed	Selective forwarding	Anomaly	Simulation	WSN	NS2	Host	Packets

the table does not include all the accessible literature. The IDS techniques are listed in the fifth column. The attack type that IDS was designed to detect is listed in the fourth column. The audit features that the system is working on are showed in the audit features column. The source of data column lists the datasets for each of the surveyed IDSs. The time line column indicates whether the IDS is a real-time or offline-based IDS or the work has been tested through simulation. The architecture column indicates the IDS architecture. The audit material specifies the type of the data collecting mechanism. Finally, the network column indicates the targeted network for which the IDS that was implemented.

13.8 Comparison among CPS, WSN, MANET, VANET, and TICS Intrusion Detection Function

Through understanding similarities and differences among different types of self-organizing network, we can increase our understanding and learn more about them. This usually involves a process of analysis, in which we compare the specific parts as well as the whole. Thus, we can decide which is more useful or valuable.

A study by Mitchell and Chen [13] presented a comparison between CPS IDSs and TICSs, while a study by Erritali and El Ouahidi [14] presented a comparison between WSN IDSs and MANET IDSs. In this chapter, we are concerned with the self-organizing network, and there are a number of important differences among SON that can influence intrusion detection functions in these networks.

As exemplified in Table 13.6, a summary of the key differences of the intrusion detection functions in CPS, VANET, MANET, WSN, and TICSs is presented. This study compares these systems based on (1) monitored components, (2) monitored events, (3) sophisticated attacks model, (4) system components, (5) system constraints, and (6) audit material nature. In the following subparagraphs, we explain these factors.

1. *Monitored components*: While a CPS IDS estimates physical properties and processes, a TICS IDS monitors computer or network machine activities. Correspondingly, a WSN IDS observes sensors collecting and transmitting data from surrounding environments, and a VANET's IDS monitors moving objects, which is the same case with MANET's IDS. However, it is possible to predict the mobility pattern in VANET, while such a feature is missing in MANET. In fact, there may be convergence and similarity between these types of IDSs from the point of view of these criteria. For example, WSN's IDS and CPS's IDS for some WSN applications are similar in monitoring physical processes.

2. *Monitored events*: The events in these systems may be different from each other. To illustrate, events in CPS environments are regularly or habitually automated as well as time-driven in a closed-loop setting [13] environments are

Table 13.6 Comparison Among TICS, CPS, WSN, MANET, and VANET IDSs

CRITERIA	TICS	CPS	WSN	MANET	VANET
Monitored components	Examine host and/or network-level user and/or machine activity, such as ftp, http, and https requests, or mail server	Observes physical processes that control physical devices behaviors, which makes specific behaviors more probable to appear more than others	Observes the sensor that collects, transmits, and receives data from the surrounding environment which makes specific specifications and behaviors to be seen more than others	Observes mobile nodes that interact with human beings such as iPad, iPod, laptop, cell phone	Monitors moving vehicles with predictable mobility pattern with the purpose of monitoring the behavior of vehicles and transportation system in road traffic
Monitored events	Monitors activities triggered by user, leading to undesired high FPR because the user's behaviors are unpredictable	Monitors repeatedly automatic activities driven by time in a feedback control setting leading to some uniformity as well as predictability of the monitored behaviors	Monitors activities triggered by environment at most, and the nodes are independent in the sense that they receive and transmit information from/to the base station	Monitors activities triggered by user with random mobility pattern making it difficult to collect audit data and to accurately characterize the normal conduct of the network, which leads to undesired high FPR because the user behaviors are unpredictable	Monitors vehicle triggered activities with a predictable mobility pattern leading to predictability for behavior monitoring

(Continued)

Table 13.6 (Continued)   Comparison Among TICS, CPS, WSN, MANET, and VANET IDSs

CRITERIA	TICS	CPS	WSN	MANET	VANET
Attacks model	Concerns at most with nonzero day attacks recalling SBD is effective	Concerns at most with highly sophisticated zero-day attacks, recalling KBD is ineffective	Highly vulnerable to security threats at different layer levels, rendering CLBD and CBD is ineffective	Very vulnerable to multiple security threats because of the features of MANET environment, rendering ABD or hybrid-based detection (HBD) is effective	Concerns with highly sophisticated attacks, zero-day attacks, rendering KBD is in effective
System components	Usually is not concerned with inheritance components; nonetheless, it is concerned with devices such as routers, switches, and so on	Often must be concerned with inherited components (also referred to in the literature as legacy components), which make BSBD an efficient technique by accurately describing the physical processes of controlling behavior of inherited components	Often is not concerned with inheritance systems and components, which makes behavior specification of the physical processes controlling these systems and components unnecessary	Often is not concerned with inheritance systems and components; furthermore, the mobility of the components is a key to be considered	Usually is not concerned with inheritance systems and components

(Continued)

Table 13.6 (*Continued*) Comparison Among TICS, CPS, WSN, MANET, and VANET IDSs

CRITERIA	TICS	CPS	WSN	MANET	VANET
System constraints	Does not have to deal with resource constrained and limited power consumption	Does not have to be concerned with resource constrained and limited power consumption. However, the privacy of information and criticalness of infrastructure are of the major issues, because the system mostly belonged to multiple organizations	Has to be concerned with resource constrained and limited power consumption. As a consequence, simple and fast detection techniques in an online as well as distributed manner concurrently should be used, in addition to the privacy of information in some of the WSN environments that are hazardous	Has to trade with limited power consumption and limited bandwidth. As a result, the limitation of the intensity of data transfer is necessary for the intrusion detection process, recalling host-based IDS is effective	Does not have to deal with resource constrained and limited power consumption since it is powered by huge batteries

(Continued)

Table 13.6 (*Continued*)   Comparison Among TICS, CPS, WSN, MANET, and VANET IDSs

CRITERIA	TICS	CPS	WSN	MANET	VANET
Audit material nature	Has multiple concentration points, where collecting, inspection, and monitoring of the audit data can be achieved	Has multiple concentration points where collecting, inspection, and monitoring of the audit data can be fulfilled in a distributed rather than centralized manner	Lacks concentration points, where collecting, inspection, and monitoring of the audit data can be accomplished. A backup process to collect all the data transmitted among the sensor nodes and the base station might be helpful	Lacks concentration points where collecting, inspection, and monitoring of the audit data can be performed. Yet developing a mobile agent module to gather the data is recommended	Lacks concentration points, where collecting, inspection, and monitoring of the audit data can be performed. However, it has other devices like RSU and GPS units that might be helpful

triggered by environments, and these share some similarities with CPS environments especially for habitual monitoring applications in WSNs. Events in MANET environments are triggered by mobile users with a random mobility pattern, while in VANET environments the events are triggered by vehicles with a predictable pattern.

3. *Sophisticated attacks model*: The reward that an adversary may gain from attacking hundreds of patients and millions of utility customers by putting them in a situation where there is a danger of loss, harm, or failure can lead to an increase in attack sophistication and the extensive use of zero-day attacks. The same situation could be applicable to VANET environments. This situation makes both CPS and VANET concerned at most with highly sophisticated zero-day attacks, while MANETs and WSNs are highly vulnerable to multiple security threats because of the broadcasting nature of these networking environments.

4. *System components*: These criteria distinguish a CPS from other systems since the majority of CPS's components consist of legacy components (inheritance components). These components have certain specifications that govern the physical processes which provide an advantage by making behavior specification based detection techniques effective mechanisms to build the IDS.

5. *Audit material nature*: This criterion distinguishes TICS from other systems and provides flexibility in the audit material process since the system has multiple concentrations which facilitate the collection, inspection, and monitoring of the audit data process. These competitive advantages are absent in other systems.

6. *Audit material nature*: This criterion distinguishes TICS from other systems and provides flexibility in the audit material process since the system has multiple concentrations which facilitate the collection, inspection, and monitoring of the audit data process. These competitive advantages are absent in other systems.

13.9 Discussion, Analysis, and Critical Problems

IDSs are a vital part of modern networking environments such as VANET, WSN, MANET, and CPS. Recent IDSs have some problems such as the ability to detect and respond in real time and the ability to identify novel or modified attacks. These are considered as the most recent challenges in the scope of IDS. In producing more desired or intended results, it is important to detect nonzero and zero-day attacks through employing a hybrid-based IDS. The following subsections conclude our survey and provide the advantages, disadvantages, and applicability of intrusion detection systems for each of the networking systems based on the detection approach.

13.9.1 CPS Environments

Behavior-based detection approaches can benefit CPS because these systems can deliver a well-defined concept of cognitive process, which can exploit its consistent behavior. Likewise, signature-based detection approaches can assist CPSs because of the minimal processing burden and because of the restrictive processing capabilities of the CPS. From a performance metrics point of view, developing metrics for the detection latency is required in evaluating CPS IDSs because the speed of detection is essential in CPS IDS. The highly large scale of CPS requires a hybrid-based intrusion detection approach rather than an autonomous approach. It is worth mentioning that the present distributed IDSs approaches are not effective in CPS, because these lack one of the important criterion that the CPS IDS requires, which is a scalable privacy protection mechanism. Generally speaking, the detection techniques have not been well studied in CPS. Hence, the expectant detection techniques must have the ability to identify real attacks from random defects, ingrained defects in the design, misconfiguration of the system devices, system faults, human errors, and software implementation bugs. Another critical problem in CPS IDSs is the design of an appropriate architecture deployment. Furthermore, developing security vulnerability and threat taxonomy for CPS is a substantial subject. By the same token, an obstruction associated with CPS IDS research is the absence of an obtainable test benchmark for comparing the performance and accuracy of the proposed solution. This is understandable from the operative perspective of CPS as a result of the sensitivity and privacy of the data. Knowledge-based designs are not efficient in the implementation of CPSs. Readers who are interested in the CPS IDSs are directed to Reference 112. In addition, there are many unresolved problems in CPS IDS fields. For instance, defining new IDS performance metrics and lifecycle metrics that comply with these two systems is a substantial issue. Most developers usually report numerical results of FPR, FNR, TPR, TNR, and detection rate. Developing metrics for the power consumption and memory usage should be considered in evaluating WSN IDSs.

13.9.2 VANET Environments

IDSs and such systems for VANETSs are still hardly explored. IDS in VANETs is a challenging task and therefore traditional intrusion detection techniques are not directly usable. To illustrate, the detection techniques must have the ability to detect bugs and fabricated information as well as identify real attacks from bugs and wrongly fabricated information. Furthermore, the highly dynamic topology and deployed applications of VANET environments add an extra burden on the design and building of such system. Cross-layer detection approaches may benefit a VANET since these can provide high performance and support real-time applications. Systems such as VANET and CPS with health care application trends require a trade-off between security and privacy issues. Providing security solutions through

IDSs would require an appropriate protection of the privacy of both the drivers and vehicle owners since these may conflict with one another in a number of situations. Furthermore, a real-time implementation with cost-effective and fast responses is recommended for VANET environments.

13.9.3 MANET Environments

Behavior-based detection approaches present a challenge for MANET networks since their profiles are unpredictable. Cluster-based IDS can work for both MANET and WSN, because of the nature of these two types of networking systems. ABD benefits MANET due to its ability to detect unknown or nonzero day attacks. There are several reasons why the IDS approaches that are proposed in Ad-Hoc networks such as MANET might not be easily implemented in WSN. Few reasons include the fact that the number of nodes in WSNs are much higher than those in ad-hoc networks, as well as the point that that sensor nodes are densely deployed, and are likely or liable to suffer from failure, and their resources are limited.

Multilayer attacks can be detected efficiently using layer-based IDS. Owing to the vulnerabilities of multilayer attacks in both WSN and MANET environments, a cross-layer intrusion detection model is required. There have been extensive studies regarding cross-layer IDS in the literature. However, one criterion of this technique is that it needs more resources, yet many research works use this technique for detection in WSN and MANET environments. This chapter surveyed some of the recently implemented works that used this technique, but many of the available works lack considering power efficiency and resource availability. Our study suggests considering these two factors in future cross-layer-based intrusion detection for WSN environments.

13.9.4 WSN Environments

Hybrid-based detection is suitable for large and sustainable WSNs because of various threats and attacks that could compromise the environment for the large-size WSN. Moreover, behavior-based detection approaches can benefit WSN because these systems can deliver a well-defined concept of cognitive process, which can exploit its consistent behavior. Likewise, signature-based detection approaches can assist WSN because of the minimal processing burden and because of the restrictive processing capabilities of the WSN. Regarding detection techniques for the case of WSN, intrusion detection techniques have been well explored, investigated, and studied. However, the credibility of the state-of-the-art simulators for WSNs has not been sufficiently examined by researchers. Furthermore, there is no tool that helps network operators to optimize the configuration of an IDS for their needs. In WSNs with mobile applications, where sensor nodes are also mobile, the usage of distributed and collaborative architectures for IDS is recommended, while in stationary applications, where there is a centralized computing unit at the BS or in the data sink, the use of centralized architecture for the IDS is recommended. Readers who are interested in the WSN IDSs are directed to refer to cited work [10]. Another critical problem in WSN IDSs is the design of an appropriate architecture deployment. Moreover, it is important to provide a benchmark that is totally suitable for research in WSN and that has been collected over a WSN environment for both mobile and stationary applications as well. Again, the same problem in CPS are applicable here which represented by many, are, unresolved problems in WSN. For instance, defining new IDS performance metrics and lifecycle metrics that comply with these two systems is a substantial issue. Most developers usually report numerical results of FPR, FNR, TPR, TNR, and detection rate. Developing metrics for the detection latency and memory usage should be considered in evaluating WSN IDSs. In addition, ABD can contribute to WSNs because their operation concept is very well defined. Thus, anomalies will quickly contradict the baseline behavior. Finally, WSNs can benefit from anomaly-based approaches because of their limited RAM storage requirement. Nevertheless, there are several weighty open issues pertinent to WSNs and MANETs that are either unresolved or not explored extensively, such as performance optimization, detection framework standardization, and reduction of design redundancy, network throughput, and power consumption. The single-layer technique is not effective for WSNs or MANETs due to the lack of a centralized infrastructure in these two environments [112] and because there are variant vulnerabilities that are related to WSN and MANET environments. Thus, a cross-layer-based design will be more effective in securing the targeted network.

13.10 Conclusion

This chapter investigated and surveyed recently proposed works regarding IDSs and their applicability to SON such as CPS, MANET, VANET, and WSN. We classified these systems based on different design approaches such as postdetection action, timeline, implementation mechanism, detection technique, architecture deployment, and audit material. Furthermore, the chapter provides a brief overview of the advantages and drawbacks of the classified detection techniques along with a sample of the recent IDSs and their implementation phase details.

In addition, we present a review of the performance metrics that are used to evaluate IDS systems. We conclude that there is a need to develop new metrics that must consider memory usage, computational power, detection latency, and power consumption in evaluating IDS performance. These metrics are important for self-organizing environments.

Additionally, we took a glance at the datasets that are used by the developers of IDSs and conclude that there is a potential absence of an efficient intrusion detection benchmark dataset that could be used for SON.

Furthermore, we explore some of the common existing attacks against SON and signify new threats in VANET and CPS environments. We concluded that many of the proposed solutions can work only with a specific attack and they are still vulnerable to unexpected attacks. Furthermore, self-organizing network researchers should focus on both exploring and preventing all possible future attacks to make self-organizing network environments secure and reliable environments.

This chapter sheds new light on a comparison among CPS, WSN, MANET, VANET, and TICS in terms of the distinguished characteristics, the security challenges, and the function of IDSs. We looked at different detection approaches and analyzed their suitability for each type of SON and whether they could handle the security challenges in these networks efficiently and effectively.

References

1. J. P. Anderson, Computer security threat monitoring and surveillance, Technical Report, James P. Anderson Company, Fort Washington, Pennsylvania, 1980.
2. M. Whitman and H. Mattord, *Principles of Information Security*, Cengage Learning, Boston, MA, 2011.
3. R. R. Mitchell III, Design and analysis of intrusion detection protocols in cyber physical systems, Doctoral Dissertation, Virginia Tech, Blacksburg, VA, 2013.
4. R. Srivastava and V. Richhariya, Survey of current network intrusion detection techniques, *Journal of Information Engineering and Applications*, **3**: 27–33, 2013.
5. H. Debar, M. Dacier, and A. Wespi, Towards a taxonomy of intrusion-detection systems, *Computer Networks*, **31**: 805–822, 1999.
6. T. F. Lunt, Automated Audit Trail Analysis and Intrusion Detection: A Survey, SRI International, Business Intelligence Program, Menlo Park, CA, 1989.
7. M. Esmaili, R. Safavi-Naini, and J. Pieprzyk, Intrusion detection: A survey, in *Proceedings of the 12th International Conference on Computer Communication on Information Highways: For a Smaller World and Better Living*, Seoul, South Korea, pp. 409–414, 1996.
8. S. Axelsson, Intrusion detection systems: A survey and taxonomy, Technical Report, Chalmers University of Technology, Sweden, 2000.
9. T. S. Sobh, Wired and wireless intrusion detection system: Classifications, good characteristics and state-of-the-art, *Computer Standards & Interfaces*, **28**: 670–694, 2006.
10. I. Butun, S. D. Morgera, and R. Sankar, A survey of intrusion detection systems in wireless sensor networks, *Communications Surveys & Tutorials, IEEE*, **16**: 266–282, 2014.
11. S. Mandala, M. A. Ngadi, and A. H. Abdullah, A survey on MANET intrusion detection, *International Journal of Computer Science and Security*, **2**: 1, 2007.
12. T. Anantvalee and J. Wu, A survey on intrusion detection in mobile ad hoc networks, in *Wireless Network Security*, Springer, New York, NY, pp. 159–180, 2007.
13. R. Mitchell and I-R. Chen, A survey of intrusion detection techniques for cyber-physical systems, *ACM Computer Surveys*, **46**(4): 55, 2014.
14. M. Erritali and B. El Ouahidi, A survey on VANET intrusion detection systems, in *Proceedings of the 2013 International Conference on Systems, Control, Signal Processing and Informatics*, Rhodes Island, Greece, 66–69, 2013.

15. C. Modi, D. Patel, B. Borisaniya, H. Patel, A. Patel, and M. Rajarajan, A survey of intrusion detection techniques in cloud, *Journal of Network and Computer Applications*, **36**: 42–57, 2013.
16. E. M. Shakshuki, N. Kang, and T. R. Sheltami, EAACK—A secure intrusion-detection system for MANETs, *IEEE Transactions on Industrial Electronics*, **60**: 1089–1098, 2013.
17. B. Kannhavong, H. Nakayama, Y. Nemoto, N. Kato, and A. Jamalipour, A survey of routing attacks in mobile ad hoc networks, *IEEE Wireless Communications*, **14**: 85–91, 2007.
18. M. Tavallaee, N. Stakhanova, and A. A. Ghorbani, Toward credible evaluation of anomaly-based intrusion-detection methods, Systems, Man, and Cybernetics, *Part C: Applications and Reviews, IEEE Transactions on*, **40**: 516–524, 2010.
19. R. G. Bace, *Intrusion Detection*, Sams Publishing, Indianapolis, IN, 2000.
20. J. Singh, L. Kaur, and S. Gupta, A cross-layer based intrusion detection technique for wireless networks, *The International Arab Journal of Information Technology*, **9**: 201–207, 2012.
21. K. Hwang, M. Cai, Y. Chen, and M. Qin, Hybrid intrusion detection with weighted signature generation over anomalous internet episodes, *IEEE Transactions on Dependable and Secure Computing*, **4**: 41–55, 2007.
22. P. Brutch and C. Ko, Challenges in intrusion detection for wireless ad-hoc networks, in *Proceedings of 2003 Symposium on Applications and the Internet Workshops*, Orlando, FL, pp. 368–373, 2003.
23. P. Uppuluri and R. Sekar, Experiences with specification-based intrusion detection, in *Proceedings of the 4th International Symposium on Recent Advances in Intrusion Detection*, Springer-Verlag, London, UK, pp. 172–189, 2001.
24. C. Ko, M. Ruschitzka, and K. Levitt, Execution monitoring of security-critical programs in distributed systems: A specification-based approach, in *Proceedings of the 1997 IEEE Symposium on Security and Privacy*, IEEE CS Press, Los Alamitos, CA, pp. 175–187, 1997.
25. Y. Ma, H. Cao, and J. Ma, The intrusion detection method based on game theory in wireless sensor network, in *2008 First IEEE International Conference on Ubi-Media Computing*, Lanzhou University, China, pp. 326–331, 2008.
26. S. Misra, P. V. Krishna, and K. I. Abraham, Energy efficient learning solution for intrusion detection in wireless sensor networks, in *2010 Second International Conference on COMmunication Systems and NETworks (COMSNETS 2010)*, Bangalore, India, pp. 1–6, 2010.
27. F. Li, N. Clarke, M. Papadaki, and P. Dowland, Behaviour profiling on mobile devices, in *2010 International Conference on Emerging Security Technologies (EST)*, Canterbury, UK, pp. 77–82, 2010.
28. F. Haddadi and M. A. Sarram, Wireless intrusion detection system using a lightweight agent, in *2010 Second International Conference on Computer and Network Technology (ICCNT)*, Washington, DC, pp. 84–87, 2010.
29. A-S. K. Pathan, *The State of the Art in Intrusion Prevention and Detection*, CRC Press, Boca Raton, FL, 2014.
30. S. J. Stolfo, W. Fan, W. Lee, A. Prodromidis, and P. K. Chan, Cost-based modeling for fraud and intrusion detection: Results from the JAM project, in *Proceedings of the DARPA Information Survivability Conference and Exposition (DISCEX'00)*, Hilton Head, SC, pp. 130–144, 2000.

31. R. P. Lippmann, D. J. Fried, I. Graf, J. W. Haines, K. R. Kendall, D. McClung et al., Evaluating intrusion detection systems: The 1998 DARPA off-line intrusion detection evaluation, in *Proceedings of the DARPA Information Survivability Conference and Exposition (DISCEX'00)*, Los Alamitos, CA, pp. 12–26, 2000.

32. M. Tavallaee, E. Bagheri, W. Lu, and A-A. Ghorbani, A detailed analysis of the KDD CUP 99 data set, in *Proceedings of the Second IEEE Symposium on Computational Intelligence for Security and Defence Applications 2009*, Piscataway, NJ, pp. 1–6, 2009.

33. V. Manjula and C. Chellappan, The replication attack in wireless sensor networks: Analysis and defenses, in *Advances in Networks and Communications: First International Conference on Computer Science and Information Technology, CCSIT 2011*, Bangalore, India, January 2–4, 2011. *Proceedings*, Part II, N. Meghanathan, B. K. Kaushik, and D. Nagamalai, editors, Springer, Berlin, Heidelberg, pp. 169–178, 2011.

34. A. Shiravi, H. Shiravi, M. Tavallaee, and A. A. Ghorbani, Toward developing a systematic approach to generate benchmark datasets for intrusion detection, *Computers & Security*, **31**: 357–374, 2012.

35. H-N. Dai, Q. Wang, D. Li, and R. C-W. Wong, On eavesdropping attacks in wireless sensor networks with directional antennas, *International Journal of Distributed Sensor Networks*, **2013**: 1–13, 2013.

36. M. Raya and J-P. Hubaux, Security aspects of inter-vehicle communications, in *5th Swiss Transport Research Conference (STRC)*, Ascona, Switzerland, pp. 1–15, 2005.

37. M. V. Mahoney and P. K. Chan, An analysis of the 1999 DARPA/Lincoln Laboratory evaluation data for network anomaly detection, in *6th international symposium on Recent Advances in Intrusion Detection*, Pittsburgh, PA, pp. 220–237, 2003.

38. C. Thomas, V. Sharma, and N. Balakrishnan, Usefulness of DARPA dataset for intrusion detection system evaluation, in *SPIE Defense and Security Symposium*, Orlando, FL, pp. 69730G–69730G-8, 2008.

39. R. Fontugne, P. Borgnat, P. Abry, and K. Fukuda, Mawilab: Combining diverse anomaly detectors for automated anomaly labeling and performance benchmarking, in *Proceedings of the 6th International Conference*, The Graduate University of Advanced Studies, Tokyo, p. 8, 2010.

40. A. Sperotto, R. Sadre, F. Van Vliet, and A. Pras, A labeled data set for flow-based intrusion detection, in *IP Operations and Management*, Springer, Berlin, pp. 39–50, 2009.

41. U. S. M. A. W. P. C. R. C. DataSets. (2015, 12-10). Cyber Research Center—DataSets. Available: http://www.usma.edu/crc/SitePages/DataSets.aspx

42. I. S. C. Reports. (2015, 1–15). Internet Security SANS ISC. Available: https://isc.sans.edu/reports.html

43. M. Karanikolas, D. Aretha, P. Kiekkas, G. Monantera, I. Tsolakis, and K. Filos, Intravenous fentanyl patient-controlled analgesia for perioperative treatment of neuropathic/ischaemic pain in haemodialysis patients: A case series, *Journal of Clinical Pharmacy and Therapeutics*, **35**: 603–608, 2010.

44. E. W. A. Kevin Bauer, D. McCoy, D. Grunwald, D C. Sicker, Dataset of received signal strength indication (RSSI) collected from within an indoor office building. Online, Available: http://www.crawdad.org/cu/rssi/20090528/

45. N. Patwari, Measured CIR (Channel Impulse Response) Data Set. Online, Available: http://crawdad.org/utah/CIR/20070910

46. T. Goodspeed and N. Filardo, RFID tracking data. Online, Available: http://crawdad.org/hope/nh_amd/20100718

47. G. K. Constantinos Kolias, A. Stavrou, and S. Gritzalis, AWID, University of the Aegean, Samos, Greece, 2015.
48. C. Kolias, G. Kambourakis, A. Stavrou, and S. Gritzalis, Intrusion detection in 802.11 networks: Empirical evaluation of threats and a public dataset, *IEEE Communications Surveys & Tutorials*, **18**: 184–208, 2016.
49. I. Almomani, B. Al-Kasasbeh, and M. AL-Akhras, WSN-DS: A dataset for intrusion detection systems in wireless sensor networks, *Journal of Sensors*, **2016**: 16, 2016.
50. R. Zuech, T. M. Khoshgoftaar, and R. Wald, Intrusion detection and big heterogeneous data: A survey, *Journal of Big Data*, **2**: 1–41, 2015.
51. M. D. Ilic, L. Xie, U. A. Khan, and J. M. Moura, Modeling of future cyber–physical energy systems for distributed sensing and control, *IEEE Transactions on Systems, Man, and Cybernetics-Part A: Systems and Humans*, **4**: 825–838, 2010.
52. A. Saatsakis and P. Demestichas, Context matching for realizing cognitive wireless network segments, Wireless personal communications, **55**(3): 407–440, 2010.
53. Simply Mod Bus: Data Communication Test Software. (2015, 8-10). Available: http://www.simplymodbus.ca/
54. N. Pereira, B. Andersson, and E. Tovar, WiDom: A dominance protocol for wireless medium access, *IEEE Transactions on Industrial Informatics*, **3**: 120–130, 2007.
55. A. F. Molisch, K. Balakrishnan, C-C. Chong, S. Emami, A. Fort, J. Karedal et al., IEEE 802.15. 4a Channel Model-Final Report, *IEEE P802*, **15**: 0662, 2004.
56. S. Han, M. Xie, H-H. Chen, and Y. Ling, Intrusion detection in cyber-physical systems: Techniques and challenges, *Systems Journal, IEEE*, **8**: 1049–1059, 2014.
57. R. Mitchell and I-R. Chen, Modeling and analysis of attacks and counter defense mechanisms for cyber physical systems, *IEEE Transactions on Reliability*, **65**: 350–358, 2016.
58. A. Cardenas, S. Amin, B. Sinopoli, A. Giani, A. Perrig, and S. Sastry, Challenges for securing cyber physical systems, in *Workshop on Future Directions in Cyber-Physical Systems Security*, Newark, NJ, p. 5, 2009.
59. J. A. Stankovic, I. Lee, A. Mok, and R. Rajkumar, Opportunities and obligations for physical computing systems, *Computer*, **38**: 23–31, 2005.
60. E. K. Wang, Y. Ye, X. Xu, S-M. Yiu, L. C. K. Hui, and K-P. Chow, Security issues and challenges for cyber physical system, *Proceedings of the 2010 IEEE/ACM International Conference on Green Computing and Communications & International Conference on Cyber, Physical and Social Computing*, Hangzhou, China, pp. 733–738, 2010.
61. H. N. Saha, D. Bhattacharyya, and P. Banerjee, A novel energy efficient and administrator based secured routing in MANET, *International Journal of Network Security & Its Applications*, **4**: 73, 2012.
62. S. K. Das, K. Kant, and N. Zhang, *Handbook on Securing Cyber-Physical Critical Infrastructure*, Elsevier, Burlington, MA, 2012.
63. A-S. K. Pathan, *Security of Self-Organizing Networks: MANET, WSN, WMN, VANET*, CRC Press, Boca Raton, FL, 2010.
64. M. Faezipour, M. Nourani, A. Saeed, and S. Addepalli, Progress and challenges in intelligent vehicle area networks, *Communications of the ACM*, **55**: 90–100, 2012.
65. M. N. Mejri, J. Ben-Othman, and M. Hamdi, Survey on VANET security challenges and possible cryptographic solutions, *Vehicular Communications*, **1**: 53–66, 2014.
66. A. Tomandl, K-P. Fuchs, and H. Federrath, REST-Net: A dynamic rule-based IDS for VANETs, in *Wireless and Mobile Networking Conference (WMNC), 2014 7th IFIP*, Vilamoura, Portugal, pp. 1–8, 2014.

67. G. Karagiannis, O. Altintas, E. Ekici, G. Heijenk, B. Jarupan, K. Lin et al., Vehicular networking: A survey and tutorial on requirements, architectures, challenges, standards and solutions, *Communications Surveys & Tutorials, IEEE*, **13**: 584–616, 2011.
68. A. Dhamgaye and N. Chavhan, Survey on security challenges in VANET 1, *International Journal of Computer Science and Network*, **2**: 88–96, 2013.
69. S. Misra, I. Zhang, and S. C. Misra, *Guide to Wireless Ad Hoc Networks*, Springer Science & Business Media, London, 2009.
70. J. L. Hill and D. E. Culler, Mica: A wireless platform for deeply embedded networks, *IEEE Micro*, **22**: 12–24, 2002.
71. M. Krämer and A. Geraldy, Energy measurements for micaz node, *Technical Report KrGe06*, University of Kaiserslautern, Kaiserslautern, Germany, 2006.
72. C. Karlof and D. Wagner, Secure routing in wireless sensor networks: Attacks and countermeasures, *Ad Hoc Networks*, **1**: 293–315, 2003.
73. N. W. Lo and H. C. Tsai, Illusion attack on VANET applications—A message plausibility problem, in *2007 IEEE Globecom Workshops*, Washington, DC, pp. 1–8, 2007.
74. M. Raya and J-P. Hubaux, The security of vehicular ad hoc networks, in *Proceedings of the 3rd ACM Workshop on Security of Ad hoc and Sensor Networks*, Alexandria, VA, 2005.
75. B. Wu, J. Chen, J. Wu, and M. Cardei, A survey of attacks and countermeasures in mobile ad hoc networks, in *Wireless Network Security*, Springer, Boston, MA, 2007, pp. 103–135.
76. M. S. Al-Kahtani, Survey on security attacks in vehicular ad hoc networks (VANETs), in *Signal Processing and Communication Systems (ICSPCS), 2012 6th International Conference on*, Springer, New York, pp. 1–9, 2012.
77. D. G. Padmavathi and M. Shanmugapriya, A survey of attacks, security mechanisms and challenges in wireless sensor networks, *arXiv preprint arXiv:0909.0576*, 2009.
78. O. Linda, T. Vollmer, and M. Manic, Neural network based intrusion detection system for critical infrastructures, in *Neural Networks, 2009. IJCNN 2009. International Joint Conference on*, Atlanta, Georgia, 2009, pp. 1827–1834.
79. H. Hadeli, R. Schierholz, M. Braendle, and C. Tuduce, Leveraging determinism in industrial control systems for advanced anomaly detection and reliable security configuration, in *Emerging Technologies & Factory Automation, 2009. ETFA 2009. IEEE Conference on*, Palma, Spain, pp. 1–8, 2009.
80. R. Mitchell and I-R. Chen, Behavior rule specification-based intrusion detection for safety critical medical cyber physical systems, *Dependable and Secure Computing, IEEE Transactions on*, **12**: 16–30, 2015.
81. B. Asfaw, D. Bekele, B. Eshete, A. Villafiorita, and K. Weldemariam, Host-based anomaly detection for pervasive medical systems, in *Risks and Security of Internet and Systems (CRiSIS), 2010 Fifth International Conference on*, Nice, France, pp. 1–8, 2010.
82. R. Mitchell and I-R. Chen, Specification based intrusion detection for unmanned aircraft systems, in *Proceedings of the first ACM MobiHoc workshop on Airborne Networks and Communications*, Hilton Head, SC, pp. 31–36, 2012.
83. Y. Yang, K. McLaughlin, T. Littler, S. Sezer, B. Pranggono, and H. Wang, Intrusion detection system for IEC 60870-5-104 based SCADA networks, in *2013 IEEE Power and Energy Society General Meeting (PES)*, Vancouver, British Columbia, Canada, pp. 1–5, 2013.

84. C-H. Lo and N. Ansari, CONSUMER: A novel hybrid intrusion detection system for distribution networks in smart grid, *IEEE Transactions on Emerging Topics in Computing*, **1**: 33–44, 2013.

85. H. R. Ghaeini and N. O. Tippenhauer, HAMIDS: Hierarchical monitoring intrusion detection system for industrial control systems, in *Proceedings of the 2nd ACM Workshop on Cyber-Physical Systems Security and Privacy*, Vienna, Austria, pp. 103–111, 2016.

86. V. Paxson, Bro: A system for detecting network intruders in real-time, *Computer Networks*, **31**(23): 2435–2463, 1999.

87. V. Schiffer, The CIP family of fieldbus protocols and its newest member-Ethernet/IP, in *Emerging Technologies and Factory Automation, 2001. Proceedings. 2001 8th IEEE International Conference on*, Juan-Les-Pins, France, pp. 377–384, 2001.

88. G. Thamilarasu, S. Mishra, and R. Sridhar, A cross-layer approach to detect jamming attacks in wireless ad hoc networks, in *Military Communications Conference, 2006, MILCOM 2006. IEEE*, pp. 1–7, 2006.

89. J. F. C. Joseph, A. Das, B-C. Seet, and B-S. Lee, Cross layer versus single layer approaches for intrusion detection in MANETs, in *Networks, 2007. ICON 2007. 15th IEEE International Conference on*, New York, NY, pp. 194–199, 2007.

90. R. O. Duda, P. E. Hart, and D. G. Stork, *Pattern Classification*, John Wiley & Sons, New York, NY, 2012.

91. G. Vigna, S. Gwalan, K. Srinivasan, E. M. Belding-Royer, and R. A. Kemmerer, An intrusion detection tool for AODV-based ad hoc wireless networks, in *Computer Security Applications Conference, 2004. 20th Annual*, Tucson, Arizona, pp. 16–27, 2004.

92. A. Mitrokotsa and C. Dimitrakakis, Intrusion detection in MANET using classification algorithms: The effects of cost and model selection, *Ad Hoc Networks*, **11**: 226–237, 2013.

93. A. Chaudhary, V. N. Tiwari, and A. Kumar, Design an anomaly-based intrusion detection system using soft computing for mobile ad hoc networks, *International Journal of Soft Computing and Networking*, **1**: 17–34, 2016/01/01, 2016.

94. G. Santhi, An efficient intrusion detection system based on adaptive acknowledgement with digital signature scheme in MANETs, in *Proceedings of the International Conference on Informatics and Analytics*, Pondicherry, India, p. 103, 2016.

95. E. J. Singh and E. N. Sharma, Wormhole attack detection by using intrusion detection system in VANET, *International Journal of Computer Networks and Wireless Communications (IJCNWC)*, **2**: 2250–3501, 2012.

96. A. Alheeti, M. Khattab, A. Gruebler, and K. D. McDonald-Maier, An intrusion detection system against malicious attacks on the communication network of driverless cars, in *Consumer Communications and Networking Conference (CCNC), 2015 12th Annual IEEE*, Las Vegas, NV, pp. 916–921, 2015.

97. M-K. Yoon, S. Mohan, J. Choi, J-E. Kim, and L. Sha, SecureCore: A multicore-based intrusion detection architecture for real-time embedded systems, in *Real-Time and Embedded Technology and Applications Symposium (RTAS), 2013 IEEE 19th*, Pittsburgh, PA, pp. 21–32, 2013.

98. N. Beigi-Mohammadi, J. Misic, H. Khazaei, and V. Misic, An intrusion detection system for smart grid neighborhood area network, in *Communications (ICC), 2014 IEEE International Conference on*, Philadelphia, PA, pp. 4125–4130, 2014.

99. H. Sedjelmaci and S. M. Senouci, An accurate and efficient collaborative intrusion detection framework to secure vehicular networks, *Computers & Electrical Engineering*, **43**: 33–47, 2015.

100. G. Bruneau, *The History and Evolution of Intrusion Detection*, SANS Institute, Virginia, USA, 2001.
101. Z. Yu and J. J. Tsai, *Intrusion Detection: A Machine Learning Approach*, vol. **3**, World Scientific, Hackensack, NJ, 2011.
102. N. Kumar and N. Chilamkurti, Collaborative trust aware intelligent intrusion detection in VANETs, *Computers & Electrical Engineering*, **40**: 1981–1996, 2014.
103. W. S. Hortos, Cross-Layer Design for Intrusion Detection and Data Security in Wireless Ad Hoc Sensor Networks, *Optics East 2007*, Boston, MA, pp. 677303–677303-16, 2007.
104. W. S. Hortos, Bio-inspired, cross-layer protocol design for intrusion detection and identification in wireless sensor networks, in *Local Computer Networks Workshops (LCN Workshops), 2012 IEEE 37th Conference on*, Florida, USA, pp. 1030–1037, 2012.
105. A. Mutazono, M. Sugano, and M. Murata, Frog call-inspired self-organizing anti-phase synchronization for wireless sensor networks, in *Nonlinear Dynamics and Synchronization, 2009. INDS'09. 2nd International Workshop on*, Klagenfurt, Austria, pp. 81–88, 2009.
106. D. E. Boubiche and A. Bilami, Cross layer intrusion detection system for wireless sensor network, *International Journal of Network Security & Its Applications*, **4**: 35, 2012.
107. C. I. Ezeife, M. Ejelike, and A. K. Aggarwal, WIDS: A sensor-based online mining wireless intrusion detection system, in *Proceedings of the 2008 International Symposium on Database Engineering & Applications*, New York, NY, pp. 255–261, 2008.
108. R. S. Sachan, M. Wazid, D. P. Singh, and R. Goudar, A cluster based intrusion detection and prevention technique for misdirection attack inside WSN, in *Communications and Signal Processing (ICCSP), 2013 International Conference on*, Cambridge, UK, pp. 795–801, 2013.
109. R. Bhatnagar and U. Shankar, The proposal of hybrid intrusion detection for defence of sync flood attack in wireless sensor network, *International Journal of Computer Science and Engineering Survey*, **3**: 31, 2012.
110. N. Alajmi and K. Elleithy, Multi-layer approach for the detection of selective forwarding attacks, *Sensors*, **15**: 29332–29345, 2015.
111. C. Kolias, V. Kolias, and G. Kambourakis, TermID: A distributed swarm intelligence-based approach for wireless intrusion detection, *International Journal of Information Security*, 1–16, 2016.
112. C. Adams, Impersonation attack, in H. C. A. van Tilborg and S. Jajodia, editors, *Encyclopedia of Cryptography and Security*, Springer US, Boston, MA, pp. 596–596, 2011.

Chapter 14

A Survey of Intrusion Detection Systems in Wireless Sensor Networks

Eleni Darra and Sokratis K. Katsikas

Contents

14.1 Introduction

A wireless sensor network (WSN) is a spatially distributed network of a few tens to thousands autonomous sensor nodes that monitor physical or environmental conditions, and cooperatively pass their data through a gateway in the network to a base station. Each sensor node is a low power device equipped with one or more sensors, a processor, memory, a power supply, a radio, and possibly, an actuator; these are typically battery operated. A WSN typically has little or no infrastructure. WSNs have great potential for many applications in scenarios such as home automation, traffic monitoring and control, military target tracking and surveillance, natural disaster relief, biomedical health monitoring, and hazardous environment exploration and seismic sensing.

WSNs often operate in hostile environments and/or monitor or control critical applications without or with little capacity for protection, supervision, or intervention. These characteristics, in conjunction with the inherent vulnerability of the broadcast nature of the transmission medium, make WSNs vulnerable to a large variety of security attacks.

In response to the need for defending against these attacks, preventive, detective, and mitigation security mechanisms for WSNs have been developed. On the detective side, intrusion detection systems (IDSs) stand out as the most prevalent and widespread defensive mechanism. An IDS is a device or piece of software that monitors the network to detect unauthorized or malicious activities, such as attacks. It is responsible for monitoring the network, determining whether an attack is taking place, and preventing destruction of the system by raising an alarm or even possibly by taking action against the identified attacker. Within the operating characteristics of a WSN, an IDS must enjoy several properties not necessarily required of IDSs operating in other computing and communication environments.

In this chapter, we survey the literature on IDSs for WSNs, with an eye toward assessing the capacity of the current state-of-the-art IDSs in terms of attack coverage on the one hand and toward identifying research gaps and proposing research directions for the future on the other.

The rest of this chapter is organized as follows: Section 14.1 discusses characteristics of IDSs for WSNs that are used to propose an appropriate multidimensional taxonomy. Section 14.3 surveys intrusion detection approaches for WSNs, and classifies them according to the taxonomy of Section 14.1. Finally, Section 14.4 concludes the chapter.

14.2 Taxonomy of IDSs for WSNs

Little if anything has changed in the conceptual process model of intrusion detection since it was first proposed in Reference 1. The model consists of four core functions:

- Monitoring
- Analysis
- Decision making
- Reporting and responding

Monitoring is the process by which an IDS gathers data originating from the WSN it defends. In WSN intrusion detection, possible data sources are the network traffic and the values of selected node parameters. Analysis is the process by which an IDS processes the available data; the results of the analysis are then used to decide whether an intrusion has occurred or not. Several analysis techniques are possible in WSN intrusion detection and different decision-making approaches have been proposed in the literature. The analysis and decision-making processes are often collectively referred to as the detection engine. The final stage of the IDS process is reporting and responding, which bear responsibility for communicating the result of the decisions made within the IDS to the outside world and for taking action to respond to intrusions. Responding functionality is not commonly found in IDSs operating in WSN environments.

Since the original paper on IDS taxonomy by Debar et al. [2], many IDS taxonomies have been proposed in the literature. Some of these have been used to classify IDSs in specific contexts, including WSNs [3–16]. In most of these surveys [5–7,12,14,15], a one-dimensional classification scheme is employed, the sole dimension is the detection technique. Fewer works [4,10,13] classify WSN IDSs using their architecture as well, whereas authors of cited work [11,16] propose full taxonomies. However, the taxonomy of Reference 11 is not specifically tailored to IDSs operating in a WSN context, whereas that of Reference 16 is designed to apply to other categories of wireless networks as well.

Despite the large number of publications on intrusion detection for WSNs and the relatively large number of surveys on the subject, it is interesting to note that the largest number of intrusion detection approaches reviewed in these surveys is 20. Additionally, with the exception of References 10 and 12, none of these surveys map IDSs to the attacks they can detect, while none report their performance metrics.

In the following, we expand and extend the work in Reference 3 to review more than 90 WSN IDS approaches that have been proposed in the literature, to classify them according to a taxonomy scheme specifically tailored to IDSs operating in a WSN context, to review their attack detection capabilities, and to discuss their reported performance and the performance metrics used.

The taxonomy we use is depicted in Figure 14.1. Its elements are discussed in the sequel.

14.2.1 Architecture

There are two architectural types of IDSs for WSNs: *individual* and *collaborative*. An individual WSN IDS resides in the WSN base station, whereas a collaborative WSN IDS consists of multiple elements distributed over the network, each one communicating with the other. The collaborative architectural type can be further divided into three categories [17]:

1. *Centralized*: Each intrusion detection element produces alerts locally, which are sent to a central server that correlates and analyzes them.
2. *Hierarchical*: The IDS is divided into several small groups, often following the division of the WSN nodes into clusters. The elements at the lowest level of the hierarchy work as detection elements, while those at higher levels act as both detection elements and alert correlators. The correlated alerts are then passed to an even higher level for further analysis.
3. *Distributed*: No centralized information gathering or analysis is taking place. All detection elements act fully autonomously, collect their own data, and make decisions locally.

14.2.2 Detection Technique

14.2.2.1 Anomaly-Based Intrusion Detection

According to Reference 18, which provides a comprehensive overview of anomaly detection as a generic technique, anomaly detection refers to the problem of finding patterns in data that do not conform to expected behavior. In the context of intrusion detection, anomaly detection techniques attempt to base the decision on whether an intrusion is occurring by monitoring specific features that collectively constitute a behavior and examining whether the observed behavior pattern is not ordinary. The ordinary can be defined with respect to the history of the test signal or

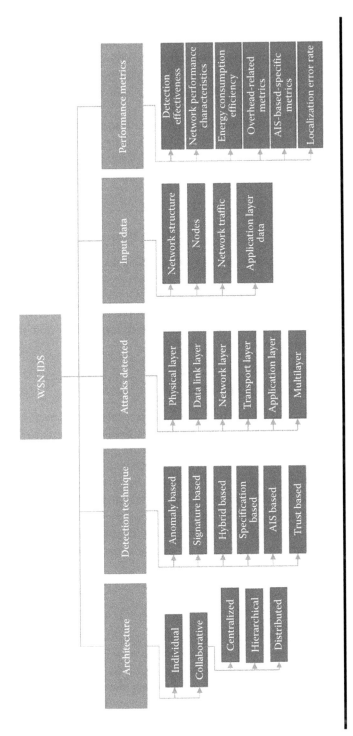

Figure 14.1 WSN IDS taxonomy.

with respect to a collection of training data that describe the ground truth. The key advantage of anomaly-based approaches is that as they monitor behavior, they can potentially detect attacks unknown at the time when the IDS was commissioned (sometimes referred to as zero-day attacks). Unfortunately, describing what is normal in terms of behavior is inherently difficult; this leads to a comparatively high level of false positives.

In Reference 19, a comprehensive survey and a taxonomy of anomaly-based IDSs based on further characteristics of such systems (e.g., method used, type of anomaly detected) are presented. However, this level of detail does not provide more insight into the present discussion; hence, we retain a one-level classification in the sequel.

14.2.2.2 Signature-Based Intrusion Detection

Signature-based intrusion detection looks for specific patterns in the data that the IDS uses to decide whether an attack is happening or not. By definition, signature-based intrusion detection can detect all known attacks whose signature is included in the system's knowledge base; this leads to low false positive rates (FPRs). On the other hand, again by definition, it is impossible to detect any attack whose signature is unknown to the system, including zero-day attacks.

14.2.2.3 Specification-Based Intrusion Detection

Specification-based intrusion detection is similar to anomaly-based detection. However, the two methods differ in that the specification-based approach exploits human expertise in developing a model of legitimate program behavior, in the form of specifications. An intrusion is detected when the system departs from this model. One major advantage of specification-based intrusion detection is a low false negative rate (FNR), because only situations that violate what a human expert previously defined as proper system behavior constitute intrusions [16]. On the other hand, specification-based intrusion detection can be prohibitively tedious and error-prone due to its reliance on the level of the human expertise employed [20].

14.2.2.4 Hybrid

In order to combine the advantages of both anomaly-based detection and signature-based detection, hybrid intrusion detection has been proposed. This consists of two detectors, one anomaly-based and one signature-based, combined in parallel or, more frequently, in series, with either one being the first in the series. The major drawback to hybrid intrusion detection is getting different technologies to interoperate successfully and, more importantly, efficiently.

14.2.2.5 Trust-Based Intrusion Detection

A trust-based IDS focuses on identifying malicious nodes rather than attacks or anomalies. Malicious nodes are identified on the basis of their reputation within the community of nodes in the WSN. The higher the reputation of a node, the more other nodes would wish to communicate with it, thus raising its reputation even further. On the other hand, a low-reputation node would eventually be practically cut off from communicating with other nodes. Several metrics have been proposed to measure (and assign) reputation, including metrics related to the quality of service (QoS) that the node provides, such as the consumption of energy and the cooperativeness of the node; and metrics borrowed from the field of social networks, such as the honesty of a node, measured by false self-reporting, trust fluctuation, and abnormal trust recommendations. The major advantage of this detection technique is that it focuses on identifying malicious nodes rather than attacks; thus, it can detect a variety of attacks, including zero-day ones. Its major disadvantage is that it is vulnerable to malicious nodes that may also employ techniques to either increase their own reputation or decrease that of normally behaving nodes.

14.2.2.6 Artificial Immune System-Based Intrusion Detection

Artificial immune systems (AISs) are a form of biologically inspired computing that attempts to imitate the behavior of the human immune system in identifying and defending against intruders. Different types of AISs have been proposed for intrusion detection within the WSN context. The most prevalent ones are systems that use the *negative selection approach*, whereby "self" is distinguished by "non-self" (intruder) by generating a set of detectors (the analogous of T cells in the human immune system) which are used for detecting anomalies; and systems using the *danger theory approach*, which suggests that a natural immune response is the result of sensing danger in the system rather than detecting foreign symptoms of the cause.

14.2.2.7 Game Theory-Based Intrusion Detection Design

Game theory is a field of applied mathematics that can be used for describing multi-person strategic decision-making scenarios. Game theory has been used in the field of intrusion detection in WSNs for analyzing the behavior of the attacker and the defender. Therefore, strictly speaking, game theory has not been used as an intrusion detection technique, but rather as a technique for informing the design of improved IDSs in terms of improved defense strategy. However, some game theory-based IDS designs utilize a detection technique to identify malicious nodes.

14.2.3 Attacks Detected

A large number of attacks against WSNs have been described in the literature, and different taxonomies have been proposed to classify them [21–32]. Criteria such as passive or active, internal or external, protocol layer used, stealthy or nonstealthy character, cryptography, or noncryptography—related nature, etc. have been used for classifying attacks against WSNs. However, the Open Systems Interconnection (OSI) layer-based classification is most commonly used, stable and unambiguous; as in References 14 and 33, we will follow this scheme in the sequel, using all attacks classified in Reference 3.

14.2.3.1 Physical Layer Attacks

In the *monitor and eavesdropping attack*, the attacker monitors the data traffic and can discover the communication content among the nodes in the sensor network [21]. When control information about the sensor network configuration is transmitted, the eavesdropping attack can act effectively against privacy.

In the *jamming attack*, a malicious node attempts to jam the frequencies of the radio links used for communication between the nodes in the network. Moreover, the adversary attempts to disrupt the operation of the network by broadcasting a high-energy signal [22–24,26,30,32].

In the *tampering attack,* also referred to as *replication attack* or *clone attack*, the attacker can alter or replace sensors and parts of computational and sensitive hardware and can also extract information such as cryptographic keys to gain unrestricted access to higher communication layers [22,23,26–28,30,31]. Any type of physical attack on sensors in the network actually constitutes a tampering attack.

In the *passive information gathering attack*, an attacker can collect unencrypted information from the sensor network. Moreover, an intruder with a well-designed antenna can easily pick off the data stream. An attacker monitors the messages which contain the physical locations of the sensor nodes and intercepts their content. In addition to the locations of sensor nodes, an adversary can thus observe the application-specific content of messages including message IDs, timestamps, and other fields [21,29,32].

In the *physical attack*, the attacker destroys one or more sensors permanently, so that the losses are irreversible [21]. For instance, the attacker can extract cryptographic secrets, tamper with the associated circuitry, modify the software in the sensors, or replace legitimate with malicious sensors under the control of the attacker. The physical attack can be further classified into two main categories according to (a) the extent of control that the attacker gains and (b) the time span during which the regular operation of a node is interrupted [27].

14.2.3.2 Data Link Layer Attacks

The *collision attack* occurs when two nodes attempt to transmit packets on the same frequency at the same time. When this happens, data will change, causing a checksum mismatch at the receiving end. The packet will then be discarded as invalid [21,22,25,30].

The *energy exhaustion attack* is similar to the collision attack but with the slight difference that a malicious node may conduct a collision attack repeatedly in order to exhaust the power in the communicating nodes [21,22,26,30].

The *unfairness attack* does not entirely prevent legitimate access to the communication channel, but could result in marginal performance degradation [26,30]. In fact, in order to ensure fairness in WSN, the use of small frames might be helpful so that any individual node might seize the channel for a shorter time. However, this would also incur some framing overhead.

Although the messages are transferred through the sensor network and the information they convey is encrypted, the communication patterns remain vulnerable to a possible *traffic analysis attack* [21,32]. Sensor activities can potentially reveal enough information to enable an adversary to cause harm to the sensor network.

In a *node malfunction attack*, the attacker forces a node to generate inaccurate data that could jeopardize the integrity of the sensor network, especially if the node is a cluster head in the network [21,32].

Random numbers are frequently used to prevent replay attacks. Unfortunately, since truly random numbers are difficult to generate, pseudorandom numbers are almost invariably used. The noise from an electronic device or the position of a pointer device is a source of such randomness. However, under certain conditions, the pseudorandom number sequence can be revealed to an attacker [24]; this constitutes a *pseudorandom number attack*.

An adversary can launch a *digital signature attack* by using a message and its associated signature to fake another message's signature [24]. Digital signature attacks are further classified into three types, namely *known-message*, *chosen-message*, and *key-only* attacks. In the known-message attack, the attacker knows the messages that the victim has signed. In the second attack, the attacker can choose a particular message to be signed by the attacker. In the key-only attack, the attacker only knows the public verification algorithm.

A *hash collision attack* tries to find two messages having the same hash. Such an attack, if successful, could be used to tamper with existing certificates, as the adversary might be able to construct a valid certificate corresponding to the hash collision [24].

A *node outage attack* occurs when an attacker manages to stop one or more nodes from functioning properly, for example, by depleting their power source [21,32].

Data aggregation is one of the most important sensor network services. In an *integration attack*, the malicious node will inject additional frames into the messages that are transferred in the network and render these messages invalid [34].

In the *camouflaged adversary* type of attack, the attacker inserts or compromises one or more nodes in the sensor network. This node(s) can hide in this network and can act as normal in order to attract and misroute packets [21].

14.2.3.3 Network Layer Attacks

The most direct attack against a routing protocol in any network is to target the routing information itself while it is being exchanged between nodes, by launching a *routing attack*. The adversary *may spoof, alter, or replay routing information* in order to disrupt traffic in the sensor network. These adversaries are able to create routing loops, attract or repel network traffic, extend or shorten source routes, generate false error messages, partition the network, increase end-to-end latency, capture sensors and turn them into inside attackers, perform random, opportunistic, and insidious attacks to evade detection, and maximize their chance of success, etc. [21,22,25,26,30,32,35,36]. By spoofing, altering, or replaying routing information, adversaries may be able to create routing loops in the network in order to increase end-to-end latency, jam normal communications, and even disable the network. In the *selective forwarding attack*, also referred to as the *message negligence attack*, a malicious node can refuse to forward certain packets and simply drop them, ensuring that they are not transmitted any further. However, neighbor nodes might start using another route. This attack is particularly effective when combined with an attack that gathers traffic through the malicious node [21,22,25,26,29,30,32].

In a *blackhole attack*, the attacker listens to the route request and then replies to the target node saying that it has the shortest path to the base station. The blackhole node can drop the packets, selectively forward those to the base station or to the next node, or even change the content of the packets [30].

In the *neglect and greed attack*, a malicious node drops packets or denies transmitting legitimate packets or gives excessive priority to the transmitted messages [26]. The dynamic source routing (DSR) protocol and the protocols that are based on it are especially vulnerable to this type of attack.

In WSNs, some nodes undertake some special responsibilities like managing cryptographic keys, making use of acquired data, maintaining a local group, etc. These nodes are called *leader nodes*. In the *homing attack*, the adversaries are attracted to these leader nodes, try to eavesdrop on their activities, and hamper the normal functioning of these leader nodes. The homing attack is especially dangerous for the location-aware routing protocols which rely on geographic information [26,30].

In the *misdirection attack*, the role of a malicious node is to direct the legitimate packets to a wrong path with no route to the intended destination [22,26,30]. Instead of sending the packets in the correct direction, the attacker misdirects those packets and this node may be victimized.

In the *sinkhole attack*, the adversary's goal is to attract nearly all the traffic from a particular area through a compromised node. The attacker makes a compromised node look attractive to surrounding nodes by forging routing information. The surrounding nodes will choose the compromised node as the next node to route their data through [21–23,25,26,29–32].

In a *Sybil attack*, a single node duplicates itself and appears in multiple locations at once. Moreover, this malicious node will send incorrect information to another node in the network [21,22,25,26,29–32]. In this type of attack, the WSN is subverted by a malicious node, which forges a large number of fake identities in order to disrupt the network's protocols. The Sybil attack targets fault tolerant schemes such as distributed storage, multipath routing, and topology maintenance.

In a *wormhole attack*, a malicious node receives packets from one location of the network, forwards them through a wormhole link, and releases them into another location [21–26,28–32]. For example, a single node is situated between two other nodes forwarding messages between the two of them. Usually, this type of attack involves two distant malicious nodes that collaborate with each other to relay packets along an out-of-bound channel available only to the attacker.

Many routing algorithms used in sensor networks require acknowledgments to be used in each communication. An attacker can spoof the acknowledgments of overheard packets destined for neighboring nodes in order to provide false information to those neighboring nodes, thus launching an *acknowledgment spoofing attack* [25,30].

In the *HELLO flooding attack*, an attacker with a high radio transmission range and processing power sends HELLO packets to a number of sensor nodes that are isolated in a large area within the network, thus leading them to believe that the malicious node is their neighbor. As a result, the victim nodes try to transmit the information through the attacking node [21,22,25,26,29–32].

The *routing loop attack* targets the information exchanged among nodes. When an attacker alters and replays routing information among nodes, false error messages are generated. Routing loops attract or repel the network traffic and increase the node-to-node latency [32].

In the *node replication/clone attack*, the attacker adds a node to an existing sensor network by copying the node ID of an existing sensor node. The replicated node is deployed arbitrarily throughout the network. These cloned sensor nodes can be installed to capture the information of the network and they can severely disrupt a sensor network's performance. The adversary can also inject false information, or manipulate the information passing through cloned nodes [21,23].

The *periodic route error attack* is a form of a denial of service (DoS) attack. A sensor node is initially physically compromised by the attacker. Next, the compromised node will proceed to broadcast route error messages to neighboring nodes. These error messages inform the neighboring nodes that the route to the base station is down. Nodes that utilized this route will lose their path to the base station and would have to repeat the process of searching for a route to the base station. This causes the affected portion of the network to be congested with packets, and also causes sleep deprivation of the affected sensor nodes [37].

In the *network partition attack*, the accessibility of nodes is denied, even though there exists a path between the nodes [38].

In the *simple broadcast flooding attack*, the attacker floods the network with broadcast messages. The false information passes through the whole network [38].

In the *simple target flooding attack*, the attacker tries to flood the network through some specific nodes [38].

The *false identity broadcast flooding attack* is similar to the simple broadcast flooding attack with the slight difference that the attacker deceives the whole network with a wrong source node ID [38].

The *false identity target flooding attack* is similar to the simple target flooding attack with the difference that the attacker deceives the whole network with a wrong source node ID [38].

In the *denial of sleep attack*, the attacker sends useless control traffic and forces the nodes to forgo their sleep cycles so that they are completely exhausted and hence stop working [39,40].

A *false node attack* involves the addition of a node by an adversary and causes the injection of malicious data. An intruder has the ability to add a node to the system that feeds false data or prevents the passage of true data. Malicious code injected in the network could spread to all nodes, potentially destroying the whole network or, even worse, taking over the network on behalf of an adversary [21,32].

In the *repetition attack*, an attacker retransmits the same message several times in the network [41].

In the *message delay attack*, an attacker retransmits a message after a defined timeout has elapsed [41].

In networks using a reputation mechanism to trust their nodes, a *slandering (Reputation Trap—RepTrap) attack* may be launched. In this attack, the attackers find some high-quality objects that have a small number of feedbacks and provide feedback such that these objects are marked as low quality by the system. Consequently, the system is led to believe that the attackers' negative feedbacks are in line with the object quality, while those from honest users are not. Therefore, the reputation of the attackers will be increased, while that of honest users will be reduced [42].

In the *sleep deprivation attack*, the malicious node makes requests to victim nodes only as often as is necessary to keep the victims awake [43].

An attack specific to WSNs using the directed diffusion routing protocol is described in Reference 44. Under this protocol, each node maintains an interest cache that records the history of received interest packets. The basic idea of the *interest cache poisoning attack,* which corrupts the routing process, is to inject fabricated interest packets to replace benign entries in the interest caches of other nodes.

14.2.3.4 Transport Layer Attacks

In the *flooding attack*, the attacker triggers multiple connection requests toward the target node, aiming at exhausting the latter's resources. This attack either blocks only the node or the link along with the node [22,23,26,30].

In the *desynchronization attack*, the connection between two end points can be interrupted by desynchronization. The adversary forges messages to one or both end points which request transmission of missed frames [22,26,30]. These messages are retransmitted, and if the adversary maintains a proper timing, she can prevent the end points from exchanging any useful information.

14.2.3.5 Application Layer Attacks

In a *path-based DoS attack*, an attacker overwhelms sensor nodes by flooding a multi hop end-to-end communication path with either replayed or injected false messages in order to waste energy resources [22,30].

In an *overwhelming attack*, the attacker deluges the network nodes with large volumes of traffic to a base station [30]. The attack consumes a large amount of network bandwidth and drains node energy.

In many deployed networks, reprogramming nodes is feasible. Sometimes the process of reprogramming is not always secure and this can allow an attacker to handle a large portion of the network and deceive the process by launching a *deluge (reprogramming) attack* [30].

In a *node subversion attack*, the sensor network can be compromised if the attacker captures a node and reveals its information including disclosure of cryptographic keys [21,32]. A particular sensor might be captured, and information (keys) stored on it might be obtained by an adversary.

In a *message corruption attack,* the attacker modifies the content of the message in the sensor network; thus, the integrity of the message is compromised [21,32].

14.2.3.6 Multilayer Attacks

Such attacks can be implemented in different ways and cannot be classified as pertaining to a particular OSI layer; hence, we classify them as multilayer attacks.

In a *man-in-the-middle attack*, the attacker connects all the potential victims together, broadcasts messages among them, making them to believe that they are talking directly to each other over a private connection. Moreover, the attacker is able to intercept all messages that the victims exchange and to force new ones [23].

A *masquerade attack* (or *fabricate information attack*) can take the form of (a) inserting messages into the network using a false identity, (b) replaying previously intercepted messages, (c) spoofing a network service, or (d) taking the address of another host or service, essentially becoming that host or service [24].

Several attacks that aim at making the services of a node or a network unavailable, regardless of the method used to achieve the aim, are collectively referred to as *DoS attacks*, or, when they are launched from multiple attackers, *distributed denial of service (DDoS) attacks*.

Likewise, attacks that result in turning a node malicious, regardless of the type of maliciousness behavior or of the method used to compromise the node, are referred to as *malicious node attacks* [45].

Table 14.1 summarizes the classification of attacks against WSNs per OSI protocol layer.

Table 14.1   Classification of Attacks against WSNs per OSI Protocol Layer

OSI Protocol layer	Attack
Physical layer	Monitor and eavesdropping attack (PL1) Jamming attack (PL2) Tampering attack (PL3) Passive information gathering attack (PL4) Physical attack (PL5)
Data link layer	Collision attack (DL1) Energy exhaustion attack (DL2) Unfairness attack (DL3) Traffic analysis attack (DL4) Node malfunction attack (DL5) Pseudorandom number attack (DL6) Digital signature attack (DL7) Hash collision attack (DL8) Node outage attack (DL9) Integration attack (DL10) Camouflaged adversary attack (DL11)
Network layer	Routing attack (NL1) Selective forwarding attack (NL2) Blackhole attack (NL3) Neglect and greed attack (NL4) Homing attack (NL5) Misdirection attack (NL6) Sinkhole attack (NL7) Sybil attack (NL8) Wormhole attack (NL9) Acknowledgment spoofing attack (NL10) HELLO flooding attack (NL11) Routing loop attack (NL12) Node replication/clone attack (NL13) Periodic route error attack (NL14) Network partition attack (NL15) Simple broadcast flooding attack (NL16) Simple target flooding attack (NL17) False identity broadcast flooding attack (NL18) False identity target flooding attack (NL19) Denial of sleep attack (NL20) False node attack (NL21)

(Continued)

Table 14.1 (*Continued*)   Classification of Attacks against WSNs per OSI Protocol Layer

OSI	
	Repetition attack (NL22) Message delay attack (NL23) Slandering (Reputation Trap) attack (NL24) Sleep deprivation attack (NL25) Interest cache poisoning attack (NL26)
Transport layer	Flooding attack (TL1) Desynchronization attack (TL2)
Application layer	Path-based DoS attack (AL1) Overwhelming attack (AL2) Deluge (reprogramming) attack (AL3) Node subversion attack (AL4) Message corruption attack (AL5)
Multilayer	Man-in-the-middle attack (ML1) Masquerade (fabricate information) attack (ML2) DoS attacks (ML3) DDoS attacks (ML4) Malicious node attacks (ML5)

14.2.4 Input Data

Every IDS relies upon the availability of audit data that are being used as input to its detection engine. In the case of IDSs operating within a WSN context, these data come from four sources, namely, the *network structure,* the *nodes,* the *network traffic,* and the *application layer data,* that is, the measurements that the network performs.

Input data related to the network structure that have been used in IDSs for WSNs include the *radio transmission range,* the *network topology,* and the *routing protocol used.*

The nodes themselves are also being used as sources of input. Specifically, *physical characteristics* such as *energy consumption characteristics* (energy consumed, energy consumption rate, and energy prediction error), and *behavior characteristics* (such as honesty, unselfishness, cooperativeness, and intimacy) that are being used to assess the trustworthiness of a node in trust-based intrusion detection have been used in intrusion detection approaches within the WSN context.

The network traffic is a major source of input data to IDSs operating in the WSN context. IDS approaches have used either the traffic itself (i.e., the *contents of data packets*) or derived *characteristics of the network traffic* (such as the data rate, the throughput, the response time, the data drop rate, the one hop delay, packet routes) input data to the detection process.

Finally, some IDS approaches for WSNs use the measurements that the WSN performs to decide whether an intrusion in the form of a malicious node is taking place. These approaches apply anomaly detection methods to detect anomalies in the *Application layer data*. If such an anomaly is detected, the IDS concludes that an intrusion is taking place.

14.2.5 *Performance Evaluation Metrics*

The most commonly used performance evaluation metrics in intrusion detection are related to the *detection effectiveness* characteristics of the IDS, expressed as the percentage of cases when a malicious node is erroneously classified as normal over all cases (FNR); as the percentage of cases when a normal node is erroneously classified as malicious over all cases (FPR); and as functions of these. Such functions are the detection rate—DR, defined as the percentage of cases when a malicious node is correctly classified as such over all malicious cases and detection accuracy— DA, defined as the inverse of FPR. The number of *malicious packets dropped* and the *number of detected intruder* nodes have also been proposed as appropriate metrics.

In addition to these metrics, frequently used ones within the WSN IDS context are the *network performance characteristics* after the introduction of the IDS as compared to the same before. These are expressed as the *packet delivery ratio*, defined as the ratio of packets that are successfully delivered to a destination compared to the number of packets that have been sent out by the sender; the *message drop ratio*, defined as the ratio between the number of messages not received to the total number of messages; the *average delay*, defined as the ratio between the sum of all packets delayed to the total number of packets received; the *network throughput*; the *energy consumption*; the *response time*, defined as the average detection cycles of correctly detected malicious cases; the misdetection ratio, defined as the ratio of misdetected cases to all detected cases, including correctly detected and misdetected cases; the *number of hops for received packets*; the *suppression rate*, defined as the ratio of number of packets received in the presence of an attack over that in its absence, etc.

Metrics related to the *energy consumption efficiency* of the IDS itself, namely, the *energy consumption* itself and the *network lifetime* (including the number of alive nodes), are also frequently used performance evaluation metrics.

Overhead related metrics, namely, the *computational overhead*, the *communication overhead*, the *storage overhead*, and the *routing overhead* induced by the introduction of the IDS, are less frequently used metrics.

Metrics specific to the AIS-based intrusion detection are the *number of activated agents* of different types. However, these do not directly relate to the performance of the IDS in terms of its capacity to detect intrusions.

Finally, the *localization error rate* has been used as a performance evaluation metric in mobile WSN intrusion detection.

14.3 IDSs for WSNs

14.3.1 Anomaly Based

A distributed anomaly detection mechanism is proposed in Reference 37 that uses a clustering algorithm to build a model of normal traffic behavior; this model of normal traffic is then used to detect abnormal traffic patterns. This approach is able to detect routing attacks. Twelve features have been identified as relevant for intrusion detection. The performance of the IDS in detecting a periodic route error attack (form of DoS), as well as passive and active sinkhole attacks, has been evaluated. In the case of the periodic route error attack, a 95% detection rate is achieved for a 5% FPR. In the case of the passive sinkhole attack, the detection rate is 70% for a 5% FPR. In the case of the active Sinkhole attack, the detection rate is 100% for a 5% FPR.

In References 39 and 40, an IDS based on an isolation table is proposed to isolate malicious nodes so as to detect intrusions in hierarchical WSNs and to estimate the effect of intrusion detection. The proposed method includes the definition of the IDS, the cluster head that monitors the member nodes, the member nodes that monitor the cluster head, and the system backing up the isolation table. The intruders are assumed to launch HELLO flooding, DoS, denial of sleep, and sinkhole and wormhole attacks against the WSN. The proposed hierarchical IDS achieves 95% of detection accuracy when the number of monitor nodes is large.

Da Silva et al. [41] proposed an IDS that is based on the inference of the network behavior obtained from the analysis of events detected by a monitor node. The IDS first acquires relevant data, and then it applies seven types of detection rules to interval, retransmission, integrity, delay, repetition, radio transmission range, and jamming. Simulation results are presented that provide performance results for the message delay, repetition and wormhole attacks, jamming attack, data alteration, and message negligence blackhole and selective forwarding attacks. The results of the proposed approach indicate a detection accuracy ranging from as low as 30% up to 100%, depending on the attack and on the IDS parameter settings.

In Reference 42, a hierarchical trust management protocol that leverages clustering to cope with a large number of heterogeneous SNs for scalability and reconfigurability, as well as to cope with selfish or malicious SNs for survivability and intrusion tolerance with vastly different social and QoS behaviors, is described. The authors address the key design issues of trust management, including trust composition, trust aggregation, and trust formation. The hierarchical trust management protocol is resilient to blackhole, sinkhole, and slandering (RepTrap) attacks. The trust-based IDS algorithm outperforms traditional anomaly-based IDS techniques in terms of detection rate, while maintaining sufficiently low false positives. The strength of the trust-based IDS algorithm is especially pronounced when FPR approaches zero: the trust-based IDS algorithm can still maintain a high detection rate (> 90%) when FPR is close to zero, a point at which the detection rate of anomaly detection-based IDS schemes drops sharply.

A distributed group-based IDS is proposed in Reference 45. This proposal partitions the sensor network into parts having the same attributes (such as sensor ID and sensor remaining power) and runs a group-based algorithm in each group. This IDS detects fabricate information (Masquerade), energy exhaustion, selective forwarding, blackhole, sinkhole, HELLO flooding, and wormhole attacks. This refined group-based intrusion detection scheme uses as performance parameters FPR and the detection accuracy.

In Reference 46, a lightweight, low energy consumption IDS has been proposed. The authors explore how ontology concept mechanisms on anomaly detection and lightweight IDS are related. Protocol and attack type packages are simulated and useful intrusion behavior features are extracted. The IDS is claimed to be capable of detecting the Sybil attack, but no evaluation regarding the detection rate or any other metric is provided.

The mechanism of detection proposed in Reference 47 uses different roles for each node: cluster headers are responsible for monitoring all common member nodes in the cluster, while the common member nodes are responsible for monitoring the cluster head. Each node executes different detection operations, depending on its role. Four kinds of agents are installed on each node. Two clustering algorithms are used in this scheme in two stages. In the first stage, a self-organizing map (SOM) neural network is used for rough clustering. The output of this algorithm is then fed to the second stage, which uses the K-means clustering algorithm to refine the clusters generated in the first stage. Attacks on the application layer have been evaluated and the simulation results show that the detection rate of the proposed scheme is much better than that of the traditional SOM, although with a little higher FPR.

Ponomarchuk and Seo [48] propose a lightweight, fast, and efficient traffic intensity-based intrusion detection method. It contains a large number of nodes, which transmit data periodically. An intrusion detection method based on the analysis of the neighbor's behavior and a thresholding technique, applied to selected parameters is presented. The approach can detect DoS attacks. According to the simulated results, FPR increases with the growing density of the network, with the increase of the packet size, or with the increase in background noise. The decrease in data rate does not affect FPR significantly, but the detection rate grows. The detection rate grows with the increase in traffic manipulation intensity or with the decrease in transmission failure probability.

In Reference 49, the concepts of self-organized criticality (SOC) and hidden Markov models (HMM) events are used to design a lightweight IDS specially designed to infer intrusions (or failures) in a WSN that measures environmental temperatures, based on anomalous measurements reports.

The IDS proposed in Reference 50 is adapted to static sensor networks; it introduces a neighbor monitoring technique known as spontaneous watchdog to optimally watch over the communications of the sensors' neighborhood in certain scenarios. A local and a global agent are utilized. The local agent audits data that come from those nodes that lie inside its node's radio range or are its neighbors and generate an alert if any node works abnormally (e.g., it floods) or if it receives a message from a node that is not present in the neighbor list. The global agent monitors the communication between neighbors. The spontaneous watchdog technique relies on the broadcast nature of sensor networks; the global agent works as a spontaneous watchdog. Attacks against the physical or logical safety of sensor nodes can be discovered if the nodes are able to know whether they are being manipulated or not.

In Reference 51, a lightweight anomaly-based intrusion detection scheme for WSNs is proposed. The IDS follows a distributed and cooperative architecture. Its main characteristic is that the nodes monitor their neighborhood and collaborate with their nearest neighbors to bring the network back to its normal operational condition when needed. This design principle is applied to detect blackhole and selective forwarding attacks by defining appropriate rules that characterize the corresponding malicious behavior. According to the first rule, if node A sends a packet to node B, then the monitoring node stores the packet in its buffer and watches to see whether B forwards it or not. According to the second rule, if the majority of the monitor nodes have raised an alert, then the target node is compromised. The threshold value for the percentage of packets dropped over a period w is set to $t = 20\%$. Above this threshold, each watchdog generates an alarm. FNR is reduced as the window length w is increased. This mechanism produces a small number of false positives and this effect is shown clearly on smaller drop probabilities.

A centralized anomaly detection mechanism called ANDES is presented in Reference 52. It tries to detect several routing protocol attacks. It consists of a collection of application data units, a collection of management information units, and a detection policy. The collection of application data units aggregate regular data. The collection of management information uses an additional management routing protocol. The detection policy works in three phases: analysis of application data, analysis of management data, and cross-checking to determine the root cause of the attack. The system is geared toward detecting blackhole, sinkhole, selective forwarding, and flooding attacks. Evaluation using a 32-node sensor testbed shows that ANDES is effective in detecting fail-stop failures and most routing anomalies with negligible computing and storage overhead. ANDES has a relatively high false positive ratio, especially in the case of flooding anomalies. Computation overheads on network nodes and storage overhead are almost negligible because ANDES is a centralized scheme. The message overhead is directly proportional to the frequency of queries and number of attributes in each response. ANDES does induce significant message overhead.

A simple graph theory-based approach that efficiently detects compromised beacon nodes is presented in Reference 53. Beacon nodes provide location information to the sensor nodes. It is assumed that an IDS agent is installed in each beacon node, which produces alerts about the maliciousness of sensor nodes. A compromised beacon node transmits false information about other nodes and degrades the performance of the routing protocol. This approach is centralized-distributed because beacon nodes generate alerts about the malicious activity. Once efficient amount of data is gathered, the proposed graph theory based detection mechanism is applied to find out whether the information is received from a reliable source or not. The evaluation metrics of the experiments are the detection rate and FPR.

In Reference 54, an intrusion detection scheme called EPIDS is presented, which is based on the energy prediction in cluster-based WSNs. The main contribution of EPIDS is to detect attackers by comparing the energy consumption of sensor nodes. Thus, sensor nodes can be managed locally by cluster heads. A rotating cluster heads policy makes it possible to elect malicious nodes as cluster heads. Adversaries can compromise any node in the network and launch DoS attacks such as selective forwarding, HELLO flooding, wormhole, sinkhole, and Sybil attack. Since malicious nodes require abnormal energy to launch an attack, the approach focuses on the energy consumption rate of nodes in order to discover the compromised nodes. The simulation results show that the HELLO flooding attack has the highest energy consumption rate 0.0333 J/s, the sinkhole attack has 0.020 J/s, the Sybil attack has close to 0.015 J/s, and the wormhole attack has close to 0.010 J/s while the selective forwarding attack has the lowest energy consumption rate 0.00297 J/s. Finally, the detection accuracy of the proposed scheme is much higher than that of others.

A distributed IDS is presented in Reference 55 that monitors the communication in the network and a criterion for the placement of intrusion detection nodes is proposed. The IDS searches for violations of that criterion to detect wormholes of length above a certain minimum value. This IDS consists of a number of intrusion detection nodes, which monitor the communications in a WSN. The intrusion detection nodes can communicate securely and do not have the same strict power and computation constraints of the sensors. Moreover, they can share their collected data and use these data to detect attacks collaboratively. The proposed IDS measures the success of detecting wormholes through simulating different wormhole lengths against different ID node ranges. The system fully detects the active wormholes if the wormhole length is two times greater than the communication range of the intrusion detection nodes. The passive wormhole will be detected if the length between two communicating nodes through the wormhole is three times greater than the communication range of the nodes. Generally, the detection ratio of this system is about 100% if the wormhole connects a pair of sensors that is not monitored by one ID node.

In Reference 56, a traffic prediction algorithm for sensor nodes which exploits the Markov model is proposed. A distributed anomaly detection scheme, called traffic prediction-based intrusion detection (TPID), is designed to detect attacks such as selective forwarding attacks and DoS attacks. In TPID, each node acts independently when predicting the traffic and detecting an anomaly. According to the simulation results, the packet loss rates of TPIDS are relatively low.

In Reference 57, a lightweight anomaly-based IDS is described. The wormhole attack, the HELLO flooding attack, and jamming and flooding attacks can be detected by the system. In contrast, the physical attack, the Sybil attack, and the sinkhole attack are difficult to detect. The collision attack is one of the attacks that yields better detection results, whereas less successful is the system for detecting the energy exhaustion attack, the selective forwarding attack, and the desynchronization attack.

A mobile agent-based hierarchical intrusion detection system (MABHIDS) for WSNs is proposed in Reference 58. The proposed scheme performs two levels of intrusion detection by utilizing as few network resources as possible. The network intrusion detection system (NIDS) and local intrusion detection system (LIDS) are involved in providing two tier security in the WSN. NIDS is installed on all cluster head nodes, whereas LIDS is based on a mobile agent. LIDS is activated whenever a cluster head node finds any suspicious nodes. The cluster head node issues LIDS for further scrutiny of malicious activities of the suspicious node in order to affirm it as a compromised node. LIDS uses the resources of the suspicious node. No performance evaluation results are presented and no attacks are mentioned as being detected by the system.

A neighbor-based detection scheme for securing sensor networks by analyzing the behavior of a node with respect to that of its neighbors is presented in Reference 59. A node detects the neighboring node as malicious if it performs abnormally with respect to the set parameters, the basic idea being that neighboring nodes should be dealing with similar network traffic and should therefore behave similarly. Hence, a node is considered malicious if its behavior differs from that of its neighbors. Selective forwarding, HELLO flooding, and jamming attacks can be detected by the neighbor-based technique. The performance metric of false negatives is at the rate of 3.76% and that of false positives is 0.28% for the HELLO flooding attack. The corresponding rates for the jamming attack are 3.36% and 4.24%, respectively.

The IDS proposed in Reference 60 combines several existing approaches. The network is divided into clusters and a cluster head is assigned to each cluster. The energy consumption remains low as the transmitted data are forwarded first to the cluster head and then to the base station. Routing attacks are mentioned as those that the system is able to detect. However, no performance evaluation results are presented.

A machine learning (ML)-based anomaly detection scheme is proposed in Reference 61, where a Bayesian classifier is used to detect anomalous nodes. Some learning samples such as throughput, packet drop ratio, and the packet average delay are used in order to inform the machine learning for intrusion detection. The detection effectiveness of the replay attack using this approach is evaluated and, according to the experimental results, it is found that when the attack strength is weak, the true positive rate is low and a higher detection rate is achieved. FPR is nearly 3%.

An extension of the learning automata-based protocol for intrusion detection (LAID) named simple LA-based intrusion detection (S-LAID) is presented in Reference 62. The extension is more efficient and energy aware, as each node functions independently without knowing the behavior of each neighboring node. S-LAID assumes that the system sampling budget of a single node is analogous to the amount of energy that the node can spend on intrusion detection during its lifetime. Moreover, the balance budget of the system is analogous to the residual sampling energy of the system. If the sampling budget of a node is exhausted, the node has no more energy that can be spent on intrusion detection tasks. In S-LAID, each node continuously samples its interface at a minimum sampling budget. If malicious packets are found and the detection rate is more than the penalty threshold, then the sampling rate is increased by a penalization function. When the detection rate is less than the penalty threshold, the sampling rate is decreased by a reward function.

In Reference 63, an anomaly-based IDS using fuzzy C-means clustering (FCM) with hierarchical network architecture was introduced. The IDS can detect the sinkhole attack, the simple broadcast flooding attack, and the periodic route error attack, which are caused by abnormal flows of data. The FCM model collects data from a cluster and sorts them as data of the same type, which should be close, or as data of different type, that should differ a lot. The cluster heads collect from all regions detection information to be conveyed to the base station for intrusion detection. The simulation results show that if FPR is less than 1.5%, the detection rate can reach 96%.

The approach described in Reference 64 is based on exchanging control packets between the sensor nodes and base station. The blackhole attack and selective forwarding attack are experimentally evaluated. In order to detect the blackhole attack, each sensor node must send the number of packets exchanged with the base station. Another cluster head forwards the control packets to the base station. All sensor nodes maintain a blackhole table, which contains identifiers of detected blackhole nodes. Each sensor node checks its blackhole table before the selection of its next cluster head; this prevents the attacker node from being selected as cluster head.

In Reference 65, an IDS based on the KNN classification algorithm is proposed. This system separates abnormal nodes from normal nodes by observing their abnormal behaviors. Moreover, data mining technology is also used to design and implement the proposed IDS. The system has three advantages: (a) the value of K for mining has little effect on the results, (b) the cutoff value used to determine the abnormal node is easy to determine, and (c) the algorithm is fast and efficient. Experimentation with the flooding attack shows that the detection rate is above 98.5%, and the false alarm rate is 4.63%, that is, relatively high. When the average detection rate is 99.0%, the average false alarm rate is 1.5%.

A lightweight, energy-efficient system is proposed in Reference 66, which uses mobile agents to detect intrusions using the energy consumption of the sensor nodes as a metric. A mobile agent moves randomly from node to node, carrying the battery status of the nodes. The battery status is used to estimate the expected power consumption based on past observations. This approach is used to detect some types of DoS attacks. Specifically, flooding and blackhole attacks cause drastic changes in the battery status of various nodes. Simulation results show that a high detection accuracy while maintaining a low FPR can be achieved. Having a high migration rate leads to a higher FPR. In contrast, using a lower migration rate leads to a very low FPR and a useful detection rate of the flooding attack.

A weighted and evidence theory-based IDS is proposed in Reference 67. A multidimensional method is adopted to collect the behavior characteristic of each evaluated node. Most of the attacks which influence one or more aspects of a node can be correctly detected. Simulation results show that this approach is more effective in detecting malicious nodes with a higher detection ratio and lower misdetection ratio compared with existing schemes.

In Reference 68, only the base station analyses the traffic and concludes whether an attack exists. This method can detect selective forwarding and blackhole attacks. The method uses support vector machines (SVMs) with a polynomial kernel or radial basis function (RBF) kernel. An IDS module at the base station monitors the bandwidth and the hop count values. When no attacker is active within the network's area, the IDS module collects data and trains the SVMs to minimize FPR. The most efficient SVM is chosen for further attack detection. The simulation results indicate that the system can detect blackhole attacks with 100% accuracy and selective forwarding attacks in which 80% of the network is ignored, with approximately 85% accuracy.

In Reference 69, a leader-based intrusion detection system (LBIDS) is proposed to detect and prevent DoS attacks in the network. In this approach, the LBIDS is deployed into cluster heads in the network. The data are forwarded to the network through the cluster heads by verifying the IP addresses of the nodes in the route and the packets to be transmitted. The simulation results show that this approach can achieve a higher detection rate, energy, and average delay in relation to the DRPGAC (dynamic random password generation and comparison) approach, which is used for preprocessing and post processing solution for abnormal activities.

The approach in Reference 70 makes use of a clustering protocol. It selects a set of cluster heads among different nodes in the network and tries to cluster the rest of the nodes with the cluster heads. The latter are responsible for the coordination among the nodes and for forwarding the collected data to the sink node after efficiently aggregating them. The cluster heads are able to detect attacks against other cluster heads in the network. Moreover, a new IDS algorithm is proposed, with dedicated procedures for secure cluster formation, periodic reclustering, and efficient cluster member monitoring and, finally, the detection of different attacks (selective forwarding, blackhole, wormhole, and Sybil attack). The algorithm defines a trust-aware leader election metric and introduces a monitoring mechanism to monitor both the cluster members and the cluster heads. It specifies a rule-based detection engine that accurately analyzes data packets and also detects signs of sensor network anomalies. Three performance metrics are considered to evaluate the algorithm: the communication overhead, the percentage reduction in the network lifetime, and the detection accuracy. The communication overhead increases smoothly as the percentage of malicious nodes increases. The same context happens to the network lifetime. Results show that as the percentage of malicious nodes inside the network increases, the reduction in the network lifetime increases.

In the approach proposed in Reference 71, an algorithm is proposed to detect the Sybil attack and to conserve energy in doing so. A comprehensive energy model is adopted that includes sensing, logging, and switching energies apart from the processing and communication energy values. Two cases are considered in order to detect the Sybil attack. In the first case, the Sybil node does not reply to the query sent by the cluster head, that is the proposed algorithm is implemented on sending and acknowledging the query data packets. In the second scenario, the Sybil node replies with the same identity and different coordinates. According to the simulation results, the network lifetime is enhanced when the proposed technique is applied to detect the Sybil node in the network.

A model-based methodology to derive a closed form solution of the WSN lifetime is proposed in References 35 and 36. This is used to design defense mechanisms against selective capture and smart attacks. One of these mechanisms is a voting-based, anomaly-based IDS operating in each node.

In Reference 72, a modified cluster-head selection algorithm has been proposed, which is based on the remaining battery life and distance. This modified cluster-head election is done on the basis of the residual battery life of candidate nodes and the geographic distance from the candidate node to the base station. The scheme works in five phases, namely, the cluster head selection; the authentication check; the detection; the information dissemination, whereby messages are transmitted; and isolation/elimination, whereby the cluster head broadcasts one encrypted message to all the nodes in the network except to the blocked node. Simulation results show the nodes detected as malicious, the number of messages transmitted and received as well as the selection of the cluster heads.

An improvised hierarchical blackhole detection algorithm is presented in Reference 73, in which each sensor node sends a control packet to one of the agents and the cluster head at the end of the transmission phase. Each control packet contains the node ID, and the number of packets sent to the cluster head. The base station compares the number of packets of each node with the amount of packets received from its agent and the cluster head. If there is any mismatch in the number of nodes, then the base station detects an eventual blackhole attack and an alarm packet will be broadcasted to all network nodes, which contains the ID of the blackhole node. All sensor nodes maintain a blackhole table, which contains the ids of already detected blackhole nodes. Each sensor node checks its blackhole table before selecting its next cluster head; this prevents a malicious node from being selected as cluster head. The node with the maximum vitality backup and neighbor to more number of nodes will be selected as a second cluster head.

A zone-based node revocation and compromise detection scheme for sensor networks is proposed in Reference 74. To succeed this as well as to provide high security, two methods are proposed. The first one is the data packet format matching (DPFM), which detects the attacker node, and the second is the divert attention attacker (DAA), which prevents nodes from compromising the zone. The simulation experiments show that the compromising node is eliminated and attacks are contained. The throughput and energy consumption is very low compared to existing systems.

Swarm intelligence (SI) is proposed in Reference 75 as one of the most effective methods that can be applied for sinkhole attack detection. There are two popular swarm intelligence methods, namely, ant colony optimization (ACO) and particle swarm optimization (PSO). ACO is a probabilistic technique for solving computational problems which can be reduced to find good paths through graphs. PSO is a stochastic optimization technique, inspired by social behavior of bird flocking or fish schooling. An enhanced particle swarm optimization (EPSO) mechanism has also been proposed, whereby hash tables are used to obtain a more accurate suspect list. The results of EPSO show that the detection rate and the packet delivery ratio are improved compared to ACO and PSO by 90.076% and 83.834%, respectively. So far as the false alarm rate, the average delay, and the message drop rate go, the results show values of 8.472%, 5.316%, and 6.958%, respectively.

In Reference 76, the energy efficient intrusion detection scheme (EEIDS) is proposed, which detects a malicious node, by comparing its actual and predicted energy consumption; a node with abnormal energy consumption is identified as malicious. A Bayesian approach is used for predicting the energy consumption of the sensor nodes and an energy efficient routing protocol called APTEEN (adaptive periodic threshold energy efficient sensor network) is used to improve the lifetime of the network. The simulation results show that EEIDS gives good network lifetime, throughput, and energy consumption and effectively detects malicious nodes.

In Reference 77, a distributed IDS is proposed to provide attack detection in wireless body area sensor nodes. Genetic algorithms are used in this approach to generate a subset of relevant network features necessary to classify attacks, with the goal of increasing the attack detection rate, while lowering false positives and energy consumption. Experiments were conducted implementing jamming attack targeted at a wireless body area network and the results show that the detection algorithm seems to have a higher detection accuracy with increase in detection rounds and using a two-point crossover compared to a single-point crossover.

In Reference 78, a lightweight, energy-efficient system is proposed which makes use of mobile agents to detect intrusions based on the energy consumption of the sensor nodes. This mobile agent collects energy readings and raises an alert if sudden changes occur. The feasibility of mobile agents used for intrusion detection in WSNs has been demonstrated. Simulation results indicate that DoS attacks, such as flooding and blackhole attack, can be detected with high accuracy that is close to 100%, while keeping the number of false-positives very low at 3.1%.

The mechanism proposed in Reference 79 resolves the selective forwarding attack and the HELLO flooding attack that are carried out by an internal attacker in a WSN. The approach authenticates nodes with a key mechanism and retraces the routing path as an evasive action from the path involved with the victim node in the form of an internal attacker in the network. Once the network is established, the routing among nodes is done by means of the shortest path algorithm. If a threat is involved along the route, the IDS detects the threat and reports it to the key server to change the private keys of the nodes and take actions by rerouting. Rerouting is achieved by leaving the private key of the affected node unaltered. Simulation results show that the reliable transmission of data and the performance of the network have been improved when the proposed system is used.

A secure data aggregation framework using a trust monitoring system (TMS) at node level and an IDS at the base station is proposed in Reference 80. Each node in the network assesses the behavior of its neighbors using their behavior in performing the network activities such as cluster head selection and data aggregation, and reports to the base station. Then the base station analyzes the received information using the IDS and reports on any malicious activities back to nodes in the network. The malicious nodes are identified and excluded from the data aggregation process. Simulation results show that the proposed framework is robust in detecting and isolating the malicious nodes, regardless of how these became malicious, and the network lifetime is improved as compared to other trust aware data aggregation methods.

Two efficient and effective anomaly detection models principal component classifier-based anomaly detection (PCCAD) and adaptive PCCAD (APCCAD) are proposed for static and dynamic environments in Reference 81. The PCCAD model is based on the one-class principal component classifier (OCPCC) to measure the dissimilarity and showed advantages in terms of low computational complexity while keeping the memory utilization fixed. Moreover, the new model showed consistent performance in terms of detection accuracy. However, FPR was increased when the training samples were not good representative of the whole data nature. To solve this, APCCAD was designed. The APCCAD model incorporates an incremental learning method that is able to track the dynamic normal changes of data streams in the monitored environment. Finally, the experimental results showed that the adaptive model has a high detection rate (100%), FPR to zero in most cases (0.09%), and FNR to 0%.

In Reference 82, a lightweight ontology-based wireless IDS (OWIDS) is proposed. The system also applies an ontology to a patrol IDS (PIDS). Patrol nodes carry knowledge of how to detect intrusion and for that reason a PIDS is used to detect anomalies via detection knowledge. According to this approach, the system gathers attack and nonattack packages to build an intrusion database to enable evaluation of anomalous transmission packages. The system then applies the ontology to construct the relationship between the wireless sensor nodes. The manager sets the threshold value of the ontology relationship to detect attacks. The OWIDS is divided into three stages: the preprocessing stage, the ontology construction stage, and the intrusion detection stage. The system records anomaly information in an isolation table. The isolation tables prevent repeated detection of an anomaly and record the error information. The experimental results show that OWIDS can reduce energy consumption. The system detection accuracy of PIDS and OWIDS is higher than 89.61%. FPR of PIDS is 13.29% and that of OWIDS is 3.77%. Moreover, the detection accuracy of PIDS is 89.61% and that of OWIDS is 96.39%. The detection accuracy of both methods for Sybil, sinkhole, blackhole, and HELLO flooding attack is larger than 90%.

14.3.2 Signature Based

In Reference 83, an intrusion framework for information sharing is developed which utilizes the hierarchical architecture to improve intrusion detection capability for all participating nodes. In this IDS architecture, every node belongs to a single cluster among the clusters that are geographically distributed across the whole network. The aim is to utilize cluster-based protocols in energy saving, reduced computational resources, and data transmission redundancy. An IDS agent is located in every sensor node. Each sensor node has two intrusion modules, called the local IDS agent and global IDS agent. Owing to the limited battery life and resources, each agent is only active when it is needed. Routing attacks such as selective forwarding, sinkhole, wormhole, HELLO flooding attack, and Sybil attack can be detected by this scheme. According to the simulation results, the probability of detection is close to 1 if the number of monitor nodes exceeds 5, regardless of the high probability of a missed detection. The probability of a false positive indicates that the number of nodes is related to the probability of false detection. Increasing the number of nodes results in an increase in the probability of a collision. Generally, the proposed scheme yields a good detection rate that exceeds 90% when the collision error is 2%–5% and the percentage of malicious nodes is under 5%.

In Reference 84, the desirable properties of a distributed mechanism for the detection of node replication attacks are analyzed. Moreover, a new randomized, efficient, and distributed (RED) protocol for the detection of node replication attacks is proposed. RED achieves a large improvement in terms of communication and computation, is more energy, memory, and computationally efficient, and detects node replication attacks with high probability. Specifically, the detection rate is more than 80%.

An IDS for sensor networks that is able to detect the sinkhole attack is proposed in Reference 85. The IDS has a distributed architecture; it is composed of identical IDS clients running on each node in the network. Krontiris et al. extend the IDS system they presented in Reference 51, so that it can detect sinkhole attacks. Simulation results show that the majority of the watchdogs were able to detect the malicious node. When the network density approaches six neighbors on average in the simulated environment, 75% of the watchdogs will identify the attacker, while for 12 neighbors the percentage increases to 88.3%. For a network density of six neighbors on average, FNR is 11%, while for 10 neighbors it is 5.3%.

The proposal in Reference 86 consists of a centralized IDS which uses multi-level dynamic tree routing. HELLO packets are being sent by the base station to all the nodes, for the purpose of setting the level of each node. Only the nodes which are neighbors of the base station receive these packets; these form the first level. The same process continues so as to set all the levels of the tree. Then the routing paths are created. This scheme can detect both the blackhole attack and the selective forwarding attack. The scheme is found to lead to improved, as compared to alternatives, packet delivery ratios in the presence of attack. However, as the number of malicious nodes increases, the packet delivery ratio decreases. It is also found that the scheme is energy efficient.

A partially distributed IDS is proposed in Reference 87 with low memory and power demands. It employs a bloom filter, which allows a reduced signature code size. A classification method to distribute attack signatures among multiple bloom filter arrays is proposed as well. The proposed method eliminates overhead messages that can account for significant energy consumption. The operation principles of this IDS are as follows: a node can communicate with its neighboring nodes at one hop distance. There are many other nodes that deliver the data packets. A packet may pass through many relay nodes to arrive at its destination node. During this procedure, the relay nodes can detect attack signatures in them using the IDS in the network layer. If the packet is fragmented, the attack signature will be divided into several packets, and relay nodes would not be able to detect attack signatures. According to the simulation results, this has reduced energy requirements, because it eliminates the overhead messages. Detection is better when a priority-based distribution mechanism is applied. An average hop distance of 6 leads to a 95% detection rate of DoS attacks being achieved with four Bloom filter arrays. When two bloom filter arrays are used, over 98% can be achieved with a 6-hop distance. Similarly, a 95% detection rate is achieved in the nonprioritized system using four Bloom filter arrays under similar conditions with an average hop distance of 10.

A cluster-based network topology is presented in Reference 88, where the network is divided into three levels: bottom, middle, and top. The bottom level consists of all sensor nodes that collect data from the environment. The second level is formed by cluster heads, which send periodically a control packet to the base station that represents the third level, which can monitor the other levels. The intrusion detection algorithm is decomposed in three phases: data collection, rules control, and intrusion detection phases. In the first phase, all sensors of the middle level send control packets to the base station. In the second phase, signature rules are applied to all received data. In the third phase, the base station detects an eventual attack based on the previous phase, and an alarm is raised to all sensor nodes. Simulation results analyze the behavior of the proposed IDS under the blackhole and selective forwarding attacks. The proposed IDS consumes 0.04 J, if there are no intruder nodes in the network, which represents 0.02% of the overall power.

The overall power is 0.03% when 10 intruder nodes have been detected in the network.

An approach for intrusion detection that employs the genetic k-means algorithm is presented in Reference 89. This algorithm is applied to differentiate between normal and abnormal intrusion behavior and to update the rule base of the IDS. The system is capable of detecting the blackhole attack with a high detection rate and low FPR.

In Reference 90, a signature-based IDS against the sinkhole attack is designed. It is based on a hierarchical topology to secure cluster-based routing protocols. This architecture allows optimizing energy consumption by reducing communication costs and minimizing the number of nodes that run their IDS. In the detection model, the IDS is activated only when an important event appears; this helps at avoiding wastage of energy. The network area is divided into a flat grid of cells to distinguish between real and fake sink nodes. The proposed IDS considers two types of sink mobility: periodic, at which the sink node calculates its new position, moves to that position and advertises it in the network, and random, at which the number of unnecessary movements made by the sink are minimized. Simulation results show that the energy consumption increases according to the increase in sinkhole attacks and the detection rate is high.

14.3.3 Hybrid

A hierarchical overlay design-based IDS is proposed in References 38 and 39, which concentrates on saving the power of sensor nodes by distributing the responsibility of intrusion detection to four levels of a hierarchy with the help of a policy-based network management system. Moreover, it follows a core defense strategy, whereby the cluster-head is the center point of defense. In this approach, the authors claim that the proposed IDS can identify known as well as unknown attacks, but they do not evaluate their proposed system.

In Reference 92, a hybrid intrusion detection system (HIDS) for heterogeneous cluster-based WSNs (CWSN) has been proposed. One of the Sensor Nodes (SN) in the CWSN serves as the cluster head (CH); this has higher capabilities than other SNs. The proposed system can detect attacks such as spoofed, altered, or replayed routing information, selective forwarding, sinkhole, Sybil, wormhole, DoS, HELLO flooding, and acknowledgment spoofing. The simulation results show that the detection rate is 99.81%, the FPR is 0.57%, and accuracy reaches 99.75%.

In Reference 93, a dynamic model of intrusion detection for WSNs has been proposed. It is a hierarchical model of IDS, based on the clustered network concept.

A hybrid, lightweight IDS for sensor networks is proposed in Reference 94. It takes advantage of a cluster-based protocol to build a hierarchical network and to provide an intrusion framework based both on anomaly and misuse techniques. This scheme can detect most of the routing attacks (selective forwarding, sinkhole, wormhole, HELLO flooding, Sybil). The authors make use of a preset threshold to estimate the probability detection of the proposed scheme. If the threshold is too small, node failure can be easily recognized as a malicious node and it increases the FPR. If the threshold is too high, it is difficult to detect the misbehaving or failed node.

Abnormal node detection in WSNs by dividing the network into a number of pairs is proposed in Reference 95. A pair is formed using an algorithm that considers different attributes of the nodes and produces the initial knowledge base for the pair, as well as for the sensor network group. The nodes in a pair have the same sensing capability and are close to each other. The abnormal node detection algorithm is scheduled to run locally for each pair and centrally for the whole sensor network. The method uses both knowledge-based and signature-based techniques to identify an abnormal node in a pair or in a group of WSNs. No results concerning the attack detection capabilities of the proposed scheme or its performance are provided, and no performance or evaluation metrics are proposed.

A hierarchical cluster-based network topology which is based on cross-layer interaction between the network, Mac, and physical layers is proposed in Reference 96. This topology divides the network into several clusters, and selects as cluster head node the one which has the greatest energy reserves in the cluster. In this proposal, the authors use the cross-layer interaction concept to detect different types of attacks on several layers of the OSI model. Once intrusion is detected, various kinds of actions (like dropping a packet, flagging a neighbor, etc.) can be taken. The system is evaluated to demonstrate its effectiveness in detecting different types of attacks, such as Spoofed, altered & replayed routing information attack, sinkhole attack, and energy exhaustion attack. The experimental simulation shows that the IDS consumes 0.118 J to detect 10 intruder nodes, which represent 0.06% of the overall network power.

In Reference 97, a policy-based IDS for hierarchical architecture is presented. In this architecture, a clustering mechanism is designed to build four-level hierarchical networks which enhance the network scalability to large geographical areas. The policy-based mechanism is a powerful approach to automating network management. The management system for intrusion detection and response system described in this chapter shows that a well-structured reduction in management traffic can be achieved by policy management. However, no experimental performance evaluation results are presented and no attack is mentioned as detectable.

A three-logic-layer architecture of IDS is presented in Reference 98; it employs the agent technology and the concept of the immunity system mechanism. It has two work modes: (a) active work mode to improve the effectiveness and intelligence for unknown attacks and (b) passive work mode to detect and defend known attacks. Three kinds of lightweight agents are designed (monitor agents, decision agents, and defense agents) in order to reduce communication overhead, computation complexity, and memory cost. The experimental results show that the system can detect the flooding, playback, and other resource exhaustion attacks.

HIDS that uses a clustering algorithm to reduce the amount of information and decrease the consumption of energy is introduced in Reference 99. The IDS uses a support vector machine (SVM), which separates data into normal and anomalous in order to detect anomalies. A misuse detection technique to determine known attack patterns is also used, with the focus being only on detecting DoS attacks. The combination of both techniques can achieve a high detection rate with a low FPR and FNR. Specifically, when the number of IDSs and sensor nodes is significant, the detection rate is 99.07% for the centralized approach and 98.39% for the distributed approach.

HIDS proposed in Reference 100 uses an anomaly-based SVM technique and a set of known attacks, which are designed to validate the malicious behavior of a target. The detection approach uses one known node, the cluster head, which forwards node packets to the base station. The proposed cluster-based architecture divides the sensors in several groups, each of which has a cluster head. Then each node belongs to only one of the clusters, which are distributed geographically across the whole network. The role of the cluster head is to reduce the energy consumption of the network as well as to increase its lifetime. Simulation results under four attacks (selective forwarding, HELLO flooding, blackhole, and wormhole) show that the combination of anomaly detection based on SVM and detection based on attack signatures allows the intrusion detection model to achieve a high detection rate (almost 98%) with a number that reduces false alarms (close to 2%).

The work presented in Reference 101 proposes an IDS framework based on multilevel clustering for hierarchical WSNs. In this framework, two types of IDS are proposed: (a) "downwards-IDS" that detects the abnormal behavior of the subordinate nodes and the effect of cluster size on the detection probability of a malicious node was evaluated and (b) "upwards-IDS" that detects the abnormal behavior of the cluster heads and the effect of the total number of monitoring nodes on the detection probability of a malicious cluster head was evaluated.

An IDS algorithm for cluster-based networks with no retransmission mechanism is proposed in Reference 102. During a time interval, each node in a cluster sends its node ID, the number of packets sent, and the number of packets received to the cluster head. When a node ID changes, the cluster head finds out the total packet drop ratio. If this is greater than some threshold, then an alarm is raised. Each node also measures the time interval between two successive receptions. For each node, there is a threshold value set for the energy consumption rate. If the energy consumed in a particular time interval is more than the threshold value, a warning message to the cluster head will be sent. If the cluster head receives the warning message repeatedly, it will raise an alarm. Simulation results examine different scenarios, such as the attack period versus packet dropped and attack period versus packet received. The proposed IDS potentially improves the performance of the network under attack in both cases.

An IDS mechanism for a network using the LEACH protocol for its routing operation is presented in Reference 103. The malicious node launches the attack by advertising that it is the nearest node to the base station to attract the packets, and alters those that pass through it. Detection efficiency metrics, such as the number of packets transmitted and received, are used to compute the intrusion ratio by the IDS agent. When the sinkhole attack is detected, the IDS agent alerts the network to stop the data transmission. The simulation results show that the proposed algorithm detects the sinkhole attack with a high detection rate and that it consumes around 2% less energy compared to MS-LEACH.

An IDS based on mobile agents, which employs classification algorithms (K-means, Naive Bayes, and SVM) to perform intrusion detection in WSNs, is proposed in Reference 104. This IDS is based on multiple mobile agents, such as a collector agent (collects the data from the wireless environment, stores it in a file, and gives it as an input to the misuse detection agent), a misuse detection agent (detects the known attacks in the network), an anomaly detection agent (detects the unknown attacks or intrusions by using the SVM algorithm), and an alert agent (used to alert the system if an attack or intrusion occurs in the network). The experimental results show that the SVM classifier is more efficient than the K-means and the Naive Bayes classifiers, with a classification rate reaching 97.4%.

An intrusion detection technique based on the trust level of neighboring nodes is proposed in Reference 105. A trust manager manages the direct and indirect trust of a node. The behavior of a node is classified as trustworthy or malicious according to the trust values and calculations obtained by the trust manager. In case the node is deemed to be trustworthiness, packet forwarding is allowed. In case of risky behavior, the trust manager informs the forwarding engine accordingly. In case of attack behavior, the attack classifier distinguishes the attack pattern and the observed node is excluded from forwarding. The proposed scheme detects the HELLO flooding attack, the jamming attack, and the selective forwarding attack by analyzing the network statistics and the node behavior. The simulation results show that the network performs better in the presence of the proposed technique and the detection rate is 0.8.

A hybrid clustering method called density-based fuzzy imperialist competitive clustering algorithm (D-FICCA) is introduced in Reference 106. This algorithm identifies data distribution anomaly profiles such as DDoS attacks and is a combination of the imperialist competitive algorithm (ICA) with a density-based algorithm and fuzzy logic for clustering. D-FICCA is evaluated using real measurements; its performance is compared against existing empirical methods, such as K-MICA, K-mean, and DBSCAN. The results show that the proposed framework achieves a detection accuracy of 87% and clustering quality of 0.99.

14.3.4 Specification Based

The IDS proposed in Reference 43 focuses at the sleep deprivation attack on networks using optimized link state routing (OLSR). The OLSR protocol is a routing protocol that uses the stability of a link state algorithm and provides the advantage of having routes immediately available when needed. Experiments on studying the impact of energy efficient OLSR under the sleep deprivation attack in terms of the packet delivery ratio and end-to-end delay metrics show that energy efficient OLSR outperforms traditional OLSR in terms of throughput, packet delivery ratio, end-to-end delay, and average node lifetime.

Islam et al. [107] describe a four-layer architecture for an IDS that uses the specification-based detection technique and constitutes a modified form of the architecture proposed in Reference 41. According to this architecture, no intrusion detection capability is implemented in the leaf level sensors. The functionality of monitor nodes is divided into three phases. In phase 1, all the leaf level sensors collect information from their environments and report it to the level 2 sensors. In phase 2, layer-based attack detection is used to detect masquerade attacks and all layer-based attacks (physical layer, link layer, network layer, and application layer). Phase 3 is used to reduce the false alarm rate. Threshold values can be defined manually or they can be adjusted based on the requirements of a particular WSN. It is claimed that the proposed architecture can reduce the number of false positive alarms, but no implementation is reported and no performance results are provided.

The system proposed in Reference 108 works in a distributed environment to detect intrusions by collaborating with the neighboring nodes. Routing attacks such as selective forwarding, sinkhole, wormhole, HELLO flooding attack, and Sybil attack can be detected by this approach. The scheme works in two modes: (a) online prevention, which allows safeguarding from those abnormal nodes that are already declared as malicious and (b) offline detection, which finds those nodes that are being compromised by an adversary during the next epoch of time. Routing attacks are the focus of this scheme. Simulation results show that the intrusion detection rate is almost 100% and FPR is below 0.06 in most cases.

In the IDS proposed in Reference 109, some nodes are called monitoring nodes, as they observe the whole network. Each monitoring node is located somewhere in the network and it can monitor the neighboring nodes as well as the entire network. The network traffic as well as the set of rules of the neighboring nodes infers the normal and the abnormal behavior of the nodes. The repetition attack is efficiently detected by the proposed IDS and the simulation results show that if there are not enough monitoring nodes, false positives could be generated.

A method called multi-protocol-oriented intrusion detection (MPOID) is proposed in Reference 110. This method can generate all the attack types for any WSN routing protocol. The routing protocol is formally described with the use of the process algebra for wireless mesh networks (AWN) language, and then all the potential attacks are classified into four classes, according to their original purpose (consume the energy, break the original route, prevent the establishment of a new route, and insert the attacker itself to the route). Even though it is claimed that MPOID can detect all types of attacks, no experimental validation of the claim is provided.

A specification-based IDS is proposed in Reference 111, tailored to detecting a variant of the wormhole attack, called camouflaging wormhole attack, in optimized link state routing (OLSR) protocol WSNs. The camouflaging wormhole attack maintains a private tunnel between two nodes and sometimes attacker nodes may drop some packets. Simulation experiments assess the performance of the proposed algorithm in terms of the packet delivery ratio, end-to-end delay, and throughput metrics. Unfortunately, these are minimally affected by the camouflaging wormhole attack, thus making detection difficult.

14.3.5 AIS Based

The objective of Reference 44 is to investigate how the dendritic cell algorithm (DCA) can detect the interest cache poisoning attack against a sensor network. The method is validated using two separate implementations: a simulation using J-sim and an implementation for the T-mote sky sensor using the TinyOS. The attack highlights a general vulnerability in sensor network protocols that rely on caches with limited capacity to keep track of state of the network. The evaluation of this attack showed that it can easily disturb the data delivery from the sensor nodes to the sink node.

In Reference 112 the proposed IDS is based on immunology theory and can detect five types of attack (route loop, jamming attack, sinkhole attack, wormhole attack, and blackhole attack). It is claimed that this IDS achieves a 100% detection rate under all these attacks. Moreover, FPRs of sinkhole and wormhole attacks are below 10%. Route loop has a medium FPR of 20%. The FPRs of blackhole and jamming attacks are 63.2% and 92.3%, respectively.

An IDS framework inspired by the human immune system (HIS) is proposed in Reference 113. It uses a decentralized and customized version of a dendritic cell algorithm, which allows nodes to monitor their neighborhood and to collaborate in order to identify an intruder. Despite reaching lower values of the true positive rate for the denial-of-sleep attack, FPRs were much smaller.

Drozda et al. [114] employed mechanisms based on AISs in order to detect WSN node misbehavior. AIS-based misbehavior detection offers a decent detection performance at a very low computational cost. In this mechanism, the system maintains a list of self-strings (normal behavior) and non–self-strings (misbehavior). No specific attacks that can be detected by the system are listed. The detection rate and FPR are being used as performance metrics.

An IDS inspired by immunology and danger theory is proposed in Reference 115. Danger theory essentially suggests that the immunizing system is not centrally responding to pathogens but it is the result of distributed information gathered from various tissues located throughout the whole body. When the components of the immune system interact with each other and invaders locally provide protection, the system cannot fail in any way. The proposed architecture defines two kinds of agents: (a) static agents that stay fixed in predetermined sensors and (b) mobile agents that are transmitted between sensors, simulating the behavior of biological cells. The agents are used to collect data in various nodes and cooperate with each other in order to detect an attack. DDoS attacks can be detected by this approach. Simulation results indicate that FNR is 40.0% and FPR is 8.23%.

An AIS based solution that employs a negative selection algorithm (NSA) is presented in Reference 116 for anomaly detection in WSNs. Anomalies including sensor network packets dropped, packets delayed, and wormholes are addressed. Simulation experiments show that the detection rate of NSA is 97.3% and FPR is ±2.6%.

The main aim of the work proposed in Reference 117 is to enhance the efficient distributed detection method by employing danger theory to detect the replica node in mobile WSNs. To achieve this, the enhanced efficient distributed detection (EEDD) algorithm is proposed. The advantages of the proposed method include (i) increased detection rate, (ii) decreased overheads, (iii) high packet delivery ratio, and (iv) low energy consumption. The proposed method is tested in a simulated environment and the experimental behavior of the EEDD is compared to that of the existing EDD; the results show that average delay, energy, overhead, and message drops are minimum with a higher packet delivery ratio value and higher detection ratio.

A cooperative multi-agent based, fuzzy AIS (Co-FAIS) to protect against attacks on wireless sensor nodes is proposed in Reference 118. Co-FAIS is implemented in the LEACH protocol and evaluates its performance in terms of recognition and defense accuracy. The defense strategy is adopted whenever a victim node receives a flooding packet beyond a certain alarm event threshold. A Co-FAIS mechanism is applied to reinforce the agent's self-learning abilities and to provide detector players with an incentive function to protect the most vulnerable sensor nodes that represent possible security threats. Experimental results show that the Co-FAIS mechanism preserves true confidence rate 98.53%, FPR 2.51%, and FNR 1.85%.

14.3.6 Trust Based

A hierarchical model based on weighted trust evaluation (WTE) is proposed in Reference 119. The basic idea is that forwarding nodes give trust values to each of the nodes in the cluster; if a node sends meaningless/wrong information (which implies that a node has been compromised or is not functioning), the forwarding node directly lowers that node's trust level. The performance metrics used to evaluate the proposed model are the response time, the detection rate, and the misdetection ratio.

A trust-based intrusion detection scheme is proposed in Reference 120, which utilizes a highly scalable hierarchical trust management protocol for clustered WSNs. A trust metric is considered taking into account QoS trust and social trust, in order to detect malicious nodes. The results collected from sensor nodes are analyzed and each cluster head applies trust-based intrusion detection to assess the trustworthiness and the maliciousness of sensor nodes in its cluster. Selective forwarding, exhaustion, and blackhole attacks can be detected by this approach.

According to the simulation results, when time (in days) is small, FNR is high. As time progresses, FNR probability drops but FPR increases.

A trust-based adaptive acknowledgment (TRAACK) intrusion-detection system for WSN is proposed in Reference 121. This IDS is based on the Kalman filter and predicts node trust. In TRAACK, the entire route is based on trust value and, as a result, an ACK is initiated on selecting packets to decrease control overhead. TRAACK is able to detect malicious nodes and avoid them in the route discovery process. On the basis of the trust value (low, medium, or high) of the entire route, "Adaptive Acknowledgement (AACK)" is initiated on chosen packets to decrease control overhead. Simulations show improved performance in the presence of malicious nodes without compromise in energy. Packet delivery ratio, routing overheads, and end to end delay are evaluated in this approach. TRAACK's routing overhead decreases by 9.41% when malicious nodes are present, the packet delivery ratio increases by 7.86% as compared to AACK, and end to end delay decreases by 18.43% as compared to AACK. The average power consumption (J/s) is lower by 4.78% as compared to AACK.

The IDS proposed in Reference 122 is called distributed trust-based intrusion detection (DTBID). It considers trust, direct trust, recommendation trust, and indirect trust in the process of intrusion detection. This approach not only considers communication behavior to detect the trustworthiness of each sensor node but some other factors of trust as well, such as energy, data trust, reliability, communication trust, etc. The system attempts to decide whether a particular node is malicious or not by comparing the subjective trust to the objective trust, which is calculated based on the actual information of each node. If these two quantities differ significantly, then the sensor node is considered to be malicious.

14.3.7 Game Theory-Based Approaches

In this section, we describe the game theoretical approaches we found in the literature review. These approaches, as mentioned in Section 14.1, have not been used as intrusion detection techniques, but rather as techniques for informing the design of improved IDSs in terms of improved defense strategy.

In Reference 123, the proposed architecture establishes an attack-defend game model, where the strategy space and payoff matrix are given to both the IDS and the malicious nodes. The results show that the average packet loss rate declines by 6% when the IDS runs. The results of simulations show a high detection rate of attacks, even in dense networks with intensive traffic flow.

The approach described in Reference 124 uses a noncooperative game theoretic framework, which can help each cluster head node decide the probability of starting up IDS service. The authors assume that only cluster head nodes run the IDS and the sensors are stationary and homogeneous. Game theory is used to simulate offensive and defensive gambling between the cluster head and the attacker and to predict the probability of the cluster head choosing to start the IDS and of the attacker launching an attack. The simulation results showed that the detection rate is at least 70% when jamming, energy exhaustion, routing, and flooding attacks are considered.

A robust stochastic game framework is proposed in Reference 125. This approach tries to model and analyze the ID problem in WSNs in the presence of uncertainty. The requirements of a practical WSN intrusion detection solution are characterized within the parameters of the framework. Unlike many ID models applicable only to specific WSN settings, this framework is based on the game equilibrium analysis and the expected robust optimal behaviors of rational players are derived and analyzed. These equilibrium behaviors provide insights into effective and efficient IDS design, as they enhance the clarity about the intruder's intent and improve situational awareness. The experimental results indicate that the proposed game model, compared to its nominal counterpart, reduces the sensitivity of the solution to data perturbations, and increases the design's stability.

In Reference 126, a game theoretic method, namely, cooperative game-based Fuzzy Q-learning (G-FQL), which adopts a combination of both the game theoretic approach and the fuzzy Q-learning algorithm to counter the DDoS attack against WSNs, is introduced. The game is a three-player one, consisting of sink nodes, a base station, and an attacker. The LEACH routing protocol is simulated in order to evaluate the performance of the proposed model. The results indicate an improvement in the attack detection rate and defense accuracy as compared to existing machine learning methods.

A repeated game model of dropping packets is presented in Reference 127. The model prevents malicious nodes from attacking by establishing a sub-game perfect periodic collusion-resistant punishment mechanism and impels sensor networks to reach a cooperative Nash equilibrium. When the malicious nodes populate less than the majority of the network nodes, the proposed model is partially collusion resistant. On the basis of rationality, malicious nodes try to impersonate their true type and show a little collaborative behavior. This game model focuses on detecting the blackhole attack. According to the results of the simulations, when one third of the nodes are malicious, the network's throughput drops to about 60%. The proposed approach succeeds in decreasing the average number of dropped packets to 0.0782 per packet. When the number of nodes increases and the percentage of malicious nodes remains invariable, the average number of packets dropped per received packet can stabilize.

In Reference 128, a game theory-based model that prevents passive DoS attacks at the network layer is proposed. The intrusion detector is located in the base station and monitors the cooperation among the nodes. If the performance of some nodes is lower than a predefined threshold, then these nodes are considered as malicious. The number of hops for received packets as well as the throughput are mentioned as metrics of trustworthiness of the nodes.

14.3.8 Experimental Performance Evaluation Setups

The performance of several, but not all, IDS proposals that we have reviewed has been experimentally assessed via simulations. Among those experimental setups mentioned in the literature, the most frequently used network simulator environment is Network Simulator 2 (ns-2) [37,56,93,96,117], an event simulator targeted at networking research. The ns-2 simulator is used to implement a sensor network that in most cases uses the AODV routing protocol [37]; many other routing protocols may be in use in a WSN (e.g., DSR; destination-sequenced distance-vector (DSDV) routing protocol; temporally ordered routing algorithm (TORA); and geographic multicast routing (GMR) protocol). The number of sensor nodes and the rectangular area used in the simulation environment varies but generally lies between 100 and 1000 nodes and 50 to 1200 m^2, respectively [51,88,98,123]. The network density is mostly chosen so that each node has a specific number of neighbors around it. The channel type, the radio-propagation model, the network interface type, the traffic model, the simulation time, the simulation area, and the transmission rate are some of the control parameters which are specified by the user and a set of output parameters [56,69,85,88].

Table 14.2 summarizes the classification of the reviewed IDSs.

14.4 Conclusion

We reviewed more than 90 approaches for intrusion detection in WSNs and classified them according to a taxonomy specially tailored to characteristics of IDSs used within the WSN context. On the basis of the findings of this survey, the following areas seem to be promising for further research in the area:

1. Powerful machine learning techniques that have been used for anomaly detection in other contexts have not been adequately researched within the context of IDS in WSNs. This may largely be due to the computational requirements of such techniques in relation to the very limited processing capabilities of WSN environments. Thus, approaches targeting to reduce the complexity of machine learning techniques, perhaps also utilizing the massively collaborative computing capability of WSN nodes, constitute an area of interesting and challenging research.

2. WSN IDSs, as all IDSs suffer from FPR curse. Several techniques have been employed in intrusion detection in other contexts to alleviate this problem, including, for example, post processing of alarms and alarm correlation. However, very little if at all consideration has been given to this problem within WSN IDSs.

3. A small portion of the attacks that have been described in the literature can be detected by WSN IDSs. Some have not been considered at all (monitor

Table 14.2 IDSs for WSNs

IDS	Year of Publication	Architecture	Detection Technique	Attack Type	Input	Performance Metrics
Adaptive Mechanism in Homogenous Clustered Sensor Networks [35,36]	2015	Distributed	Anomaly based	NL1, NL24	■ Energy consumption characteristics	■ Detection effectiveness
Intrusion Detection for Routing Attacks [37]	2006	Distributed	Anomaly based	NL7, NL14	■ Traffic characteristics	■ Detection effectiveness ■ Energy consumption efficiency
Isolation Table-Based IDS [39,40]	2009 & 2010	Hierarchical	Anomaly based	NL7, NL9, NL11, NL20, ML3	■ Energy consumption characteristics ■ Traffic characteristics	■ Network performance characteristics ■ Energy consumption efficiency
Decentralized ID in WSNs [41]	2005	Distributed	Anomaly based	PL2, NL2, NL3, NL4, NL9, NL22, NL23	■ Energy consumption characteristics ■ Traffic characteristics	■ Detection effectiveness

(Continued)

Table 14.2 (Continued)   IDSs for WSNs

IDS	Year of Publication	Architecture	Detection Technique	Attack Type	Input	Performance Metrics
Hierarchical Trust Management for WSNs [42]	2012	Hierarchical	Anomaly based	NL3, NL7, NL24	■ Energy ■ Node behavior characteristics	■ Detection effectiveness
IDS for Power-Aware OLSR [43]	2015	Distributed	Specification based	NL25	■ Energy consumption characteristics	■ Network performance characteristics
Cache Poisoning in Sensor Networks [44]	2010	Distributed	AIS based	NL26	■ Packets	■ Network performance characteristics
Group-Based IDS in WSNs [45]	2008	Distributed	Anomaly based	ML2, DL2, NL2, NL3, NL7, NL9, NL11	■ Traffic characteristics	■ Detection effectiveness
Ranger IDS for WSNs [46]	2010	Distributed	Anomaly based	NL8	■ Traffic characteristics	■ Energy consumption efficiency

(Continued)

Table 14.2 (Continued) IDSs for WSNs

IDS	Year of Publication	Architecture	Detection Technique	Attack Type	Input	Performance Metrics
Multi-Agent and Refined Clustering Method [47]	2009	Distributed	Anomaly based	NL1, AL1, AL4	▪ Traffic characteristics ▪ Energy consumption characteristics	▪ Detection effectiveness
ID Based on Traffic Analysis [48]	2010	Distributed	Anomaly based	ML3	▪ Traffic characteristics	▪ Detection effectiveness
Self-Organized-Based IDS for WSNs [49]	2003	Distributed	Anomaly based	ML5	▪ Application layer data	▪ Detection effectiveness ▪ Energy consumption efficiency
Applying IDSs to WSNs [50]	2006	Distributed	Anomaly based	NL25	▪ Traffic characteristics	▪ No performance evaluation
Toward ID in WSNs [51]	2007	Distributed	Anomaly based	NL2, NL3	▪ Traffic characteristics	▪ Detection effectiveness ▪ Network performance characteristics
ANDES [52]	2007	Centralized	Anomaly based	NL2, NL3, NL7, TL1	▪ Application layer data ▪ Network topology ▪ Traffic characteristics	▪ Detection effectiveness ▪ Communication overhead

(Continued)

Table 14.2 (Continued) IDSs for WSNs

IDS	Year of Publication	Architecture	Detection Technique	Attack Type	Input	Performance Metrics
A Framework for Identifying Compromised Nodes [53]	2008	Centralized	Anomaly based	PL5	■ Traffic characteristics	■ Detection effectiveness
An Energy Prediction Approach [54]	2012	Distributed	Anomaly based	NL2, NL7, NL8, NL9, NL11	■ Energy consumption characteristics	■ Detection effectiveness ■ Network performance characteristics
Distributed Detection of Wormhole Attacks [55]	2010	Distributed & Cooperative	Anomaly based	NL9	■ Traffic characteristics	■ Detection effectiveness ■ Communication overhead
Traffic Prediction Model [56]	2012	Distributed	Anomaly based	ML3	■ Traffic characteristics	■ Detection effectiveness ■ Network performance characteristics
Lightweight Anomaly ID in WSNs [57]	2007	Distributed	Anomaly based	PL2, PL5, DL1, DL2, NL2, NL7, NL9, NL8, NL10, NL11, TL1, TL2	■ Energy consumption characteristics ■ Traffic characteristics ■ Network structure ■ Packets	■ Detection effectiveness

(Continued)

Table 14.2 (Continued)   IDSs for WSNs

IDS	Year of Publication	Architecture	Detection Technique	Attack Type	Input	Performance Metrics
Mobile Agent-Based Hierarchical IDS [58]	2012	Hierarchical	Anomaly based		■ Packets	■ No performance evaluation
Neighbor-Based ID for WSNs [59]	2010	Distributed	Anomaly based	PL2, NL2, NL11	■ Packets	■ Detection effectiveness
A Full Approach for ID in WSNs [60]	2007	Hierarchical	Anomaly based	NL1	■ Traffic characteristics	■ No performance evaluation
An Anomaly Detection Scheme [61]	2009	Distributed	Anomaly based	NL1	■ Traffic characteristics	■ Detection effectiveness ■ Network performance characteristics
Energy Efficient Learning Solution [62]	2010	Distributed	Anomaly based		■ Energy consumption characteristics ■ Traffic characteristics	■ Detection effectiveness
Fuzzy C-means Clustering Approach [63]	2009	Distributed	Anomaly based	NL7, NL14, NL16, TL1	■ Traffic characteristics	■ Detection effectiveness

(Continued)

Table 14.2 (Continued)   IDSs for WSNs

IDS	Year of Publication	Architecture	Detection Technique	Attack Type	Input	Performance Metrics
Hierarchical Energy Efficient IDS for Blackhole Attacks [64]	2013	Hierarchical	Anomaly based	NL2, NL3	■ Energy consumption characteristics ■ Traffic characteristics	■ Detection effectiveness
KNN-Based Approach [65]	2014	Distributed	Anomaly based	TL1	■ Traffic characteristics	■ Detection effectiveness
Energy Consumption-Based IDS for WSNs [66,78]	2013&2015	Distributed	Anomaly based	NL3, TL1	■ Energy consumption characteristics	■ Detection effectiveness
An IDS for Cluster-Based WSNs [67]	2013	Distributed	Anomaly based	TL1, ML3	■ Energy consumption characteristics ■ Traffic characteristics	■ Detection effectiveness
Detecting Selective Forwarding Attacks Using SVM [68]	2007	Centralized	Anomaly based	NL2, NL3	■ Traffic characteristics	■ Detection effectiveness

(Continued)

Table 14.2 (Continued)   IDSs for WSNs

IDS	Year of Publication	Architecture	Detection Technique	Attack Type	Input	Performance Metrics
A Leader-Based IDS in Heterogeneous WSNs [69]	2015	Distributed	Anomaly based	ML3	■ Packets	■ Detection effectiveness ■ Network performance characteristics
A Novel Anomaly Detection Algorithm for WSNS [70]	2015	Distributed	Anomaly based	NL2, NL8, NL3, NL9	■ Traffic characteristics	■ Detection effectiveness ■ Communication overhead ■ Energy consumption efficiency
A Sybil Node Detection Algorithm for IDSs in WSNs [71]	2014	Centralized	Anomaly based	NL8	■ Traffic characteristics ■ Energy consumption characteristics	■ Detection effectiveness
Detecting Malicious Nodes in WSNs [72]	2015	Hierarchical	Anomaly based	PL3	■ Traffic characteristics	■ No performance evaluation
An Improvised Hierarchical Blackhole Detection Algorithm in WSNs [73]	2015	Hierarchical	Anomaly based	NL3	■ Packets ■ Network topology	■ Energy consumption efficiency

(Continued)

Table 14.2 (Continued)   IDSs for WSNs

IDS	Year of Publication	Architecture	Detection Technique	Attack Type	Input	Performance Metrics
Detecting and Revocation the Compromised Node in Zone-Based WSN [74]	2014	Distributed	Anomaly based	ML5	■ Packets	■ Network performance characteristics ■ Detection effectiveness ■ Energy consumption efficiency
Detecting Sinkhole Attack using Enhanced Particle Swarm Optimization Technique [75]	2016	Distributed	Anomaly based	NL7	■ Traffic characteristics	■ Detection effectiveness ■ Network performance characteristics
Bayesian Energy Prediction Approach [76]	2015	Distributed	Anomaly based	NL2, NL7, NL8, NL11	■ Energy consumption characteristics	■ Network performance characteristics ■ Energy consumption efficiency

(Continued)

Table 14.2 (Continued)   IDSs for WSNs

IDS	Year of Publication	Architecture	Detection Technique	Attack Type	Input	Performance Metrics
Genetic Algorithm-Based IDS for Wireless Body Area Networks [77]	2015	Distributed	Anomaly based	PL2	■ Traffic characteristics	■ Detection effectiveness ■ Computational overhead ■ Energy consumption efficiency
Localized DoS Attack Detection Architecture [79]	2016	Distributed	Anomaly based	NL2, NL11	■ Traffic characteristics	■ Detection effeteness ■ Network performance
Secure Data Aggregation and Intrusion Detection in WSNs [80]	2015	Distributed	Anomaly based	ML5	■ Packets ■ Application layer data ■ Energy consumption characteristics	■ Energy consumption efficiency
Adaptive Anomaly Detection for WSNs [81]	2014	Distributed	Anomaly based	ML5	■ Application layer data	■ Detection effectiveness ■ Energy consumption efficiency

(Continued)

Table 14.2 (Continued) IDSs for WSNs

IDS	Year of Publication	Architecture	Detection Technique	Attack Type	Input	Performance Metrics
Patrol IDS for WSNs [82]	2014	Distributed	Anomaly based	NL3, NL8, NL7, NL11	■ Packets	■ Detection effectiveness ■ Energy consumption efficiency
A lightweight ID framework [83]	2010	Hierarchical	Signature based	NL2, NL7, NL8, NL9, NL11	■ Traffic characteristics	■ Detection effectiveness ■ Energy consumption efficiency
Distributed Detection of Clone Attacks in WSNS [84]	2011	Distributed	Signature based	NL13	■ Packets	■ Detection effectiveness ■ Computational overhead ■ Energy consumption efficiency ■ Storage overhead
ID of Sinkhole Attacks in WSNs [85]	2007	Distributed	Signature based	NL7	■ Traffic characteristics	■ Detection effectiveness
Selective Forwarding-Based IDS for Secure WSN [86]	2013	Centralized	Signature based	NL2, NL3	■ Energy consumption characteristics ■ Traffic characteristics	■ Detection effectiveness

(Continued)

Table 14.2 (*Continued*)   IDSs for WSNs

IDS	Year of Publication	Architecture	Detection Technique	Attack Type	Input	Performance Metrics
Partially Distributed IDS for WSNs [87]	2013	Distributed	Signature based	ML3	■ Traffic characteristics	■ Detection effectiveness ■ Storage overhead ■ Energy consumption efficiency
Centralized IDS based on Misuse Detection [88]	2015	Centralized	Signature based	NL2, NL3	■ Traffic characteristics	■ Network performance characteristics ■ Energy consumption efficiency ■ Detection effectiveness
ID in WSN using Genetic K-Means Algorithm [89]	2014	Distributed	Signature based	NL3	■ Packets	■ Detection effectiveness
IDS against Sinkhole Attack in WSNs with Mobile Sink [90]	2015	Hierarchical	Signature based	NL7	■ Traffic characteristics	■ Detection effectiveness ■ Energy consumption efficiency

(*Continued*)

Table 14.2 (Continued)   IDSs for WSNs

IDS	Year of Publication	Architecture	Detection Technique	Attack Type	Input	Performance Metrics
Hierarchical Design-Based IDS for WSN [38,91]	2008&2010	Hierarchical	Hybrid	PL2, PL3, DL1, NL20	■ Traffic characteristics	■ No performance evaluation
Hybrid IDS of Cluster-Based WSNs [92]	2009	Hierarchical	Hybrid	NL1, NL2, NL7, NL8, NL9, NL10, NL11, ML3	■ Packets	■ Detection effectiveness ■ Energy consumption efficiency
A Dynamic Model of IDS in WSNs [93]	2008	Hierarchical	Hybrid	ML5	■ Energy consumption characteristics ■ Traffic characteristics	■ Network performance characteristics ■ Energy consumption efficiency
Hybrid IDS for WSNs [94]	2007	Hierarchical	Hybrid	NL1	■ Packets	■ Detection effectiveness
Pair-Based Approach [95]	2008	Distributed	Hybrid		■ Traffic characteristics ■ Energy consumption characteristics	■ No performance evaluation

(Continued)

Table 14.2 (Continued) IDSs for WSNs

IDS	Year of Publication	Architecture	Detection Technique	Attack Type	Input	Performance Metrics
Cross-Layer IDS for WSNs [96]	2012	Hierarchical	Hybrid	DL2, NL1, NL7	■ Traffic characteristics	■ Detection effectiveness
Policy-Based approach in WSNs [97]	2012	Distributed	Hybrid		■ Traffic characteristics	■ No performance evaluation
SAID [98]	2006	Distributed	Hybrid	DL2	■ Traffic characteristics	■ Detection effectiveness ■ Communication overhead ■ Network performance characteristics ■ Energy consumption efficiency
Novel Hybrid IDS for Clustered WSN [99]	2011	Distributed	Hybrid	ML3	■ Packets	■ Detection effectiveness ■ Energy consumption efficiency

(Continued)

Table 14.2 (Continued)   IDSs for WSNs

IDS	Year of Publication	Architecture	Detection Technique	Attack Type	Input	Performance Metrics
A Global Hybrid IDS for WSNs [100]	2015	Distributed	Hybrid	NL2, NL3, NL9, NL11	■ Traffic characteristics ■ Packets	■ Detection effectiveness ■ Energy consumption efficiency
Multi-Level Clustering for Hierarchical WSNs [101]	2015	Hierarchical	Hybrid	ML3	■ Traffic characteristics	■ Detection effectiveness
Intelligent IDS in WSNs [102]	2014	Distributed	Hybrid	DL2	■ Packets ■ Traffic characteristics	■ Network performance characteristics
ID Algorithm on LEACH Protocol [103]	2015	Distributed	Hybrid	NL7	■ Packets	■ Detection effectiveness ■ Network performance characteristics ■ Energy consumption efficiency
IDS in WSN Based on Mobile Agent [104]	2014	Distributed	Hybrid		■ Traffic characteristics	■ Detection effectiveness

(Continued)

Table 14.2 (Continued)   IDSs for WSNs

IDS	Year of Publication	Architecture	Detection Technique	Attack Type	Input	Performance Metrics
Neighbor Node-Based IDS for WSN [105]	2015	Distributed	Hybrid	PL2, NL2, NL11	■ Traffic characteristics	■ Detection effectiveness
D-FICCA [106]	2014	Distributed	Hybrid	ML3	■ Application layer data ■ Network topology	■ Detection effectiveness
A Hierarchical IDS in WSNs [107]	2010	Hierarchical	Specification based	ML2	■ Traffic characteristics	■ Detection effectiveness
A Novel ID Framework for WSNs [108]	2013	Distributed	Specification based	NL2, NL7, NL8, NL9, NL11	■ Traffic characteristics ■ Packets	■ Detection effectiveness
A Collaborative Approach for IDS on WSNs [109]	2010	Distributed	Specification based	NL22	■ Energy consumption characteristics ■ Traffic characteristics	■ Detection effectiveness
MPOID [110]	2015	Distributed	Specification based	PL4, NL2, NL3, ML1, ML3	■ Packets	■ No performance evaluation

(Continued)

Table 14.2 (Continued)   IDSs for WSNs

IDS	Year of Publication	Architecture	Detection Technique	Attack Type	Input	Performance Metrics
IDS for Camouflaging Wormhole Attack in OLSR [111]	2015	Distributed	Specification based	NL9	■ Traffic characteristics	■ Network performance characteristics
Immunity-Based IDS [112]	2008	Distributed	AIS based	PL2, NL3, NL7, NL9, NL12	■ Traffic characteristics	■ Detection effectiveness
Danger Theory Immune-Inspired Technique [113]	2012	Distributed	AIS based	NL20	■ Traffic characteristics	■ Detection effectiveness ■ Energy consumption efficiency
AIS-Based Misbehavior Detection [114]	2007	Distributed	AIS based		■ Traffic characteristics	■ Detection effectiveness
A DDoS−Aware IDS Model [115]	2009	Distributed	AIS based	ML4	■ Traffic characteristics	■ Number of activated agents
Anomaly Detection Using Immune-Based Bioinspired Mechanism [116]	2015	Distributed	AIS based	NL9	■ Traffic characteristics	■ Detection effectiveness

(Continued)

Table 14.2 (Continued)   IDSs for WSNs

IDS	Year of Publication	Architecture	Detection Technique	Attack Type	Input	Performance Metrics
Clone Detection Using EEDD [117]	2015	Hierarchical	AIS based	NL13	■ Traffic characteristics	■ Detection effectiveness ■ Network performance characteristics ■ Routing overhead ■ Node localization error rate ■ Energy consumption efficiency
Co-FAIS [118]	2014	Distributed	AIS based	ML4	■ Traffic characteristics ■ Energy consumption characteristics	■ Detection effectiveness
Weighted Trust Evaluation Method [119]	2008	Hierarchical	Trust based	ML5	■ Application layer data	■ Detection effectiveness

(Continued)

Table 14.2 (*Continued*)   IDSs for WSNs

IDS	Year of Publication	Architecture	Detection Technique	Attack Type	Input	Performance Metrics
Trusted-Based ID in WSNs [120]	2011	Hierarchical	Trust based	DL2, NL2, NL3	■ Energy consumption characteristics ■ Node behavior characteristics	■ Detection effectiveness
Trusted-Based IDS with Adaptive Acknowledgment for WSNs [121]	2016	Distributed	Trust based	NL3	■ Node behavior characteristics	■ Network performance characteristics ■ Routing overhead ■ Energy consumption efficiency
Distributed Trust-Based ID Approach in WSNs [122]	2015	Distributed	Trust based	ML5	■ Node behavior characteristics	■ Detection effectiveness
Payoff Matrix approach [123]	2009	Hierarchical	Game theory based	NL1, NL2, NL7, NL8, NL9	■ Traffic characteristics	■ Detection effectiveness ■ Network performance characteristics

(Continued)

Table 14.2 (Continued)   IDSs for WSNs

IDS	Year of Publication	Architecture	Detection Technique	Attack Type	Input	Performance Metrics
ID Method Based on Game Theory in WSNs [124]	2008	Distributed	Game theory based	PL2, DL2, NL1, TL1	■ Energy consumption characteristics	■ Detection effectiveness
Game-Theoretic Framework for Robust Optimal ID in WSNs [125]	2014	Distributed	Game theory based	ML5	■ Traffic characteristics	■ Detection effectiveness
Cooperative Game Theoretic Approach [126]	2014	Distributed	Game theory based	ML3	■ Traffic characteristics	■ Detection effectiveness ■ Energy consumption efficiency
Game-Theoretical Model in WSNs [127]	2010	Hierarchical	Game theory based	ML5	■ Energy consumption characteristics ■ Traffic characteristics	■ Network performance characteristics
A Repeated Game Theory Approach [128]	2007	Centralized	Game theory based	ML3	■ Node behavior characteristics ■ Energy consumption characteristics	■ Detection effectiveness ■ Network performance characteristics

and eavesdropping attack, unfairness attack, traffic analysis attack, node malfunction attack, pseudorandom number attack, digital signature attack, hash collision attack, node outage attack, integration attack, camouflaged adversary attack, homing attack, misdirection attack, network partition attack, simple target flooding attack, false identity broadcast flooding attack, false identity target flooding attack, false node attack, overwhelming attack, deluge attack, and message corruption attack). Others have received minimal attention (tampering attack, passive information gathering attack, physical attack, collision attack, acknowledgment spoofing attack, routing loop attack, node replication attack, periodic route error attack, simple broadcast flooding attack, slandering attack, sleep deprivation attack, interest cache poisoning attack, desynchronization attack, path-based DoS attack, node subversion attack, man-in-the-middle attack, masquerade attack, and DDoS attack); and others only limited attention (neglect and greed attack, denial of sleep attack, repetition attack, and message delay attack). Only a few attacks (jamming attack, energy exhaustion attack, routing attack, selective forwarding attack, blackhole attack, sinkhole attack, wormhole attack, HELLO flooding attack, DoS attack, and malicious node attack) have attracted considerable interest. Furthermore, the majority of WSN IDSs have been designed to detect one or at best a few attacks, with a limited number claiming capability of detection of multiple attacks and very few having demonstrated multiple attack capability. The design of WSN IDSs with demonstrable multi-attack detection capability therefore remains a promising area of research.

4. Research into issues related to the evaluation of performance of WSN IDSs seems to be needed and promising. Indeed, no comprehensive studies comparing WSN IDS approaches with regard to their performance are available. Moreover, whereas many performance evaluation metrics have been proposed, their relevance and possible cross-correlations have not been assessed. Furthermore, no widely used and publicly available test bed platform and/or synthetic dataset that can be used for "standardized" WSN IDS performance evaluation against set benchmarks has been made available in the literature. Last, save approaches relying on application layer data, no WSN IDS approach has been evaluated against real-world WSN data, as such datasets are not publicly available.

References

1. D. Denning, An intrusion—detection model, *IEEE Transactions on Software Engineering*, SE-13(2): 222–232, 1987, doi: 10.1109/TSE.1987.232894.
2. H. Debar, M. Dacier, and A. Wespi, Towards a taxonomy of intrusion detection systems, *Computer Networks*, 31(9): 805–822, 1999.

3. E. Darra and S. Katsikas, Attack detection capabilities of intrusion detection systems for wireless sensor networks, in *4th International Conference on Information, Intelligence, Systems and Applications (IISA)*, Piraeus, Greece, pp. 1–7, 2013.

4. H. Alsafi and S. Basamh, A review of intrusion detection system schemes in wireless sensor network, *Journal of Emerging Trends in Computing and Information Sciences*, 4(9): 688–697, 2013.

5. K. Goyal, N. Gupta, and K. Singh, A survey on intrusion detection in wireless sensor networks, *International Journal of Scientific Research Engineering & Technology*, 2(2): 113–126, 2013.

6. O. Can and O. Sahingoz, A survey of intrusion detection systems in wireless sensor networks, in *6th International Conference on Modeling, Simulation and Applied Optimization (ICMSAO)*, Istanbul, Turkey, pp. 1–6, 2015.

7. T. Bhattasali and R. Chaki, A survey of recent intrusion detection systems for wireless sensor network, in *Advances in Network Security and Applications*, India, Springer CCIS, pp. 268–280, 2011.

8. M. Gupta, G. Sharma, and R. Signh Jadon, A review on intrusion detection approaches in wireless sensor network, in *National Conference on Security Issues in Network Technologies*, Banmore, Gwalior, 2012.

9. Z. Li and G. Gong, http://cacr.uwaterloo.ca/techreports/2008/cacr2008-20.pdf, Online. Available: http://cacr.uwaterloo.ca/techreports/2008/cacr2008-20.pdf. Accessed Aug 21, 2016.

10. A. Farooqi and F. Khan, Intrusion detection systems for wireless sensor networks: A survey, in *Communication and Networking*, Springer CCIS, Korea, pp. 234–241, 2009.

11. I. Butun, S. Morgera, and R. Sankar, A survey of intrusion detection systems in wireless sensor networks, *IEEE Communications Surveys & Tutorials*, 16(1): 266–282, 2014.

12. N. Alrajeh, S. Khan, and B. Shams, Intrusion detection systems in wireless sensor network: A review, *International Journal of Distributed Sensor Networks*, 9(5): 7, 2013.

13. H. Soliman, N. Hikal, and N. Sakr, A comparative performance evaluation of intrusion detection techniques for hierarchical wireless sensor networks, *Egyptian Informatics Journal*, 13: 225–238, 2012.

14. M. Rassam, M. Maarof, and A. Zainal, A survey of intrusion detection schemes in wireless sensor networks, *American Journal of Applied Sciences*, 9(10): 1636–1652, 2012.

15. M. Singh, K. Babbar, and K. Jain, A survey on intrusion detection system in wirelsess sensor networks, *International Journal of Wireless Communications and Networking Technologies*, 3(3): 40–43, 2014.

16. R. Mitchell and I.-R. Chen, A survey of intrusion detection in wireless network application, *Computer Communications*, 42: 1–23, 2014.

17. A. Patel, M. Taghavi, K. Bakhtiyari, and J. Junior, An intrusion detection and prevention system in cloud computing: A systematic review, *Journal of Network and Computer Applications*, 36: 25–41, 2013.

18. V. Chandola, A. Banerjee, and V. Kumar, Anomaly detection: A survey, *ACM Computing Surveys*, 41(15): 1–58, 2009.

19. M. Bhuyan, D. K. Bhattacharyya, and J. K. Kalita, Network anomaly detection: Methods, systems and tools, *IEEE Communications Surveys & Tutorials*, 16(1): 303–336, 2014.

20. N. Stakhanova, S. Basu, and J. Wong, On the symbiosis of specification-based and anomaly based detection, *Computers and Security*, 29: 253–268, 2010.

21. D. Padmavathi and M. Shanmugapriya, A survey of attacks, security mechanisms and challenges in wireless sensor networks, *International Journal of Computer Science and Information Security (IJCSIS)*, 4(1&2): 1–9, 2009.

22. M. Ahmed, X. Huang, and D. Sharma, A taxonomy of internal attacks in wireless sensor network, *International Journal of Electrical, Computer, Energetic, Electronic and Communication Engineering*, 6(2): 203–206, 2012.

23. R. Dubey, V. Jain, R. Thakur, and S. Choubey, Attacks in wireless sensor networks, *International Journal of Scientific & Engineering Research*, 3(3): 1–4, 2012.

24. T. Lupu, Main types of attacks in wireless sensor networks, in *International Conference in Recent Advances in Signals and Systems*, Wisconsin, USA, pp. 180–185, 2009.

25. C. Karlof and D. Wagner, Secure routing in wireless sensor networks: Attacks and countermeasures, *Ad Hoc Networks Journal*, 1: 293–315, 2003.

26. A. S. Pathan and C. Hong, Security attacks and challenges in wireless sensor networks, in *Encyclopedia on Ad Hoc and Ubiquitous Computing: Theory and Design of Wireless Ad Hoc, Sensor, and Mesh Networks*, World Scientific Publishing, Singapore, pp. 397–426, 2009.

27. A. Becher, Z. Benenson, and M. Dornseif, Tampering with motes: Real-world physical attacks on wireless sensor networks, in *3rd International Conference on Security in Pervasive Computing (SPC)*, York, UK, pp. 104–118, 2006.

28. Z. Benenson, P. Cholewinski, and F. Freiling, Vulnerabilities and attacks in wireless sensor networks, in *Wireless Sensors Networks Security, Cryptology & Information Security Series (CIS)*, IOS Press, Amsterdam, pp. 22–43, 2008.

29. K. Sharma and M. Ghose, Wireless sensor networks: An overview on its security threats, *International Journal of Computer Applications (IJCA)*, 1(8): 42–45, 2010.

30. H. Chaudhari and L. Kadam, Wireless ensor networks: Security, attacks and challenges, *International Journal of Networking*, 1(1): 4–16, 2011.

31. W. Zhu, J. Zhou, R. Deng, and F. Bao, Detecting node replication attacks in wireless sensor networks: A survey, *Journal of Network and Computer Applications*, 35: 1022–1034, 2012.

32. T. Zia and A. Zomaya, Security issues in wireless sensor networks, in *International Conference on Systems and Networks Communications (ICSNC '06)*, Tahiti, p. 40, 2006.

33. Y. Wang, G. Attebury, and B. Ramamurthy, A survey of secuirty issues in wireless sensor networks, *IEEE Communication Surveys and Tutorials*, 8: 2–23, 2006.

34. D. Xiao, C. Chen, and G. Chen, Intrusion detection based security architecture for wireless sensor networks, in *IEEE International Symposium on Communications and Information Technology (ISCIT)*, Beijing, China, pp. 1412–1415, 2005.

35. H. Al-Hamadi and I.-R. Chen, Adaptive network defense management for countering smart attack and selective capture in wireless sensor networks, *IEEE Transactions On Network And Service Management*, 12(3): 451–466, 2015.

36. H. Al-Hamadi and I.-R. Chen, Integrated intrusion detection and tolerance in homogeneous clustered sensor networks, *ACM Transactions on Sensor Networks*, 11(3): 2015.

37. C. Loo, M. Ng, C. Leckie, and M. Palaniswami, Intrusion detection for routing attacks in sensor networks, *International Journal of Distributed Sensor Networks*, 2(4): 313–332, 2006.

38. M. Mamun and A. Sultanul Kabir, Hierarchical design based intrusion detection system for wireless ad hoc sensor network, *International Journal of network Security & Its Applications*, 2(3): 102–117, 2010.

39. R.-C. Chen, C.-F. Hsien, and Y.-F. Huang, An isolation intrusion detection system for hierarchical wireless sensor network, *Journal of Networks*, 5(3): 335–342, 2010.

40. R.-C. Chen, C.-F. Hsien, and Y.-F. Huang, A new method for intrusion detection on hierarchical wireless sensor networks, in *3rd International Conference on Ubiquitous Information Management and Communication (ICUIMC)*, Suwon, Korea, pp. 238–245, 2009.

41. A. Da Silva, M. Martins, B. Rocha, A. Loureiro, L. Ruiz, and H. Wong, Decentralized intrusion detection in wireless sensor networks, in *1st ACM International Workshop on Quality of Service & Security in Wireless and Mobile Networks (Q2SWinet)*, Montreal, Quebec, Canada, pp. 16–23, 2005.

42. F. Bao, I.-R. Chen, M. Chang, and J.-H. Cho, Hierarchical trust management for wireless sensor networks and its applications to trust-based routing and intrusion detection, *IEEE Transactions on Network and Service Management*, 9(2): 169–183, 2012.

43. C. Dutta and U. Biswas, Intrusion detection system for power-aware OLSR, in *International Conference on Computational Intelligence & Network (ICRCICN)*, Kolkata, India, 2015.

44. C. Wallenta, J. Kim, P. Bentley, and S. Hailes, Detecting interest cache poisoning in sensor networks using an artificial immune system, *Applied Intelligence*, 32: 1–26, 2010.

45. G. Li, J. He, and Y. Fu, Group-based intrusion detection system in wireless sensor networks, *Computer Communications*, 32(18): 4324–4332, 2008.

46. R.-C. Chen, Y. Huang, and C.-F. Hsieh, Ranger intrusion detection system for wireless sensor networks with Sybil attack based on ontology, in *Proceedings of the 10th WSEAS International Conference on Applied Informatics and Communications, and 3rd WSEAS International Conference on Biomedical Electronics and Biomedical Informatics*, Taipei, Taiwan, pp. 176–180, 2010.

47. W. Huai-Bin, Y. Zheng, and W. Chun-Dong, Intrusion detection for wireless sensor networks based on multi-agent and refined clustering, in *WRI International Conference on Communications and Mobile Computing*, Yunnan, China, pp. 450–454, 2009.

48. Y. Ponomarchuk and D. Seo, Intrusion detection based on traffic analysis in wireless sensor networks, in *19th Annual Wireless and Optical Communications Conference (WOCC)*, Shanghai, China, 2010.

49. S. Doumit and D. Agrawal, Self-organized criticality and stochastic learning based intrusion detection system for wireless sensor networks, in *Military Communications Conference*, Boston, MA, pp. 609–614, 2003.

50. R. Roman, J. Zhou, and J. Lopez, Applying intrusion detection systems to wireless sensor networks, in *IEEE Consumer Communications & Networking Conference (CCNC)*, Las Vegas, USA, pp. 640–644, 2006.

51. I. Krontiris, T. Dimitriou, and F. Freiling, Towards Intrusion Detection in Wireless Sensor Networks, in *13th European Wireless Conference*, Paris, France, pp. 1–10, 2007.

52. S. Gupta, R. Zheng, and A. Cheng, ANDES: An anomaly detection system for wireless sensor networks, in *IEEE International Conference on Mobile Ad hoc and Sensor Systems*, Pisa, Italy, pp. 1–9, 2007.

53. Q. Zhang, T. Yu, and P. Ning, A framework for identifying compromised nodes in WSNs, *ACM Transaction Information System Security (TISSEC)*, 11(3): 1–35, 2008.

54. W. Shen, G. Han, L. Shu, J. Rodrigues, and N. Chilamkurti, A new energy prediction approach for intrusion detection in cluster-based wireless sensor networks, in *The 2nd International Conference on Green Communications and Networking*, Gandia, Spain, pp. 1–12, 2012.

55. R. Graaf, I. Hegazy, J. Horton, and R. Safavi-Naini, Distributed detection of wormhole attacks in wireless sensor networks, *Ad Hoc Networks, LNCS*, 28(1): 208–223, 2010.

56. H. Zhijie and W. Ruchuang, Intrusion detection for wireless sensor network based on traffic prediction model, *International Conference on Solid State Devices and Materials Science*, 25: 2072–2080, 2012.

57. H. Chen, P. Han, X. Zhou, and C. Gao, Lightweight anomaly intrusion detection in wireless sensor networks, in *Pacific Asia Workshop*, PAISI, Chengdu, China, pp. 105–116, 2007.

58. S. Khanum, M. Usman, and A. Alwabel, Mobile agent based hierarchical intrusion detection system in wireless sensor networks, *International Journal of Computer Science Issues (IJCSI)*, 9(1): 101–108, 2012.

59. A. Stetsko, L. Folkman, and M. Vashek, Neighbor-based intrusion detection for wireless sensor networks, in *6th IEEE International Conference on Wireless and Mobile Communications*, Valencia, Italy, 2010.

60. A. Strikos, A full approach for intrusion detection in wireless sensor networks, *School of Information and Communication Technology*, KTH-Royal Institute of Technology, Sweden, 2007.

61. Z. Xiao, C. Liu, and C. Chen, An anomaly detection scheme based on machine learning for WSN, in *1st International Conference on Information Science and Engineering*, Nanjing, China, pp. 3959–3962, 2009.

62. S. Misra, P. Krishna, and K. Abraham, Energy efficient learning solution for intrusion detection in wireless sensor networks, in *Second International Conference on Communication Systems and Networks*, Bangalore, India, pp. 423–428, 2010.

63. T. Wang, Z. Liang, and C. Zhao, A detection method for routing attacks of wireless sensor network based on fuzzy C-means clustering, in *The Proceeding of 6th International Conference on Fuzzy Systems and Knowledge Discovery*, Tianjin, China, pp. 445–449, 2009.

64. S. Athmani, D. Boubiche, and A. Bilami, Hierarchical energy efficient intrusion detection system for black hole attacks in WSNs, in *World Congress on Computer and Information Technology (WCCIT)*, Sousse, Tunisia, pp. 1–5, 2013.

65. W. Li, P. Yi, Y. Wu, L. Pan, and P. Li, A new intrusion detection system based on KNN classification algorithm in wireless sensor network, *Journal of Electrical and Computer Engineering*, 2014: Article ID 240217, 2014.

66. M. Riecker, S. Biedermann, R. El Bansarkharni, and M. Hollick, Lightweight energy consumption-based intrusion detection system for wireless sensor networks, *International Journal of Information Security*, 14(2): 155–167, 2015.

67. X. Deng, An intrusion detection system for cluster based wireless sensor networks, in *16th International Symposium on Wireless Personal Multimedia Communications (WPMC)*, Atlantic City, NJ, 2013.

68. S. Kaplantzis, A. Shilton, N. Mani, and A. Sekercioglu, Detecting selective forwarding attacks in wireless sensor networks using support vector machines, in *3rd International Conference on Intelligent Sensors, Sensor Networks and Information (ISSNIP)*, Melbourne, Australia, pp. 335–340, 2007.

69. D. Udaya Suriya Rajkumar and R. Vayanaperumal, A leader based intrusion detection system for preventing intruder in heterogeneous wireless sensor network, in *IEEE Bombay Section Symposium (IBSS)*, Mumbai, India, pp. 1–6, 2015.

70. A. Balakrishnan and Rino PC, A novel anomaly detection algorithm for WSN, in *5th International Conference on Advances in Computing and Communications*, Kochi, India, 2015.

71. A. Babu Karuppiah, J. Dalfiah, K. Yuvashri, S. Rajaram, and A.-S. K. Pathan, A novel energy-efficient Sybil node detection algorithm for intrusion detection system in wireless sensor networks, in *3rd International Conference on Eco-friendly Computing and Communication Systems*, Surathkal, Mangalore, India, pp. 95–98, 2014.

72. S. Das and A. Das, An algorithm to detect malicious nodes in wireless sensor network using enhanced LEACH protocol, in *International Conference on Advances in Computer Engineering and Applications (ICACEA)*, Ghaziabad, India, pp. 875–881, 2015.

73. A. Babu Karruppiah, J. Dalfiah, K. Yuvashri, and S. Rajaram, An improvised hierarchical black hole detection algorithm in wireless sensor networks, in *International Conference on Innovation Information in Computing Technologies (ICIICT)*, India, 2015.

74. D. Udaya Suriya Rajkumar and R. Vayanaperumal, Detecting and revocation the compromised node in zone—based wireless sensor network using a two stage approach, in *Sixth International Conference on Advanced Computing (ICoAC)*, Chennai, India, pp. 10–16, 2014.

75. G. Keerthana and G. Padmavathi, Detecting sinkhole attack in wireless sensor network using enhanced particle swarm optimization technique, *International Journal of Security and its Applications*, 10(3): 41–54, 2016.

76. S. Shivaji and A. Patil, Energy efficient intrusion detection scheme based on Bayesian energy prediction in WSN, in *Fifth International Conference on Advances in Computing and Communications*, Kakkanad, Kochi, pp. 114–117, 2015.

77. G. Thamilarasu, Genetic algorithm based intrusion detection system for wireless body area networks, in *3rd IEEE International Workshop on Security and Forensics in Communication Systems*, Larnaka, Cyprus, pp. 160–165, 2015.

78. M. Riecker, S. Biedermann, and M. Hollick, Lightweight energy consumption based intrusion detection system for wireless sensor networks, in *Symposium on Applied Computing*, Coimbra, Portugal, pp. 1784–1791, 2013.

79. C. Anand and R. Granamurthy, Localized DoS attack detection aArchitecture for reliable data transmission over wireless sensor network, *Wireless Personal Communications*, 90(2): 847–859, 2016.

80. P. Vamsi and K. Kant, Secure data aggregation and intrusion detection in wireless sensor networks, in *IEEE International Conference on Signal Processing and Communication (ICSC)*, Noida, India, pp. 127–131, 2015.

81. M. Rassam, M. Maarof, and A. Zainal, Adaptive and online data anomaly detection for wireless sensor systems, *Knowledge-Based Systems*, 60: 44–57, 2014.

82. C.-F. Hsien, R.-C. Chen, and Y.-F. Huang, Applying an ontology to a patrol intrusion detection system for wireless sensor networks, *International Journal of Distributed Sensor Networks*, Hindawi Publishing Corporation, 2014: 14, 2014.

83. T. Hai, E.-N. Huh, and M. Jo, A lightweight intrusion detection framework for wireless sensor networks, *Journal on Wireless Communications & Mobile Computing*, 10(4): 559–572, 2010.

84. M. Conti, R. Di Pietro, L. Manchini, and A. Mei, Distributed detection of clone attacks in wireless sensor networks, *IEEE Transactions on Dependable and Secure Computing*, 8(5): 685–698, 2011.

85. I. Krontiris, T. Dimitriou, T. Giannetsos, and M. Mpasoukos, Intrusion detection of sinkhole attacks in wireless sensor networks, in *Third International Workshop in*

Algorithmic Aspects of Wireless Sensor Networks (ALGOSENSORS), Wroclaw, Poland, pp. 150–161, 2007.

86. B. Bains and R. Vaid, Selective forwarding based intrusion detection system for secure wireless sensor network, *International Journal of Computer Applications*, 77(13): 20–26, 2013.

87. E. Cho, C. Hong, S. Lee, and S. Jeon, A partially distributed intrusion detection system for wireless sensor networks, *Sensors*, 13(12): 15863–15879, 2013.

88. F. Hidoussi, H. Toral-Cruz, D. Boubiche, K. Lakhtaria, A. Mihovska, and M. Voznak, Centralized IDS based on misuse detection for cluster based wireless sensors networks, *Wireless Personal Communications*, 85(1): 207–224, 2015.

89. G. Sandhya and A. Julian, Intrusion detection in wireless sensor network using genetic K-means algorithm, in *IEEE International Conference on Advanced Communication Control and Computing Teclmologies (ICACCCT)*, Ramanathapuram, India, pp. 1–4, 2014.

90. M. Guerroumi, A. Derhab, and K. Saleem, Intrusion detection system against Sink-Hole attack in wireless sensor networks with mobile sink, in *12th International Conference on Information Technology—New Generations*, Las Vegas, Nevada, pp. 307–313, 2015.

91. S. Saha, M. S. Islam, M. S. Hossen, and M. S. Mamun, A Novel Overlay IDS for Wireless Sensor Networks, in *International Conference of Wireless Applications and Computing*, ISBN: 978-972-8924-62-1, 2008.

92. K. Yan, S. Wang, and C. Liu, A hybrid intrusion detection system of cluster-based wireless sensor networks, in *Proceedings of the International Multi Conference of Engineers and Computer Scientists*, Hong Kong, pp. 18–20, 2009.

93. G. Huo and X. Wang, DIDS: A dynamic model of intrusion detection system in wireless sensor networks, in *International Conference on Information and Automation*, Zhangjiajie, China, pp. 374–378, 2008.

94. T. Hai, F. Khan, and E.-N. Huh, Hybrid intrusion detection system for wireless sensor networks, in *International Conference on Computer Science and Applications (ICCSA)*, San Francisco, USA, pp. 383–396, 2007.

95. K. Ahmed, A. Shihavuddin, K. Ahmed, M. Shirajum Munir, and M. Anwar Asad, Abnormal node detection in wireless sensor network by pair based approach using IDS secure routing methodology, *IJCSNS International Journal of Computer Science and Network Security*, 8(12): 339–342, 2008.

96. D. Boubiche and A. Bilami, Cross layer intrusion detection system for wireless sensor network, *International Journal of Network Security & Its Applications (IJNSA)*, 4(2): 35–52, 2012.

97. M. Islam Mamun, K. A. Sultanul, M. Sakhawat Hossen, and M. R. Hayat Khan, Policy based intrusion detection and response system in hierarchical WSN architecture, in *Proceeding in International Conference on Wireless Communication, Vehicular Technology, Information Theory and Aerospace & Electronic Systems Technology*, Aalborg, Denmark, 2012.

98. J. Ma, S. Zhang, Y. Zhong, and X. Tong, SAID: A self-adaptive intrusion detection system in wireless sensor networks, in *7th International Conference on Information Security Applications (WISA)*, Korea, pp. 60–73, 2006.

99. H. Sedjelmaci and M. Feham, Novel hybrid intrusion detection system for clustered wireless sensor network, *International Journal of Network Security & Its Applications (IJNSA)*, 3(4): 1–14, 2011.

100. Y. Maleh, A. Ezzati, Y. Qasmaoui, and M. Mbida, A global hybrid intrusion detection system for wireless sensor networks, in *The 5th International Symposium on Frontiers in Ambient and Mobile Systems (FAMS 2015)*, London, UK, Vol. 52, pp. 1047–1052, 2015.

101. I. Butun, I.-R. Ra, and R. Sankar, An intrusion detection system based on multilevel clustering for hierarchical wireless sensor networks, in *SENSORS*, Busan, South Korea, pp. 28960–28978, 2015.

102. A. Sardar, R. Sahoo, M. Singh, S. Sarkar, J. Singh, and K. Majumder, Intelligent intrusion detection system in wireless sensor network, in *3rd International Conference on Frontiers of Intelligent Computing: Theory and Applications (FICTA)*, Bhubaneswar, India, pp. 707–712, 2014.

103. R. Sundararajan and U. Arumugam, Intrusion detection algorithm for mitigating sinkhole attack on LEACH protocol in wireless sensor networks, *Hindawi Publishing Corporation, Journal of Sensors*, 2015: 12, 2015.

104. Y. El Mourabit, A. Toumanari, A. Bouirden, H. Zougagh, and R. Latif, Intrusion detection system in wireless sensor network based on mobile agent, in *Second World Conference on Complex Systems (WCCS)*, Agadir, Morocco, pp. 248–251, 2014.

105. S. Sajjad, S. Bouk, and M. Yousaf, Neighbor node trust based intrusion detection system for WSN, in *6th International Conference on Emerging Ubiquitous Systems and Pervasive Networks (EUSPN)*, Berlin, Germany, Vol. 63, pp. 183–188, 2015.

106. S. Shamshirband, A. Patel, N. Anuar, L. Kiah, V. Rohani, D. Petrovic, S. Misra, and A. Khan, D-FICCA: A density-based fuzzy imperialist competitive clustering algorithm for intrusion detection in wireless sensor networks, *Measurement*, 55: 212–226, 2014.

107. M. Islam, R. Khan, and D. Bappy, A hierarchical intrusion detection system in wireless sensor networks, *IJCSNS International Journal of Computer Science and Network Security*, 10(8): 21–26, 2010.

108. A. Farooqi, F. Khan, J. Wang, and S. Lee, A novel intrusion detection framework for wireless sensor networks, *Personal and Ubiquitous Computing*, 17(5): 907–919, 2013.

109. M. de Sousa Lemos, L. Barroso Leal, and R. Holanda Filho, A new collaborative approach for intrusion detection system on wireless sensor networks, in *Novel Algorithms and Techniques in Telecommunications and Networking*, USA, pp. 239–244, 2010.

110. Q. Guo, X. Li, Z. Feng, and G. Xu, MPOID: Multi-protocol oriented intrusion detection method for wireless sensor networks, in *17th International Conference on High Performance Computing and Communications (HPCC), 7th International Symposium on Cyberspace Safety and Security (CSS), 12th International Conference on Embedded Software and Systems (ICESS)*, New York, USA, pp. 1512–1517, 2015.

111. C. Dutta and U. Biswas, Specification based IDS for camouflaging wormhole attack in OLSR, in *IEEE 23rd Mediterranean Conference on Control and Automation (MED)*, Torremolinos, Spain, pp. 960–966, 2015.

112. Y. Liu and F. Yu, Immunity-based intrusion detection for wireless sensor networks, *International Joint Conference on Neural Networks (IJCNN)*, Hong Kong, pp. 439–444, 2008.

113. H. Salmon, C. de Farias, P. Loureiro, L. Pirmez, S. Rosseto, P. de A. Rodrigues, R. Pirmez, F. Delicato, and L. de Costa Carmo, Intrusion detection system for wireless sensor networks using danger theory immune-inspired techniques, *International Journal of Wireless Information Networks*, 20(1): 39–66, 2012.

114. M. Drozda, S. Schaust, and H. Szczerbicka, Is AIS based misbehavior detection suitable for wireless sensor networks?, in *IEEE Wireless Communications and Networking Conference*, Hong Kong, pp. 3128–3133, 2007.

115. M. Zamani, M. Movahedi, M. Ebadzadeh, and H. Pedram, A DDoS-aware IDS model based on danger theory and mobile agents, in *International Conference on Computational Intelligence and Security*, Beijing, China, pp. 516–520, 2009.

116. R. Rizwan, F. Khan, H. Abbas, and S. Chauhdary, Anomaly detection in wireless sensor networks using immune-based bioinspired mechanism, *Hindawi Publishing Corporation, International Journal of Distributed Sensor Networks*, 2015: 1–10, 2015.

117. L. Sindhuja and G. Padmavathi, Clone detection using enhanced EDD (EEDD) with danger theory in mobile wireless sensor network, *International Journal of Security and its Applications*, 9(4): 185–202, 2015.

118. S. Shamshirband, A. Patel, N. Anuar, L. Kiah, V. Rohani, D. Petrovic, S. Misra, and A. Khan, Co-FAIS: Cooperative fuzzy artificial immune system for detecting intrusion in wireless sensor networks, *Journal of Network and Computer Applications*, 42: 102–117, 2014.

119. I. Atakl, H. Hu, Y. Chen, W.-S. Ku, and Z. Su, Malicious node detection in wireless sensor networks using weighted trust evaluation, in *The Symposium on Simulation of Systems Security (SSSS)*, Ottawa, Ontario, Canada, pp. 52–60, 2008.

120. F. Bao, I.-R. Chen, M. Chang, and J.-H. Cho, Trust-based intrusion detection in wireless sensor networks, in *Proceedings of IEEE International Conference on Communications (ICC)*, Kyoto, Japan, pp. 1–6, 2011.

121. G. Rajeshkumar and K. Valluvan, An energy aware trust based intrusion detection system with adaptive acknowledgement for wireless sensor network, *Wireless Personal Communications*, 1–15, 2016.

122. A. Dhakne and P. Chatur, Distributed trust based intrusion detection approach in wireless sensor network, in *International Conference on Communication, Control and Intelligent Systems (CCIS)*, Mathura, India, pp. 96–101, 2015.

123. R. Dong, L. Liu, J. Liu, and X. Xu, Intrusion detection system based on payoff matrix for wireless sensor networks, in *3rd International Conference on Genetic and Evolutionary Computing*, 2009, doi: 10.1109/WGEC.2009.87.

124. Y. Ma, H. Cao, and J. Ma, The intrusion detection method based on game theory in wireless sensor network, in *First International Conference on Ubi-Media Computing*, China, pp. 326–331, 2008.

125. H. Moosavi and F. Bui, A game-theoretic framework for robust optimal intrusion detection in wireless sensor networks, *IEEE Transactions on Information Forensics and Security*, 9(9): 1367–1379, 2014.

126. S. Shamshirband, A. Patel, N. Anuar, L. Kiah, V. Rohani, D. Petrovic, S. Misra, and A. Khan, Cooperative game theoretic approach using fuzzy Q-learning for detecting and preventing intrusions in wireless sensor networks, *Journal of Engineering Applications of Artificial Intelligence*, 32: 228–241, 2014.

127. M. Estiri and A. Khademzadeh, A game-theoretical model for intrusion detection in wireless sensor networks, in *IEEE 23rd Canadian Conference on Electrical and Computer Engineering (CCECE)*, Calgary, Alberta, Canada, pp. 1–5, 2010.

128. A. Agah and S. Das, Preventing DoS attacks in wireless sensor networks: A repeated game theory approach, *International Journal of Network Security*, 5(2): 145–153, 2007.

Chapter 15

Intrusion Detection and Tolerance for 6LoWPAN-Based WSNs Using MMT

Vinh Hoa La and Ana R. Cavalli

Contents

15.1 Introduction

There are today about 15 billion devices on the Internet of Things (IoT) and there would be 50 billion connected devices by 2020 according to a report by Cisco and DHL [1]. As a subdomain, wireless sensor networks (WSNs) are more and more widely used in various fields, for example, to monitor physical and environmental conditions in areas where human access is seemingly obstructed. Smart cities are also another application domain based on the collaboration of a number of WSNs. However, researchers observed three principal difficult challenges in designing and implementing a secure WSN:

- *The vulnerable characteristics of wireless communication nature*: For example, eavesdropping, unauthorized access, spoofing, replay and denial of service (DoS) attacks, etc.
- *The severely constrained resources of sensor devices*: Typical WSNs are composed of a large number of low-power tiny sensors and actuators. These nodes have typically limited energy lifetime, slow embedded processors, severely constrained memory, and low-bandwidth radios. For example, Waspmote [2], the modern open-source sensor device distributed by Libelium,* contains simply a 14-MHz microprocessor, 3.3–4.2 V battery voltage, 8 kB SRAM, 128 kB flash memory, and 4 kB EEPROM to save sensed data and to run an operating system and application programs. These resource constraints limit the degree of encryption, decryption, and authentication that can be deployed; thus, the concepts *security* and *WSNs* sound likely contradictory.
- *Additional physical security risks*: WSNs are commonly deployed in inaccessible terrains or unattended and even hostile environments to sense data or to observe the occurrence of certain events. They can self-organize into an ad hoc-style wireless network that collects and forwards sensor data to an information sink (e.g., a base station acting as a gateway to the wired network).

Having been standardized by IETF, 6LoWPAN-based WSNs consist of low-power objects equipped with sensors. They use IEEE 802.15.4 as the physical layer standard. However, they are exposed to various types of security threats due to the intrinsic characteristics and the lack of security considerations in the design of protocols for them. The failure of nodes may result in network partition,

---

* http://www.libelium.com/

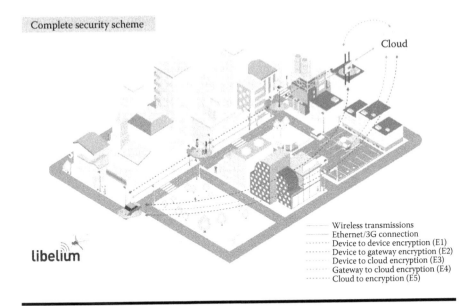

Complete security scheme

Cloud

Wireless transmissions
Ethernet/3G connection
Device to device encryption (E1)
Device to gateway encryption (E2)
Device to cloud encryption (E3)
Gateway to cloud encryption (E4)
Cloud to encryption (E5)

libelium

Figure 15.1 A complete example of the security scheme proposed by Libelium.

decreasing the cover ratio, reducing the availability of the sensor network, and even causing entire network failure. We therefore need an adapted monitoring tool that takes into account the particular characteristics of 6LoWPANs (e.g., resource constraints).

Nevertheless, the research on IoT in general and on 6LoWPANs in particular has so far mainly focused on how to make the concepts realistic and practical. In the other words, most of the research works have been focusing on standardizing the communication protocols, ameliorating the performance of the IoT systems, optimizing the resource consumption, etc. Security is always considered an important issue but it is difficult to achieve thoroughly because it seems contradictory with the system's performance.

Additionally, the research on IoT security concentrates mostly on *designing* secure communication protocols, light encryption, and authentication. For example, Figure 15.1 displays the complete security scheme proposed by Libelium that deals with common security issues, including *access control (privacy)*, *authentication, data confidentiality, data integrity, data freshness (avoiding packet injection)*, and *nonrepudiation*. In general, there are so far several security propositions for 6LoWPAN-based IoT:

- *Hop-by-hop security*: TinySec, Minisec, ContikiSec, IEEE802.15.4 security mechanism, WSNSec
- *End-to-end approach*: WSN-ETESec

Recently, there are more and more research works on monitoring in general and intrusion detection in particular for IoT/WSNs. However, some existing approaches are still at the design level and not implemented yet. Some others focus only on the routing problem (e.g., Foren6) or seemingly affect the performance of the systems (e.g., SVELTE). Therefore, security monitoring with minimum influence on the running system is the topic that we study in this chapter. We also propose the concept of "intrusion tolerance" as an open issue that will probably be playing an important and innovative role in future WSNs.

Initially, traditional battery-powered networks or low-bit-rate networks (e.g., 802.15.4) were considered incapable of running Internet Protocol (IP) due to their typical characteristics:

■ *Limited processing capability*: From 8-bit clock speed processors
■ *Low memory capacity*: From a few kilobytes of RAM with a few dozen kilobytes of ROM/flash memory
■ *Low power*: From a few dozen of milliamperes
■ *Short range*: Normally from 10 to 100 m

A majority of local area networks (LANs) and wide area networks (WANs) are running IP. As a result, 6LoWPAN has been designed to work on top of 802.15.4 networks as an adaptation layer, which makes the layer 2 compatible with layer 3 routing and internetwork technology. 6LoWPAN supports uniquely IPv6 (no IPv4 support available) and is promising to allow low-power and lossy devices connecting to other IP-based networks, without intermediate entities such as translation gateways or proxies. This success will enable reusing existing IP-based technology, including tools for monitoring, diagnostic, and management.

6LoWPAN standards [3] are basically completed. Figure 15.2 demonstrates the protocol stack of a 6LoWPAN, including the following standardized protocols [4]:

■ *6LoWPAN*: IPv6 over low-power WPAN (RFC 4919, August 2007) [5]

TCP/IP protocol stack				6LoWPAN protocol stack	
HTTP		SSL/TLS	Application	CoAP	DTLS
TCP	UDP	ICMP	Transport	UDP	ICMP
IP			Network	RPL	
			(adaptation)	6LoWPAN	
Ethernet MAC			Data link	IEEE 802.15.4. MAC	
Ethernet PHY			Physical	IEEE 802.15.4. PHY	

Figure 15.2   6LoWPAN protocol stack in comparison with TCP/IP.

- *RPL*: IPv6 routing protocol for low-power and lossy networks (RFC 6550, March 2012) [6]
- *CoAP*: Constrained application protocol (RFC 7252, June 2014) [7]
- *DTLS*: Datagram transport layer security (RFC 6347, January 2012) [8]

6LoWPAN standards are still being complemented to satisfy routing needs and to extend to other link layer technology.

A 6LoWPAN typically includes devices realizing a combined work: collecting the physical or environmental parameters and sending them to real-world applications. The most seemingly popular devices are wireless sensors, although a 6LoWPAN is not necessarily composed of sensor nodes only, but also actuators. Figure 15.3 illustrates an example of typical WSN/IoT solutions proposed by Libelium. Data collected from sensors can be stored in a local or external database, which will be queried by cloud-based applications.

However, in reality, sensors are usually affected by noises, misconfigurations, and other malicious nodes. In addition, 6LoWPAN devices are themselves unreliable due to various reasons, namely, uncertain radio connectivity, battery drain, device lockups, and physical tampering. Security-concerning designs (e.g., encryption and decryption) are limited due to restricted resources of devices. These make them vulnerable to failures and attacks. Network monitoring and anomaly detection therefore become essential.

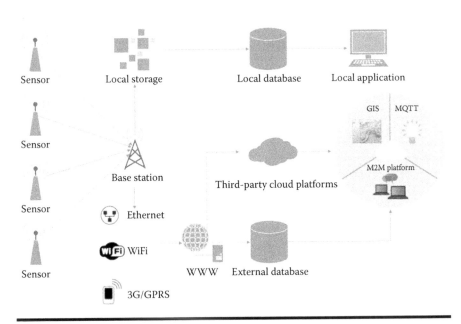

Figure 15.3   An example of the actual application of WSN/IoT.

15.2 Montimage Monitoring Tool

15.2.1 Overview

Montimage Monitoring Tool (MMT) is our monitoring tool [9,10] that allows capturing and analyzing network traffic in both online and offline manners. MMT supports network traffic inspection by extracting necessary attributes and referring to a set of security rules. Figure 15.4 illustrates the architecture of MMT. MMT uses Deep Packet/Flow Inspection (DPI/DFI) techniques and consists of three principal modules:

- *MMT-Extract* enables the extraction of network protocol fields of not only offline structured traffic (e.g., PCAP files) but also real-time online network traffic passing by an interface. It is possible to build a new plugin for the addition of new protocols and the parsing of proprietary structured data. In practice, this module permits monitoring different applications, systems, or

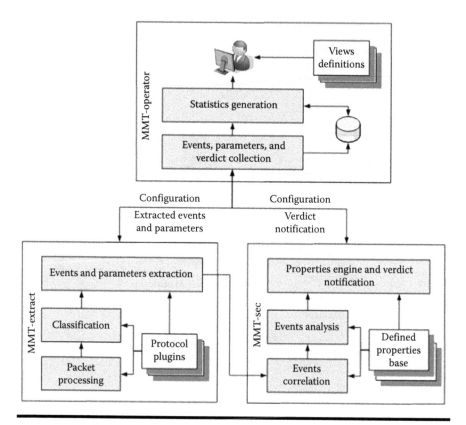

Figure 15.4   MMT global architecture.

networks. In the case of application, the input can be the exchanged messages or events log.

■ *MMT-Security* contains security rules written in XML that refer to both expected and unexpected behaviors. MMT-Security model is inspired from linear temporal logic. Different rules can be correlated in order to detect security incidents. Rules in XML provide the advantage of simple and straightforward structure verification. A property is an IF ⟨*context*⟩ THEN ⟨*trigger*⟩ relation. The trigger is checked if and only if the context is valid. If the trigger is found valid, then the property is satisfied. Otherwise, the property is violated. Embedded functions can also be added to preprocess the data input before passing to MMT-Security rules.

■ *MMT-Operator* allows a graphical user interface, which is customizable to display the result.

The typical position of MMT for monitoring an organization's network (TCP/IP network) is between the network router and the (outside) Internet. It plays a role similar to a firewall filtering and analyzing the passing traffic. For example in Reference 10, MMT is implemented in the gateway of a wired LAN to detect ARP spoofing attack. The detection rules are described in Figure 15.5. Nonetheless, MMT can be installed in an individual local host to listen to traffic passed by one or several interfaces. Web administrators can also integrate MMT in their web servers to inspect incoming requests before processing them.

Compared to existing intrusion detection techniques, the originality of MMT is that MMT is not based on only pattern matching (i.e., signature-based) as SNORT and does not require writing executable scripts as in BRO. MMT is a flexible

```
<property value="THEN" delay_max="10" property_id="1"
      type_property="ATTACK"
      description="IPv4 address conflict detection (RFC5227). Possible ARP spoofing.">
<operator value="THEN" delay_max="1">
<event value="COMPUTE" event_id="1"
      description="ARP who has request"
      boolean_expression="(ARP.OPCODE == 1)"/>
<event value="COMPUTE" event_id="2"
      description="ARP reply MAC address"
      boolean_expression="((ARP.OPCODE == 2)&amp;&amp;
      (ARP.SRC_PROTO == ARP.DST_PROTO.1))"/>
</operator>
<event value="COMPUTE" event_id="3"
      description="ARP reply but with different MAC address"
      boolean_expression="(((ARP.OPCODE == 2)&amp;&amp;
      (ARP.SRC_PROTO == ARP.DST_PROTO.1))&amp;&amp;
      (ARP.SRC_HARD == ARP.SRC_HARD.2))"/>
</property>
```

Figure 15.5   MMT security rules to detect ARP spoofing attack.

solution that can integrate pattern matching, statistics, and machine learning [11] techniques, depending on the actual problem. MMT property rules are descriptive and straightforward. They can be written and added to describe normal/abnormal behaviors. Furthermore, MMT is open for developer to add new plugins in order to deal with new structured input as well as to preprocess the attributes before analyzing them in the module MMT-Security. This is the main reason why we have adapted MMT to inspect 6LoWPAN traffic that is so far not covered by other monitoring solutions.

15.2.2 MMT Adaptation for 6LoWPAN-Based WSNs

Having been standardized by IETF, 6LoWPAN-based WSNs consist of low-power objects equipped with sensors. They use IEEE 802.15.4 as the physical layer standard. However, they are exposed to various types of security threats due to the intrinsic characteristics and the lack of security designs. The failure of nodes may result in network partition, decreasing the cover ratio, reducing the availability of the sensor network, and even causing entire network failure. An adapted monitoring tool that takes into account the particular characteristics of 6LoWPANs (e.g., resource constraints) is therefore a need.

Nevertheless, to our knowledge, there has not been any official monitoring solution for such kind of networks yet. The initial propositions concentrate only on routing issues and they are likely impossible to allow a deep inspection on the network traffic. We aim to fulfill this mission. Indeed, we have adapted our original version of MMT, which has been working well over TCP/IP networks [9,10]. Our goal is to consider not only theoretical topology of the network but also ready-to-use elements in network traffic to monitor itself (i.e., passive monitoring). Avoiding creating additional traffic, which is costly in 6LoWPAN, is an important priority throughout our work. We validate MMT integrated with new 6LoWPAN plugins over a real test-bed in analyzing real-world 6LoWPAN traffic. Experimental results prove the applicability of our tool, which can be useful for both research community and industrial companies.

Attempting to adapt MMT for 6LoWPANs, we have built several 6LoWPAN plugins in addition to the original version working properly over TCP/IP networks. These plugins take into consideration the encapsulation and header compression mechanisms of the 6LoWPAN standard. Attributes and protocols can thus be recognized and extracted for being analyzed. To the best of our knowledge, existing monitoring tools and intrusion detection system (IDS) (e.g., Suricata, SNORT) have not provided any official support to IEEE 802.15.4 or 6LoWPAN yet.

Figure 15.6 presents an example of a packet captured while nodes were exchanging topology information for routing. It should be noted that there are three different header structures corresponding to IEEE 802.15.4 ACK packets, IEEE 802.15.4 DATA Unicast packets, and IEEE 802.15.4 DATA Multicast packets. The field "Frame Control" plays the role of their identifier.

```
▾IEEE 802.15.4 Data, Dst: Broadcast, Src: 00:00:00_00:00:00:00:03
 ▸Frame Control Field: Data (0xd841)
  Sequence Number: 131
  Destination PAN: 0xabcd
  Destination: 0xffff
  Extended Source: 00:00:00 00:00:00:00:03 (00:00:00:00:00:00:00:03)
  FCS: 0xaa9b (Correct)
▾6LoWPAN
▾Internet Protocol Version 6, Src: fe80::200:0:0:3 (fe80::200:0:0:3), Dst: ff02::1a (ff02::1a)
▾Internet Control Message Protocol v6
```

Figure 15.6 **A sample captured packet with IEEE 802.15.4 fields.**

Our plugins aim to cover all possible structures of packets. Figure 15.7 briefly resumes our plugins and their supporting protocols (i.e., packet structures) at the time of writing this chapter. They include already-done ones (black boxes), almost-done and under-tested ones (dark gray boxes), and a to-be-done one (light gray box). For the moment, we are mostly focusing on routing control packets that can identify efficiently the network's state. For the long-term goal, we would like to verify other protocols in higher layers, especially security-related protocols (e.g., DTLS).

Actually, building a new plugin for any structured data/traffic/event log is a feasible task. Researchers and industrial network administrators can build the plugins themselves, taking into consideration their own interesting data to extract. Montimage provides supporting tools to create skeletons for new plugins based on predefined attributes, which are in need of being extracted.

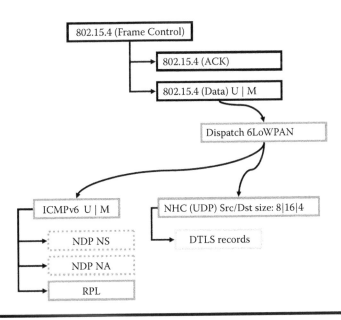

Figure 15.7 **List of MMT plugins corresponding to supported protocols.**

15.3 Intrusion Detection for 6LoWPAN-Based WSNs Using MMT

15.3.1 Detection Methodology and Algorithm

In this section, we summarize our methodologies and algorithms to detect anomalies. The learning phase is realized by utilizing supervised learning approach, that is, we knew the label (normal or abnormal) of the audit traffic before learning it. More specifically, we propose two detection algorithms, one based on statistical learning and another based on information theory (entropy).

15.3.1.1 Misbehaving Node Detection Algorithm Based on Statistical Learning

We suppose that s is a sink node (i.e., base station node, gateway) and n_i is the ith sensor node. For a node n_i at the moment t, $W_i(t)$ denotes the weight of the link between n_i and s. Depending on real-world case study and requirements, W_i can be defined and calculated differently. Later in this section, we present a specific case study in which we define the necessary time duration for a packet traveling from one node to another as the link weight between them.

In general, our detection algorithm consists of two phases: *learning phase* and *monitoring phase*.

1. *Learning phase*: We assume that $W_i(t) \sim N(\mu_i, \sigma_i^2)$, that is, W_i is distributed normally with mean μ_i and variance σ_i. $N(\mu, \sigma^2)$ is the normal (or Gaussian) distribution in probability theory [12].

   According to *3-sigma rule*, approximately 95% and 99.7% of values drawn from a normal distribution lie correspondingly within two and three standard deviations σ away from the mean μ. This percentage increases according to the gap away from the mean. In case of 7σ, the percentage approaches up to 99.99999999974%. In other words, the probability that X is within $[(\mu - 7\sigma), (\mu + 7\sigma)]$ is high up to 0.9999999999974.

   In the learning phase, we assume that every node functions normally. This phase should be performed right after the sensor network starts operating. In fact, multiple attempts in learning phase could be useful to identify "the most common normal status of the network," thus, to determine the best values for μ_i and σ_i for the node n_i. We then define $[(\mu_i - \varepsilon_i), (\mu_i + \varepsilon_i)]$ as the promising interval that W_i should lie within. ε_i is a customizable parameter, which defines the frontier between normal and abnormal behaviors. Its value is generally from 3σ to 7σ.

2. *Monitoring phase*: In this phase, we listen to the network and calculate $W_i(t)$ for every node. We evaluate whether a node n_i is normal or abnormal by comparing $W_i(t)$ with μ_i defined in the learning phase.

■ *Step 1: Malicious path identification*—Let S_i be the state of n_i, $S_i(t) = 0$ if n_i operates normally at the moment t. Otherwise, $S_i(t) = 1$ and there must be (a) misbehavior node(s) somewhere. Hence, $S_i(t)$ is deduced as follows:

$$S_i(t) = \begin{cases} 0 \text{ if } W_i(t) \in [(\mu_i - \varepsilon_i), (\mu_i + \varepsilon_i)], \\ 1 \text{ otherwise} \end{cases} \quad (15.1)$$

It is worth noting that the fact that $S_i(t) = 1$ does not lead to the conclusion that n_i is malicious. The problem can also come from another sensor node within the path from n_i to the sink node s. Our mission then is to identify a misbehavior node that we know definitely within the path from n_k to s. This is the goal of step 2.

■ *Step 2: Misbehavior node identification*—Suppose that we are (passively) monitoring in real time a 6LoWPAN-WSN and suddenly we witness the occurrence of the event "$S_k = 1$." Thus, there must be a malicious node within the path $s \rightarrow n_1 \rightarrow n_2 \rightarrow \cdots \rightarrow n_k$.

Thanks to the learning phase, we have already known $S_1, S_2, \ldots, S_{k-1}, S_k$. Then we have the following logic deduction:

$\exists j \in \mathbf{N}, j \geq 1 \mid (S_i = 0 \text{ for every } i \in \{0, 1, \ldots, j-1\}) \bigwedge (S_j = 1)$

The sink node s is considered as S_0 ($S_0 \equiv s$).

Evidently, n_i is legitimate for $i \in \{0, 1, \ldots, j-1\}$ and n_j is logically the first misbehavior node detected. We continue to test the other nodes, including n_{j+1} until n_k:

$s \rightarrow \cdots \rightarrow n_j \rightarrow n_{j+1} \rightarrow \cdots \rightarrow n_k$.

$S_0 = 0 \rightarrow \cdots \rightarrow S_j = 1 \rightarrow S_{j+1} = 1 \rightarrow \cdots \rightarrow S_k = 1$.

We define α as the difference (i.e., delay) between the link weight calculated in reality and the expected (predicted) link weight: $\alpha_i = W_i - \mu_i$, and β as the additional link cost to the neighbor caused by the node n (Figure 15.8) ($\beta = 0$ if and only if n is normal). α is directly calculated thanks to known values of W and μ, while β would be deduced indirectly.

Obviously, Equation 15.1 equals to the two following expressions:

$$S_i(t) = \begin{cases} 0 \text{ if } |\alpha_i(t)| > \varepsilon_i, \\ 1 \text{ otherwise} \end{cases} \quad (15.2)$$

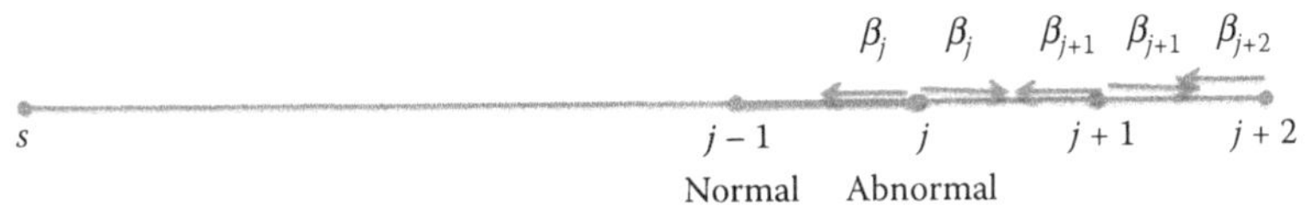

Figure 15.8 Additional link cost to the neighbor.

$$S_i(t) = \begin{cases} 0 \text{ if } \beta_i(t) = 0, \\ 1 \text{ otherwise} \end{cases} \tag{15.3}$$

Thus,

$$(\alpha_{j+1} - \alpha_j) = (W_{j+1} - W_j) - (\mu_{j+1} - \mu_j) \tag{15.4}$$

Obviously, $|\alpha_j| > \varepsilon_j$ and $|\alpha_{j+1}| > \varepsilon_{j+1}$. μ_j and μ_{j+1} were derived from the learning phase. W_j and W_{j+1} are calculated in real-time monitoring.

Because all nodes from the sink node to n_{j-1} function normally, α_j exists principally as a result of the communication delay between n_{j-1} and n_j. In other words,

$$\alpha_j \approx \beta_j \tag{15.5}$$

In Equation 15.4, $(\mu_{j+1} - \mu_j)$ is the weight of the link $n_j \rightarrow n_{j+1}$ in the normal condition (in theory); $(W_{j+1} - W_j)$ is the one in the under-monitored condition (in practice). The right side is thus the additional cost caused over the link $n_j \rightarrow n_{j+1}$, that is, $\beta_j + \beta_{j+1}$.

Therefore,

$$(\alpha_{j+1} - \alpha_j) = \beta_j + \beta_{j+1} \tag{15.6}$$

Because of Equations 15.5 and 15.6, β_{j+1} can be inferred as

$$(\beta_{j+1} \approx \alpha_{j+1} - 2 * \alpha_j) \tag{15.7}$$

We have achieved identifying the status of n_{j+1}.
Continuously, now we are testing n_{j+2}.
Similar to Equation 15.4, we have

$$(\alpha_{j+2} - \alpha_{j+1}) = (W_{j+2} - W_{j+1}) - (\mu_{j+2} - \mu_{j+1}) \tag{15.8}$$

The right side of Equation 15.8 is the additional cost caused over the link $n_{j+1} \rightarrow n_{j+2}$, that is, $\beta_{j+1} + \beta_{j+2}$.
To sum up,

$$\beta_{j+2} = \alpha_{j+2} - \alpha_{j+1} - \beta_{j+1} \tag{15.9}$$

All elements in the right side are disclosed, thus Equation 15.9 gives us the condition to determine whether n_{j+2} is normal or not.
Similarly, we repeat the aforementioned steps to verify the status of the rest:

$$\{n_{j+3}, n_{j+4}, \ldots, n_k\}$$

15.3.1.2 Anomalies Detection Based on Information Theory

This subsection aims to take information theory into consideration in order to provide a theoretical base for the learning phase of our framework. These measures can be defined and calculated from extracted attributes. They can be useful to describe the characteristics of an audit dataset, define a suitable detection model, as well as evaluate the performance of the model.

1. *Entropy*: Entropy [13] is an important concept measuring the uncertainty (or impurity) of a collection of data items. Let X be the collection, including N classes of data items $x_i (i = 1,2,\ldots,N)$. The entropy of X is defined as

$$H(X) = H(x_1, x_2, \ldots, x_N) = -\sum_{i=1}^{N} P(x_i) * \log P(x_i) \qquad (15.10)$$

where $P(x_i)$ is the probability of x_i in X for $i = 1,2,\ldots,N$. The "purer" dataset has a smaller entropy, that is, the class distribution is skewer. The smallest possible value of the entropy is 0 in case the dataset has only one class of items, that is, there is no uncertainty because we know for sure every item belongs to this unique class. When the data are more "impure," the uncertainty increases, and the entropy value is bigger.

   In the context of this subsection (anomaly detection), we use entropy to measure the regularity of the data input. For example, a trace file can be translated to a set of events $E = \{e_1, e_2, \ldots, e_N\}$. The high regularity refers to the fact that many events are repeated and they will likely appear again in the future. Additionally, if a system works in the mode of *duty cycle* and frequently (e.g., WSNs in which sensors periodically send sensed data), its regularity is seemingly stable. This assumption is assessed in the experimental section.

2. *Conditional entropy*: The conditional entropy of the dataset X given the dataset Y is defined as

$$H(X|Y) = -\sum_{\substack{i=\overline{1,N} \\ j=\overline{1,M}}} P(x_i, y_j) * \log P(x_i|y_j) \qquad (15.11)$$

where x_i, y_j are classes of data items of X and Y, respectively ($i = \overline{1,N}, j = \overline{1,M}$), $P(x_i, y_j)$ is the joint probability of x_i and y_j, and $P(x_i|y_j)$ is the conditional probability of x_i given y_j.

   This concept can be used to measure temporal or sequential characteristics of complex audit datasets corresponding to temporal user, program, and network activities. For the intrusion detection point of view, this is usable for detecting complicated attacks demanding the correlation of different events.

3. *Other measures*: In addition to the two aforementioned measures, there exist several concepts that can be taken into consideration, namely, *relative conditional entropy, information gain and classification*, and *information cost*. They can be useful for building an anomaly detection model as well as evaluating it.

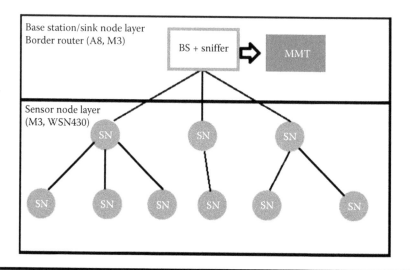

Figure 15.9 Hierarchical architecture of the 6LoWPAN-WSN in our experiment.

15.3.2 Test-Bed Description

In order to illustrate the monitoring performed by MMT, we deployed a real 6LoWPAN-WSN using the open platform provided by FIT-IoT lab.* Figure 15.9 depicts the hierarchical architecture of our network acting as the proof of concept. The BRs (Border Routers) play the role of sink nodes equipped with a sniffer, which allows capturing live traffic. In the context of this work, we used A8 and M3 nodes to deploy BR nodes. M3 and WSN430 nodes[†] were utilized to implement sensor nodes. Our nodes were running Contiki as the operating system. Sniffers were integrated with BRs to capture and pass the network traffic to MMT. Extracted attributes were stored in a local database and further computations would be performed to detect the problems.

The network deployment was realized step by step as follows:

■ Selecting available sensor nodes in FIT-IoT test-bed for the experiments: M3/A8 node as the BR node and M3/WSN430 nodes as sensor nodes.
■ Starting *tunslip6*[‡] to bridge the BRs to the front-end network.
■ Loading suitable firmware for each node: BR firmware with sniffer integrated and sensor node firmware with HTTP server code included.

---

* https://www.iot-lab.info/
[†] Hardware information about nodes: https://www.iot-lab.info/hardware/
[‡] https://www.iot-lab.info/tutorials/build-tunslip6/

■ Booting nodes and starting the experiments. Sensor nodes would periodically (every 10 s) send sensed data to their corresponding BR. MMT would take the traffic captured by sniffers as the input.

15.3.3 Experimental Results

15.3.3.1 Case 1: Statistical Learning

1. *Performance evaluation with offline traffic (PCAP files)*: First, we assessed the processing speed of MMT in the cases of different sizes of the network (i.e., the number of nodes). In each case, the sniffer recorded the traffic passing by the BR for 5 min and saved as a PCAP (packet capture) file. MMT would analyze the PCAP files and extract all attributes that we had defined by the plugins. Figure 15.10 summarizes the results. Evidently, the more nodes we inserted to the network, the more traffic they generated and the more time MMT required to process. However, MMT has indeed shown a promising processing rate, which is always around 420 Mbps. The processing rate is calculated as follows:

$$\text{Processing rate (Mbps)} = \frac{\text{traffic volume (kB)} * 8}{1024 * \text{average execution time}} \quad (15.12)$$

No. of nodes	Traffic (kB)	Processing time (ms)	Processing rate (Mbps)
5	47	0.87	422
10	118	2.21	417
15	235	4.38	419
20	393	7.33	418
25	648	12.14	417
30	1038	19.35	419
35	1334	24.87	419
40	2096	39.08	419

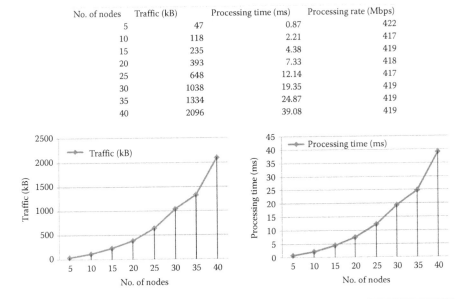

Figure 15.10 Volume of traffic and processing time depending on the size of network.

This processing rate introduces MMT as a potential candidate for monitoring even big networks consisting of hundreds or thousands of connected objects.

2. *Real-time monitoring and response delay*: Second, we validated and compared our solution's performance with Foren6, which is one of the first and the most well-known open-source debugging tools for IoT, while detecting abnormal activities triggered by some misbehaving nodes in the network. In fact, similar to MMT, Foren6 permits passively capturing 6LoWPAN traffic and renders the network state in a graphical user interface. Although it is able to detect abnormal activities in routing, it is mainly used to reconstruct a visual and textual representation of the network (i.e., network troubleshooting). There is, for the moment, no specific Foren6-based application for detecting security violations. In our experiments, we created abnormal activities by modifying the firmware loaded to a number of nodes and forcing them to delay the message-forwarding process or even sometimes avoid forwarding messages (selective forwarding attack). These misbehaving nodes would affect all downstream nodes that use them as the forwarder to reach the BR.

In order to detect these behaviors, we applied our own detection algorithm, which is explained in detail in Section 15.3.1.1. The general idea is to calculate the *travel time* of each packet coming from the BR to each node or vice versa. This task was realized by extracting suitable attributes in packets. The *travel time* would act as the link weight mentioned in the algorithm. We saved the results when the network was functioning properly and when the aforementioned abnormal activities were performed.

In case of proper conditions, we observed every node and extracted two attributes *time stamp* and *MAC address* of every packet coming in and out. After that, based on those values stored, we calculated the *travel time* related to each node (i.e., the necessary time for a packet delivered from this node to the BR or vice versa). We considered this as a random variable and statistically analyzed them in using RStudio.[*] We received the same result regardless of using MMT or Foren6. Figure 15.11 presents the histogram of two PDFs (probability density functions) and two CDFs (cumulative distribution functions) of the *travel time* of packets concerning a sample node. Line (a) corresponds to the case where we took into account 5 min of monitoring, while line (b) corresponds to the case of 10 min. We witnessed that it is likely normally distributed (Gaussian distribution [12]). For both two cases, the *mean* (i.e., expectation) of the distribution is approximated at 380 ms. We call this observation as *the learning phase*.

In case of abnormal activities added to the network, we repeated that procedure (extraction and calculation) and compared received values with the ones derived from the learning phase. As seen in the case above, if we suppose

---

[*] https://www.rstudio.com/

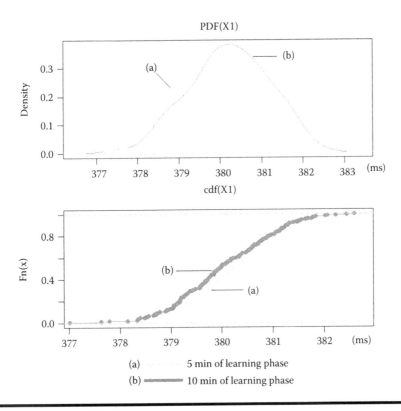

Figure 15.11 Probability density functions and cumulative distribution functions of the *travel time*.

$X_i(t)$ is the random variable representing the *travel time* related to the node n_i and the sink node s at the moment t, $X_i(t)$ can be dealt as a Gaussian distribution: $X_i(t) \sim N(\mu_i, \sigma_i^2)$, where μ_i is the mean and σ_i is the variance.

Thanks to *3-sigma rule*, we could then define $[(\mu_i - \varepsilon_i),(\mu_i + \varepsilon_i)]$ as the promising interval that X_i should lie within at whatever moment. ε_i should be customizable and generally between 3σ and 7σ. Each occurrence of the event when we witness a value fall outside this interval should trigger the alert about an abnormal activity. In such case, our detection algorithm would be applied to determine the misbehaving node.

While performing experiments with MMT and Foren6, we successfully detected the evil nodes that were loaded with malicious firmware. Table 15.1 depicts the detection delay of MMT and Foren6 depending on the network size. In those experiments, we fixed the number of malicious nodes equivalent to 20% of the total nodes in the network. We witnessed that the demanding processing time to identify misbehavior nodes are strikingly increased with the number of nodes (both normal and abnormal ones). This delay consists of

Table 15.1 Comparison of Detection Delay between MMT and Foren6 (in Seconds)

No. of Nodes	No. of Malicious Nodes	MMT (s)	Foren6 (s)
5	1	13.87	13.43
10	2	31.9	32.14
15	3	48.54	49.11
20	4	66.58	64.87
25	5	84.61	85.22
30	6	108.22	110.56
35	7	128.92	131.94
40	8	152.57	155.15

the time for extracting the attributes, calculating variables, and performing the detection algorithm to determine malicious entities. It was growing with the number of nodes and paths under test because the number of computations used for the algorithm increases correspondingly. In any case, we observed basically the same performance for both MMT and Foren6.

3. *Real-time monitoring and the detection algorithm performance*: Third, in order to evaluate the influence of the threshold ε_i to the accuracy of our algorithm, we repeated the experiments with different thresholds in counting the number of false-positives and false-negatives. In the framework of this research, we did not observe any false-negative. Figure 15.12 illustrates the false-positive and accuracy rate related to an observed node N_i in function with the threshold ε_i. It is computed as follows:

$$\text{False\_positive rate (\%)} = \frac{\text{number of false\_positives} * 100}{\text{number of detection}} \quad (15.13)$$

Since there was no false-negative:

$$\text{Accuracy rate} = 100 - \text{false\_positive rate (\%)} \quad (15.14)$$

Indeed, we observed a very good accuracy when the threshold ε_i is bigger than $3 * \sigma_i$. These results validated once again the *3-sigma rule*.

4. *Algorithm extension*: Although the accuracy witnessed in the aforementioned experiments was high, we recognized that certain abnormal nodes detected were not really valuable. Sensor nodes are weak and sometimes fall to failure

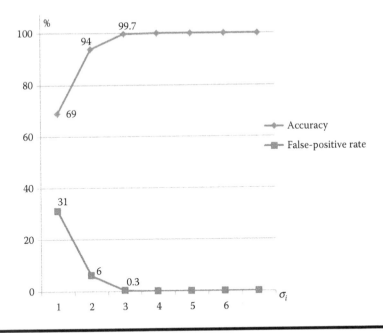

Figure 15.12 **Proposition's false-positive and accuracy rate in function with the threshold ε_i.**

but only momentarily and then come back to the normal state. The temporary fault state should be tolerated. To deal with this issue, we attempted to replace the momentary state in the monitoring phase by a more long-lasting state. X_i would not be calculated at a single moment t but as an average value in the period from $t - \tau$ to t, where τ is the observation duration and predefined based on the characteristics of the network. We performed our experiments with $\tau = 1$ min.

Evidently, the detection algorithm with and without the presence of τ must have different performance in terms of response delay (i.e., processing time) due to the additional time for querying other older values of X_i and for extra calculations (Table 15.2, AVG means *average (of)*). Experiments demonstrated that our solution allowed processing network traffic with bit

Table 15.2 **Solution's Average Processing Time and Throughput**

	Without τ	With τ
AVG (processing time) (ms)	383	528
Throughput (Kbps)	104	76

rate up to 104 Kbps (without τ) and 76 Kbps (with τ), which is sufficient for 6LoWPAN-based WSNs.

In conclusion, looking to some sample events occurring in some specific moment is not enough for a thorough security monitoring. Instead, monitoring should be a continuous process taking into account the history and the sequence of events.

5. *Related work*: There are actually several malicious and abnormal node detection schemes proposed in the literature for WSNs in general and recently for 6LoWPAN specifically. As a result of energy issues, most of them are based on a distributed model, using either neighbor coordination or clustering. For example, Curiac et al. [14] proposed an autoregression technique to detect malicious node. They saved past and present values provided by each sensor as the input of an autoregressive predictor to estimate an expected value. If the received value is too different from the expected one, the related sensor node must be questionable. The similar point of this work to our one is that we both predict an expected range based on the received one, that is, anticipate the future from the past. Falling outside from this range signifies an abnormal behavior. However, Reference 14 specifically copes with suspicious nodes sending malicious data; in other words, it cares more about the content of message rather than other aspects of the network, for example, delay, bit rate, and packet loss rate, which is the main concern taken in our work.

Atakli et al. [15] proposed another scheme to detect the compromised node using *weighted-trust evaluation*. The authors utilized a clustered topology for their hierarchical WSN network and built their detection scheme based on *weighted-trust evaluation*. They divided their network into three layers, including AP (access point), FN (forwarding node), and SN (sensor node) layers. FNs assigned a *weighted trust* to each SN and an algorithm was proposed to update this value based on what FNs receive from their SN. Nonetheless, Reference 15 presents simply some preliminary results derived from some simulations and the performance and the scalability of the solution are still a problem that the authors left as their future work. As an improvement, Seo Hyun Oh et al. [16] proposed another scheme using *dual-weighted-trust evaluation* to reduce the misdetection rate while maintaining comparable performance. Although *weighted trust* is very close to our idea in using *link weight* represented by *packet's travel time*, FNs in our case are also sensor devices that are not powerful enough to perform computations.

15.3.3.2 Case 2: Information Theory

The main idea of these experiments is to monitor the entropy value of the system (6LoWPAN-based WSNs) and see if it can be useful to design an anomaly detection model. Similar to the experiments in Section 15.3.3.1, we deployed 6LoWPAN-based WSNs, but this time, we cared only about routing packets. For each packet, we extracted the set of attribute consisting of *source's MAC address, destination's MAC address, timestamp*, and *type of routing packet*. So far, we defined five different routing packet types: *RPL DIS, RPL DIO, RPL DAO, Neighbor solicitation,* and *Router Advertisement*. An event e_i is defined as a triplet ⟨source's MAC address, destination's MAC address, type of routing packet⟩. We analyzed the traffic and recorded events received and then calculated the entropy of the set of all received events as a temporal variable.

First, we performed five experiments on the networks of 10 nodes. We monitored the entropy of the set of received events in approximately 40 min (from the booting of sensor nodes). The topology of the networks in five experiments were fixed but we loaded BR firmware to five different nodes. The sniffer was always located in the BR node.

As displayed in Figure 15.13, after the first 2 or 3 min of increasing very fast, the entropy became quite stable and slightly oscillated around a convergence value.

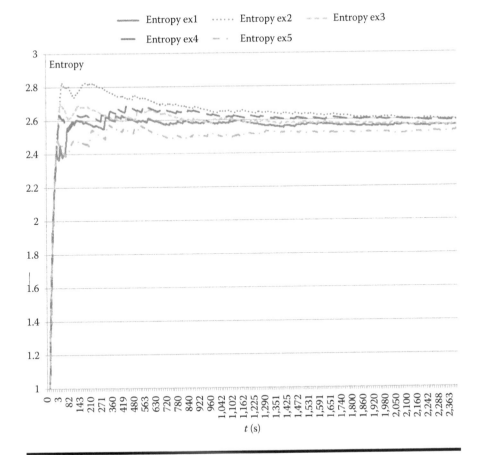

Figure 15.13   Entropy monitoring of 10 nodes under normal condition.

Figure 15.14 Entropy monitoring of 30 nodes under normal condition.

We witnessed this in all five experiments regardless of the fact that the convergence values are a little bit different among the five cases. We also observed that the more the BR node was located at the center of the network (i.e., the more symmetric the topology is), the smaller the entropy became (i.e., the purer the set of events is).

Second, we repeated the experiments another three times but on the networks of 30 nodes. We received similar results (Figure 15.14). Moreover, we noticed that the entropy regarding the 30-node networks is higher than the one regarding the 10-node networks. This is obviously understandable because the larger systems easily become more impure than the smaller systems.

Third, we performed another two experiments on the networks of 30 nodes. However, we rebooted the BR several times to see how the entropy variable reacted. Indeed, it reacted like we restarted the experiment (ex4 and ex5 in Figure 15.15).

Finally, we injected some routing attacks to the networks. Figure 15.16 (ex6) depicts the results when we forced some nodes to perform the *flooding attack* and the *selective forwarding attack* [17]. An almost immediate augmentation of the entropy is noticed at the moment of the attack. When the attack is terminated, the entropy returned to the stable state at a point higher than before the attack. This can be explained by the fact that our model had also taken into account the attack and thus the taken events are more impure.

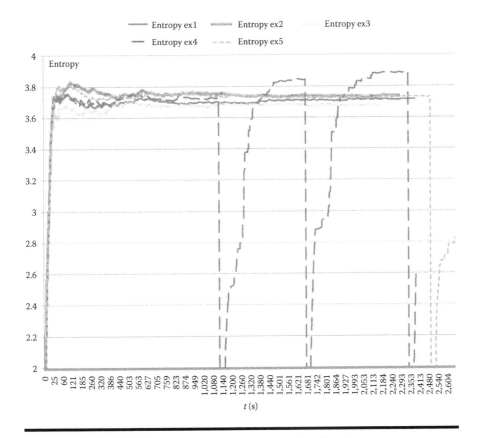

Figure 15.15   Entropy monitoring of 30 nodes under rebooting.

In short, we acknowledged the usability of entropy as a metric to monitor the 6LoWPAN-based WSNs. It can be a good candidate for other systems where we need to define the normal states. From our point of view, the link weight can be the metric providing the local view of the network, while the entropy can provide a global view of the whole system. As the future work, we would like to apply other supervised learning machine algorithms, for example, neural networks, support vector machines, and decision trees, to automatically learn the valuable metrics. The unsupervised learning should also be considered if we do not have labeled training data.

15.3.4 Other Existing Solutions

15.3.4.1 6LoWPAN Troubleshooting With Foren6

As far as we know, there have not been many monitoring tools for IoT in general and for 6LoWPAN-based WSNs in particular. Foren6 is seemingly the most well-known

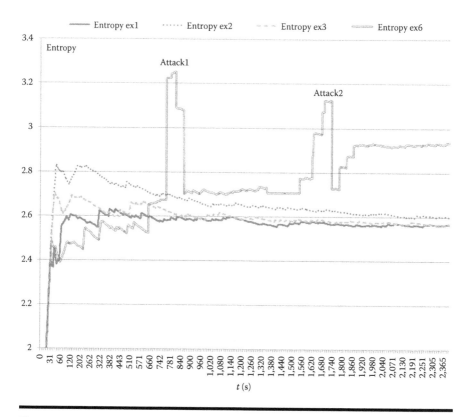

Figure 15.16 Entropy monitoring of 10 nodes under attacks.

one, which is compared in the previous section. However, the current version of Foren6 mainly focuses on visualizing the network topology and analyzing routing issues. Meanwhile, MMT is an extendable monitoring tool that allows adding plugins to define new input as well as writing rules describing both wanted and unwanted behaviors from the input. This flexibility makes MMT open to different types of input as well as to be able to adapt to different scenarios.

15.3.4.2 SVELTE: Real-Time Intrusion Detection in the Internet of Things

Regarding research works dealing with the security of 6LoWPAN objects, SVELTE [18] has been presented as the most well known among very few intrusion detection tools working over such small devices. SVELTE consists of three main centralized modules, including lightweight modules and mini-firewalls deployed in SNs, and a central modules called 6Mapper located in BRs. 6Mapper collects the routing information thanks to their "little" collaborators located in SNs. Experiments have been carried out by the authors and their team to evaluate SVELTE. In comparison with our approach, SVELTE is more active and creates additional traffic to realize their goal. We attempt to passively monitor the network based on the network's traffic to avoid additional costs that might hamper 6LoWPAN.

15.4 Intrusion Tolerance: An Open Issue

For a long time, people conducted much research on security mechanisms to prevent or detect intrusions and attacks. However, attacks are more and more sophisticated and thus, they are difficult to be captured. A system may fail to complete its mission whenever a successful attack occurs and it may be impossible to recover quickly. In the past few years, the research community has started to spread the issue of attack tolerance that will allow a system strong enough to tolerate attacks.

Intrusion or attack tolerance of a system is generally understood as the capability to continue to function properly with minimal degradation of performance, despite intrusions or malicious attacks [19]. In terms of networks, this concept means the ability to maintain the overall connectivity and diameter of the network as nodes are removed.

Figure 15.17 indicates features that should be integrated to a modern system or network. To gain intrusion/attack tolerance, systems employ redundancy, diversity, and reconfiguration to remove unwanted intrusions and recover the normal state. From our point of view, *intrusion tolerance* can be the next step of *intrusion detection*. In other words, if a system detects some malicious signs, it can react by taking into effect a number of adaptive countermeasures to "tolerate" the attacks (to continue to provide the service with the minimal influence of performance).

Specifically in WSNs, this task can be done by changing the network routing protocol in reacting to a detected intrusion. Indeed, the idea of intrusion-tolerant routing for WSNs first appeared around the middle of the last decade; however, there has been very little research on this topic. Given that an intrusion-tolerant routing protocol would definitely cost some extra resource of the network devices in terms of computing and energy, applying it even under the normal conditions seems expensive. Therefore, the idea is that a universal routing protocol (e.g., RPL) is used until the intrusion detector discovers some malicious signals inside the network, then an intrusion-tolerant routing protocol substitutes. It will be maintained till the intrusive elements are eliminated or neutralized.

INSENS is to date the most well-known intrusion-tolerant routing protocol for WSNs. ITSRP is another proposition that is much less popular but still deserves to be considered because it takes into account the problem of "energy consumption" that is a very critical issue of WSNs. These two approaches were studied in our paper work [21] in which we provided a comparative evaluation and proposed several suggestions to improve the protocol's effectiveness and performance. As a future work, we would like to implement and trigger these two protocols in response to

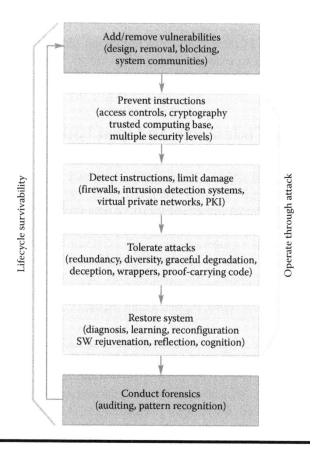

Figure 15.17 Security features for modern systems. (Adapted from Raytheon, Intrusion-tolerant systems, *Technology Today Journal*, 2007.)

intrusive activities to appraise the pros and cons of each proposition in reality. The results will be, in our opinion, useful for many critical IoT systems, such as those in the military domain.

15.5 Conclusions

In summary, this chapter proposes MMT as the monitoring tool for WSNs using 6LoWPAN technologies that allows detecting intrusions and attacks. To the best of our knowledge, it is the first monitoring tool dealing with the traffic in the 6LoW-PAN environment. As a demonstration, we implemented a real test-bed to assess the tool in using different methodologies based on statistical learning and information theory (entropy). Promising preliminary results prove our approaches' applicability and extensibility. In comparison with other traffic analyzer and intrusion detection system for 6LoWPAN devices, our approach provides a more thorough view of the network with lighter effect to its performance. Intrusion tolerance will be, from our perspective, an innovative concept that should be integrated into future WSNs. It could be achieved by design (intrusion-tolerant routing protocols) and/or by performing adaptive reactions, enabling tolerance of the intrusive elements.

References

1. J. Macaulay, L. Buckalew, and G. Chung, Internet of Things in logistics, *DHL Trend Research*, 1(1): 1–27, 2015.
2. Libelium Comunicaciones Distribuidas S.L. Waspmote Datasheet. Product description, 2014.
3. O. Hersent, D. Boswarthick, and O. Elloumi, *The Internet of Things: Key Applications and Protocols*, Wiley, Chichester, West Sussex, 2012.
4. M. Rita Palattella, N. Accettura, X. Vilajosana, T. Watteyne, L. Alfredo Grieco, G. Boggia, and M. Dohler, Standardized protocol stack for the internet of (important) things, *IEEE Communications Surveys and Tutorials*, 15(3): 1389–1406, 2013.
5. X. Ma and W. Luo, The analysis of 6LoWPAN technology, in *Computational Intelligence and Industrial Application, 2008. PACIIA '08. Pacific-Asia Workshop on*, vol. 1, pp. 963–966, Dec 2008.
6. J. P. Vasseur, N. Agarwal, J. Hui, Z. Shelby, P. Bertrand, and C. Chauvenet, RPL: The IP routing protocol designed for low power and lossy networks, *Internet Protocol for Smart Objects (IPSO) Alliance*, April 20, 2011.
7. D. Karaman, N. Gozuacik, M. O. Alagoz, H. Ilhan, U. Cagal, and O. Yavuz, Managing 6LoWPAN sensors with CoAP on Internet, in *Signal Processing and Communications Applications Conference (SIU), 2015 23th*, Malatya, Turkey, pp. 1389–1392, 2015.
8. S. Raza, D. Trabalza, and T. Voigt, 6LoWPAN compressed DTLS for CoAP, in *Distributed Computing in Sensor Systems (DCOSS), 2012 IEEE 8th International Conference on*, pp. 287–289, 2012.
9. V. H. La, R. Fuentes, and A. R. Cavalli, Network monitoring using MMT: An application based on the user-agent field in HTTP headers, in *2016 IEEE 30th International Conference on Advanced Information Networking and Applications (AINA)*, pp. 147–154, 2016.
10. B. Wehbi, E. Montes de Oca, and M. Bourdelles, Events-based security monitoring using MMT tool, in *IEEE Fifth International Conference on Software Testing, Verification and Validation (ICST), 2012*, pp. 860–863, 2012.
11. J. Pokhrel, B. Wehbi, A. Morais, A. Cavalli, and E. Allilaire, Estimation of QoE of video traffic using a fuzzy expert system, in *Consumer Communications and Networking Conference (CCNC), 2013 IEEE*, Las Vegas, USA, pp. 224–229, Jan 2013.
12. C. M. Bishop, *Pattern Recognition and Machine Learning (Information Science and Statistics)*, Springer-Verlag New York, Inc., Secaucus, NJ, USA, 2006.
13. C. E. Shannon, A mathematical theory of communication, *SIGMOBILE Mobile Computing and Communications Review*, 5(1): 3–55, 2001.

14. D.-I. Curiac, O. Banias, F. Dragan, C. Volosencu, and O. Dranga, Malicious node detection in wireless sensor networks using an autoregression technique, in *International Conference on Networking and Services (ICNS '07)*, Athens, Greece, pp. 83–83, 2007.

15. I. M Atakli, H. Hu, Y. Chen, W.-S. Ku, and Z. Su, Malicious node detection in wireless sensor networks using weighted trust evaluation, in *Proceedings of the 2008 Spring Simulation Multiconference*, Ottawa, Ontario, Canada, pp. 836–843, 2008.

16. S. Hyun Oh, C. O. Hong, and Y.-h. Choi, A malicious and malfunctioning node detection scheme for wireless sensor networks, *Wireless Sensor Network*, 2012(March): 84–90, 2012.

17. I. Grand, E. Nancy, and T. Nancy, A taxonomy of attacks in RPL-based internet of things, *International Journal of Network Security*, 18(3): 459–473, 2016.

18. S. Raza, L. Wallgren, and T. Voigt, SVELTE: Real-time intrusion detection in the internet of things, *Ad Hoc Networks*, 11(8): 2661–2674, 2013.

19. F. Wang, R. Uppalli, and C. Killian, Analysis of techniques for building intrusion tolerant server systems, in *Military Communications Conference, 2003. MILCOM '03. 2003 IEEE*, vol. 2, pp. 729–734, Oct 2003.

20. Raytheon, Intrusion-tolerant systems, *Technology Today Journal*, (2): 12, 2007.

21. V. Hoa La and A. R. Cavalli, A comparative evaluation of two intrusion-tolerant routing protocols for wireless sensor networks, in *10th International Conference on Broadband and Wireless Computing, Communication and Applications (BWCCA 2015)*, Krakow, Poland, pp. 6–12, Nov 2015.

Chapter 16

Security Concerns in Cooperative Intelligent Transportation Systems

Konstantinos Fysarakis, Ioannis Askoxylakis,
Vasilios Katos, Sotiris Ioannidis, and Louis Marinos

Contents

16.1 Introduction

Smart vehicles will be an important segment of the imminent Internet of Things (IoT)-enabled world. Modern vehicles already feature many embedded electronics that monitor and control their subsystems, to enhance passenger comfort and safety, achieve energy-efficient operation, and maximize vehicle lifetime. Superior safety features can help avoid many accidents, and they are the focus of various governmental initiatives worldwide, which define stricter regulations (such as the COMMISSION DIRECTIVE 2008/89/EC enforcing daytime running lights in new vehicles). This push is expected to intensify, leveraging the benefits of intelligent transportation systems (ITS), and advanced features like early braking, road lane departure warnings, and prompt emergency response services. This trend is also evident in the relevant guidelines and policies, such as the "Policy orientations on road safety 2011–2020" European Union (EU) program [1].

To be able to support this range of sophisticated features, modern vehicles are equipped with an assortment of embedded computing devices, sensors, actuators, and communication interfaces. While a Boeing 787 Dreamliner aircraft requires about 6.5 million lines of software code to operate its avionics and onboard support systems, a modern car may already feature over 20 million lines of code, with predictions that cars will require 200 million to 300 million lines of software code soon [2].

While software is the main area of innovation and value in modern vehicles, the added complexity comes at a cost. Experts predict that the cost of software and electronics, already at 35%–40% in some vehicles today, may reach 80% for certain types of vehicles in the future; moreover, over 50% of car warranty costs can be attributed to electronics and their embedded software [3]. In February 2014, carmaker Toyota recalled 1.9 million hybrid cars around the world following the discovery of faulty software in the car's hybrid-control system [4]. The software glitch could cause the hybrid system to shut down while the vehicle is being driven, resulting in the loss of power and the vehicle coming to a stop.

Moreover, security-related incidents are a tangible threat. Exploiting vulnerabilities in the vehicle's electronics can allow the remote control of vehicle components; an attacker can control turn off the lights or even control the brakes while on the move [5]. More recently, hackers managed to successfully control Fiat/Chrysler production vehicles over the Internet; by exploiting vulnerabilities on the UConnect system, they were able to apply the brakes, kill the engine and even take control of the steering. In response, the company urged owners to update their cars' software to patch the identified vulnerabilities [7].

Considering that, as estimated, there are over 250 million vehicles already roaming EU roads alone [8], the potential for improvement in passenger safety and corresponding reduction in loss of human lives, as well as the potential for new types of massive, distributed smart vehicle and infrastructure-based attacks, which may endanger human lives, are equally high.

Moreover, many of the promised enhanced services of this new interconnected world rely on the location of the vehicle and its driver, a private-sensitive information in nature that gives rise to significant privacy concerns.

Still, securing the various heterogeneous hardware and software platforms and networks in the ITS ecosystem is a challenging task. While security is necessary in various aspects of the smart vehicle-related information and communications technology (ICT) deployments, many aspects of efficient ITS operations rely on very low latency (especially safety related ones) and other quality of service (QoS) characteristics which often limit the applicability of complex security primitives. Therefore, the proposed solutions should consider and work around these limitations.

There is currently no EU policy or requirements on security for transport [9,10]. In addition, methodologies and tools for the assessment of combinations of physical and cyber risks are scarce and offer only limited guidance for the transport sector on how to assess these risks. Yet such combinations of risks are expected to increase [11,12]. This highlights the need for the development of transport specific tools to assist in the analyses of risks due to combined physical and cyber-attacks and especially on interdependent and dependent land transport systems. Efforts should focus on providing a methodology for multi-hazard risk analyses in C-ITS interdependent/dependent systems. These analyses should be based on a detailed and credible vulnerability assessment, including an extended and detailed analysis of how combined attacks can provide cascading effects and how hazards can propagate throughout the C-ITS deployments. Moreover, the effectiveness of these efforts should be enhanced by introducing a method for analyses to cope with the required level of detail and extent of the problem and a scalable framework that can be extended to include additional systems and threat scenarios.

Motivated by the above, this work presents the current security landscape in the cooperative intelligent transportation system (C-ITS), examining both vehicle-to-vehicle (V2V) and vehicle-to-infrastructure (V2I) interactions. Moreover, future directions are highlighted in the context of providing a holistic intrusion detection, prevention, and mitigation approach for C-ITS deployments, as they constitute an integral part of the critical cyber-physical systems that it is urgent to protect.

The chapter is organized as follows: Section 16.2 presents the motivation behind this effort and key background information on the technologies involved in a C-ITS environment, and Section 16.3 provides a comprehensive overview of the current threat landscape in the field; Section 16.4 presents some key pointers to providing a future-proof and comprehensive approach to intrusion detection, prevention, and mitigation in the context of C-ITS deployments, while Section 16.5 concludes this work with some important points for future efforts.

16.2 Background

There is already a consensus in the academia and the industry alike that the introduction of smart vehicles, smart road infrastructure, and the associated services will significantly reduce accidents, and will provide more efficient and environmentally friendly transportation for everyone. Moreover, real-time monitoring of the vehicle's state and the driver's behavior will allow public entities, logistics organizations, and other businesses to minimize the vehicle investment risks and promote strategies for increasing productivity and safety while reducing transportation and staff costs. Government regulations are decisive motivators of pertinent research efforts.

The European Commission defines new regulations for vehicle safety, such as the eCall system [13], which will become mandatory for every vehicle moving in the European Union by 2018. This emergency service dictates that when an accident occurs, the vehicle should automatically relay essential information (its location, its direction and speed before the crash, number of passengers, etc.) to appropriate public safety answering points (PSAP). By providing early notification and allowing efficient coordination of the emergency services, it is expected to decrease the response time to such incidents by 50% in rural areas and 40% in urban areas, drastically reducing the number of deaths and the severity of injuries for the thousands of people involved in road accidents every year [1].

The United Kingdom aims to minimize road deaths in business-owned vehicles; starting in 2008, road death is considered an unlawful killing, enabling seizing of the company's records and bringing prosecutions against directors who fail to enforce safe driving policies. Therefore, fleet management is now imperative for organizations owning a significant number of vehicles. Automotive legislation also necessitates the production of more eco-friendly vehicles, a target partly achieved by subsystems monitoring the vehicles' operation in real time, triggering adjustments to engine parameters.

On the basis of the above stimuli, the integrated electronics increase with every vehicle generation, and are expected to rise steeply with the introduction of smart and, eventually, self-driving vehicles. This "intelligence" will also enable a variety of novel services that everyone will enjoy, from end-users (e.g., parents lending the family vehicle to their teenager) to private and public entities operating vehicle fleets (logistics, car-rental, governments, rescue services, etc.).

A modern vehicle may already utilize over 80 built-in microprocessors, typically interconnected via the controller area network (CAN bus). These microprocessors are tasked with providing advanced safety systems, emission monitoring, and in-car commodities [14] which aim to enhance passenger comfort and safety, also protecting the vehicle's subsystems by providing early warning of failures and/or adjusting their operation accordingly. Typically, electronic control units (ECU) manage and interconnect the distinct systems [15], and the infotainment infrastructure provides enhanced facilities, like navigation, to passengers [16]. A rough sketch of a modern vehicle's typical elements (color-grouped by ECU function) and the network architecture that interconnects them can be seen in Figure 16.1.

Newer vehicle generations will take this further, supporting communication with other vehicles (vehicle-to-vehicle and V2V communications), the road infrastructure (vehicle-to-Infrastructure, V2I communications), and backend systems

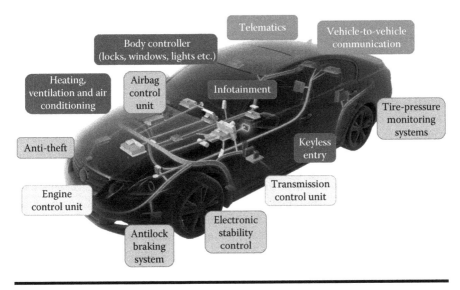

Figure 16.1 A modern vehicle's typical electronics, color-grouped by ECU function.

providing a number of enhanced services (e.g., vehicular cloud computing services or sophisticated car insurance services). Such prototype deployments are already under assessment in the EU and the United States (USA); for example, the UMTRI Safety Pilot [19].

In ITS environments, V2V interactions typically rely on the instantiation of vehicular ad-hoc networks (VANETs), a sub-type of self-organized, large-scale mobile ad-hoc networks (MANETs), with single-hop and/or multi-hop and broadcasting or multicasting communications. Since VANETs feature vehicles as the mobile nodes, they come with all the associated intricacies compared to typical MANETs; for example, not as resource-constrained as a sensor, high speed and large-scale mobility, highly dynamic contact between numerous nodes, and privacy concerns. Nevertheless, a fully featured ITS deployment, or cooperative intelligent transportation system (C-ITS), as it may be referred to, is not limited to communications between vehicles but also includes other heterogeneous devices and the services and applications that run on top of those. In this context, VANETs are only part of the communication infrastructure of the ecosystem. A C-ITS features a multi-communication model that, in addition to communications between vehicles, also features communications between other forms of transport (e.g., trains, buses, motorcycles, and bicycles) and even other objects such as flying drones and other autonomous systems (referred to as vehicle-to-everything communications, or V2X). Other types of communications can include pedestrians (vehicle-to-pedestrians, V2P) as well as interactions with the infrastructure and other parts of the road network (vehicle-to-infrastructure, V2I), such as fixed road-side units

(RSUs) and mobile RSUs. Moreover, communication between infrastructure entities is needed (infrastructure-to-infrastructure, I2I), for example, a smart traffic light communicating with a smart road lamp, as well as the presence of backend systems (e.g., for traffic management). The RSUs will typically also feature direct communication with the backend infrastructure, via a backbone network. One or more trusted authorities (TAs), or certificate authorities (CAs), can also be present at the backend for the registration, issuance, and validation of certificates of the involved entities, an integral part of vehicular public key infrastructure (VPKI) setups. Other than the above entities, another important element in the C-ITS landscape are the services themselves, including enhanced version of existing services (e.g., tolling) as well as novel transport-related services (e.g., safety systems, fleet management, and travel planning) and the associated infrastructures that support them. Consequently, a C-ITS may also involve a variety of service providers (e.g., fleet management and leasing companies), ICT systems and communication networks that enable the corresponding applications, as well as the data generated and operated upon in the context of these services. The coordination and integration of said services aims to bring major social and economic benefits, by maximizing the benefits of transportation to both commercial users and the general public, leading to greater transport efficiency, minimized environmental impact, and increased safety.

In terms of communications, various heterogeneous networking technologies are proposed in this environment, such as WiFi IEEE 802.11p, IEEE 1609 WAVE (a higher layer standard based on the IEEE 802.11p), WiMAX IEEE 802.16 (for wider area communications), Bluetooth, and cellular (e.g., LTE). For vehicles, the IEEE 802.11p-based dedicated short-range communications (DSRC, [20]) set of protocols and standards is widely accepted for one or two-way short to medium-range communications, being adopted in the United States, EU (where it coexists with ITS-G5 [21]), Japan, and other major markets, though discrepancies and incompatibilities exist between the variants [22]. In parallel to the above efforts, an important competing (or complementing, in some cases) approach, and the focus of numerous academia and industry-based research efforts, is the introduction of 5th generation (5G) networks, which aim to be applicable to a variety of vertical applications, including cellular vehicle-to-everything (C-V2X) interactions that will pave the way for fully autonomous driving [23].

To enable these enhanced features, vehicles must typically feature onboard units (OBUs) supporting the communication technologies required for V2V, V2I, and V2X communications. Moreover, one or more computing platforms are required, controlling all the sensors, cameras and radars/LIDARs needed (especially in the case of autonomous vehicles), a positioning system (GPS), electronic license plates (ELPs), as well as an event data recorder (EDR), a tamper-proof device acting as the vehicle's "black box," recording all critical events and operations [24].

This heterogeneous networking infrastructure enables all parts of the C-ITS to share information, improving decision making and enabling the provision of novel, enhanced types of services. These enhanced services may include enhanced

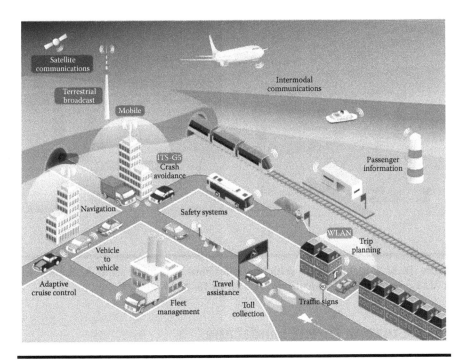

Figure 16.2 High-level view of a C-ITS ecosystem. (Adapted from European Telecommunications Standards Institute (ETSI), EN 302 663 V1.2.1, Intelligent Transport Systems (ITS); Access Layer Specification for Intelligent Transport Systems Operating in the 5 GHz Frequency Band, July 2013, http://www.etsi.org/ deliver/etsi_en/302600_302699/302663/01.02.01_60/en_302663v010201p.pdf. © European Telecommunications Standards Institute 2017. Further use, modification, copy and/or distribution are strictly prohibited.)

vehicle insurance models, whereby the driving behavior, distance, and areas travelled directly affect the insurance fees, providing novel usage-based insurance (UBI) schemes, and the associated pay as you drive (PAYD) and pay how you drive (PHYD) models. The real-time data link between road infrastructure and vehicles (types of which can be seen in Figure 16.2, presenting a view of a C-ITS deployment) can also enable numerous other services. The said services promise to reduce wait times and increase the efficiency of transport, by providing up-to-date information on mobile roadworks, wrong-way driver and pedestrian alerts, remaining red and green light times, and by enabling features such as the dynamic coordination green light phases, intelligent parking space management, and priority for emergency vehicles and public transport [25–27]. Figure 16.3 presents some indicative applications. While the potential and impact of these new cooperative traffic communications will reach its full potential with autonomous vehicles, considerable benefits can be earned even in these early stages.

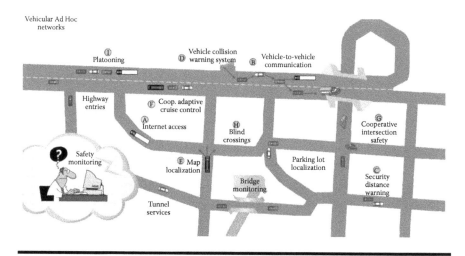

Figure 16.3 Possible smart vehicle applications. (Adapted from A. Boukerche et al., *Computer Communications,* **31(12): 2838–2849, 2008.)**

16.3 Current Threat Landscape

Aiming to provide an overview of the current security landscape in the area, the next subsections identify key *assets* (i.e., anything that has value and must be protected), the *threats* to these assets (i.e., anything that poses some danger to an asset), the *adversaries* (i.e., agents who wish to abuse, damage, or otherwise compromise the assets), and the types of *attacks* that these adversaries may use to realize the said threats. A full risk assessment is not the aim of this work; nevertheless, any of the available risk assessment methodologies [29] could be followed by involved stakeholders, if the scope of the evaluation is limited to their specific use cases and applications. Instead, the application-specific security requirements and challenges will be presented to complete the view of the landscape.

16.3.1 Assets

While it is not possible to list all assets that may be present in a complex, full-scale C-ITS deployment, some key assets that must be considered and protected include the smart vehicles and other transport means (e.g., mass transit), the RSUs, the operators' and services providers' infrastructure, the communications infrastructure as well as the various human assets and their personal data. A mind map of key assets is presented in Figure 16.4.

The key asset domains include the smart vehicles and other smart transport entities, the smart road infrastructure (e.g., RSUs or the road itself), the communication infrastructures, the backend services, and the existing (potentially enhanced) and novel services and applications offered in the context of the C-ITS ecosystem.

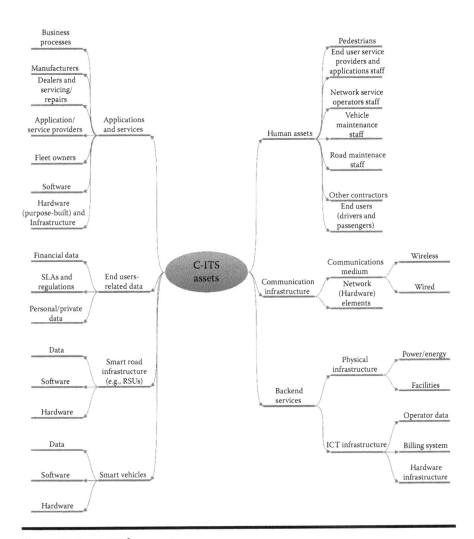

Figure 16.4 C-ITS key assets.

Finally, an important domain is that of the human assets and the data associated with them (often private sensitive in nature). With respect to the smart entities comprising the C-ITS ecosystem, the most common one is the smart vehicles themselves. Specific assets that can be found in such vehicles include the hardware components (e.g., EDR, GPS receiver, ECUs, antennas, and ELPs), the software components (e.g., infotainment and operating systems), and the associated data (e.g., recorder on the EDR). The same types of assets can be found in the smart road infrastructure entities, such as the RSUs, as each comes with its own set of hardware (e.g., traffic lights) and software (e.g., firmware) components as well as the relevant data required (e.g., settings) or produced (e.g., logs) during its operation. The communication infrastructure domain includes all types of heterogeneous networking technologies that can potentially be found in a C-ITS environment (e.g., for V2V or V2I interactions), including their hardware (e.g., network devices) and the communication medium itself. Two other significant domains are those of the backend services that the operator and/or owner of the C-ITS infrastructure is responsible for (including the physical infrastructure and ICT-based assets such as the billing system or the associated data), and the enhanced applications and services that are enabled by (and form an integral part of) the C-ITS. The latter domain may feature a wide range of assets, from smart vehicle manufacturers, dealers and servicing, the fleet owners (e.g., logistics companies), the application and enhanced service providers themselves, and the business processes (e.g., the distribution monitoring processes), the software (e.g., application), and the hardware (e.g., purpose-built hardware modules that monitor driving patterns for vehicle insurance purposes) used to enable the said services. Since most of the above are intelligent systems, they operate upon and produce data. Thus, arguably the most important domain in this context is the end-user (e.g., driver)-related data, such as the financial data, the relevant service level agreements (SLAs) and regulations, and, of course, the private sensitive data pertaining to each user (e.g., her location or her driving habits). Finally, a very important domain is that of the human assets, that is, the individuals who work in, use and/or benefit from the C-ITS in any way. The end users (be it drivers or passengers of private and public transport vehicles) are a key asset in this category and related assets include the users' physical (e.g., items and vehicles owned by the users) and virtual (partly the user data already mentioned above, but also data stored on his/her personal devices, the associated software purchases, virtual currency, settings, etc.) properties. Other human assets involved in the C-ITS context are the pedestrians who walk on the smart road or even interact with smart vehicles (e.g., via their mobile phones or, in the future, via sensors on their smart clothing). Other individuals in this domain are the staff of the various businesses and service providers involved in C-ITS deployments, such as the service/application provider companies, road maintenance and contractors' staff, as well as personnel responsible for servicing the various vehicles. Moreover, in every one of these cases, an important asset in this category is the health of all involved individuals that could, for example, be harmed if smart vehicles crash.

16.3.2 Adversaries

When considering actors who may try to damage or otherwise hinder the operation of individual assets or the smart vehicle ecosystem, the following types of adversaries are identified:

■ *Eavesdroppers*: Passive attackers who monitor the network, capture traffic, and passively gather information, for example, to track individual vehicles. These may vary from invasive insurance companies trying to monitor their clients' driving habits beyond the agreed service terms, burglars trying to figure out when someone is away from home to break in or just curious individuals trying to see if it is possible to track their neighbor.

■ *(Infrastructure) Insiders*: Attackers who work for the stakeholders managing the infrastructure or the associated services (e.g., smart road contractors or a service operator) and who, for example, exploit their access to the management of these assets to launch an attack.

■ *Greedy drivers*: Vehicle drivers who will try to exploit the systems to maximize their own gains, disregarding the convenience or even safety of other drivers, by, for example, informing neighboring vehicles that the road ahead is blocked, causing the other vehicles to choose an alternate course, thus clearing the road ahead and enabling a faster trip to his/her destination.

■ *Pranksters*: Individuals who have no specific goal other than to become famous via an attack that makes the headlines or goes viral or individuals who are just bored and try to meddle with the vehicles or the smart vehicle infrastructure.

■ *Malicious attackers*: Adversaries whose sole aim is to damage a specific asset or disrupt the normal operation of the whole system. These could include individuals seeking revenge from a specific person, competitors trying to harm a service operator, a contractor or vehicle manufacture, terrorists who wish to harm human lives, or even state actors who wish to disable the transport infrastructure of a country they are hostile to, and disrupt supply lines, for example, during a combined physical and cyber-attack.

■ *Unintentional attackers*: Drivers or employees managing the infrastructure who accidentally cause damage to an asset by, for example, damaging critical sensors on the vehicle or erroneously setting up a smart roadside sign.

16.3.3 Threats and Associated Attacks

As with any ICT system, C-ITS deployments are exposed to a number of threats and associated attacks. Threats can typically be classified as deliberate (e.g., a malicious attacker trying to disable a vehicle) or accidental (e.g., a roadworks employee erroneously setting up or disabling a roadside unit). Deliberate threats can be further categorized as passive (e.g., monitoring communications) or active (e.g., tampering with the contents of a message). Individual threats include but are not limited to network-specific ones, such as message deletion, message modification, message forgery/fabrication, and message replay, but also more generic threats such as information disclosure, denial of service (DoS), repudiation, identity spoofing and physical damage, as well as application-specific ones, such as tracking.

The potential attacks launched by the adversaries to realize the various threats can be classified [30] as follows:

■ *Passive versus active*: *Passive* attacks include techniques such as monitoring, passive tracking, and noninvasive network reconnaissance, while *Active* attacks involve invasive techniques such as traffic modifications and packet injections.

■ *Malicious versus rational*: *Malicious* attackers typically do not aim to gain a specific benefit but focus on damaging or disrupting the normal operation of an asset or the system in general (e.g., by destroying a roadside unit or jamming the network), thus having no limitations on the means that may be used to accomplish their target. *Rational* attackers, on the other hand, aim to gain a specific benefit (e.g., to avoid toll charges or to arrive faster at their destination), and thus have limited tools to accomplish their goal and their actions and intents are easier to predict.

■ *Intentional versus unintentional*: *Intentional* attacks are those attacks that are carried out on purpose, with the intent to achieve a specific goal. *Unintentional* attacks are those that are, for example, the result of negligence, mishandling, or inappropriate training on behalf of some users (e.g., road maintenance personnel defining wrong settings on a smart roadside unit). Unintentional attacks may also include the results of natural (environmental) disasters.

■ *Insider versus outsider*: Insider attacks are those carried out by entities belonging to the ITS deployment, that is, having valid credentials and access to (a subset) of the assets and services; these could include a driver, a vehicle, an employee working for a contractor or an operator, acting, for example, in a malicious, unauthorized, or unpredicted manner. Outsider attacks are those launched by external to the system entities, such as an intruder or a remote hacker.

■ *Local versus extended*: Local attacks are limited to a specific area (e.g., an area covered by a compromised RSU or a jammer). Extended attacks cover a large area, affecting more areas and entities across the network; this enables more sophisticated attack techniques, such as continuous tracking and wormhole attacks.

It is not feasible to list all current and future attacks that adversaries may launch against the C-ITS deployment, exploiting known and unknown security weaknesses that such a complex and heterogeneous ecosystem may have. VANETs share most of the security weaknesses identified in MANETs; still, the former have the advantage of increased resources and almost no energy restrictions, thus excluding some attacks. The infrastructure itself is also vulnerable as it mostly relies on typical ICT systems. Moreover, all assets, including vehicles and RSUs, are subject to physical attacks, though detailing these is beyond the scope of this work.

The expected targets of the attacks include safety-related applications, traffic optimization applications, payment-based applications, and the end user's privacy. Some basic attack types and ones that are expected to occur more often include

Snooping is a passive attack where an attacker monitors and possibly captures all network communications, hoping to retrieve sensitive data (e.g., location and number of vehicles) or even confidential data that may be sent unencrypted.

Traffic analysis is a passive attack against the privacy (and potentially the confidentiality) of the users. It involves the analysis and correlation of aggregated (eavesdropped) network traffic to extract useful information about its operation, retrieve sensitive information, or profile legitimate users and entities to, for example, learn the driving habits and daily schedule of a specific individual.

(Location) Tracking is an attack against the privacy of the end users [31]. It involves tracking the unique ID of the vehicle if these are exposed (e.g., in direct interactions with other vehicles) and/or involves attempts to correlate temporary IDs that the vehicles may use to obfuscate their identity. Compromised vehicles and RSUs, as well as the centralized monitoring of the vehicles' location from a backend system (enabling more sophisticated attacks that e.g., calculate vehicle trajectories), can be used to enhance the efficacy of the location tracking.

Message replay is a trivial active attack that involves rebroadcasting/injecting a message that was previously sent by a legitimate entity and which the attacker had captured, such as a message that informs vehicles in the vicinity about a crash or broadcasts a specific vehicle ID. It can typically be launched by anyone, even entities not having legitimate access to the network.

Denial of service (DoS) and distributed DoS (DDoS) are attacks that target the availability of the system. These types of attacks are typically easy to launch and very dangerous in the context of C-ITS deployments, as, for example, an autonomous vehicle may not get a timely warning for an accident on the road ahead. One type of such attacks is *spamming* which is a technique used to flood the network with high volumes of messages to disable nodes (e.g., a vehicle's OBU or an RSU) that cannot handle such amount of data, thus launching either a DoS [32] or DDoS [33] attack. *Jamming* [34] is a DoS attack launched by transmitting a signal that disrupts communications at the physical level; the signal-to-noise ratio (SNR) of the channel is decreased significantly, making communication impossible. Depending on the power of the jamming signal's transmission, it may affect a limited or larger area. Jamming attacks can also focus on disabling the global positioning system (GPS), which is a critical part of most C-ITS applications. *Greedy attack* [35] refers to malicious nodes not respecting the channel access method, for example, minimizing its wait time, thus gaining faster access but also causing collision problems that do not allow other nodes to access the medium, producing delays. *Blackhole attack* [36], in its basic form, is an attack whereby a malicious node disrupts the routing process by constantly advertising that it is available to route data, but then refuses to forward the packet it receives. If the malicious node additionally advertises good routing characteristics, thus attracting more traffic, this can be categorized as a *Sinkhole attack* [37].

Message suppression is a more sophisticated and subtle class of attacks than all out DoS; it involves selectively delaying or dropping certain packets, possibly also stealthily manipulating the routing process (both from a network topology as well as physical location perspective). In this category of attacks, *selective forwarding* [38] refers to an attack whereby the attacker only forwards specific packets (e.g., of a specific application or node), dropping the rest. This causes significant degradation or even DoS for the affected services. For example, a prankster could choose to drop messages that carry congestion-related information, but forward the rest; this would not harm the overall operation of the VANET, but affected drivers will have to unnecessarily wait in traffic. *Timing attacks* [39] affect the delivery of time-critical messages; malicious nodes do not forward time-critical messages (e.g., safety-related alerts) immediately, but only after some delay, and the neighboring nodes do not receive it on time. Thus, for example, other vehicles are only informed of an accident that happened ahead when it is already too late for them to brake on time. A *grayhole attack* [40] is a more complex variant of the Blackhole and Sinkhole attacks mentioned above, the attacker selectively removing data related to applications vulnerable to packet loss, affecting the operation of the service and associated application for all (or specific) legitimate users. A *wormhole attack* [41] is a cooperative attack whereby two or more malicious or compromised nodes, which are far from each other, collaborate to deceive their neighboring, legitimate nodes to believe that the malicious nodes are close together. This creates a tunnel, transmitting data to and from distant parts of the network, disrupting the multicast and broadcast operation of the network. The malicious entities can communicate via the legitimate VANET medium (in band attack) or via their own separate channel (out of band attack).

Message tampering refers to an active attack against integrity, whereby an attacker tampers with the content of legitimate messages. Moreover, in *Message fabrication* attacks, adversaries are expected to broadcast fabricated false information, either to damage some of the assets by causing, for example, some vehicles to brake immediately and others to speed up, causing a crash (malicious attack) or to gain benefits, by broadcasting, for example, messages that make neighboring vehicles change path, so that the adversary can get faster to his/her destination (a rational attack). Some examples of these attacks include *bogus information,* which refers to attackers (typically rational drivers) transmitting fake information on the network to trick their peers and benefit from it, for example, by making sure that all vehicles are diverted and the road ahead is free and thus faster to traverse. *Masquerading* is an attack whereby an attacker hides her true identity and pretends to be another valid entity, possibly producing false messages that appear to come from a legitimate source, gaining immediate benefits (e.g., by pretending to be an emergency vehicle and get priority) or paving the way for other attacks (e.g., Blackhole). A sophisticated version of this is the *Sybil attack* [42], where the malicious entity assumes multiple identities simultaneously, using them to gain a disproportionally large influence on routing protocol, compromising and effectively controlling the network. *GPS spoofing* is an attack on the reported position of an entity, typically achieved by broadcasting localization signals that are stronger than the one received by the GPS satellites, thus misleading the GPS receivers in the area [43]. The repercussions of such a successful attack are significant, considering that ITS applications rely heavily on the accurate and authentic location of vehicles and other means of transport for some of their safety features as well as the services they provide. *Illusion attack* [44] refers to an attack that involves having adversary-controlled sensors transmit erroneous information (e.g., about traffic conditions) to the other entities in the vicinity. These may be external sensors (e.g., a fake RSU or a set of sensors emulating a vehicle) or the sensors of a compromised legitimate vehicle which are made to report fake sensing data. The latter case is harder to defend against, as a legitimate vehicle has no problem authenticating itself and its messages to the system and the other entities it communicates with. A successful illusion attack can cause car accidents, traffic jams, and a decrease in VANET performance in terms of bandwidth utilization.

Repudiation and accountability evasion attacks are another important concern in C-ITS environments. Repudiation refers to a sender/receiver denying having sent/received a message, respectively. In its most innocent form, this may cause the legitimate entities to have to retransmit the said message. Nevertheless, more mischievous attacks could involve malicious attackers issuing, for example, safety-related messages that cause other vehicles to crash, and then repudiating having sent these messages. Besides, a rational attacker could exploit potential vulnerabilities in the nonrepudiation and accountability mechanisms to erase all evidence implicating them in road accidents that they may have caused [45].

Man-in-the-middle refers to a relatively broad category of attacks, whereby an attacker inserts herself between legitimate communicating parties [46]. Around C-ITS deployments, this attack can be launched on various fronts, as we have numerous types of interactions (V2V, V2I, I2I, etc.) that adversaries will try to insert themselves into. Thus, an instance of such an attack would be to have a malicious vehicle intercept, tamper with, and then re-broadcast the messages transmitted between other neighboring vehicles, while in another case a malicious base station deployed as a roadside unit could be used to tamper with the communications between vehicles and legitimate roadside units nearby.

Vehicle attacks are also an important concern, as an obvious target for attackers is the smart vehicle platform to either damage the specific vehicle itself or to gain access to the C-ITS network, enabling them to use it as a stepping stone to launch some of the attacks listed here. Various types of vehicle-specific attacks can be identified. These include *CAN bus attacks*, which rely on directly interfacing with the vehicle's CAN bus, allowing access to (and control of) all the vehicle's subsystems, including critical ones such as steering and braking, as already demonstrated by researches [5,17]. A common interface for gaining access to the CAN bus is the on-board diagnostics (OBD) port that all vehicles have, and which is trivial to gain access to. *Infotainment System attacks* are another type of such attacks that rely on compromising the infotainment system of the vehicle. These attacks can even be launched remote ("over the air"), as modern vehicles' infotainment systems feature Internet access (typically via cellular networks) to inform, entertain, support, and protect the driver. Vulnerabilities in the infotainment system can potentially allow attackers to, for example, remotely rewrite its firmware, consequently gaining access to the CAN bus (with which the infotainment typically interfaces). The feasibility of this approach has already been demonstrated in production vehicles [6]. *Malware* is also a concern in this context. Infotainment systems are nowadays becoming open to installing third-party applications, even having specific "application stores," much like any tablet or smart phone. The possibility of installing malicious software (as already happening in, e.g., smart phones) poses a significant security risk that, depending on the type of malicious software installed, may even raise safety concerns for the driver and the passengers [47]. Considering that there are also software components on other vehicles and the infrastructure nodes as well, it is possible that malicious software spreads to these entities as well, amplifying the attacks impact.

Physical attacks are attacks that physically damage one or more of the involved entities (e.g., vehicles or roadside units), partially limiting the operation or completely disabling their target. These could be carried out on purpose (e.g., vandals) or be accidental (e.g., vehicle colliding with a roadside unit).

Unintentional attacks, software faults, and component failures are attacks typically taking place by human entities participating in the C-ITS ecosystem, such as smart vehicle users or maintenance personnel erroneously setting up their vehicles or roadside units, respectively. Moreover, failures in the dependability of some components or faults in the software used throughout the smart platforms comprising the C-ITS are to be expected; such occurrences could lead to loss of the availability of certain services, communications in general, or even pave the way for other attacks.

Some of the above mentioned types of attacks are visualized in Figure 16.5.

16.3.4 Security Requirements

The intricacies of C-ITS applications impose several strict requirements that prohibit the adoption of some existing security mechanisms and that make the

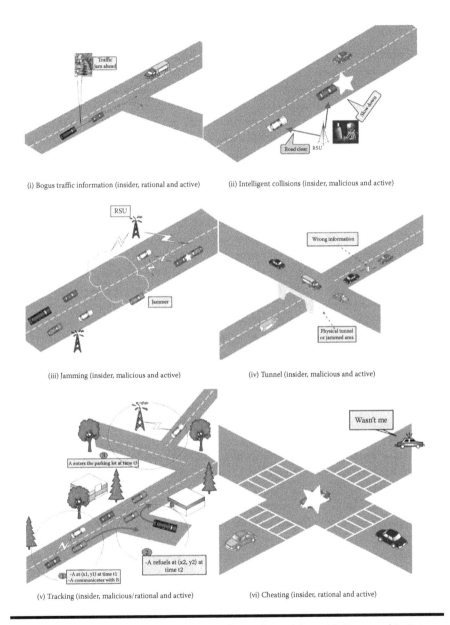

(i) Bogus traffic information (insider, rational and active)

(ii) Intelligent collisions (insider, malicious and active)

(iii) Jamming (insider, malicious and active)

(iv) Tunnel (insider, malicious and active)

(v) Tracking (insider, malicious/rational and active)

(vi) Cheating (insider, rational and active)

Figure 16.5 Some types of VANET attacks. (Adapted from M. Raya and J.-P. *Journal of Computer Security*, 15(1): 39–68, 2007; B. Parno and A. Perrig, *Workshop on Hot Topics in Networks*, no. 4, Philadelphia, PA, pp. 1–6, 2005; M. A. Razzaque, S. A. Salehi, and S. M. Cheraghi, *Security and Privacy in Vehicular Ad-Hoc Networks: Survey and the Road Ahead*, Springer, Berlin, pp. 107–132, 2013.)

design of alternative mechanisms a challenging task. Some important requirements that must be considered in this context include

- *Confidentiality*: Information stored or in transit is only accessible to legitimate, authorized entities.
- *Message integrity*: Message contents have not been altered, intentionally (e.g., by attacker) or unintentionally (e.g., fault in storage device or transmission).
- *Availability*: All assets, such as communications network and services, must be available for authorized entities to use; thus, the system should operate reliably and be able to withstand DoS attacks.
- *(Entity and Data) Authentication*: Ensuring that entities are identified properly and have valid credentials, thus gaining access to the parts of the assets they are authorized to. Moreover, message authentication ensures that a message can be trusted, that is, comes from a recognized sender and its contents have not been altered; thus, it is safe to act upon them (e.g., immediately engage the brakes as an accident just happened ahead).
- *Privacy/anonymity*: Vehicle and driver location is private sensitive information and should be protected. Access to location-based services should be provided in an anonymized way.
- *Traceability*: Entities (drivers, vehicles, and RSUs) that abuse the network and/or are otherwise compromised should be detectable.
- *Revocation*: Mechanisms should be in place to revoke the credentials of malicious or compromised entities (e.g., vehicles), thus disabling them.
- *Nonrepudiation*: Providing proof that a transaction took place, preventing the sender and/or the receiver from denying having taken part in the said transaction; thus, for example, drivers will not be able to avoid liability for the accidents they have caused.
- *Data consistency*: The plausibility of messages should be examined, considering similar messages generated by the same or neighboring entities in a relatively close space and time.
- *Mobility*: The adopted mechanisms should be able to accommodate the high mobility of smart vehicles and the highly dynamic and fleeting interactions between involved entities.
- *Real-time constraints*: Given the very high speeds that vehicles may travel at, and the time-critical actions that must take place for safe, smart, and autonomous vehicle operation (e.g., emergency braking or collision avoidance), key safety-related communications, verification, and processing should take place in real time or near real time.

16.3.5 Security Challenges

Smart vehicles and the associated infrastructure pose significant security challenges that must be addressed to motivate the adoption of these technologies and associated service (Figure 16.6 aggregates this information). These challenges include

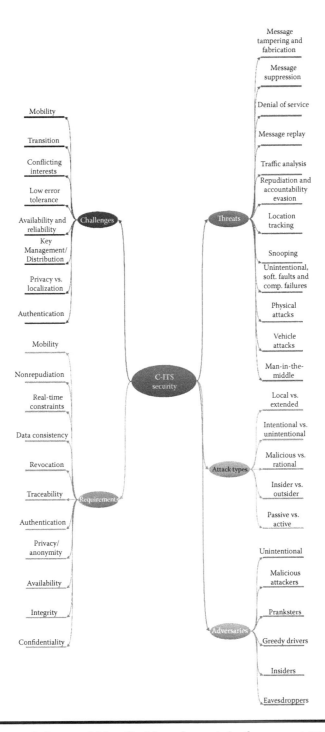

Figure 16.6 Mind map of identified key elements in the current VANET threat landscape.

Authentication: C-ITS communications require to have efficient and reliable authentication of all participating entities, and to guarantee the authenticity of exchanged messages. This allows, for example, a vehicle to act with confidence on a message received that dictates the need to brake immediately and, in general, helps prevent masquerading and other spoofing attacks. Moreover, these mechanisms can enable law enforcement to reliably deter, prevent, or detect such attacks. In cases where attacks do take place, the strong entity and message authentication and authenticity mechanisms that prohibit, for example, repudiation, can act as forensic evidence for pursuing legal action.

Privacy versus localization: Many vehicular applications rely on accurate localization and, moreover, it is essential to achieve efficient and reliable authentication of all participating entities and to guarantee the authenticity of exchanged messages. Nevertheless, schemes that compromise the users' privacy, for example, by having a single key linked to each vehicle or individual who is exposed during network authentication, cannot be accepted. Furthermore, there are discrepancies in the laws (privacy-related or otherwise) between the countries where a vehicle may be sold, and these must be considered as well in any adopted solution.

Availability and reliability: Critical interactions in vehicular applications will require highly reliable, always available, and real-time (or near real-time) communications. The challenge to successfully address these requirements is exacerbated by the type of unreliable, constantly changing links between the involved entities, and the guarantees that also need to be considered, for example, with respect to message authenticity.

Mobility: Vehicular networks are self-organized, with each vehicle interacting with numerous new entities (vehicles, other means of transport, RSUs, etc.) on just a single path (e.g., from home to work), often for just a few seconds. The distances and speeds involved are in the scale of kilometers and kilometers/hour, respectively. Techniques that rely on previous interactions with the nodes (e.g., reputation-based routing systems) or complex, time-consuming interactions are thus not applicable.

Key management and distribution: Key distribution is an important primitive for secure protocols, while key management is a challenge in all large-scale, dynamic deployments. Expecting the prevalence of VPKI solutions, several important decisions need to be made, such as where and how the keys should be installed (e.g., factory or by governments), also considering the intricacies of each approach, such as the numerous vehicle manufacturers that exist, as well as the differences in legislation between countries. Moreover, the use of keys in the adopted security protocols in a way that does not compromise users' privacy should also be considered.

Low error tolerance: Considering safety issues and the involvement of human lives, techniques that rely on detecting attacks may often be inappropriate; it is essential to be able to prevent attacks, because a postattack detection may be of limited use if the attack has already resulted in, for example, an accident that endangered or harmed the passengers. Moreover, the employed mechanisms should work with minimal chances of error; the large number of vehicles and lives involved means that even a slight chance of error or failure may translate to hundreds if not thousands of lives lost.

Conflicting interests: Vehicle manufacturers, service providers (e.g., providing enhanced insurance services or personalized advertising), governments (including law enforcement), and consumers often have conflicting requirements and interests. For example, governments and some service providers may want accurate location and driving behavior tracking, but consumers will reject such an approach. On the other hand, consumers may require that EDRs are only accessible by them and, possibly, the vehicle manufacturer, but the authorities would also want access to the EDR for investigation purposes. These differences should be reconciled, and incentives should be provided to all involved parties, if the adoption of C-ITS technologies is to progress rapidly.

Transition: At the first stages of introducing smart vehicles, the clear majority of vehicles will be traditional ones, or ones with limited "smartness." Moreover, the smart infrastructure's deployment will be equally sparse. Therefore, the techniques adopted should be able to effectively operate under these restrictions, increasing their efficacy and benefits as the technologies become more widely adopted.

16.4 Intrusion Prevention, Detection, and Mitigation

To effectively protect the C-ITS assets from the threats identified above, the provision and integration of an assortment of protection mechanisms and other countermeasures will be necessary. Indeed, this has been the focus of numerous research efforts, as already aggregated in a variety of published surveys in the field (e.g., [30,48–58]). Nevertheless, in the context of intrusion detection, prevention, and mitigation techniques for smart vehicle deployments, relatively few efforts can be identified, and, for the most part, detection mechanisms are focused on VANET-specific intrusion detection mechanisms (e.g., [59–62]), or refer to the prompt detection of road incidents [63–65], such as vehicle collisions. What is even more rare are approaches that consider the topic of C-ITS security as a characteristic use case of critical cyber-physical systems (e.g., [66]), applying equivalent high-level principles and providing comprehensive frameworks as is more common for other cyber-physical applications.

Identifying this gap and the necessity for such an approach, in this subsection we consider C-ITSs as a class of critical cyber-physical systems, highlighting a holistic approach to the intrusion detection, prevention and mitigation of these C-ITS deployments, as well as the potential benefits that such an approach may yield if applied appropriately.

16.4.1 Monitoring, Detection, and Response

Nowadays every Internet connected system is subject to cyber-attacks. An attack can be initiated from anywhere in the world and the target could be either a critical infrastructure or a client endpoint. The paths that the attacks are following to achieve their target are quite complex and the attacks are getting more and more sophisticated. There was, and is, an urgent need for sufficient network traffic data collection to combat such incidents.

Approaches of collecting data include collecting and sharing firewall and intrusion detection system (IDS) logs coming from heterogeneous sources. Examples of such sources include the Internet Storm Centre from the SANS Institute [67], DShield [68], and MyNetWatchman [69]. In the same direction, software antivirus companies have deployed their own data collection environment, such as DeepSight Early Warning Services from Symantec [70] and X force threats alert system from IBM [71]. These approaches are quite interesting but fall short when having to deal with cyber-attacks; they operate well in isolation but could provide results that are more valuable in a combined manner instead of being isolated.

Indeed, there is a huge amount of data widespread across the Internet and a proactive intelligent gathering and knowledge management mechanism is needed to find more data about an attack and thus maximize the effectiveness of the countermeasures taken. Identifying how cyber-attacks are performed will reveal their intelligence and help in identifying threat activities earlier. Having that in mind, effective approaches combine various innovative real-time data collection and gathering mechanisms and frameworks [72–76] to get the maximum output from their combined results. The main objective of the above process will be to correlate and combine the traffic gathered from various sources and appropriately enrich them to gain the maximum value of the data collected and thus identify potential cyber-attacks. A side effect target of the above will be to provide a knowledge management framework for cyber security-related information [77–79].

To enable the above, information from various levels should be captured. The monitoring levels should include host level data (for all the ICT, road infrastructure, and vehicle assets involved), at the software and hardware level (e.g., for malware and hardware tampering, respectively), as well as network level data from the heterogeneous networks that are expected to be present in the context of a C-ITS deployment.

While most of the related topics have been investigated extensively when considering the typical ICT systems of the C-ITS ecosystem, the same cannot be said for the VANET part of the said ecosystem. The parameters that should be considered when designing and implementing the associated mechanisms (including security mechanisms) for VANETs are [80]: *node speed* (ranging from 0 to over 200 km/h on highways), *movement patterns* (which are very different for a dense city road and a highway), *node density* (from traffic jams to isolated rural roads), and *node heterogeneity* (considering the different types of vehicles and their varied capabilities—an aspect significantly exacerbated in the context of C-ITS environments where RSUs and other smart devices are present). The above factors and their variations make it challenging to address the different ways information is disseminated, as well as the related latency and priority requirements (e.g., for the aggregation of monitoring data from the vehicles), and research efforts aim to define some practical approximations. For example, in the context of incident detection, research [81] has shown that, for a 5-minute sensor collection interval and a 30-second sensing interval, data accuracy will satisfy the requirement that vehicle detection is realized within ± 1 vehicle for 90% of all 5-minute intervals; occupancy will be measured within $\pm 1\%$ at 25% occupancy, volume within ± 1 vehicle/min at 2,000 vehicles/hour and speed within ± 2–4 mph 95% of the time. These and similar metrics should be considered also in the context of security (and privacy) applications, possibly producing additional, more security application-focused metrics as well. The spatio-temporal density variations affect VANET protocols and architecture; as such, both short-term and long-term variations impose critical challenges on VANET protocol design [82]. Thus, designing VANET monitoring mechanisms that are robust under all circumstances (i.e., from highly sparse networks to highly dense networks of vehicles) is challenging. This issue is exacerbated when considering the various levels of market penetration that are expected to be observed until the use of VANETs dominates transportation. In general, the efficient, scalable, and reliable data collection in VANETs is, in most scenarios, an open issue. Researchers are still investigating ways to efficiently collect large amounts of vehicle data without overloading the network, avoiding communication collisions (common in dense traffic conditions), by exploiting techniques such as data spatial correlations to reduce information redundancy and improve communication efficiency [83].

Moreover, when monitoring at the network layer (i.e., VANET traffic), the communication patterns typically found in such networks must be considered, ensuring that critical ones are monitored in an appropriate manner. These patterns typically are [80] *Beaconing* (i.e., continuous update of information among all neighboring nodes, e.g., update on current position), *Geobroadcast* (i.e., immediate distribution of information in a larger area, e.g., to inform of vehicle crash ahead), *Unicast routing* (i.e., transport of data through the ad hoc network to a specific destination, such as another vehicle or an RSU), *Advanced information dissemination* (i.e., dissemination of information among vehicles enduring a certain time, capable of bridging network partitions and prioritizing information (e.g., to inform vehicles that arrive later in time/where not previously reachable), and *Information aggregation* (whereby communicated data are processed and merged by network nodes and not simply forwarded, such as when multiple vehicles detected a traffic jam).

When considering the design of an efficient threat response mechanism that is relied upon, two main characteristics: produce accurate results and produce them instantly. Two large categories exist while speaking about various cyber-defense response mechanisms: the prevention mechanisms and the reaction mechanisms. Some of the previous preventing [84–86] mechanisms were too complex and costly to be applied; thus, reactive cyber-defense mechanisms had to be developed and deployed. In the latter case of reactive mechanisms, the information system operators can detect potential malicious activities and perform defense actions against them in an efficient manner. Current state-of-the art in response mechanisms include host/network level IDS [87–89], which, in the context of C-ITS, can be augmented with VANET-specific IDS schemes [60–62]. In addition to the above, other existing IDS techniques could be ported to the C-ITS environment, such as swarm intelligence ones, which are a bio-inspired family of methods [90]. Moreover, in the context of a C-ITS, there is a significant presence of smart devices, such as smartphones and tablets of vehicle passengers and pedestrians, and infotainment systems of modern vehicles (that are not unlike a tablet, often with their own application store). Thus, the integration of IDS systems designed for such types of hosts (such as a host/cloud-based IDS for smartphones [91]) would offer a more holistic protection of the C-ITS ecosystem. Another important protection tool is the use of security information and event management (SIEM) platforms; many vendors have developed and deployed their own solutions [92]. Other categories of tools that assist system and network operators fight against organized spam email and malware-spreading campaigns, include botnet detection tools and traffic anomaly detection tools [93,94]. Furthermore, these could be augmented with equivalent tools designed specifically for the VANET section of the C-ITS, such as tools designed to detect vehicle misbehavior [95], malicious data injection [96], and DoS attacks [97].

An effective solution should capitalize on those tools to provide accurate and instant results to the network operators. This can be performed by a combination of such tools, efficient information exchange mechanisms through appropriate SIEM mechanisms, and correlation mechanisms able to eliminate the large portion of false positives produced, also integrating frameworks designed to monitor and manage, in real time, smart vehicles and their associated subsystems [98,99]. The design proposed should react in real time because effort will be provided to design the system in a way that it will eliminate human factors from the identification, analysis, and response process cycle.

16.4.2 Analysis

After an event has been detected, it is important to analyze the incident not just to mitigate the attack (often, it is too late if automated response mechanisms have failed), but also to provide valuable feedback for the adaptation and improvement of the systems to avoid similar future occurrences.

During the last years, a lot of effort has been devoted from the security research and industry community in the collection, analysis, and detection of malicious network traffic that is represented by malicious software and malicious user activities. Various mechanisms and methods have been developed and deployed for inspecting malicious software in its various forms (e.g., embedded code in documents, executable files mobile applications). These methods are categorized into three types: static analysis, dynamic analysis, and a hybrid approach. Static analysis, the most basic approach to understanding the malicious software, includes the process of disassembling the binary code of an executable. The disassembling process produces signatures and indicative patterns. Some of the forms of these signatures and patterns include byte strings [100–102], identified instruction traces [103,104], and control flow graphs [105]. Dynamic analysis is an alternative to the static analysis and is performed under controlled environments (either sandboxes like Anubis [106], TEMU [107], or other instrumented frameworks). The hybrid analysis approach is a mixture of the previous two solutions that is trying to combine their advantages and eliminate their weaknesses. A representative framework of that approach is the Reanimator [108]. Moreover, there are tools that allow users to upload a captured malware sample and check whether it is malicious or not. Such tools include VirusTotal [109], VX Heaven [110], Zoo [111], and VxCage [112]. Those are web-based tools while collabREate [113], CrowdRE [114], and BinCrowd [115] are client-based tools. Those sources could also be active parts of the analysis once they offer a Web service interface to the community. In the context of C-ITS, these above are not only usable in the context of the ICT systems, but also the infotainment and other embedded computers present on the vehicles themselves, which, as already mentioned, are expected to become the target of malware attacks.

All the above include ways to analyze identified captured data, but they are capable enough to provide enough insights about the attacks. Current attacks are no longer isolated and a mixture of information is needed to eliminate false positives and produce valuable results. Malware economy [116,117] is flourishing and new analytic mechanisms are needed to understand the nature of the attacks, the reasons, and the potential targets. Events that in the past were forming a single standalone attack nowadays are just a small part of the complex mosaic of a sophisticated cyber-attack and malicious campaigns. Several steps are needed to enable correlation of the events that are taking place in different places, different network layers, and different times. Future approaches should aim to address the above issue by (i) identifying data that can be correlated (a level of preprocessing on some of them could also be performed, e.g., padding mechanisms including, e.g., the insertion of the country of origin of the attacker and/or the time difference between attacker and victim, etc.). The second step should include the (ii) integration of the data in a high-performance big data analytics engine (e.g., Hadoop), and (iii) apply high-performance analysis to extract results in near real time. The whole process could be empowered by applying the rating mechanism to the data collected depending on the sources that contribute those data.

16.4.3 Incident Management and Adaptation

Addressing many kinds of hazards, including physical and cyber-attacks on transport infrastructures and systems of critical nature for people, requires careful investigation of new security risks and threats due to the increased interconnection among the impact of physical hazards and cyber-attacks. To improve cyber security, stakeholders must act in synergy and proactively, instead of being isolated and reactive to cyber-attacks. To this direction, there is a need for holistic approaches providing end-to-end security, also considering the very specific requirements of the highway control systems, traffic and bridge/tunnel control, lighting systems, and other such critical parts of a C-ITS ecosystem.

The latest standards and solutions for cyber security management involve multidisciplinary methods, including: user-driven design and experience, big data analytics, visualization of cyber security analytics, incorporation of human behavior when designing cyber security technologies, predictive analytics for situational threat detection. Moreover, security incident management is a continuous process that cycles through five key stages [118]: (1) planning and preparation, (2) detection and reporting, (3) assessment and decision, (4) responses, (5) lessons learned. The lifecycle is typically supported by the combination of various tools, such as firewalls, intrusion detection systems and other monitoring tools, incident tracking (ticketing systems), SIEMs, real-time streaming analytics, information security operation centers, computer emergency response teams (CERTs), and information extraction from security mailing lists and forums. On an individual basis, some tools are mature, but their integration within and across organizations poses challenges. The complexity increases in current information systems, involving mobile and/or cloud services, with several actors or security domains. Yet, owing to existing business and regulatory requirements, solutions that address these challenges will be increasingly in demand, creating a real business opportunity for those capable of offering an integrated end-to-end (E2E) solution for incident management. We have pinpointed four key aspects that must be addressed to create a modern E2E incident management framework:

1. Better information sharing
2. Better data collection and analysis capabilities
3. Better awareness, trust, and transparency
4. Better mapping of granularity and overlaps

Effective cyber security management solutions can help reverse the imbalance of intelligence capabilities of the attackers versus network and service infrastructures under attack, by developing an intelligence-driven, dynamically configurable, adaptive and evolvable security management framework to enable the monitoring, information sharing, runtime adaptation, and incident response of network and service infrastructures. Researchers should aim to develop a comprehensive, yet transportation-specific, approach to assure the security and the integrity of existing and emerging connected and interdependent cyber-physical transportation installations, driving the adoption of better cybersecurity management solutions in critical transport infrastructures. An objective of this work is to develop complete, end-to-end, security-by-design solutions adjusted to the very specific requirements of C-ITS. In this context, efforts can also benefit from related research projects, such as CIPSEC [119] and NECOMA [120], which have investigated how to combine state-of-the-art network monitoring and detection capabilities with information sharing means, keeping stakeholders up to date with cyber-threats and attack attempts.

16.4.4 *Information Sharing*

New and increasingly significant security breaches are reported practically every day. For most companies, it is no longer a matter of whether they will be attacked, but rather how long ago they were attacked. Enterprises and cloud providers alike face a constant barrage of threats and attacks. They all have a distinct need to understand the types of incidents that peers and technology partners are experiencing. In this environment, sophisticated organizations understand that the difference between a minor incident and massive breach often comes down to the ability to quickly detect, contain, and mitigate an attack. Unfortunately, evidence suggests the opposite, despite a growing number of security tools and solutions at our disposal. Based on attacks observed during 2014, 75% of attacks spread from Victim 0 to Victim 1 within one day (24 hours), while over 40% hit the second organization in less than an hour, highlighting the need to close the gap between sharing speed and attack speed [121].

A key reason that the delta between compromise and detection is growing, is the increasing sophistication of attackers. Once an exploit is shown to be effective, it is often quickly disseminated via a number of underground channels. For example, immediately after the target breach [122], 18 other companies were attacked using the same methods. Yet despite this disturbing trend, owing to a longstanding and pervasive corporate reluctance to share information, companies are understandably hesitant to externally disclose any information until they fully understand the incident. The only obvious approach to protect themselves against legal/market/reputational risks while instantly sharing information is probably through anonymity.

Information Sharing and Analysis Center (ISAC) model[*] requires sending sensitive data to a trusted third party thus revealing the identity of the company affected. The Snowden incident undermined the little trust within the market making sharing with trusted third parties undesirable today. This lack of trust highlights the need for a trusted sharing method, in which (i) company and end-user identity is

---

[*] National Council of ISACs (https://www.nationalisacs.org/).

not known, (ii) incident data submission is quick and simple, (iii) rapid analysis of data is performed, (iv) alerts are sent in minutes, and (v) the ability to anonymously discuss attacks and share solutions is provided.

Researchers should aim to design an incident sharing program that supports the C-ITS market stakeholder and the society at large in the following ways: (i) enable sharing: share meaningful C-ITS incident data safely, easily, and early in the response process to leverage external expertise during remediation efforts and provide early warning to help others reduce their own exposure; (ii) expand expertise: collaborate with skilled security to analyze attack indicators, develop defensive strategies, and decrease time to mitigation; and (iii) provide context and support decision-making: avoid duplication of effort and benefit from what others have already learned. Once an incident report is shared, the system should provide a near real-time correlation with reports supplied by other vetted members. If similarities are discovered, members can be alerted and provided with the related reports that contain additional attack indicators and mitigation advice. Members might also decide to collaborate in other ways, such as joining the response efforts.

16.5 Conclusions

Regarding security incidents and cyber-attacks on typical ICT deployments, the vulnerabilities and attack classes (such as distributed denial of service, malware, and phishing attacks) to most information technologies, such as computers and servers, have been researched over the past years and are well understood by security researchers. Although the specifics of the attacks and potential consequences can vary with each type of attack, the basic structures and general mitigations for these attacks are known. However, with the introduction of ITS and C-ITS systems and cyber-physical innovations in general, the vulnerabilities, the resulting mitigating factors, and the potential consequences of cyber-attacks still need further research.

Addressing many kinds of threats, such as the ones identified earlier in this work, including physical and cyber-attacks on transport infrastructures and systems of a critical nature for people, requires careful investigation of new security risks and threats due to the increased interconnection among the impact of physical hazards and cyber-attacks. Moreover, as combined physical and cyber-attacks on interconnected transport systems are expected to become more mainstream, there is a need for comprehensive approaches that will orchestrate cyber security, communications, intrusion detection and response solutions, guided by the expected sequence of incidents at linked subsystems to maximize the level of protection.

Thus, to effectively improve the security of C-ITS deployments, research must not only focus on specific security primitives and technological building blocks, but also act in synergy and proactively, instead of being isolated and reactive to cyber-attacks. To this direction, complete, end-to-end and secure-by design solutions, adjusted to the intricacies of C-ITS systems, are needed. An intelligence-driven, dynamically configurable, adaptive, and evolvable security management framework can enable the monitoring, information sharing, runtime adaptation and incident response of network and service infrastructures, reversing the imbalance of intelligence capabilities between the attackers and the network and service infrastructure operators.

Further efforts should focus on providing a comprehensive, yet transportation-specific, approach to assure the security and the integrity of existing and emerging connected and interdependent cyber-physical C-ITS deployments, driving the adoption of the said technologies and associated services that have the potential to significantly reduce the number of road victims, improve our everyday lives, and introduce an assortment of new services and business models.

References

1. European Commission, Jeanne Breen Consulting, Road Safety Study for the Interim Evaluation of Policy Orientations on Road Safety 2011–2020, Feb 2015, https://ec. europa.eu/transport/road_safety/sites/roadsafety/files/pdf/study_final_report_february_ 2015_final.pdf
2. S. Ramesh, Software's Significant Impact on the Automotive Industry, Frost & Sullivan's Automotive Practice, October 2008, https://www.frost.com/sublib/display-market-insight.do?id = 145328674
3. R. N. Charette, This Car Runs on Code, IEEE Spectrum, February 2009, http:// spectrum.ieee.org/transportation/systems/this-car-runs-on-code
4. Reuters, Toyota to recall 1.9 million Prius cars for software defect in hybrid system, February 2014, http://www.reuters.com/article/us-toyota-recall-idUSBREA1B1B920140212
5. K. Koscher et al., Experimental security analysis of a modern automobile, in *Proceedings of the IEEE Symposium on Security and Privacy*, Berkeley/Oakland, CA, pp. 447–462, 2010.
6. A. Greenberg, Hackers Remotely Kill a Jeep on the Highway—With Me in It, Wired, July 2015, https://www.wired.com/2015/07/hackers-remotely-kill-jeep-highway/
7. Jeep owners urged to update their cars after hackers take remote control, The Guardian, July 2015, https://www.theguardian.com/technology/2015/jul/21/jeep-owners-urged-update-car-software-hackers-remote-control
8. European Automobile Manufacturers' Association, 2015–2016 Automobile Industry Pocket Guide, June 2015, Brussels, Belgium.
9. European Commission (EC), Commission Staff Working Document on Transport Security, Brussels, Belgium, SWD, 2012, 143 Final.
10. ENISA, Cyber Security and Resilience of Intelligent Public Transport, European Union Agency for Network and Information Security, 2015, www.enisa. europa.eu/publications/good-practices-recommendations
11. US Department of Homeland Security (DHS), The Future of Smart Cities: Cyber-Physical Infrastructure Risk, National Protection and Programs Directorate, Office of Cyber and Infrastructure Analysis, 2015.
12. Ferris, J., Terrorist Attack Shows Vulnerability in Critical Infrastructure, 2014, http://dailysignal.com/2014/02/19/terrorist-attack-shows-vulnerability-critical-infrast ructure/

13. Economic Commission for Europe, *Telematic applications: eCall HGV/GV, additional data concept specification*, September 2011, http://www.unece.org/fileadmin/DAM/trans/doc/2011/dgwp15ac1/INF.30e.pdf

14. J. A. Cook et al., Control, computing and communications: Technologies for the twenty-first century model T, *Proceedings of the IEEE*, 95(2): 334–355, 2007.

15. A. Doshi, B. T. Morris, and M. M. Trivedi, On-road prediction of driver's intent with multimodal sensory cues, *IEEE Pervasive Computing*, 10(3): 22–34, 2011.

16. K. Lee et al., AMC: Verifying user interface properties for vehicular applications, in *Proceeding of the 11th Annual International Conference on Mobile Systems, Applications, and Services—MobiSys '13*, Taipei, Taiwan, pp. 1–12, 2013.

17. S. Checkoway et al., Comprehensive experimental analyses of automotive attack surfaces, system, in *SEC'11 Proceedings of the 20th USENIX Conference on Security*, San Francisco, CA, pp. 6–6, August 8–12, 2011.

18. G. Leen and D. Heffernan, Expanding automotive electronic systems, *Computer (Long Beach, Calif)*, 35(1): 88–93, 2002.

19. U.S. Department of Transportation, Connected Vehicle Infrastructure Deployment Considerations: Lessons Learned from Safety Pilot and Other Connected Vehicle Test Programs, May 2014, http://www.roadsbridges.com/sites/default/files/Deployment_Considerations_report_06_02_2014_v1.pdf

20. D. Jiang and L. Delgrossi, IEEE 802.11p: Towards an international standard for wireless access in vehicular environments, in *Proceedings of the IEEE Vehicular Technology Conference*, Marina Bay, Singapore, pp. 2036–2040, 2008.

21. European Telecommunications Standards Institute (ETSI), EN 302 663 V1.2.1, Intelligent Transport Systems (ITS); Access Layer Specification for Intelligent Transport Systems Operating in the 5 GHz Frequency Band, July 2013, http://www.etsi.org/deliver/etsi_en/302600_302699/302663/01.02.01_60/en_302663v010201p.pdf

22. M. Horani, V2X—Emerging Technologies Comparison in the United States, Europe, and Japan, AutomotiveWorld.com, October 2012, https://s3.amazonaws.com/automotiveworld/presentations/V2V + - + Car2Car + Comparison.pdf

23. A. Osseiran et al., Scenarios for 5G mobile and wireless communications: The vision of the METIS project, *IEEE Communications Magazine*, 52(5): 26–35, 2014.

24. J. P. Hubaux, S. Capkun, and J. Luo, The security and privacy of smart vehicles, *IEEE Security and Privacy Magazine*, 2(3): 49–55, 2004.

25. European Telecommunications Standards Institute (ETSI), Technical Specification 103 301 V1.1.1, Intelligent Transport Systems (ITS); Vehicular Communications; Basic Set of Applications; Facilities layer protocols and communication requirements for infrastructure services, 2016.

26. Sitraffic ESCoS—The cooperative traffic system for the digital road of the future, Siemens Mobility, https://www.mobility.siemens.com/mobility/global/SiteCollectionDocuments/en/road-solutions/urban/trends/sitraffic-escos-en.pdf

27. European Commission, A European strategy on Cooperative Intelligent Transport Systems, a milestone towards cooperative, connected and automated mobility, Communication from The Commission to The European Parliament, The Council, The European Economic and Social Committee and The Committee of The Regions, COM(2016)766 final, 30 Nov. 2016, http://ec.europa.eu/energy/sites/ener/files/documents/1_en_act_part1_v5.pdf

28. A. Boukerche, H. A. B. F. Oliveira, E. F. Nakamura, and A. A. F. Loureiro, Vehicular ad hoc networks: A new challenge for localization-based systems, *Computer Communications*, 31(12): 2838–2849, 2008.

29. Deliverable 5.2.1: Currently established risk-assessment methods, TRESPASS EU Project, FP7, October 2014, https://www.trespass-project.eu/sites/default/files/Deliverables/D5_2_1.pdf

30. M. Raya and J.-P. Hubaux, Securing vehicular ad hoc networks, *Journal of Computer Security*, 15(1): 39–68, 2007.

31. K. Sampigethaya, M. Li, L. Huang, and R. Poovendran, AMOEBA: Robust location privacy scheme for VANET, *IEEE Journal of Selected Areas in Communications*, 25(8): 1569–1589, 2007.

32. S. Roselinmary, M. Maheshwari, and M. Thamaraiselvan, Early detection of DOS attacks in VANET using attacked packet detection algorithm (APDA), in 2013 *International Conference on Information Communication and Embedded Systems, ICICES 2013*, Chennai, India, pp. 237–240, 2013.

33. S. Szott and M. Natkaniec, Security and cooperation in wireless networks: Thwarting malicious and selfish behavior in the age of ubiquitous computing, *IEEE Communication Magazines*, 47(3): 8–8, 2009.

34. O. Punal, C. Pereira, A. Aguiar, and J. Gross, Experimental characterization and modeling of RF jamming attacks on VANETs, *IEEE Transactions on Vehicular Technology*, 64(2): 524–540, 2015.

35. M. N. Mejri and J. Ben-Othman, Detecting greedy behavior by linear regression and watchdog in vehicular ad hoc networks, in *2014 IEEE Global Communications Conference, GLOBECOM 2014*, Austin, TX, December 8–12, 2014, pp. 5032–5037.

36. V. Bibhu, K. Roshan, K. B. Singh, and D. K. Singh, Performance analysis of black hole attack in VANET, *International Journal of Computer Network Information Security*, 4(11): 47–54, 2012.

37. D. G. Padmavathi and M. D. Shanmugapriya, A survey of attacks, security mechanisms and challenges in wireless sensor networks, *International Journal of Computer Science and Information Security*, 4(1): 1–9, 2009.

38. B. Xiao, B. Yu, and C. Gao, CHEMAS: Identify suspect nodes in selective forwarding attacks, *Journal of Parallel Distributed Computing*, 67(11): 1218–1230, 2007.

39. J. T. Isaac, S. Zeadally, and J. S. Cámara, Security attacks and solutions for vehicular ad hoc networks, *IET Communications*, 4(7): 894, 2010.

40. Y. Guo, S. Schildt, and L. Wolf, Detecting blackhole and greyhole attacks in vehicular delay tolerant networks, in *2013 Fifth International Conference on Communication Systems and Networks (COMSNETS)*, Bangalore, India, pp. 1–7. doi: 10.1109/COMSNETS.2013.6465569, 2013.

41. Y.-C. H. Y.-C. Hu, A. Perrig, and D. B. Johnson, Wormhole attacks in wireless networks, *IEEE Journal of Selected Areas in Communications*, 24(2): 370–380, 2006.

42. J. R. Douceur, The Sybil attack, in P. Druschel, F. Kaashoek, and A. Rowstron (eds), *Peer-to-Peer Systems*. IPTPS 2002. Lecture Notes in Computer Science, vol. 2429. Springer, Berlin, Heidelberg, 2002, https://link.springer.com/chapter/10.1007/3-540-45748-8_24.

43. T. Nighswander, B. Ledvina, J. Diamond, R. Brumley, and D. Brumley, GPS software attacks, in *Proceedings of the 2012 ACM Conference on Computer Communications and Security—CCS '12*, Raleigh, NC, p. 450, 2012.

44. N.W. Lo and H.C. Tsai, Illusion attack on VANET applications—A message plausibility problem, in *2007 IEEE Globecom Workshops*, Washington, DC, pp. 1–8, 2007. doi: 10.1109/GLOCOMW.2007.4437823.

45. T. Leinmuller, R. K. Schmidt, E. Schoch, A. Held, and G. Schafer, Modeling roadside attacker behavior in VANETs, *2008 IEEE Globecom Workshops*, New Orleans, LO, pp. 1–10, 2008. doi: 10.1109/GLOCOMW.2008.ECP.63

46. M. S. Al-kahtani, Survey on security attacks in Vehicular Ad hoc Networks (VANETs), in *2012 6th International Conference on Signal Processing and Communication Systems*, Gold Coast, Australia, pp. 1–9, 2012.

47. McAfee, Caution: Malware Ahead. An analysis of emerging risks in automotive system security, http://www.mcafee.com/cn/resources/reports/rp-caution-malware-ahead.pdf

48. B. Parno and A. Perrig, Challenges in securing vehicular networks, *Workshop on Hot Topics in Networks*, no. 4, Philadelphia, PA, pp. 1–6, 2005.

49. M. A. Razzaque, S. A. Salehi, and S. M. Cheraghi, *Security and Privacy in Vehicular Ad-Hoc Networks: Survey and the Road Ahead*, Springer, Berlin, pp. 107–132, 2013.

50. R. G. Engoulou, M. Bellaïche, S. Pierre, and A. Quintero, VANET security surveys, *Computer Communications*, 44: 1–13, 2014.

51. I. Studnia, V. Nicomette, E. Alata, Y. Deswarte, M. Kaaniche, and Y. Laarouchi, Survey on security threats and protection mechanisms in embedded automotive networks, in *Proceedings of the International Conference on Dependable Systems and Networks*, Atlanta, GA, 2013.

52. G. Yan, D. Wen, S. Olariu, and M. C. Weigle, Security challenges in vehicular cloud computing, *IEEE Transactions on Intelligent Transportation Systems*, 14(1): 284–294, 2013. doi: 10.1109/TITS.2012.2211870.

53. L. Bariah, D. Shehada, E. Salahat, and C. Y. Yeun, Recent advances in VANET security: A survey, in *IEEE 82nd Vehicular Technology Conference, VTC Fall 2015—Proceedings*, Boston, MA, 2016.

54. S. Al-Sultan, M. M. Al-Doori, A. H. Al-Bayatti, and H. Zedan, A comprehensive survey on vehicular Ad Hoc network, *Journal of Network and Computer Applications*, 37: 380–392, 2014, https://doi.org/10.1016/j.jnca.2013.02.036.

55. M. Zhao, J. Walker, and C. C. Wang, Challenges and opportunities for securing intelligent transportation system, *IEEE Journal on Emerging and Selected Topics in Circuits and Systems*, 3(1): 96–105, 2013.

56. X. Lin, R. Lu, C. Zhang, H. Zhu, P.-H. Ho, and X. Shen, Security in vehicular ad hoc networks, *IEEE Communications Magazine*, 46(4): 88–95, 2008.

57. M. N. Mejri, J. Ben-Othman, and M. Hamdi, Survey on VANET security challenges and possible cryptographic solutions, *Vehicle Communications*, 1(2): 53–66, 2014.

58. M. Whaiduzzaman, M. Sookhak, A. Gani, and R. Buyya, A survey on vehicular cloud computing, *Journal of Network and Computer Applications*, 40(1): 325–344, 2014.

59. J. Hortelano, J. C. Ruiz, and P. Manzoni, Evaluating the usefulness of watchdogs for intrusion detection in VANETs, in *IEEE International Conference on Communications Workshops, ICC 2010*, Cape Town, South Africa, 2010.

60. N. Bißmeyer, C. Stresing, and K. M. Bayarou, Intrusion detection in VANETs through verification of vehicle movement data, in *2010 IEEE Vehicular Networking Conference*, Jersey City, NJ, pp. 166–173, 2010. doi: 10.1109/VNC.2010.5698232.

61. N. Kumar and N. Chilamkurti, Collaborative trust aware intelligent intrusion detection in VANETs, *Computers and Electrical Engineering*, 40(6): 1981–1996, 2014.

62. A. Tomandl, K.-P. Fuchs, and H. Federrath, REST-Net: A dynamic rule-based IDS for VANETs, in *2014 7th IFIP Wireless and Mobile Networking Conference (WMNC)*, Vilamoura, Portugal, pp. 1–8, 2014.

63. J. Wang, X. Li, S. S. Liao, and Z. Hua, A hybrid approach for automatic incident detection, *IEEE Transactions on Intelligent Transportation Systems*, 14(3): 1176–1185, 2013.

64. M. Abuelela, S. Olariu, and G. Yan, Enhancing automatic incident detection techniques through vehicle to infrastructure communication, in *Proceedings of the IEEE Conference on Intelligent Transportation Systems, ITSC*, Beijing, China, pp. 447–452, 2008.

65. M. Abuelela and S. Olariu, Automatic incident detection in VANETs: A bayesian approach, in *VTC Spring 2009—IEEE 69th Vehicular Technology Conference*, Barcelona, pp. 1–5, 2009. doi: 10.1109/VETECS.2009.5073411.

66. N. H. Ab Rahman, W. B. Glisson, Y. Yang, and K. K. R. Choo, Forensic-by-Design Framework for Cyber-Physical Cloud Systems, *IEEE Cloud Computing*, 3(1): 50–59, 2016.

67. SANS, The SANS Internet storm center, http://isc.sans.org

68. DShield. DShield distributed intrusion detection system, http://www.dshield.org

69. myNetWatchman network intrusion detection and reporting, http://www.mynetwatchman.com

70. Symantec. Deepsight early warning services, https://www.symantec.com/content/en/us/enterprise/fact_sheets/b-deep-sight-early-warning-service_DS-21227151.en-us.pdf

71. IBM X-Force Threat Intelligence, https://www-03.ibm.com/security/xforce/

72. P. Baecher, M. Koetter, T. Holz, M. Dornseif, and F. C. Freiling. The nepenthes platform: An efficient approach to collect malware, in *Proceedings of 9th International Symposium on Recent Advances in Intrusion Detection (RAID'06)*, Hamburg, Germany, September, 2006, pp. 165–184.

73. FP6-NoAH, European network of affined honeypots, http://www.fp6-noah.org/

74. The mwcollect alliance, http://alliance.mwcollect.org

75. Nepenthes malware collection, http://nepenthes.mwcollect.org/

76. The Honeynet Project, The honeynet project, http://www.honeynet.org/misc/project.html

77. CYBEX, The cybersecurity information exchange framework (x.1500), http://www.itu.int/en/itu-t/studygroups/com17/pages/cybex.aspx

78. STIX, Structured threat information expression, http://stix.mitre.org/

79. X-ARF, Network abuse reporting 2.0, http://www.x-arf.org/index.html

80. E. Schoch, F. Kargl, and M. Weber, Communication patterns in VANETs, *IEEE Communications Magazine*, 46(11): 119–125, 2008.

81. E. Parkany and C. Xie, *A Complete Review of Incident Detection Algorithms & Their Deployment: What Works and What Doesn't*, The New England Transportation Consortium, Storrs, CT2005, p. 120.

82. F. Bai and B. Krishnamachari, Spatio-temporal variations of vehicle traffic in VANETs, in *Proceedings of the Sixth ACM International Workshop on VehiculAr InterNETworking—VANET '09*, Beijing, China, p. 43, 2009.

83. C. Liu, C. Chigan, and C. Gao, Compressive sensing based data collection in VANETs, in *2013 IEEE Wireless Communications and Networking Conference (WCNC)*, Shanghai, China, pp. 1756–1761, 2013.

84. D. Bell and L. J. LaPadula, Secure computer systems: Mathematical foundations, *Data Base*, 1(MTR-2547 Vol. I): 513–523, 1973.

85. D. D. Clark and D. R. Wilson, A comparison of commercial and military computer security policies, in *IEEE Symposium on Security and Privacy*, Oakland, pp. 184–194, 1987.

86. D. F. C. Brewer and M. J. Nash, The Chinese Wall security policy, *Proceedings of the IEEE Symposium and Security and Privacy*, pp. 206–214, 1989. https://snort.org/

87. Snort IDS, https://snort.org/

88. Suricata IDS, http://suricata-ids.org/

89. Surfnet, http://ids.surfnet.nl

90. C. Kolias, G. Kambourakis, and M. Maragoudakis, Swarm intelligence in intrusion detection: A survey, *Computer Security*, 30(8): 625–642, 2011.

91. D. Damopoulos, G. Kambourakis, and G. Portokalidis, The best of both worlds. A Framework for the synergistic operation of host and cloud anomaly-based IDS for smartphones, in *Proceedings of the Seventh European Workshop on System Security—EuroSec '14*, New York, NY, pp. 1–6, 2014.

92. https://en.wikipedia.org/wiki/Security_information_and_event_management#Vendor_products

93. Taasera, https://www.taasera.com/products

94. Flowtraq, http://www.flowtraq.com/corporate/botnet-and-anomaly-detection/

95. S. Ruj, M. A. Cavenaghi, Z. Huang, A. Nayak, and I. Stojmenovic, On data-centric misbehavior detection in VANETs, in *IEEE Vehicular Technology Conference*, IEEE Press, San Francisco, pp. 1–5, 2011.

96. K. L. Morrow, E. Heine, K. M. Rogers, R. B. Bobba, and T. J. Overbye, Topology perturbation for detecting malicious data injection, in *Proceedings of the Annual Hawaii International Conference on System Sciences*, Hawaii, pp. 2104–2113, 2011.

97. K. M. A. Alheeti, A. Gruebler, K. D. McDonald-Maier, and A. Fernando, Prediction of DoS attacks in external communication for self-driving vehicles using a fuzzy petri net model, in *IEEE International Conference on Consumer Electronics (ICCE)*, Taiwan, pp. 502–503, 2016.

98. K. Fysarakis, G. Hatzivasilis, C. Manifavas, and I. Papaefstathiou, RtVMF: A secure real-time vehicle management framework, *IEEE Pervasive Computing*, 15(1): 22–30, 2016.

99. K. Fysarakis, G. Hatzivasilis, I. Askoxylakis, and C. Manifavas, RT-SPDM: Real-time security, privacy and dependability management of heterogeneous systems. In: T. Tryfonas and I. Askoxylakis (eds), *Human Aspects of Information Security, Privacy, and Trust*. HAS 2015. Lecture Notes in Computer Science, vol. 9190. Springer, Cham, Switzerland, 2015.

100. D. Kirat, L. Nataraj, G. Vigna, and B. S. Manjunath, SigMal: A static signal processing based malware triage, in *Proceedings of the 29th Annual Computer Security and Application Conference—ACSAC '13*, New Orleans, LA, pp. 89–98, 2013.

101. K. Griffin, S. Schneider, X. Hu, and T. Chiueh, Automatic generation of string signatures for malware detection. In: E. Kirda, S. Jha, and D. Balzarotti D. (eds), *Recent Advances in Intrusion Detection*. RAID 2009. Lecture Notes in Computer Science, vol 5758. Springer, Berlin, Heidelberg, 2009.

102. R. Perdisci, A. Lanzi, and W. Lee. Classification of packed executables for accurate computer virus detection, *Pattern Recognition Letters*, 29(14): 1941–1946, 2008.

103. R. W. Lo, K. N. Levitt, and R. A. Olsson, MCF: A malicious code filter, *Computer Security*, 14(6): 541–566, 1995.

104. M. Christodorescu and S. Jha, Static analysis of executables to detect malicious patterns, in *SSYM'03 Proceedings of the 12th Conference on USENIX Security Symposium*, vol. 12, Berkeley, CA, pp. 12–12, 2003.
105. D. Caselden, A. Bazhanyuk, M. Payer, S. McCamant, and D. Song, HI-CFG: Construction by binary analysis and application to attack polymorphism, in *Lecture Notes in Computer Science (including subseries Lecture Notes in Artificial Intelligence and Lecture Notes in Bioinformatics)*, 2013, 8134 LNCS, 164–181.
106. U. Bayer, C. Kruegel, and E. Kirda, TTAnalyze: A tool for analyzing malware, in *15th Annual Conference on European Institute for Computer Antivirus Research*, Hamburg, Germany, pp. 180–192, 2006.
107. D. Song, D. Brumley, H. Yin, J. Caballero, I. Jager, M. G. Kang, Z. Liang, J. Newsome, P. Poosankam, and P. Saxena, BitBlaze: A new approach to computer security via binary analysis, in *Lecture Notes in Computer Science (including subseries Lecture Notes in Artificial Intelligence and Lecture Notes in Bioinformatics)*, 2008, 5352 LNCS, 1–25.
108. P. M. Comparetti, G. Salvaneschit, E. Kirdai, C. Kolbitsch, C. Kruegel, and S. Zanero, Identifying dormant functionality in malware programs, in *Proceedings of the IEEE Symposium on Security and Privacy*, San Jose, CA, pp. 61–76, 2010.
109. Google, Virustotal, https://www.virustotal.com/
110. Vxheaven, http://vxheavens.com/
111. M. Quintans, Zoo, http://zoo.mlw.re
112. C. Guarnieri, Vxcage, https://github.com/botherder/vxcage
113. C. Eagle and T. Vidas, CollabREate, http://www.idabook.com/collabreate/
114. T. Werner and J. Geffner. CrowdRE, https://www.crowdstrike.com/blog/crowdre-alpha-release/
115. Zynamics, BinCrowd, http://www.zynamics.com/bincrowd.html
116. C. Kanich, C. Kreibich, K. Levchenko, B. Enright, G. M. Voelker, V. Paxson, and S. Savage, Spamalytics: An empirical analysis of spam marketing conversion, in *Proceedings of the 15th ACM Conference on Computer and Communications Security, CCS '08*, New York, NY, pp. 3–14, 2008.
117. B. Stone-Gross, M. Cova, L. Cavallaro, B. Gilbert, M. Szydlowski, R. Kemmerer, C. Kruegel, and G. Vigna, Your botnet is my botnet: Analysis of a botnet takeover, in *Proceedings of the 16th ACM Conference on Computer and Communications Security*, ACM, Denver, CO, pp. 635–647, 2009.
118. ISO/IEC 27035-1:2016, Information Technology—Security Techniques—Information Security Incident Management—Part 1: Principles of Incident Management, http://www.iso.org/iso/home/store/catalogue_ics/catalogue_detail_ics.htm?csnumber = 60803.
119. Enhancing Critical Infrastructure Protection with innovative SECurity framework (CIPSEC), H2020-DS-2015-1 EU-funded research project, http://www.cipsec.eu/.
120. Nippon-European Cyberdefense-Oriented Multilayer Threat Analysis (NECOMA), FP7 EU-funded research project, http://www.necoma-project.eu/
121. Verizon, *Data Breach Investigations Report*, 2015, http://www.verizonenterprise.com/verizon-insights-lab/dbir/
122. TechCrunch, Target Says Credit Card Data Breach Cost It $162M In 2013-14, https://techcrunch.com/2015/02/25/target-says-credit-card-data-breach-cost-it-162m-in-2013-14/

Index